DESIGN OF DEVICES AND SYSTEMS

DESIGN OF DEVICES AND SYSTEMS

~ *Third Edition* ~
Revised and Expanded

William H. Middendorf
Richard H. Engelmann
University of Cincinnati
Cincinnati, Ohio

CRC Press is an imprint of the
Taylor & Francis Group, an informa business

Library of Congress Cataloging-in-Publication Data

Middendorf, William H.
  Design of devices and systems / William H. Middendorf, Richard H. Engelmann. -- 3rd ed., rev. and expanded.
    p.   cm.
  Includes bibliographical references and index.
  ISBN 0-8247-9924-0 (hardcover : alk. paper)
  1. Engineering design. I.Engelmann, Richard H.
II. Title.
TA174.M529   1997
620' .0042--dc21   97-41590
                                                                                  CIP

The publisher offers discounts on this book when ordered in bulk quantities. For more information, write to Special Sales/Professional Marketing at the address below.

Copyright ©1998 by MARCEL DEKKER, INC. All Rights Reserved.

Reprinted 2009 by CRC Press

Neither this book nor any part may be reproduced or transmitted in any form or by any means, electronic or mechanical, including photocopying, microfilming, and recording, or by any information storage and retrieval system, without permission in writing from the publisher.

MARCEL DEKKER, INC.
270 Madison Avenue, New York, New York 10016
http://www.dekker.com

Current printing (last digit):
10 9 8 7 6 5 4 3 2 1

PREFACE

The authors of the Third Edition joined the faculty of the Department of Electrical Engineering at the University of Cincinnati within three months of each other, and collaborated on a number of academic projects during the ensuing years. In addition, we were co-editors of the *Handbook of Electric Motors*, published by Marcel Dekker, Inc., in 1995. In the early years of our careers, we both taught design courses at various times, but subsequently Dr. Middendorf became much more heavily involved with instruction in the design area because of his wide-ranging consulting work in the electromechanical product design field. Most of his consulting work was concentrated in the design of small circuit breakers, including production, testing, and approval by Underwriters Laboratories (UL). This work led on into investigations into the field of insulation.

During the same time frame, Professor Engelmann did consulting work in mechanical and electromechanical design, but with a much different emphasis than that of Dr. Middendorf. Most of his consulting work involved the design of large one or two of a kind machines, including deformation generators for rolling reinforcing rod and the drive systems and controls for large lathes, machines used in the steel industry for turning as-cast rolls for rolling mills. He was also retained for a number of years for the design of small automatic control systems, including both analog and digital systems.

We both recognized early on the importance of feasibility studies, of patents, the need to be able to create alternative designs and to invent, the usefulness of various models, the fact that much of a design project requires that informed decisions must be made at every turn, and that all of these facets of a design project required knowledge of optimum design procedures, reliability, and testing. Both of us were also sought after as experts in product liability cases (in one case on opposite sides of the same case), reinforcing our recognition of the importance of standards and of the ability to provide clear and concise testimony. All of this experience has shaped the content of this work. As might be expected, the experience of the senior author (Dr. Middendorf) has been the dominant factor in this work because the first two editions were his work alone, but this edition includes examples drawn from the experience of Professor Engelmann as well.

PREFACE

At about the time that the first edition of *Design of Devices and Systems* appeared, the Accreditation Board for Engineering and Technology released its 1986-1987 document describing, among other topics, the requirements for the design component of engineering curricula. Those requirements were well met by the coverage of topics in that first edition. The second edition went a step farther, incorporating suggestions made by students who were assigned to write reviews of each chapter. It also added a significant amount of updated material, especially in the chapters on reliability and computer-aided design. This edition extends those and other topics, and includes information on modern search methods for relevant standards and patents in any field of interest.

The material of this book has been used for a year-long course at the senior level. Although many of the concepts do not preclude an earlier presentation and the mathematical preparation required is probably well in hand by the mid-point of an engineering student's studies, it is principally the maturity of the students and the advantage of relating the material to the high-level technology presented by other senior-year courses that make the late presentation desirable. In addition, many of those students have been employed in part-time or summer engineering positions, and our own students have been employed by engineering departments in industry during their required cooperative work/study experience. They have thus brought their own experiences to class discussions, and those diverse experiences have challenged, modified, and broadened our own understanding of the design field.

We have thus had the advantage of access to the academic attitudes that push design toward ever-higher scientific and technological levels. We have also had the advantage of access to the industrial attitudes that recognize the need for design procedures that result in an entire population of acceptable models, many times with a range of ratings, and always at an acceptable price.

We want to thank William Singer for reviewing the legal aspects of product liability in Chapter 2. Chapter 4 was carefully reviewed by Dean Shupe, Professor Emeritus of Mechanical Engineering, and his suggestions for improvements were much appreciated. We also wish to thank Kurt Grossman of Wood Herron & Evans for his invaluable assistance with Chapter 5 on Patents, and Albert Klosterman, Thomas Sigafoos, and Dan Hassler of Structural Dynamics Research Corporation (SDRC) for their assistance in providing the background material for the discussions in Chapter 11 on the use of computers in design and in Chapter 12 on optimization.

William H. Middendorf
Richard H. Engelmann

INTRODUCTION

Most of the study of devices and systems by undergraduate engineering students involves analysis. Most of the engineering activity related to devices and systems in industry involves design and production. This dichotomy is an unavoidable result of the fact that the ability to design depends heavily on the ability to analyze, and the sequence of courses in a typical engineering curriculum is therefore dictated by the way in which the student develops. The importance of the design function in formal engineering education has been the subject of much discussion in the past twenty years or so, and professional and academic accrediting organizations have confirmed the need for the study of design by typically requiring one-half year of a four-year baccalaureate curriculum to be devoted to it. That is, a bridge from analysis to design has now been built into undergraduate engineering programs. Almost all of the design content occurs in the junior and senior years because it is not until this stage of a student's education that the analysis capability has been developed to a sufficiently high level.

The most important reason for studying design, however, is that it demonstrates to the student how the many seemingly diverse topics of an engineering program must be synergized to enhance his or her technical skills. *Design is interdisciplinary.* It calls for the ability to recognize situations in which certain theories or techniques can be used and to develop problem-solving methods to fit a particular situation.

Design is best learned by being faced with a variety of problem situations while working under the direction of a "chief engineer" or a mentor. However, in preparation for that kind of experience, the student needs to learn the types of problem situations likely to be encountered and must learn suitable techniques for solving those types of problems. That is, the student needs to acquire a set of tools, and, just like the craftsman, must learn which tools to make use of in a particular situation. The objective of this book is to introduce the student to a variety of tools or methods that will allow the engineer to obtain useful answers to the myriad of questions that product designers encounter. We have attempted to weave these techniques into an order that allows the student to feel that he or she is encountering the first real design experience without the necessity of inventing each step as the project proceeds. There is no suggestion that the list of techniques is complete; it is not and cannot be complete. Wherever possible, the student is directed to other references that will stimulate the inquiring mind to find more suitable or easier methods of solving the problem at hand.

This book begins with the concept of the life cycle of a product. This is introduced in the first chapter along with a five-step design process that seems to fit most products well. The process described is not a linear, once-through, process. Numerous feedback paths are shown; these are necessary because of the extensive interaction that must take place among the various departments of any successful manufacturer. Although the term "concurrent engineering" is not used, it will be recognized that the basic philosophies of the process shown and that of concurrent engineering are in harmony.

The need for safe products that truly satisfy the needs of the intended purchaser is emphasized throughout. Whenever specific advice can be given to avoid difficulty, this is done. As an example, the second chapter lists twenty ways to reduce product liability risk. Specifications are determined by need analysis and considerations of safety, again stressing the total product life from production to discard, and the early determination of feasibility motivates a treatment of engineering economics in sufficient depth that cost analysis and investment evaluation are seen to be of importance to the designer.

No product will be on the market for more than a few years unless it is improved on a regular basis. Product improvement based on new concepts or new manufacturing techniques is encouraged by the treatment of inventions and patents. Improvements rarely result from dramatic breakthroughs in technology, but most often occur as the result of incremental improvements. It is true, of course, that in a large device, such as a household appliance, or in a small system, a number of incremental improvements may be made in various subassemblies, giving the appearance of having resulted from technological breakthroughs, but this is only rarely the case.

Design is carried out by developing models of the proposed product, and for the designer of the future no preliminary model will be more important than the mathematical model. One long-neglected mathematical approach that is given much attention in Chapter 7 is that of dimensional analysis. A model can be made to represent a single product having a range of ratings by use of a π-term. It can also be shown that most new products can be represented by products already in existence, acting as pseudomodels, if corresponding π-terms can be suitably modified. This knowledge allows the engineering designer to obtain some reliable data on a new design early in the project. The ultimate expansion of the idea is, of course, computer-aided engineering, in which information is drawn from a data bank and can be used to exercise models of a product prior to detailed design.

INTRODUCTION

Design depends heavily on decision making. Traditional decision making techniques, CPM and PERT, have been included, but these are chiefly useful when making decisions on scheduling of various activities so as to expedite a project. On the other hand, the precedence matrix provides a tool for deciding which elements of a device or system to design or specify first, thus avoiding as much as possible the necessity of making any more assumptions about other elements than absolutely necessary. General techniques are also discussed for making decisions under conditions of certainty, risk, and uncertainty.

In the end, the designer must arrive at details having sufficient specificity that the device or system can be built. The five-step design procedure of Chapter 1 is expanded in Chapter 9 as a two-dimensional morphological matrix for projects at the community and system levels, and is then further expanded to a morphological cube for interdisciplinary projects at those levels. The objective is to create a structure by which projects of any complexity can be resolved into constituent subsystems and components and designed by synthesis or repeated analysis.

Although design now requires more of the engineer than ever before, it is also true that the engineer has more accurate descriptions of various phenomena, more training in mathematics, and vastly improved computational aids. These aids range from relatively small programs, such as those described early in Chapter 11 for use on personal computers (PC's) to expert systems and to the workstations used with comprehensive computer-aided engineering (CAE) systems described later in that chapter. It is also important that the student understand that CAE enables the designer to make use of powerful simulation techniques and to relate designs very closely to manufacturing processes.

Optimum design is an important tool because it provides one additional constraint on which to make decisions, thus making the decision process a bit easier. If a mathematical model is developed, the additional constraints provided by optimization can greatly facilitate the entire design process. Any design resulting in the use of less energy or material, providing an easier method of manufacture or resulting in increased profit—even if the improvement is but a few percent—can be the margin of success needed by a manufacturer.

Every acceptable design requires an acceptable level of reliability. Ways by which reliability can be assessed and improved are discussed in Chapter 13. Because reliability assessment generally relies on testing procedures and many products, if tested to failure, would be out-of-date before reaching that point, accelerated life tests and their design and limitations are discussed. Important considerations when doing testing are the difficulties inherent in drawing meaningful conclusions if there are only a few samples to test, or how to handle a testing situation in which some tests must be discontinued.

Although deferred to the last two chapters, human factors engineering and the art of design in engineering are important topics. They have been deferred to the final portion of the book because we believe that the student can better appreciate their importance after studying the principal steps of the design process. It cannot be emphasized too strongly, however, that they need to be considered very early in a design if a safe and yet aesthetically satisfying product is to be manufactured.

Reinforcement of the material in class has been done in at least two ways. One is to have students select case studies from a library of published engineering cases, either those published by the American Society of Engineering Education or by others. Each student's in-class presentation gives an opportunity to discuss how these case projects could have been carried out in the light of the material covered in the course or how the case shows use of the techniques that have been explored.

The second reinforcement method is by having the student select a product to which he or she will apply relevant techniques from the book or from references. The product may be a variation of a commercial product or it may be a concept that the student wishes to explore. It must obviously be simple enough that the student can make realistic and practical decisions. Product liability analysis, need analysis, feasibility studies, and patent and standard searches are among the design aspects that the student pursues as exercises.

CONTENTS

Preface   iii
Introduction   v

1   DESIGN AND THE ENGINEER   1

     1.1   Design   4
     1.2   The Engineer as a Designer   14
     1.3   Ethics   17
     1.4   Summary   25
           References   26
           Review and Discussion   26
           Practice Projects   28

2   PRODUCT LIABILITY   31

     2.1   Product Liability Evolution   31
     2.2   Foreseeable Use and Misuse   36
     2.3   Reducing Product Liability Risk   40
     2.4   The Importance of Design Review and Checking   62
     2.5   The Engineer as Expert Witness   64
     2.6   Summary   71
           References   72
           Review and Discussion   72
           Practice Projects   73

3	**NEED ANALYSIS AND SPECIFICATIONS**	76
	3.1 Need Analysis for Total Product Life	77
	3.2 Determining What the Customer Wants	79
	3.3 Determining the Restrictions of Groups with Authority	80
	3.4 Determining Company Restrictions	89
	3.5 Needs Analysis	90
	3.6 Summary	94
	References	95
	Review and Discussion	95
	Practice Projects	96

4	**THE FEASIBILITY STUDY**	97
	4.1 Physical Realizability	98
	4.2 A Physical Realizability Study	101
	4.3 Company Compatibility	109
	4.4 Economic Decision Analysis	110
	4.5 Cost Estimating	132
	4.6 What Determines Selling Price?	136
	4.7 Summary	139
	References	139
	Review and Discussion	140
	Practice Projects	141

5	**PATENTS**	146
	5.1 Why the Designer is Interested in Patents	147
	5.2 Structure of a Patent	151
	5.3 Patent Searches: The Classification System	159
	5.4 Basics of Patent Law	165
	5.5 Computer Software	169
	5.6 Other Intellectual Property	171
	5.7 Summary	174
	References	175
	Review and Discussion	175
	Practice Project	176

6	**ALTERNATIVE DESIGNS AND INVENTIONS**	177
	6.1 Alternative Designs	177
	6.2 What Can Be Learned from Great Inventors of the Past	179

CONTENTS

	6.3	A Theory of Invention	184
	6.4	Blocks to Creativity	187
	6.5	What Kind of Person Invents?	189
	6.6	Identifying the Strategy	191
	6.7	How to Improve Your Ability to Invent	194
	6.8	Methods to Stimulate Invention	197
	6.9	Summary	208
		References	208
		Review and Discussion	209
		Practice Projects	210
7	**MODELS**		212
	7.1	Sketches and Drawings	213
	7.2	Block Diagrams	216
	7.3	Network Models	218
	7.4	Mathematical Models	221
	7.5	Physical Models	224
	7.6	Combination of Models	225
	7.7	Use of Dimensional Analysis	231
	7.8	Summary	253
		References	253
		Review and Discussion	254
		Practice Projects	255
8	**DECISIONS**		264
	8.1	Elements of Every Decision	265
	8.2	Types of Decision Problems	266
	8.3	Decisions Under Certainty	267
	8.4	Decisions Under Risk	301
	8.5	Decisions Under Uncertainty	308
	8.6	Combination of Models	312
	8.7	Summary	318
		References	318
		Review and Discussion	319
		Practice Projects	319
9	**THE DESIGN OF SYSTEMS**		327
	9.1	What Is a System?	328
	9.2	The System Designer	329

	9.3	Bringing the System into Focus	330
	9.4	System Design	332
	9.5	Summary	338
		References	338
		Review and Discussion	339
		Practice Projects	339
10	**DETAILED DESIGN OF DEVICES AND SYSTEMS**		**340**
	10.1	Evolution, Repeated Analysis, and Synthesis	340
	10.2	Design by Repeated Analysis	342
	10.3	Design by Synthesis	348
	10.4	System Design; Reduction to Components	359
	10.5	Summary	367
		References	368
		Review and Discussion	368
		Practice Projects	369
11	**PRODUCT DESIGN USING COMPUTERS**		**374**
	11.1	Hardware	374
	11.2	Software	376
	11.3	Computer Integrated Manufacturing (CIM)	389
	11.4	Computer-Aided Design/Computer-Aided Manufacturing	392
	11.5	Expert Systems	396
	11.6	Comprehensive CAE Systems	399
	11.7	Summary	406
		References	407
		Review and Discussion	407
		Practice Projects	408
12	**OPTIMUM DESIGN**		**411**
	12.1	Preliminary Considerations	412
	12.2	The General Optimization Problem	414
	12.3	Optimization With Only a Criterion Function	416
	12.4	Optimization With Functional Constraints	420
	12.5	Optimization With Regional Constraints	426
	12.6	Optimization With Functional and Regional Constraints—Lagrange Multipliers Extended	430
	12.7	Searching	436
	12.8	Optimization Using Comprehensive CAE Programs	442

CONTENTS

	12.9	Summary	444
		References	445
		Review and Discussion	445
		Practice Projects	446

13 RELIABILITY 452

	13.1	The Nature of Failures	453
	13.2	Basic Relationships	456
	13.3	Chance Failures	457
	13.4	Early Failures	461
	13.5	Wearout	464
	13.6	Stress/Strength Interference	467
	13.7	Calculation of Product Reliability	470
	13.8	Other Statistical Distributions	475
	13.9	Design Considerations	478
	13.10	Reliability Growth Modeling	481
	13.11	Human Factors	485
	13.12	Summary	486
		References	486
		Review and Discussion	487
		Practice Projects	488

14 ACCELERATED LIFE TESTING 491

	14.1	Theory of Accelerated Testing	492
	14.2	Testing a Few Samples	499
	14.3	Goodness of Fit; Censored Data	503
	14.4	Evaluating Proposed Improvements	508
	14.5	Summary	516
		References	517
		Review and Discussion	517
		Practice Projects	518

15 HUMAN FACTORS ENGINEERING 520

	15.1	Human-Machine Interactions	521
	15.2	Anthropometry	522
	15.3	Bones and Muscles; Force and Work	525
	15.4	Speed and Accuracy	529
	15.5	Designing to Avoid Human Errors	530
	15.6	Illumination	532

	15.7	Information Feedback from the Product	534
	15.8	Fatigue, Boredom, and Vigilance	538
	15.9	Including Human Factors in Design	540
	15.10	Summary	541
		References	541
		Review and Discussion	542
16	**THE ART OF DESIGN**		**543**
	16.1	Design for Production	543
	16.2	Industrial Design	551
	16.3	Summary	555
		References	555
Index			557

1

DESIGN AND THE ENGINEER

Engineering is frequently viewed as a problem-solving discipline. In engineering colleges, first principles derived from the sciences, such as Newton's Laws of Motion, are applied to well-defined problems. These problems are usually stated in such a way that it is possible to use the given data to formulate equations that the student can easily recognize. As the complexity of the problem increases, equations are required that describe phenomena in greater detail.

For example, deflection of an electron beam in the cathode ray tube shown diagrammatically in Figure 1.1 may be analyzed by use of the tube geometry and the equation

$$D = E_d L d / 2 E_a A$$

where E_a = the accelerating voltage of the electron gun,
A = the separation between the deflecting plates, and
the other variables and parameters are defined in the figure.

The electric field between the deflecting plates is assumed to be confined entirely to the parallelepiped having the deflection plates as its upper and lower surfaces.

If the velocity of the electron is low, the ratio of the electron charge to its mass may be assumed to be that of the electron at rest; if the frequency of the voltage applied to the deflecting plates is low, the voltage may be assumed to be constant during the electron's flight time through the deflecting field. If the electron velocity is sufficiently high, the mass must be corrected using the Lorentz equations.

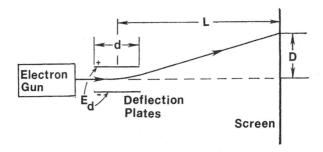

Figure 1.1 Diagrammatic representation of a cathode ray tube.

As the frequency of the voltage applied to the deflection plates is increased, a frequency will be reached at which there is enough change in the deflecting voltage during the flight time that the voltage E_d in the equation above must be represented by a time-dependent function. The simple form of the equation, which was derived from more basic equations, no longer holds. One of the first lessons to be learned from this example is that engineering is not simply problem solving, although that is an important part of the work that engineers do. *Engineering requires a great deal of decision making.* Are the electron velocity and frequency of the deflecting voltage both low enough that the equation will yield an accurate result? If not, can one expect the result to be close enough to be acceptable (in some agreed upon sense) if the mass is not corrected and the flight time between the deflection plates is used to average the deflecting voltage? Because the electron velocity as the electron leaves the deflecting plates is the vector sum of the velocity with which it left the electron gun and the transverse velocity added by the deflecting voltage, there will be a second-order effect on the electron mass. Should this be taken into consideration, or can it be neglected? One sees that, even in this apparently simple analysis problem, there are a number of decisions which must be made. In some cases, the engineer may find it desirable to use experimentation to arrive at a decision.

The antithesis of analysis is synthesis. In an ideal world, if one were given a sufficient set of functional specifications for a new device, such as the energy supply to be used, the nature of the inputs, and the desired outputs, synthesis procedures would lead directly to the final form of the device and to the manufacturing specifications. In the real world, there are very few synthesis procedures which lead directly to the final result. Even in such areas as the design of electrical networks, although there are procedures which lead directly to a network configuration (including all parameter values) which will produce the desired response, one cannot say that the design is completed until the effect on the response of tolerances of the various components has been investigated. That is, in the real world, functional specifications lead to the final form of the device and to the manufacturing specifications by an iterative process generally referred to as engineering design.

In this iterative process, large amounts of information must be obtained, much of it not in the form of precise equations or numerical data, and there are numerous decisions to be made, far more than in the cathode ray tube analysis problem used as an example above. The iterative process, as a glance at Figure 1.2 on page 11 will show, is a multiloop process in which some decisions will take the designer back to earlier parts of the process. The fact that much of the work requires decisions cannot be stressed too highly.

In addition to the order of magnitude difference in the number of decisions that must be made, there is a psychological difference which becomes obvious very early to the neophyte designer. In the case of analysis—say of the cathode ray tube above—there is a single *right* answer to the value of D given the original data. We may be willing to allow for some error by neglecting certain aspects of the problem, such as the mass correction and the fringing of the electric field, but we know that there *is* a right answer, and only one.

By contrast, there is *never* a single right answer when doing engineering design. That point may be illustrated by considering any common household appliance. If one looks at refrigerators, one finds first of all a wide variety of sizes. Even if one settles on a given size, the prospective purchaser is still faced by a wide range of choices: Freezer on top or on the bottom? Self-defrosting or not? Single door or double door? With or without a built-in ice maker? If it has an ice maker, what is the shape of the ice "cubes" that it produces? All of the refrigerators we see on the showroom floor perform the basic functions the customer is seeking, but all of them are different. There are no "right" answers, although occasionally there may be one which is clearly "wrong" and another which is far superior. From this discussion, one can fairly conclude that design is an imperfect process, and that characterization of the end result as a "good" or a "bad" design is often the result of a subjective evaluation.

In spite of the uncertainty with which the designer is faced, if one considers design and the function of the designer in general terms, one quickly finds that there are a number of questions to consider, and that answers to these equations will allow for an orderly approach to design. We need to investigate:

   What is engineering design?
   What does a designer do?
   Is there some plan of action that will help a designer to do his or her job in an efficient manner? That is, can the decisions be made in an orderly manner? How does one avoid pitfalls?
   What must a designer know to work effectively? Where does one obtain that information?

> During the course of the design process, the designer must work with others, both inside one's own company and with engineers and others outside the company. Are there ethical questions which might arise during these interactions?

A good designer is a constant learner. The answers to these questions will be discussed in general terms in this chapter, and the volume as a whole will lead the designer toward more complete answers. If at all possible, the new designer should make an effort to learn from more knowledgeable designers. For example, which choices are more likely to result in a higher level of performance or greater acceptability to the consumer?

1.1 DESIGN

What is engineering design? The introductory discussion can lead rather naturally to various definitions of engineering design. One definition that seems to cover the topic in a reasonably concise manner is as follows:

> Engineering design is an iterative decision-making activity whereby scientific and technological information is used to produce a system, device, or process which is different in some degree from what the designer knows to have been done before and which is intended to meet human needs at an acceptable cost.

To carry out this activity, the designer uses various structured techniques and scientific principles to make decisions regarding the selection of materials and components, as well as decisions regarding their placement and interconnections, in order to form a system or device that satisfies a set of specified or implied requirements. This simple statement, which relates "design" to the determination of what materials to use in which relationships, is true for products as diverse as a microprocessor, a space shuttle, or a building crane. It says, in effect, that design is driven by materials, manufacturing techniques (the interconnections), and economics.

Design engineers are essentially decision makers! It would be desirable to make all decisions with certainty: For example, deciding on the thickness of a beam after calculating all stresses in the beam with complete knowledge of the environment in which it will be used. Where possible, this is the way in which design should be done. However, many decisions must be made with incomplete knowledge; for instance, there may be some uncertainty as to the precise value of the modulus of elasticity of the beam material, and the loads on the beam may themselves be random in magnitude as well as in location along the beam. In those cases in which the exact nature of the variables is not known, the design process becomes less rigorous than is desirable and the designer must introduce

DESIGN AND THE ENGINEER

some of the elements of art and heuristic decision making. The most talented designers must be prepared to be scientific, intuitive, and creative, depending on the problem at hand and the depth or shallowness of the knowledge available. The final chapter in this book addresses some of these points.

1.1.1 Scientific and Engineering Methods

The definition of design above tells what design is, but it gives no clue as to how a design is produced. A method that can be used as a model is the "scientific method." The scientific method evolved as a logical and orderly way of relating phenomena observed experimentally to the mental activity leading to theories which explain the experimental phenomena. It is an iterative method, the steps listed below constituting one cycle of inquiry.

1. Observe a phenomenon and record data concerning it.
2. Postulate a theory to explain the phenomenon.
3. Develop and conduct an experiment to test the theory.
4. Draw conclusions as to the validity of the theory.

In the early industrial period, most products evolved slowly with step-by-step improvements being tested in the field, that is, by the user. For the most part, "design" was synonymous with "drawing." Intuition was frequently the guide used to decide if a particular part was strong enough or if it had the proper spatial relationship to another part. The drawing or drawings constituted a model which indicated how parts would move and whether they would fit.

As products became more complicated and the need for new products accelerated, it became necessary to develop techniques which would lead in a predictable manner to successful designs. Moreover, owners and managers of manufacturing plants began to perceive that by developing components which would enhance the quality of their product they would simultaneously be rewarded by the increased demand created by satisfied customers. The industrial laboratory, which began to appear about 100 years ago, was a natural outgrowth of that perception. In time, these organizations developed the "engineering method." Like the scientific method, this is an iterative method, each cycle consisting of the following steps:

1. Establish a set of specifications.
2. Develop a design concept.
3. Use physical and mathematical models to test the design concept.
4. Draw conclusions as to whether the design satisfies the specifications.

This is the design procedure stated in its most abbreviated and general terms.

The steps listed constitute the irreducible minimum number which completely describe the activity of design. If the design does not meet the specifications, the process is repeated, utilizing the information gained in the first attempt. Although it is possible to return to the first step and to modify the specifications, it is more probable that the design concept will be changed or that the models will be improved. Performance after the second cycle is again compared to the specifications, and the process is repeated as necessary until the design is acceptable.

Design is an iterative process, requiring numerous decisions.

1.1.2 Levels of Design Problems

Design problems vary in complexity. The design of a wall bracket on which to hang a coat and hat is trivial when compared to the design of a microprocessor. Yet both are designed for production by a manufacturer and application by others. The wall bracket will be used in a building; the microprocessor may be used as part of a control system. Thus both devices are components, but it is obvious that there are major differences in the technical levels required for the design and for manufacture.

Jones [1] identifies four levels of complexity of design in another way. His categories reflect the global nature (or lack thereof) of the design problem rather than the technical difficulty. The levels of design in increasing order of complexity are: The component level, the product level, the system level, and the community level.

Stevenson [2], in agreement with Jones, gives the following example of these levels:

> Consider a metropolitan area with a great deal of traffic congestion as people move to and from work. The problem is how to move people expeditiously. It might be solved by building more highways, or by building some sort of mass transit system, or by some combination of these. For a mass transit system a sub-problem involves deciding the transportation means to use: Subway or underground tube; busses, including exclusive highway bus lanes; monorail; etc. Routes to be followed form another part of the problem. This is design at the community level. Having decided to build an underground tube system, for instance, many things still must be decided, such as the tractive equipment and passenger cars to use, and whether underground trackage should be on the level or be built with a series of up and down grades. This is design at the systems level. Having decided, perhaps, that a new type of passenger car will be required, its design occurs at the product level, within limitations established at higher levels. From here we can

move to the design of a new set of trucks for the car, at the component level. This is all designing, at different levels.

This point of view, that engineering design varies in its influence from the lowest level, where a component need only operate satisfactorily with other parts of a product, to the highest level, where the design influences a whole community, is now well accepted.

Design procedures and engineering aspects of community level projects and system level projects are essentially identical. At the system level and below, these decisions are usually made only by engineers. For instance, in Stevenson's example above, the decisions as to what kind of traction equipment to use, whether to keep all tunnels at nearly the same level, whether standard passenger cars are to be used, and so forth, would probably be made by engineers. However, in this same example, the decision as to mass transit versus additional highways, although requiring engineering input, would probably be strongly affected by political considerations, public relations, and the inclinations of elected officials.

At the other end of the list of levels, there is an analogous situation. Whether something is a product or a component depends on the designer's point of view. To the manufacturer of a silver-tungsten contact, it is a product, but to the manufacturer of circuit breakers, that very necessary part is a component. Even though the designers view the specific item differently, the procedures used by engineers to develop a new contact are not essentially different from those used to design a circuit breaker.

One significant engineering difference in design procedures occurs between the product design level and the system design level. A product is a device in which essentially all major parts will be made by a single manufacturer and are thus under complete control of the designer. A system, on the other hand, involves a significant number of parts which are not all made to the designer's specifications. When using parts supplied by others, the designer must work with the input-output characteristics of the component he or she chooses from those available and must make certain that all parts and components are compatible.

Even this situation is somewhat blurred. There are few products made that do not include some off-the-shelf components. Philosophically, however, the differences between product design and system design are sufficiently great that separate treatments are warranted. Which category is appropriate for a given development depends on the extent to which the complete unit is made in-house from raw material or is made by assembling purchased parts. This dichotomy is reflected in this volume's title, where "devices" is intended to include components and products and "systems" is intended to include design at the community level. To avoid the longer expression, "devices and systems," the term "products" is frequently used to designate anything that is produced.

1.1.3 A Design Procedure

Design is an especially perplexing function for the inexperienced engineer. Most formal undergraduate training is involved with learning analytical procedures. In analysis, the engineer predicts the performance of a given device or system with stated initial conditions and stated driving forces. There is one and only one "right" answer. In design, on the other hand, the engineer must "predict" the product that is capable of the required performance in a variety of conditions, and there are a multitude of "right" answers. However, the ability to analyze is a prerequisite to the ability to design.

A successful designer usually follows some sort of pattern in responding to the challenge of a design problem. This pattern may have evolved over many years, and the designer may not be able to put it into precise words. Nevertheless, the pattern is a part of his or her method of operation. It is reasonable, then, that the series of activities by which the designer is most apt to complete a design assignment successfully should be discussed in great detail. However, just as there are benefits to be derived from such a discussion, there is also a *caveat*. No one can outline the steps of a single design process that will be correct for every situation. By the very nature of design, the process used will be different, depending on the type of system or device to be designed, the state of the art, the supporting personnel and equipment available, the number of units likely to be made, and so on. If little scientific information is available about the product, an intuitive approach may be necessary. At the other extreme, products which have been made for many years, such as steam boilers, electric motors, and automobiles, have a well-documented history and may also have well-defined mathematical procedures for use in some steps of the design process. The point is that the design procedure presented below should be looked on as a general outline of the steps to be taken. In a given situation, some of the steps will be completely unnecessary. In other situations, some of the steps shown may burgeon into major parts of the design activity.

Who actually carries out the design function? Historically, this function has been carried out by a single engineer or possibly by a small group of engineers. That engineer or group of engineers has had the complete responsibility for insuring that all of the necessary steps are carried out, as well as the responsibility for the necessary interactions with others in the company. Some of these interactions were not carried out very well, and on many occasions others in the company would be unsympathetic to requests for information from the designer. For example, manufacturing personnel tend to concentrate on the immediate problems of keeping production lines running with the minimum of down time, and are not very amenable to looking at a new design, especially in its early stages of development, from the standpoint of manufacturability. Many times the attitude would be that the designer should complete the design and bring it to manufacturing only when the design is complete. "At that point, we'll make some suggestions as to how to modify the design so that it's easier to manufacture." The purchasing department has its own set of problems, and does not want to look into the cost of purchased components, especially since the component may not even be used in the final design. Designers were frequently looked on by others in a company as a necessary evil.

The process just described is a sequential process, and it is clear that such a procedure leads to a cycle time for a new design which will be increased in length because of the redesign forced by the manufacturing personnel or by the lead time for purchasing needed components. Not only is total time increased, but costs are obviously higher because of redesign. They may also be higher because it might be necessary to use higher priced components in order to obtain acceptable delivery times. Finally, the length of the entire cycle may become so long that competitors have reached the market with a comparable product before you have prototypes to demonstrate for potential customers.

Because of the disadvantages of the sequential process, in recent years attention has been turned toward a process known as concurrent engineering, often referred to simply by the initials CE. Salamone [3] states that "The simplest definition of Concurrent Engineering (CE) is the simultaneous development of product and process." He points out that there are three main reasons for the evolution from the sequential process of design to the concurrent engineering approach. These are: The rapid pace of development of technology which forces a company to update its products on a timely basis in order to retain or expand market share, the design cycle compression consequent on this need to remain competitive, and the emergence of personal computers and workstations which allowed access to large databases and to computer-aided design tools, such as wire-frame and three-dimensional solid object representations of possible designs. The effect of the concurrent engineering philosophy can be seen in the late 1995 announcement by Toyota that it had set a goal of one year for bringing a new car from design to assembly line, a process for which a three-year time frame was once considered to be a major accomplishment.

The sequential design process has few paths that carry one back to earlier parts of the process, and many of those come very late in the cycle. The manufacturing division that says "Bring us your final design and we'll tell you what needs to be done to improve manufacturability" is an example of a feedback path which comes far too late. By contrast, concurrent engineering requires extensive interaction among various groups within a company and some who are outside, and requires that feedback occur continually throughout the entire process. The block diagram shown in Figure 1.2 shows possible feedback loops.

This figure shows activities grouped under five main headings. The three blocks off the sides of the five-block core indicate that, after working through each of the five major activities of the design process, consideration should be given to returning to one or more of the earlier activities with the objective of applying new information to those phases of the project. On occasion, iteration within one of the five major blocks is appropriate. The major blocks are:

1. Problem Definition. This is the initiating point of the design process. Often the engineering personnel recognize that a product improvement or an entirely new product is warranted. Perhaps information obtained from service calls indicates that a certain component is failing prematurely, or technological advances in the field make evident the need for a device with improved performance, or concern for safety or product liability considerations indicate that a safer machine should replace the present design. Frequently the request for a new product will come from the sales department. Sales representatives, having day-to-day contact with customers, often get suggestions for products which may have sales potential, or for improvements in existing products. No matter where or by whom the need is identified, it should be subjected to market research to evaluate the sales potential. That is, when a project is assigned to a design engineer, the potential sales volume should already have been determined.

    It is also necessary for engineering personnel to determine what technology is available to improve the product, and whether the in-house manufacturing facilities are capable of producing the product. Often, of course, the product is similar to those already being produced by the company and answers to such concerns are easily resolved. The designer should also determine how competitors make similar products that fill the same need. This is frequently done by purchasing a competitor's product and subjecting it to a series of tests. Some years ago, one of the authors (RHE) was visiting an electric motor manufacturer, which shall be identified as Company A, and was shown a 7-1/2 hp motor and a 10 hp motor, both of which were built by Company B. The nameplates were identical except for horsepower ratings and horsepower-related variables. Company A's personnel had opened up the two motors and discovered that they were identical!

2. Problem Evaluation. It is at this stage that the designer needs to practice the art of asking the right questions, because the design task is rarely described to the design engineer in complete detail. Time must be allotted to determine precisely what is wanted. Often, the first information received is an idealized description of a proposed product. This description should not be ignored; instead it must be translated into what can actually be accomplished. This is best done by considering what everyone who will come into contact with the product will expect of it. Beginning with the manufacturing organization itself, there are certain qualities the administration wants the product to have, such as ease of manufacture of parts and ease of assembly of the product. The seller of the product wants, among other things, a well-protected, attractive product. In turn, the consumer expects a product that meets or exceeds applicable industry standards. The product

DESIGN AND THE ENGINEER

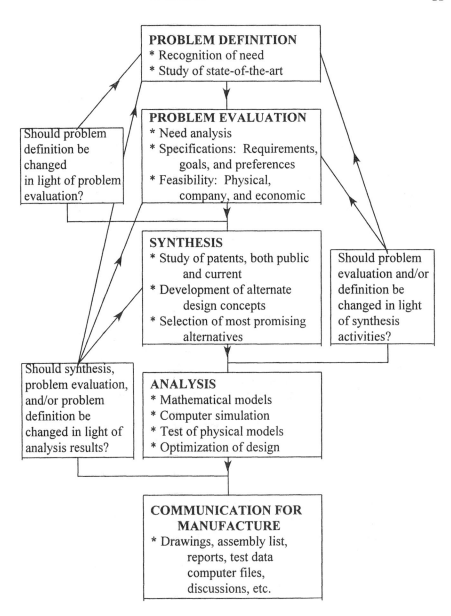

Figure 1.2 A design procedure.

must be safe to use and cause no undue problems in disposal after it has served its purpose.

In order to determine all that is expected of the product, a needs analysis must be made. The important result of this activity is a set of specifications, goals, and preferences that describe the functional requirements and physical limitations of the product.

When specifications are determined, there is sufficient information to make a preliminary study of whether the product should be made or not. Is it feasible? In the early design stages, the analysis will be questionable, but at least it will give some indication of probable success or failure. As the project proceeds, the analysis should be repeated when data are more exact. With each repetition, confidence in the results will grow. *If it becomes apparent at any point that the project cannot be successful, the project should be terminated at once. You will never have spent less time and money on the project than at that point.*

3. Synthesis. The most creative step in the design procedure is the development of a number of tentative solutions, any or all of which could lead to a workable design. Often this step leads to inventions, for which patent protection should be considered.

During this phase, a patent study should be made because a patent gives sole rights for a limited time to make, sell, or use the device or process described by the claims of the patent. Both time and money will be wasted if a product is developed that infringes on a current patent. An efficient patent search procedure as well as basic facts of patent law will be described in a later chapter.

After generating several approaches, one of the design concepts is usually chosen for further development. The selection of which concept to pursue must be done to some extent in a subjective manner. The designer's own experience and that of others in the company will have a large effect. Intuition may play a role. Sometimes tests on very crude models may be necessary to obtain sufficient data to help make a choice. It is almost inevitable that the designer will be required to make a decision based on incomplete information. Given the uncertainty with which the designer is faced at this stage, it may seem preferable to develop several design concepts to the final form; in ordinary business environments such a multiple development is much too expensive to be considered.

4. Analysis. Once the decision has been made to proceed with a certain design concept, it should be modeled, both mathematically and physically, to determine the operating characteristics in all foreseeable situations and environments. Most important at this step is to compare the results of the analyses and tests to the specifications, goals, and preferences developed in the second step. In some cases, perceived needs must be revisited in light of information accumulated from the tests and analyses. Limitations on what can be done are typically much clearer at this point than when the

specifications, goals, and preferences were originally determined. It is not at all unusual to find that some specifications will be exceeded, others will be met only conditionally, and some will not be met at all. Likewise, some components may turn out to be difficult to manufacture or expensive to purchase. At this point, the designer must consider all possibilities to optimize the system or device following a set of criteria that have been agreed on by representatives of all affected company departments.

If the product is to be evaluated by a standards writing agency such as Underwriters Laboratories, Inc., if it will require approval by a government agency, or if it is being manufactured for a single customer, models should be submitted to the appropriate organization before taking the next step.

5. Communication for Manufacture. It is at this stage that precise information about every part must be transmitted to all departments involved in making the product. Ideally, the information will be so complete that all who are involved have precise information as to correct materials, dimensions and tolerances, surface conditions, and assembly sequences. Departments affected by the introduction of a new product include purchasing, tool engineering, quality control, manufacturing, and marketing. (In some companies, there will be still others.) Information is usually transmitted by means of bills of materials, drawings, parts lists, and test reports.

These five numbered steps should be carried out in the order given, but a number of steps should be repeated as the design proceeds, as indicated by the side blocks in Figure 1.2. For example, after the design concept is selected it is prudent to repeat the search for patents as insurance that the concept selected is not in conflict with a valid patent. Moreover, at this stage a patent search may reveal related patents which were overlooked during the earlier search. As another example, after a preliminary attempt to verify that the design meets the specifications, it may be desirable to return to the analysis of the device or system and improve the mathematical or physical model. Such an improvement may lead to new values for the optimum design. In following the design procedure, one is constantly looking for opportunities to improve the quality of the design and to sharpen one's ideas on what should be done. *Design is an iterative process.*

As the design moves forward, designers must avail themselves of the expertise of others within the organization. These include engineers and technicians, especially those with special training or experience in areas different from that of the designer, and appropriate manufacturing, quality control, purchasing, and marketing personnel. One-on-one interaction is best for specific questions, but design review by a group from time to time will uncover points that have escaped the designer's attention. It is important that this not be a *pro forma* review; questions raised for which the designer has no answer or only an incomplete one must be pursued by the designer to resolve the uncertainties that prompted the questions. It is also important to recognize that the review group may provide very useful insights which can be used to improve the design. The designer should also insure that contributions by the review group are recognized by those to whom the designer reports.

Each step of the design process has been described as briefly as possible so that the overall plan is understood before considering each step in more detail. The remainder of this book presents information and techniques needed to carry out the design procedure. In all of this, the reader should think of himself or herself as the "engineer" or the "designer," and the organization for which the product is being designed as the "company," "manufacturer," or "organization."

1.2 THE ENGINEER AS A DESIGNER

One of the important advantages of studying engineering is the number of different occupational activities available to those who are knowledgeable in this field. These occupations cover a broad spectrum, from research, which usually requires a highly developed ability to use mathematical concepts and the ability to devise and implement meaningful experiments, to sales, which requires a highly developed ability to work with people and to teach others what a product's capabilities are. Somewhere in the middle of the spectrum is the occupation of engineering design, which requires a measure of skill in both mathematics and sales, as well as the ability to make acceptable decisions based on incomplete information. Perhaps because engineering design is in the middle of the occupational spectrum, or perhaps because there is so much design work to be done, a large number of engineers find themselves either partially or fully engaged in design.

1.2.1 Duties of the Designer

The designer's responsibility to produce the prototype or model from which all the units to be offered to consumers are fashioned calls for extreme accuracy. A mistake in the prototype can leave its mark on the whole population of subsequent similar devices. A 1994 example is a widely used microprocessor chip which was found to produce erroneous results for certain rare, but still important, calculations. In spite of the need for accuracy and correct decisions, the designer must make most of his or her judgments in the face of uncertainty. As the project begins, judgments may need to be based on experience with only loosely related products. Occasionally, the designer is involved with a basically new item and no previous experience is available to serve as a guide. Even as the project nears the production stage, final judgments are usually based on a small group of models, models which are usually expensive and are frequently hand-made.

Sometimes, because of the special care given to these models, they are not truly representative of the product as it will be manufactured, and the production models will not perform as well as anticipated. Should this occur, a decision must be made as to whether the reduced performance can be accepted or whether the cause or causes of the performance reduction must be identified and corrected.

On the other hand, production units may outperform the models, performing so much better than the model that some manufacturing specifications may be relaxed or less expensive components used. One of the authors (RHE) was once asked to look into the possibility of rewinding a small motor with an insulation suitable for a higher temperature than the insulation in the motor as purchased. The motor, being used in a model, was overheating. The overheating problem was not identified until the first of the production units was being fabricated, and a larger motor could not be substituted because of space limitations. The decision was made to proceed with the production run, and install the rebuilt motors as they became available. Fortunately, before the first motor was rebuilt, it was found with the first completed production units that the original motor would perform satisfactorily. The model, it was determined, had bearings that were not of as high a quality as the production units, and the overheating was due to the energy loss in the original bearings.

In any case, if the product does not perform satisfactorily for the consumer, it is said to be a poor design. Designers must accept the responsibility for producing an acceptable design despite manufacturing inaccuracies, variations in material, variations in conditions of use, and the like.

An unacceptable design can also result if the designer considers the operation of the product only under ideal conditions. An example of this is a washing machine produced around 1950 that rinsed clothes by introducing water through the center of a spinning basket containing the washed clothes. Water passed through the clothes by centrifugal force. Perhaps it worked well in the laboratory, but in areas where the water supply contained a significant level of minerals the clothes acted as a filter. White shirts came out of the wash with brown blotches. The manufacturer responded quickly. An external inlet filter was made available within several months after the washing machine was introduced, and the next model was provided with a built-in filter.

Some modern devices are extremely sensitive to environmental influences. This is particularly true of solid-state devices. The use of voltage surge arresters to protect personal computers is well known. Voltage dips, however, can also cause problems. For example, a problem occurred with a residential water softener, placed on the market in 1987. The softener has a sensor in the mineral tank that sends an electrical signal to a solid state control when the mineral concentration exceeds a threshold value. The signal was stored to initiate regeneration at the time set for this operation, usually at night. This new control was advertised as keeping the water within a narrow range of softness and reducing the amount of salt used. Unfortunately, it was tested in an environment that did not replicate that in which it was placed in service.

Some purchasers of the new model found that the salt usage increased tremendously over that with the previous model and that the water stayed extremely soft. Because the new units were within the warranty period when this occurred, dealers were subjected to multiple trouble calls. After some months of replacing parts and making adjustments to no avail, it was found that these units were being triggered by the starting transient (voltage disturbance) of motors connected to the same circuit as the softener. Putting the water softener on a separate supply circuit provided sufficient isolation from the motor transients that proper operation was restored. Certainly the presence of refrigerators, freezers, air conditioners, and power tools on circuits serving residential water softeners should have been anticipated, and the costly burden of learning about the problem from the customers could have been avoided, as could the cost of running separate circuits to the softeners. Moreover, one questions whether it would not have been better to modify the control circuitry to block the line transients rather than to install a separate circuit.

Designers must also consider the product at times other than when it is in use. For example, toy manufacturers have been subjected to loss and breakage of parts of toys by maltreatment while on the dealer's display shelves. The designer, realizing that the customers' habits are not easily changed, must design stronger parts that are captive, if possible, and must seek to improve the packaging. Products may be undesirable not of themselves but because of some condition that exists during the life cycle. Certain plastics will yield inferior molded parts if the raw material is not stored properly before molding. If the designer realizes that the organization for which he or she works does not have proper storage facilities, a different material should be selected.

The product is especially liable to events during manufacture that can render it defective. Presumably, the manufacturer's quality control program will find and eliminate defective products. However, quality control programs are never completely successful. The designer must make choices to ensure that the particular manufacturing facility involved—with its skills and deficiencies—can manufacture the product to acceptable quality standards. Nothing can be gained by designing a device that performs acceptably only if all parts are perfect. The tolerances that are allowable for satisfactory operation must be known, and the designer must be certain that the facility can manufacture the parts within those limits.

Products are invariably consumed in one way or another. Most of the consumption is caused by wearing of parts or degradation resulting from environmental conditions. With certain products, improvements occur so rapidly that a given model may be obsolescent long before it has worn out; personal computers and dedicated word processors are prime examples. When the product is judged by the consumer to be no longer efficient or desirable, it will be discarded. Usually, part of the product can be recycled to reusable material through a salvage operation in such a way that no harm is caused to people or to the environment. Other parts of a product are simply waste that cannot be rebuilt, repaired, or salvaged. Even final disposition of the device when it is no longer useful should be considered during the design process. For example, the tragic deaths by suffocation of children who played hide-and-seek in discarded refrigerators which had latches that could not be operated from the inside impelled designers to abandon such latches in favor of the magnetic seal on refrigerator doors.

Development of new materials and new manufacturing processes motivate design improvements. Designers play a key role in keeping their companies competitive by introducing improvements that stretch the capability of the manufacturer and keep the company current. Designers should be leaders, not followers. It is not enough to wait for the factory to install a new machine or process. There is no motivation until there is an obvious need, and the design can create that need. For example, automated assembly manufacture is much more likely to occur if products are designed for that manufacturing philosophy rather than attempting to use robots to replace workers in an assembly operation designed for humans.

The stages of the life of a product are production, distribution, consumption, and retirement. Products must be designed so that all four stages can be compatible with the needs of modern society.

1.3 ETHICS

New engineering graduates are often totally unprepared for problems or situations that will test their professionalism and ethics. Furthermore, published codes of ethics, such as that published by the Professionalism and Ethics Committee of the National Council of Examiners for Engineering and Surveying or the Institute of Electrical and Electronics Engineers (IEEE), may be of little value in gray area ethical situations. Consequently, the engineer must cultivate an awareness of the importance of creating, from a solid philosophical background, his or her own rules of ethical behavior for guidance in these situations. This can be done by recognizing that ethics and ethical theories must and can be learned, just as engineering theories are learned. The basic difference is that engineering theories are based on the behavior of physical objects, while ethical theories are based on models of ideal behavior of human beings. Ideal behavior of an engineer includes:

1. Using his or her knowledge and skill principally for the enhancement of human welfare,
2. Being honest and impartial in all dealings with the public, employers, and employees, and

3. Striving to increase the competence and prestige of the engineering profession.

It must also be remembered that the autonomy of professionals depends on how that group is perceived by the public. This means that the engineer must not only act ethically but also that the engineer must pay great attention to the appearance of his or her actions. An action that appears to be unethical should not be undertaken, regardless of the true ethical situation.

Because ethical behavior is superior to—that is, goes beyond—legal requirements, adherence to this deportment represents a voluntary commitment. The basis of any profession is its ethical underpinning, and every professional must either accept existing ethical guidelines or work actively to have them changed.

The purpose of this discussion is not to suggest how engineers should respond to every ethical situation, but rather to serve as a warning that design engineers are especially likely to find themselves faced with such problems. Some of the reasons for this are that designers specify materials and components that can represent considerable profit to a supplier or commission to a salesperson; design engineers often supervise technicians and other personnel who may propose ideas that are then incorporated into a design, ideas for which they deserve recognition; and engineers must design products that on the one hand are capable of returning reasonable profits to the manufacturers and on the other hand are safe and reliable for consumers.

Some actions are clearly unethical. Examples of such actions include:
- Accepting a consulting job for which the engineer is not qualified.
- Choosing a component or equipment based on friendship.
- Listing oneself inappropriately as the principal inventor on a patent application.
- Nonuniformly enforcing engineering employment contracts.
- Publicly alleging that an employer's product is unsafe ("whistle blowing") for reasons other than public safety considerations.
- Failing to provide opportunities for the professional and technical development of engineers and technicians working under his or her supervision.
- Accepting gratuities from suppliers.
- Inaccurately evaluating former employees seeking employment elsewhere.
- Awarding an unwarranted merit raise to an aggressive employee to avoid confrontation.

There have been a few well-publicized cases in which engineers have publicly accused their employers of proceeding with the development of unsafe products or projects. The ethical problems arising during the development of San Francisco's Bay Area Rapid Transit (BART) system [4] is an example. This is a high-density rail system designed to operate between San Francisco and

Oakland, California. The first step was taken in 1951 with the creation of the San Francisco Bay Area Rapid Transit Commission. In June of 1961 consultants submitted final plans for the system. It began operation in 1972. However, in 1971 three BART employees became concerned about the system's Automatic Train Control (ATC) subsystem.

They claimed, among other things, that use of a digital system increased complexity over that of an analog system to point of making it less reliable, and that the variety of frequencies used required the receivers to have a much wider bandwidth than the normally used single "train frequency" would have required. The engineers finally went public with their concerns and all three were discharged. The BART management apparently felt that the problem with the ATC could be solved and that such action on the employees' part was premature. The consequence was charges and countercharges, with the dismissed employees claiming that BART management blackballed them in a number of instances. On October 2, 1972, a train crashed through the barrier at the Fremont Terminal, which the system critics took as "proof" of the danger described by the dismissed engineers. However, the referenced article (1974) points out that at that time the system had already provided 200 million passenger-miles without fatality or serious injury.

This complicated case has many nuances beyond what can be described here. The point to this brief synopsis is that such situations do occur and to suggest that whistle blowing must be a last resort. Any engineer taking this action acts unethically if his or her accusations are motivated by the desire for notoriety or to be a hero or martyr. The engineer's motivation must be to protect the public's safety, health, or welfare from an employer who refuses to reduce significant and correctable hazards.

Unfortunately, the judgment on blowing the whistle usually cannot be made until many years after the act. The consequences of the individual's decisions are serious both to the individual and to those who might be harmed if actual defects are tolerated. These are most difficult decisions to make, and they require both a sound philosophical base and the courage to act ethically. Fortunately, the legal activity of recent years concerning product liability, which is the subject of the next chapter, has impelled manufacturers to be much more safety conscious than formerly.

Although safety problems are important and questions concerning safety do arise, you are more likely to experience other ethical problems, ones which arise without notice. You may be asked to reveal confidential information about a former employee, put under pressure by factory personnel to approve a part that is outside tolerance in order to keep production going, invited to use a condominium at an ocean beach owned by a would-be supplier, or approached by an acquaintance to accept remuneration in return for specifying a certain material. If you have considered what your ethical posture should be, you will respond to such situations in a way that is consistent with the kind of person you want to be; otherwise you may respond in a way that you will regret later.

There are many ways to express a set of guidelines to help you keep in mind what the profession would consider ethical behavior if faced with situations such as these. However, the best way to develop a posture of ethics is to be aware of the need to discuss such problems. To help clarify ethical behavior, three case studies are offered for possible class discussion. The first [5] is fictitious, the second is true, and the third [6] is hypothetical.

1.3.1 Exercises in Ethics

A. An electronics manufacturing concern, which we shall call Amptronic, Inc., was anxious to announce a new electronic component that would replace a combination of several traditional components. So it commissioned its technical public relations department to invite a dozen key editors and reporters from the trade and business press to a large-scale presentation at its plant in the Midwest.

Although the device was being made only in very small quantities on a pilot line when the press conference was being planned, Amptronic's engineers and production people were confident that there would be only routine problems in getting the device into full production. Consequently, in their invitation, the press was told they would see the device being manufactured in volume quantities using high-speed production and automatic test equipment.

Unfortunately, in the weeks before the press meeting, some unexpected snags were encountered. Barely enough good devices could be produced to fill engineering sample requests, although the demand for them was high and the initial reactions from customers could not have been better. The evening before the scheduled demonstration, the plant manager confided to the vice president of public relations, under whose aegis the press arrangements had been orchestrated, that the yield on the new device had been so low that the production line was down. Furthermore, he said, if it were started up, most of the devices would register as failures at the automatic test station.

The public relations vice president quickly consulted with the director of engineering, and they decided upon a plan. The next morning, the press group was ushered into the factory, and after viewing preliminary assembly operations, was guided to the automatic test station, where they watched about 20 devices rapidly undergo tests, all of them passing with flying colors. The tour leader at that point herded the reporters quickly into the conference room, where coffee was served and company executives answered questions. One reporter asked what the current rate of production was, and was told by an Amptronic representative that the automated line operated at a rate of 1000 devices per hour. That would equate to a throughput of 40,000 devices per week per shift, the reporter observed, and he was told that was correct.

What neither he nor any of the other press people knew was that the 20 devices they saw being tested were the only good devices available. The Amptronic people had intentionally timed the press tour so that the "window" on the automatic test operation was only 20 devices wide. In fact, the test equipment technician had been advised that if the press group did not move on promptly, he was to shut down the equipment after the 20th device had passed through and announce that the scheduled morning coffee break was about to begin. The reporters and editors subsequently wrote news articles that cited Amptronic as producing their significant new device in large quantities.

Did this deception by Amptronic represent an unethical practice? Suppose that the trouble on the production line was not easily solved and, for some months, deliveries to customers who had designed the new devices into their equipment were held up? On the other hand, suppose the problems were corrected quickly, and shipments to customers began promptly within a few days? Would this make the hoax palatable?

Some colleagues to whom we posed this fictitious case felt that it was trivial. "This kind of poker is played frequently," one felt. Some even suggested simpler ways to keep the line running for the demonstration. "Simply adjust the test equipment so that all the bad devices read good," was one suggestion. Others felt that similar deceptions are practiced internally as often as they are on customers or the press. Anyone who is frightened enough of the boss, or who wants to impress the boss, might be a party to such a ploy, some felt.

Nevertheless, there is reason not to dismiss the case lightly. It has many of the elements of a classic ethics problem. It begins with a "white lie." The company's reputation may be lost if the problems are not corrected quickly. Customers can be affected. There is collusion on the part of employees at several levels, and those in the highest levels of management may be among those misled. There a number of serious questions which can be posed.

For example, had you been the director of engineering, would you have suggested or condoned the ploy? On the other hand, had you been a senior engineer reporting to the director of engineering and were aware of the intention to deceive the press, would you have felt compelled to take some action? Would you have objected strongly? Asked for a transfer? Or would you have followed some other course?

B. On January 28, 1986, the space shuttle Challenger exploded 72 seconds after liftoff and set the stage for an unparalleled governmental investigation of engineering practice and managerial decisions. The investigative commission was headed by former Secretary of State William Rogers and convened on Capitol Hill in order to question the agencies involved in determining launch feasibility: NASA, Rockwell International, and Morton Thiokol. The explosion of the shuttle Challenger occurred because decision makers took a risk with a highly complex system which was being subjected to environmental conditions that had not been investigated. The risk included the lives of the astronauts and a civilian.

The shuttle rocket booster, produced for NASA by Morton Thiokol, is composed of four interlocking segments, as shown in Figure 1.3. The seals be-

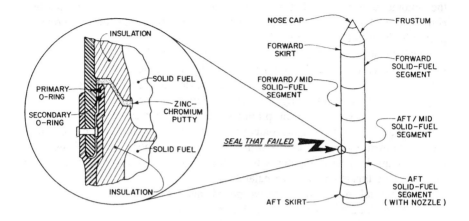

Figure 1.3 The NASA shuttle rocket with detail showing the design of the seals.

tween the interlocking rocket casings consist of three protective layers: A strip of zinc-chromium putty, a primary rubber gasket, and a secondary rubber gasket, both gaskets being O-rings. The seams between the rocket segments must be sealed to ensure that the hot gases do not escape through these seams while the rockets are firing.

The double set of O-rings was initially put in place to fulfill the NASA requirement of redundancy in design. Beginning with the second shuttle flight in November 1981, more than 30 seals had been partially eroded by rapid compression on the inside of the boosters. In 10 of these seals, hot gases were able to penetrate the primary barrier, the strip of zinc-chromium putty, and the primary rubber gasket, and to deposit some soot just beyond the secondary rubber gasket (the final barrier) [7]. Also, rapid pressurization of the booster segments caused the gaskets to form an ineffective seal because the interlocking joints widened by 0.04 to 0.06 inch, as shown in Figure 1.4. Such deformation and documented failures meant that the redundancy requirement of the seals was not being met. Instead of correcting the defect, NASA waived the redundancy requirement for Morton Thiokol in March 1983 because "...poor performance of the gaskets was identified in numerous documents as a 'budget threat,' not a safety hazard" [7]. On top of the obvious deficiency of the rubber gaskets, another concern came into play on this cold January morning. The boosters had been tested down to 47 F, but it was estimated that the joints would probably be between 27 and 29 F at launch. The concern was that the rubber rings would seal less effectively at lower temperatures, thus permitting hot gases to escape through the seams of the boosters.

DESIGN AND THE ENGINEER   23

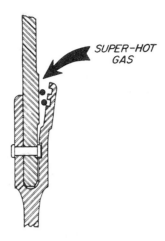

Figure 1.4 Rocket seals at failure, exaggerated for clarity.

As the Rogers Commission conducted its investigation into the cause of the Challenger's break-up, it discovered a decision-making process that it termed "flawed" and "a serious deficiency." Warnings of deep-seated problems with the rocket booster seals were voiced as early as July 1985 when Richard Cook, budget analyst for NASA, stated in a memo that, unless the O-ring problem was corrected, a catastrophic failure could result. Soon to follow was a warning to management from Roger Boisjoly, Morton Thiokol's ring expert, that the dependability of the seals was "a jump ball" and that the seals might be the cause of a catastrophic failure [8].

The decision to launch on January 28 was reached by a process that began the previous evening with a caucus of Morton Thiokol engineers and a teleconference involving Morton Thiokol management and decision makers at the Marshall Space Flight Center in Alabama. Fourteen engineers at Morton Thiokol unanimously recommended postponement of the launch. The concerns of the engineers were forwarded to Allan McDonald, Director of Solid Rocket Motor Projects for Morton Thiokol. Allan McDonald then called Robert Lund, Vice President for Engineering at Thiokol, to communicate the "very serious" findings and urge an analysis of the safety of the seals at low temperatures. McDonald also relayed this information to Stanley Reinartz, NASA manager of the Shuttle Projects Office. Reinartz did not inform Arnold Aldrich (Space Shuttle Manager at the Johnson Space Center) until he had "a full understanding of the situation" or, in other words, after a final decision was made in the teleconference.

Robert Lund, in the teleconference with the Marshall Space Flight Center and Morton Thiokol management and engineers, stated that unless the temperature reached 53 F, "I don't want to fly" [8]. The Marshall Space Flight Center, appalled at Thiokol's reasoning behind the no-go stance, asked them to reexamine their data and prove that the Challenger should *not* be launched. The rocket boosters had never been launched or tested at such low temperatures; therefore Thiokol's data were inconclusive and they could not prove that the Challenger should *not* be launched. Following this contorted line of reasoning, Morton Thiokol's management gave the Marshall Space Flight Center the go-ahead for a launch. The Space Flight Center notified Arnold Aldrich at the Johnson Space Center of the launch approval given by Thiokol, without mentioning the reservations expressed by Thiokol's engineers.

The supplier of the space shuttle (the orbiter), Rockwell International, never gave NASA the clearance to launch. Rockwell warned NASA's top management (Arnold Aldrich) directly that they "could not assure it was safe to fly" because of the fear that ice would damage the shuttle's tiles during launch. But to override Rockwell's assessment, NASA did its own technical evaluation and did not encounter the same concerns.

Discuss this tragedy from an ethical point of view. It may help to make a chart showing each person's sphere of influence and responsibility. Who should have stopped the launch?

C. In April 1990, an item appeared on the "Spectral Lines" page of the *IEEE Spectrum*, titled "Ethical Judgment" [6]. The entire text follows:

"Consider the following case study. It is based on a hypothesis of the National Society of Professional Engineers' (NSPE) Board of Ethical Review.* We have made it more germane to our profession by identifying the principals as electrical engineers.

"A municipality issues a request for proposals for a city-wide emergency communications system. In response, an engineering department head in a large consulting firm assigns the proposal to three of his staff engineers. Soon after the proposal is submitted, the city learns who actually developed the designs, and a city official approaches the three engineers, hoping to contract with them directly, independent of the consulting firm. The engineers inform their department head of these overtures, resign from the firm, and enter into negotiations with the city.

"The NSPE ethics board poses the question: Was it unethical for the three engineers to take the action they did?

"No, it was okay, the board concludes, referring to the NSPE Code of Ethics, which states that the public interest should be put ahead of the interests of the engineers or the consulting firm. The client—in this case, the city—should be able to select the engineer(s) of its choosing.

"One solution that the city could have required as a condition of its acceptance of the bid, that the three engineers be assigned to the project, is not mentioned. However, the board implies that the decision

---

*Engineering ethics: You be the judge, *Engineering Times,* January 1990, p. 3.

DESIGN AND THE ENGINEER 25

might be different if the three engineers gained any 'particular and specialized knowledge' while working for the consulting firm. But such information, the board says, must 'approach being proprietary in nature.'

"No mention is made of the implications of the three engineers' development of the proposal on 'company time,' using the consulting firm's facilities. Also, no ethical problem is apparently seen in the consulting firm's loss of the opportunity to bid on the contract in question, or in how the consulting firm would have profited from assigning the engineers, during the time they spent developing the proposal, to another, ongoing project.

"The NSPE board begs the question of whether the city official was unethical in approaching the engineers directly. Nor is the likelihood of the consulting firm bidding on the next project to come along from the city discussed.

"What do you think? Responses will be published in a future issue."

Was the NSPE board right? What do you think about the ethical stature of the three engineers, or of the city official who approached them? Did someone gain here at the expense of others?

1.4 SUMMARY

This chapter describes design in several ways, first by definition, then by a pragmatic statement that emphasizes the selection and positioning of materials in space. Another way of describing design is by comparing it to the scientific method to emphasize its closed nature, which leads from functional specifications as the first step to manufacturing specifications as the last step. More explicitly, the activities of design can be grouped under five general titles and these can be further broken down into an orderly list of tasks, as shown in Figure 1.2.

The chapter also indicates that, although the complexity of projects may vary from those that involve politics, massive funding, and major system design to those that involve a simple device where strength or shape may be the only concern, the design procedures are essentially the same.

Attention is focused on the responsibility of the engineer to develop designs that meet the needs of all persons who will be affected by the product during its existence, from manufacture to the time after it has been discarded. This responsibility is based not merely on the employer's desire for customer satisfaction but also on the ethical principle of caring for the safety of the public. The fact that the designer must consider manufacturing, distribution, consumption, and retirement needs will recur often throughout the text.

The design procedure given in this chapter is expanded in Chapter 9 for system design. It is also the basis for the content of this volume, although for pedagogical reasons the topics are not given here in the same order as they are normally used in product development as shown in Figure 1.2. Chapters 1 through 4 prepare the designer for the first two steps in design: The definition and evaluation of the problem. Device and system synthesis command the most attention and are treated in Chapters 5, 6, 9 through 11, 15, and 16. Device and system analysis are most closely related to Chapters 7, 13, and 14. Optimization is obviously the sole topic of Chapter 12. Chapter 8, on decisions, is not listed as belonging to any one procedure step because it is a cardinal element of all of the steps in the design process. The last step of the procedure shown in Figure 1.2, communication for manufacture, is not discussed in detail because the methods used depend principally on the product and the procedures of the manufacturing company. This activity is usually performed by technicians under the direction of the designer.

REFERENCES

1. Jones, J. C. *Design Methods, the Seeds of Human Future.* Wiley-Interscience, New York, 1970, pp. 30-32.
2. Stevenson, E. N., Jr. Education for engineering design involvement, *Engineering Design Graphics Journal,* 1973, 37(1): 39-42.
3. Salamone, Thomas A. *What Every Engineer Should Know About Concurrent Engineering,* Marcel Dekker, Inc., New York, 1995.
4. Friedlander, G. D. The case of three engineers vs BART, *IEEE Spectrum,* 1974, 11(10): 69-76.
5. Christensen, D. An exercise in ethics. *IEEE Spectrum,* 1981, 18(1): 35.
6. Spectral Lines, *IEEE Spectrum,* 1990, 27(4): 19.
7. Smith, R. J. Shuttle inquiry focuses on weather, rubber seals, and unheeded advice, *Science,* Feb. 28, 1986, pp. 909-911.
8. Megnuson, E. A serious deficiency, *Time,* March 10, 1986, pp. 38-42.

REVIEW AND DISCUSSION

1. What is *your* definition of design?
2. Which step of the scientific method and of the engineering method requires the most creativity?
3. Which step of the scientific method and of the engineering method requires the best understanding of scientific principles?
4. Give an example of an engineering project that includes all levels of design problems.

DESIGN AND THE ENGINEER 27

5. The steps of the design procedure were selected for breadth of meaning. List some specific activities necessary for design that would fall within or take place between the 5 steps of the procedure.
6. Many products are sold to tradespeople (e.g., water heaters, electrical panelboards, furnaces, air conditioners, and so forth) who then install products in homes or commercial establishments. Which stage of the life cycle (distribution or consumption) would you consider to be appropriate during installation? If it depends upon the product or the extent of the installation, can you devise a rule to provide consistent treatment?
7. How would you arrange for a design review?
8. You are the engineer in charge of maintenance for a large manufacturer. As such, you have the responsibility to maintain a stock of replacement equipment so that down time in case of equipment failure is minimized. A sizable number of motors used in the plant are on equipment that requires 9 horsepower at full load. You become aware by accident of the identical nature of the 7-1/2 hp and 10 hp motors produced by Company B, as described on page 10 above. It is obvious that the motors must all be 10 hp motors, and that the 7-1/2 hp rating is used merely so that the company can have a complete line. Is it ethical to order 7-1/2 hp motors for this application, even though the application requires 9 hp? Suppose one of the 7-1/2 hp motors fails during the warranty period. Would it be ethical to expect Company B to replace that motor at their expense?
9. Assume that you are an electrical engineer. You have been asked to be an expert witness in an energy theft case involving a home heated electrically by resistors in baseboard units (a heating method that is notorious for being very expensive). The charge is that the homeowner would remove the electric meter (which was not sealed) from its socket and reinstall it upside down, thus making it run backward. By doing this for part of each month, he could keep his electric bill down to a minimum. His defense is that, after two years of paying high bills, he installed a wood stove and used the stove to heat the house in succeeding winters. As an electrical engineer, you have no knowledge of the methods of analysis of domestic heating methods, although you know that a thermal energy balance can be run on the system consisting of the wood stove as the source and the energy sinks consisting of the roof, outside walls, windows, and doors. You are faced with the choice of learning the pertinent analysis methods and applying them to the situation at hand, or recommending that the attorney in the case retain an engineer skilled in the type of analysis required. The first choice means that you will spend time, for which you will of course bill the attorney, to acquire a skill that you lack; the second choice means that you will forego consulting income, which you need at the moment because you have a child in college and tuition costs are high. What do you do? Is there a third alternative?

PRACTICE PROJECTS

1. Choose a product with which you are very familiar to be used in carrying out many of the procedures given in this book. The product chosen should be a rather simple product. It may be something you believe is needed, such as a new tool or measuring device, or an improvement on something you use, for example, a better toaster or a more convenient automobile jack.
2. An engineer is developing a new product that is to be tested by Underwriters Laboratories (UL) and, if successful, listed by them to bear the UL label. The tests are to be performed at the engineer's laboratory but witnessed by an engineer from the UL office near Chicago. The engineer is nervous about the product passing the set of tests, which are always rigorous and are based on standards which are not easily met. Discuss the following questions with a group of fellow students.
   a) Is it ethical for the engineer to select a sample that appears to be one of the better ones?
   b) Is it ethical for the engineer to use special material in the sample to be tested (e.g., a better grade of insulation)?
   c) Is it ethical to select meters that are within calibration but that read current, voltage, and so forth that will be favorable to passing the tests?
   d) Is it ethical for the engineer to arrange a temperature test to take advantage of periodic cooling drafts?
3. An engineer is developing a product and has three samples made for tests. During the test one fails in a way that would result in injury if anyone were near. The other two pass. The engineer does not determine why the one sample failed and does not report the failure to his or her supervisor or include it in the written report.
   a) Did the engineer violate ethics by not determining why the one sample failed?
   b) Did the engineer violate ethics by not informing his or her supervisor?
   c) Did the engineer violate ethics by not including the mishap in the report?
   d) Who, if anyone, is harmed by those omissions?
4. A government agency plans the construction of a bridge. It retains a consulting engineer to design the entire structure. An engineer who is a sales representative of Firm A, which produces and sells prestressed concrete bridge members, contacts the consulting engineer and requests that he consider using Firm A's product. The engineer of Firm A indicates that the firm will provide the design of the structure incorporating its product at no charge to the consulting engineer, and that this design will be performed by licensed professional engineers. Is it ethical for the engineer employed

DESIGN AND THE ENGINEER 29

by Firm A to make such an offer? Is it ethical for the consulting engineer to accept such an offer?

5. Suppose you hired a young engineer after graduation to design new products. He did not work out well in that he left important details of his designs unresolved, drawings that he made (or that he was responsible for) contained an unusual number of errors, and he had a much higher rate of absenteeism than other employees in the department. Some of the absenteeism was obviously caused by sickness but most of it he explained as a severe headache, back problems, and other ailments that are impossible to verify. He finally left to go into business unrelated to engineering. Recently, the chief engineer of a local manufacturer called to ask about the former employee, who is now applying for a job as a project engineer there, but did not give you as a reference. After a number of questions that did not bring out any of the points above he asked, "What else can you tell me about him?" Is it ethical to reply that there is nothing else significant that you can add and terminate the conversation? Is it ethical to give the details stated above? What would you do? Would your answers be different if the engineer had listed you as a reference?

6. It is considered unethical for an engineer to serve as an expert witness in a lawsuit on a contingent fee basis. "Contingent fee" means that the witness would get a percentage of the damages awarded the plaintiff. Discuss the reasons why the legal profession, which uses this compensation method, holds it to be proper for its members but improper for expert witnesses.

7. Suppose you have been employed by Company A for 5 years. During that time you have had annual salary reviews. Your supervisor has made it clear that your work has been satisfactory but not brilliant. You may rationalize a bit, but privately you agree. There have been no patents, no publications, and no excitingly new ways of doing things. On the other hand, the products you designed show good engineering judgment, and mistakes have not been embarrassing. During the reviews your supervisor has consistently raised your salary to stay on the median curve published by the Engineers Joint Council. Although everyone would like more money, you feel that your supervisor has treated you fairly. You notice that a competitor of your company, Company B, is running an ad in the newspaper for an engineer with your experience. You find out that they really need someone as your counterpart in that company has died. You realize you will be in an excellent position to improve your salary. Discuss the ethics of applying for the job. Then answer the next question.

8. Suppose you are the chief engineer of a medium-sized company with full authority to hire and fire all engineering personnel. You have an engineer, Mr. A, who has been working for you for 5 years. His performance has been satisfactory but not brilliant. Suddenly you hear that an exceptional engineer, Ms. B, from a competitor, has quit her job and is available. The thought occurs to you that if you fire Mr. A you could hire Ms. B and

strengthen your department. Your budget does not permit you to hire Ms. B as an extra employee. Discuss the ethics of taking such action.

9. Suppose that in your job as a design engineer you are visited by a salesperson who informs you that your company's biggest competitor is one of his customers, and engineers there are working on a product similar to the one you are designing. It would be helpful to you to know details of the competitor's product. The thought occurs to you that you could suggest to the salesperson that you will specify the material he is selling if he freely discusses what he knows about the competitor's design. You feel a little embarrassed about bringing up the subject, but that is beside the point. Discuss the ethics of taking such action.

10. An engineer and his wife have decided to move to Florida and are building a home there that is scheduled to be completed in about six months. He has not disclosed these plans to his employer. The engineer has just been offered a promotion by his present employer as supervisor of a team of engineers on a new project, which will require 18 months to complete. Obviously, it will not be in the company's interest to have to change managers one-third of the way through the project. On the other hand, the new title will be a valuable entry on his resume. What would you do?

11. An engineer on a business trip has taken a taxi to meet with a client quite some distance from the airport. The company will reimburse him for expenses. The taxi driver volunteers that he will leave the space for the fare blank on the receipt. What should the engineer say to the driver?

12. An expert witness is retained for a product liability case which involves considerable loss. He has been ill and is using medication that makes him drowsy. There are several hundred-page depositions to read. He charges against his retainer at $100 per hour. The case will not come to trial for several months, so there is no need to read the material during this illness. Is it ethical for well-paid professional people to do client's work when their mental faculties are significantly impaired?

13. The cost of a traveling salesman to visit a prospective customer is about $150 per "call." On average, a salesman can make about four calls a day. However, if midday can be made productive by taking a customer to lunch and discussing business details, the salesman can often increase the calls to five a day. Whether he can do this depends on the travel time between calls and the ability to make the additional appointment. In view of these facts, is it ethical for an engineer to accompany the salesman for a "working lunch" as the guest of the salesman's company?

14. Repeat problem 13 but add the fact that the engineer has already ordered the equipment that this salesman sells from a competitor.

2

PRODUCT LIABILITY

Ethics demands that engineers hold paramount the safety, health, and welfare of the public in the performance of their professional duties. This admonition to protect the public has, during the twentieth century, become a *de facto* legal requirement, a gradual change in attitude having taken place since the early years of the century.

Although one goal of manufacturers is and always has been to produce the best possible device or system from a benefit/cost point of view, the possibility of injury or loss of property has become a major factor in the decision to accept a given design. The present status of the legal means by which product safety is enforced, that is, product liability, is best explained by a review of cases that have established new principles and that have been confirmed either by a state supreme court or by the United States Supreme Court.

2.1 PRODUCT LIABILITY EVOLUTION

In order to understand the discussion of product liability evolution, it is necessary to understand the legal terminology. The more important terms are as follows:

Privity is a direct successive relationship between two parties, for example, between a vendor and a buyer.

Plaintiff is the person (or organization) who (which) initiates legal action for redress of injury or loss of property.

Defendant is the person (or organization) alleged by the plaintiff to have done a wrongful act.

Landmark case is a case in which the decision was significantly different from that which would have been anticipated from previous interpretations of the law and in which the decision was later upheld when appealed to a state supreme court or to the United States Supreme Court.

Tort is any wrong or injury for which the aggrieved can seek redress by legal action.

Warranty is any promise from a vendor to a purchaser.

Implied warranty is the promise of commercial worthiness inherent in a product by virtue of its being offered for sale.

Disclaimer is an express or implied denial or renunciation of certain things in question.

Express warranty is any affirmation of fact or promise made by the seller to the buyer that relates to the goods and that influenced the sale, any description of the goods that influenced the sale, and any sample or model that influenced the sale.

Strict liability means that a manufacturer or seller is held accountable for injury or loss of property caused by a defect in the product without regard to proof that the defect was caused by negligence of the manufacturer.

Foreseeable use or misuse means an activity involving the product that can be anticipated by thoughtful consideration of the functions it can provide. For example, an aluminum ladder can be used as a scaffold although this would change the stresses in the side rails from a compression mode to a flexural mode, and may cause the ladder to collapse.

Trier of fact is usually the jury; however a judge presides and renders the verdict if a jury trial is not requested.

Deposition is the taking of a witness's testimony under oath at a pretrial meeting. Counsels for both sides, the witness, and a court reporter are present. The court reporter records the proceedings and provides a written transcript. Some depositions are videotaped.

Ultimate issue is the disagreement between plaintiff and defendant that brought the case to court.

For many years manufacturers insulated themselves from product liability loss by invoking the legal theory that redress for harm required privity between the plaintiff, that is, the ultimate consumer or user, and the defendant, that is, the manufacturer. Since products are typically distributed through a wholesaler or retailer, the manufacturer argued that the required privity was absent. Early in the twentieth century exceptions to the privity doctrine began to develop in cases in which the product was found to be inherently dangerous to life or health. The death knell of the privity doctrine in negligence cases was sounded by the landmark case of *MacPherson v. Buick Motor Co.,* in 1916.

MacPherson was driving a new automobile when one of its wheels fell off, causing injury. The automobile manufacturer's defense was based on lack of privity and the fact that the defective axle was a purchased item, having been made by another manufacturer. Judge Benjamin Cardozo, who later became a U. S. Supreme Court Justice, held that the manufacturer's liability did extend to the user because "if the nature of a thing is such that it is reasonably certain to place life and limb in peril when negligently made, it is a thing of danger; and if to the element of danger there is added knowledge that the thing will be used by persons other than the purchaser, then the manufacturer of the thing of danger is under a duty to make it carefully." This famous decision put aside the notion that the duty to safeguard life and limb, when the consequences of negligence may be foreseen, grows out of contract and nothing else. One vital requirement for the plaintiff remained, however, namely, the need to prove that the manufacturer was negligent.

Proving negligence is often difficult because the plaintiff is usually not an expert in the manufacturing techniques of the defendant's industry. Furthermore, if the plaintiff contributed to causing his or her own injury, in many jurisdictions recovery usually cannot be obtained under the negligence doctrine. Under the aggressive attack of the plaintiffs' attorneys, the requirement of negligence began to erode. At first, this simply involved bringing suit under a different doctrine, that of breach of implied warranty. A warranty, unlike negligence, is not a tort concept but a contract concept. As such, negligence on the part of the manufacturer does not have to be proved.

Despite its apparent simplicity as an effective basis of recovery for the plaintiff, especially as compared with the negligence argument, implied warranty was still encumbered with the contract concept of privity. In 1960 the privity doctrine was again struck down, but this time for implied warranty cases, by the landmark case of *Henningsen v. Bloomfield Motors, Inc.* Shortly after purchasing a new car from Bloomfield Motors, Ms. Henningsen was driving at a speed of approximately 20 miles per hour when she suddenly heard a loud noise under the hood. Simultaneously, the steering wheel spun in her hands, and the car veered sharply to the right, crashing into a wall. Witnesses corroborated her testimony. The front of the car was so badly damaged that it was impossible to determine if negligence on the part of the manufacturer was involved to any degree. However, suit was successfully brought under breach of implied warranty of merchantability and fitness, even though there was no privity of contract between Henningsen and the manufacturer. The court stated that "where the commodities sold are such that if defectively manufactured they will be dangerous to life or limb, then society's interests can only be protected by eliminating the requirement of privity between the maker and dealers and the reasonably expected ultimate consumer."

In addition to striking down the privity doctrine in an implied warranty case, the court declared the standard automobile disclaimer, advanced as a defense, to be invalid on the basis of ambiguity and because of the inequitable bargaining position of the consumer versus the automobile industry.

After *Henningsen v. Bloomfield,* it became apparent that every product placed on sale would have a warranty, implied if not expressed, that it is safe unless specific warnings are provided to indicate otherwise. Also, any defective product that could be dangerous subjects the manufacturer to liability even if care has been exercised in producing it or a disclaimer of liability has been issued in advance. This case made clear that "the burden of losses consequent upon use of defective articles is borne by those who are in a position to either control the danger or make an equitable distribution of the losses when they do occur," that is, the manufacturer.

At about the same time, a major change in legal doctrine of tort resulted from the continuing effort of plaintiffs' attorneys and the courts to insure that the injured product user could have legal recourse at least equal to the technical defense capabilities of the manufacturer. A landmark case (*Greenman v. Yuba Power Products, Inc.*, 1962) recognized strict liability in tort for the sale of a dangerously defective product. The circumstances that brought about the decision in this case were that the plaintiff Greenman was seriously injured when a piece of wood that he was turning on a popular combination lathe, saw, and drillpress machine flew out of the machine and struck him on the forehead. He brought suit for damages against both the retailer and the manufacturer, alleging negligence and breach of warranty in selling a product improperly designed because of inadequate set screws.

The trial court ruled that there was no evidence of negligence and found for the plaintiff based on breach of warranty. Upon appeal, the California State Supreme Court criticized the warranty basis of the trial court's decision as being unnecessary. The Supreme Court stated that "a manufacturer is strictly liable in tort when an article he placed on the market, knowing that it is to be used [by the ultimate consumer] without inspection for defects, proves to have a defect that causes injury to a human being." Continuing, the court also stated that "the purpose of such liability is to insure that the costs of injuries resulting from defective products are borne by the manufacturers that put such products on the market rather than by the injured persons who are powerless to protect themselves. Sales warranties serve this purpose fitfully at best."

The Greenman case was the culmination of two historical trends in product liability cases. One was the increasing tendency of the courts to view their mission as one of providing compensation to injured parties as opposed to assessing fault. The other trend was to provide a tort action for the plaintiff free of defendant defenses based on questions of negligence, privity, disclaimers, warranties, and so forth. This development of legal theory is shown in Figure 2.1.

Thrusting the burden on those most able to pay—the so-called "deep pocket"—without regard to fault hardly seems to make for equity under the law. At best, it seems apparent that the courts, in adopting the rule of strict tort, were attempting to correct what they perceived to be a social problem rather than to remedy some deficiency in the law. In any event, strict tort is the liability theory likely to be used in the great majority of states. The addition of punitive damages to the possible outcomes of product liability cases compounds the potential financial liability of the manufacturer. Engineers must recognize these facts and act accordingly.

The ultimate in manufacturers' liability would be absolute liability; that is, the mere fact of injury would be sufficient proof for damages to be awarded to a

PRODUCT LIABILITY

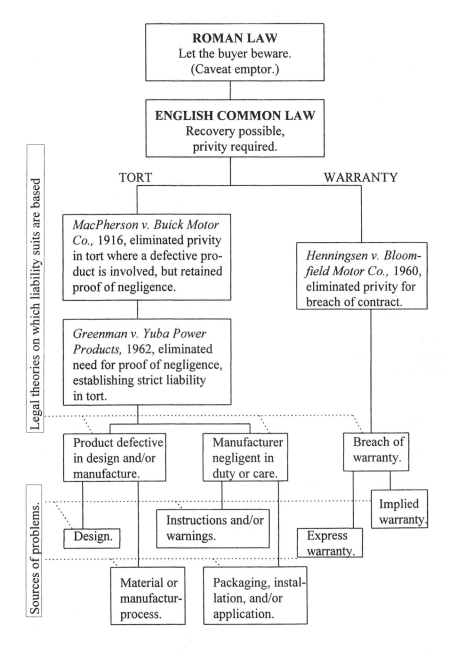

Figure 2.1. Product liability history, current legal theories, and sources of liability problems.

plaintiff injured by a product. This relief from identifying the product defect is also part of another doctrine called *res ipsa loquitur*, Latin for "the thing speaks for itself." This doctrine is designed to permit a jury to infer existence of a defect when there is no specific evidence of it but where a defect is the most reasonable inference to be drawn from the facts of the case. Most courts treat *res ipsa loquitur* as a form of circumstantial evidence. It is an evidentiary doctrine and not a rule of strict or absolute liability. It permits a court to draw a factual inference when that is the most logical explanation.

A good example of this doctrine is represented by the verdict rendered in the case of *May v. Columbian Rope Co.* The plaintiff was using a rope and truss arrangement on a scaffold. The rope, which was one-half inch in diameter, three-strand, manila line, broke and May fell from the scaffold. The plaintiff's case was built on the fact that the rope was new and the truss did not damage the rope. However, the plaintiff did not name a specific defect as the cause of the accident. The defense explained the testing procedures used in rope manufacture and questioned the alleged "newness" of the rope and the way it was used. The defense also suggested that May might well have contributed to his own misfortune.

The court said simply that "the thing speaks for itself"; that is, the rope broke, caused the accident, and injured the plaintiff. May recovered damages.

This theory is often applied in medical malpractice suits. For example, if a sponge is left inside a patient after an operation, the mere fact that it is there speaks of negligence. However, as threatening as *res ipsa loquitur* appears to be, it is the authors' experience that it is not often invoked successfully because there are relatively few accidents in which one product or one organization can unequivocally be assigned full blame. Typically, two or three adverse factors are involved in product-related injuries. For example, many products are installed by contractors who are independent of the manufacturer. When an accident occurs, the question usually arises whether it was caused by a defect in the product or by faulty installation. This uncertainty is enough to negate the claim that the thing speaks for itself.

2.2 FORESEEABLE USE AND MISUSE

In addition to the landmark cases, there are other important cases from which fundamental lessons may be learned. First is that the product must be designed and manufactured with its total life cycle and total use environment in mind, that is, from the moment the product leaves the factory until it no longer exists, both under normal conditions and in adverse conditions. The second principle is that the manufacturer is responsible not only for accidents related to the intended use of the product, but also for those caused by other foreseeable uses and misuses. The word "foreseeable" is a key legal word.

PRODUCT LIABILITY

Perhaps the potential accident easiest to overlook is one that takes place before the product is ever put to use. What can happen while the product is still in its package? Consider the following case. A door manufacturer was manufacturing doors with a large opening in the top half for later installation of glass. The doors were packed together, making a stack about 42 inches high, with all of the openings at the same end of the stack. The stack was covered with cardboard for protection, leaving the ends of the stack exposed, with two steel bands to secure the cardboard. The only marking on the cardboard were the words "fine doors." Some of these doors were loaded aboard ship for sale abroad.

A longshoreman carrying a 100-lb sack of flour walked across the doors and fell through the void area, sustaining injuries as he fell. In the ensuing trial, the manufacturer argued that this was a clear case of abuse. However, the Ninth Circuit Court of Appeals affirmed the judgment that the injury was caused solely by the manufacturer by the way in which the product was packaged. The manufacturer should have known that it is customary for stevedores to walk on material already loaded. Damages were paid because of failure to foresee that a product can cause injury even while in transit.

Another easy-to-overlook aspect of product life is the physical changes that take place in the materials that compose the product due to normal use and environment. A case in point is that of a farmer who suffered the loss of one eye when a piece of metal chipped off a hammer he was using. The hammer was a forged-head carpenter's hammer; it was being used to drive a pin into a clevis to connect a manure spreader to a tractor. Both sides agreed that there were no metallurgical flaws in the hammer when it left the manufacturer. However, a process known as work hardening occurs whenever metal is squeezed, struck, or bent. Also, it was agreed that as metal hardens it is more likely to break into chips when striking an object harder than itself. At the time of the accident, the hammer had work hardened to a Rockwell hardness of C-52, while the clevis pin had a hardness of C-57. Thus, the mere use of the hammer and the coincidence of the clevis pin being harder than the hammer formed the necessary conditions for an accident to occur.

The defense lawyer argued that the hammer was not being used as intended and that the farmer should have used a ball peen hammer for this job. However, it was pointed out that the farmer bought the hammer at the local hardware store and that this type of hammer is known to be used for all types of tasks. The court found in favor of the farmer on the basis that the work-hardening concept is well known to metallurgists, that the manufacturer's records showed chipped hammers had been returned to the factory for replacement, and that it was the manufacturer's duty to foresee that the hammer might be used in the way the farmer had used it.

The important lesson is that all materials change with age and environment. Electrical insulations will deteriorate, metals will oxidize, lubricants will undergo chemical changes that alter their properties, and so forth. In thinking about what types of accidents can occur during the use or foreseeable misuse of a product, one should include in the scenario all of the adverse changes that typically affect the materials composing the product.

It is also necessary to guard against accidents to a person who may not even be the principal user of a product, but who is taking advantage of an attractive way to misuse it. Such a misuse is that of a little girl who attempted to get on the seat behind the operator of a riding lawn mower. The mower blades were well guarded for 270 degrees around the front and the sides, but the rear was exposed. One of the girl's feet was badly mangled when she fell off the mower. The case was settled out of court for a sizable sum before it came to trial.

In considering all foreseeable circumstances, operator neglect must be given due concern. For example, in one case a self-propelled lawn mower was left unattended, but with the engine running. The design of the control lever was such that it shifted from "disengage" to "engage" because of vibration from the engine. The mower started to move, struck the plaintiff, and caused serious injury to his right foot.

Another case involving an unattended product was that of a vaporizer that overheated after the water boiled away, starting a fire. The defense claimed that this was misuse of the product in that the plaintiff failed to follow the manufacturer's instructions not to allow the water to boil away. However, the court held that it was too much to expect a consumer to use a vaporizer properly all of the time. Furthermore, since an inexpensive automatic cutoff switch would have obviated the need for perfect user compliance with the manufacturer's instructions, failure to include this protection constituted a design error.

In another vaporizer case the unit was designed with a loose-fitting filler cap so that high internal steam pressure could not develope. A 3-year-old child knocked the vaporizer over, the hot water spilled out, and the child suffered third-degree burns from the hot water. The defect in design was alleged to be the fact that the cap was loose-fitting. A cap with molded threads and small holes for the escape of the steam would have served to avoid high internal pressures and would have prevented the hot water from escaping.

In another case, when the engine of his car stopped suddenly, a motorist looked under the hood to see if he could find any obvious problem. It was after dark, and the motorist struck a match so that he could see the battery; gas from the battery exploded. The plaintiff charged defective design and an inadequate warning label. (Warning labels are discussed later.) There was a substantial award, but it is not known which allegation carried more weight with the court.

A case with fatal results was that of an electrician, working for a manufacturer of various machines for industrial applications. While checking the operation of the sliding inlet and discharge gates of an industrial mixer before shipment, he put his head inside the mixer drum through the access door opening so that he could see what the gates were doing as he pressed the

PRODUCT LIABILITY

appropriate pushbuttons on the control station, which was located close enough to the access door that he could reach it easily with his head in the drum. The access door, when it was closed, was held closed by a 4 or 5 inch threaded rod with a knob at one end; the rod threaded into a bracket on the outside of the mixer drum, and at the end of its travel it closed a pair of contacts in a limit switch that was mounted on the bracket. When the limit switch was closed, the control circuitry for the drive motor for the mixer was energized. With the door open, the limit switch contacts were open, removing power to the control circuitry for the drive motor. The drive START and STOP buttons on the control station were immediately adjacent to those for gate operation. With his head in the drum, the electrician could not see the control station. He apparently pressed the start button believing it to be a gate operating button, and the drive motor started, decapitating him.

On examination, it was found that the mounting holes for the limit switch on the bracket and the hole into which the limit switch plunger was intended to move when the threaded rod was backed out had tolerances that could allow the limit switch plunger to bind in the hole. If this misassembly occurred, the threaded rod could close the contacts, but the internal spring in the limit switch was not strong enough to return the plunger to its normal position when the rod was backed out, leaving the drive control circuitry energized. This mechanical problem was one error of design. A second error was the design of the control station; buttons that were distinctively different for different functions could have avoided this fatality. Moreover, the limit switch had a pair of normally closed contacts which could have been used to energize a "Drive Off" lamp, and/or the normally open contacts could have been used to energize a "Drive Power On" lamp connected in parallel with the drive control circuitry.

No lawsuit was ever filed in this case because of the worker's compensation laws of the state in which the fatality occurred.

A case that points up the need for product designers to be familiar with variations from the norm in size and shape of people was decided by a Federal District Court in Florida. An award of over $5 million was made to a high school student injured during a football game. The plaintiff alleged that the rear edge of his helmet pressed against his neck as he was tackled, causing permanent paralysis. Considering the different head and neck shapes of football players, it is probably impossible to design a helmet that is proper for all players. As a result, most high school athletic personnel have taken much-needed steps to ensure proper choice of equipment for each player.

The importance of variations from the norm in size became evident in late 1996 and on into 1997 when there was a considerable amount of discussion in the media concerning injuries and deaths resulting from deployment of air bags in automobiles. These were deployments that occurred as a result of low-speed collisions. Although about two-thirds of these deaths were those of children, the remainder were those of small-size adults. It became evident as more and more information became available that the air bag had been designed for medium-size adults, and that those engineers who had issued warnings about the potential danger to small-size adults or children had been ignored, or the warnings had been disregarded because of cost factors. There is some discussion of anthropometric data in Chapter 15.

The deaths caused by the lack of internal latches on household refrigerator were referred to in the previous chapter. Certainly no one would expect the manufacturer to provide an internal latch, but no one foresaw the fact that children would use inoperative or unused refrigerators as hiding places. As many as 15-20 deaths were reported annually. However, in 1956 Public Law 930 was enacted, requiring that refrigerator doors be made so that they can be opened from the inside by 15 pounds of force or less. The magnetic gasket closure on later models of refrigerators was the result.

A seemingly completely innocuous household item that has caused deaths of children between the ages of 7 months and 4 years is the pull cord used on many window coverings. These children became entangled in the cords while playing with them, and died by strangulation. Such products now come with a Consumer Product Safety Alert that warns the purchaser of this possibility.

There seems to be an endless list of cases whose verdicts have been startling enough to be described in the literature, and human beings seem to exhibit an unlimited level of ingenuity in misusing almost every product ever made. Articles in newspapers, popular magazines, and technical journals should be monitored so that the designer is aware of pitfalls into which others have fallen and the misuses of products of which the human being is capable. The law is an ever-changing discipline.

2.3 REDUCING PRODUCT LIABILITY RISK

With strict liability as a basis for product liability actions, it may appear that little can be done to prevent legal action against a manufacturer's products. Nonetheless, there are many actions that the engineer and management can take to lessen the likelihood of product liability suits, or to reduce the probability of such a suit being successful.

One of the main countermeasures is adherence to principles that will ensure design and production of products that are as safe as possible. Pursuit of this policy will protect both the manufacturer and the public. There must be both a broadening of the design perspective and a more professional approach to the design process.

Traditionally, design engineering has revolved around strictly technical considerations, such as strength of materials, machine elements, electrical considerations, manufacturability, and cost. In an era when the manufacturer is extremely vulnerable to product liability action, this traditional approach must be broadened to include more attention to standards and codes, human factors engineering, hazards analysis, express and implied warranties, warnings and labels, failure analysis, and design review.

For some companies this will require substantial changes in design procedures. There is no doubt that in the past many products were designed by persons who had little or no formal engineering education. These products were "designed" simply by specifying what appeared to be a product having reasonable strength, reasonable electrical insulation, reasonable heat transfer properties, and so forth. Many such designs were verified by running tests on one model made by one craftsperson, a model which was not at all representative of the same product when made on a production line.

When undertaking a policy of ensuring design and production of products that are as safe as possible and thus to lessen the risk of product liability, the first step is for the top management of the company to formulate policy statements concerning safe design. The next step is to select a person having a strong interest in product safety, and to vest that person with sufficient authority that he or she can reject unsafe designs. The person selected must have sufficient knowledge and experience to be able to recognize potentially unsafe designs and to suggest alternate approaches. The person selected should also be someone who is well respected by company personnel in general. Next, a formal listing of actions to be taken by management to promote product safety should be developed. Some of these actions may place constraints on marketing, purchasing, or other departments. They may also require an additional expenditure of time and effort by engineers, technicians, and nontechnical personnel. However, without sufficient backing and commitment by company management, the necessary actions will very likely not be carried out.

2.3.1 Twenty Guidelines to Reduce Product Liability Risk [1]

Reducing product liability risk requires that steps be included in the design process which increase safety where possible and which warn against hazards that cannot be eliminated. Some of these activities are included in the engineer's formal training. Others, however, may be new to the practicing engineer. An example is the preparation of warnings and labels.

With a strong commitment by management, the climate can be set for everyone in the company to become conscious of the need to give product safety its proper priority. In this climate, the engineer responsible for developing a product should carry out the following activities. To aid in their recall, they are given in the order in which they normally occur during product development.

1. Include safety as a primary specification in identifying needs during all phases of the product's existence. Engineers are accustomed to determining in detail what product is needed to meet competition and what characteristics (functional, physical, aesthetic) the product must have.

These specifications are principally set by the intended operation or use of the product. The product must be safe during that stage, but also during distribution and retirement. Even after it is discarded, the product must not poison streams, explode, entrap, and so on.
2. Design to a nationally recognized standard, if such a standard exists. Be aware of the requirements of standards elsewhere in the world if the product is to be sold abroad. Standards such as those promulgated by the Canadian Standards Association (CSA), the International Electrotechnical Commission (IEC), and the International Standards Organization (ISO) should be obtained, studied, and their requirements adhered to.
3. Only those materials and components that are known to have sufficient quality and a small enough standard deviation from the norm to satisfy the requirements in a consistent manner should be selected. Characteristics of materials change with the immediately surrounding environment. Of two materials being considered, one having a greater yield strength at 25 C may have the smaller yield strength at the normal operating temperature. One sensor of two being considered may have a significantly higher reliability than the other. The current-carrying capacity of insulated wire is reduced as ambient temperature increases. Some of the necessary data appears in reference books, such as handbooks. Some appears in manufacturer's publications, which should be available as part of a company library or, if not there, as part of one's personal collection. Some appears in the technical literature. Some may not be available in any publication that can be found, even after a diligent search, and may have to be developed by suitable testing.
4. Apply accepted analysis techniques to determine if all electrical, mechanical, and thermal stress levels are well within published limits. Present-day engineering education is made up largely of courses in analyzing various physical situations. This guideline merely reminds the design engineer to use those skills that were acquired in such courses, and to the fullest extent. Mastery of all courses in engineering curricula is not only desirable but necessary. You can never be sure that the material in a given course will never be used.
5. Test the device using accelerated aging tests (Chapter 14). If possible, use a recognized test, that is, one specified in the standard or standards applicable to the product being developed. If there is no established or recognized test, develop an accelerated test that truly represents in-use conditions at elevated stress levels. Designing such a test is indeed difficult and usually evokes dissenting opinions as to its value. However, the test is certainly valuable if it uncovers a defect in the design. Even if the test shows no defects, it is still valuable in disputing testimony of design defects if the product should be involved in a liability suit.
6. Conduct a design review that includes persons knowledgeable about the distribution, installation, and use of the product, about manufacturing prob-

lems that can arise to lower the quality of the product, and dangers to persons and to the environment after discard. When a review is conducted, the designer must be careful not to take negative comments personally. The point of the review is to improve the design, and that means that the designer must take advantage of the experience and knowledge of the members of the review team that the designer may not possess. The members of the team will see features which are to be praised, but they will also see features which they may question. They are, in a sense, to act as "the devil's advocate." For example, if the door manufacturer mentioned earlier had had a design review team that included a packaging engineer and if a representative of the marketing department had pointed out that some of their product was transported by ship, the accident which occurred when a stevedore fell into the void where glass was later to be installed might have been avoided. Section 2.4 relates a dramatic failure of design review.

7. Make a failure and hazards analysis of the product for each stage of product life. (See Section 2.3.2 for more detail.)
8. Make a worst-case analysis of the product assuming the material characteristics, part dimensions, and so forth will simultaneously take on values at the tolerance extremes that are most detrimental to your product's performance. Although it is highly unlikely that all parts will simultaneously assume the worst values, you should ascertain that the product is not hazardous if this unlikely condition is even approached.
9. If possible, submit your product to an independent testing laboratory for evaluation and approval (or listing). Some laboratories engaged in this work are described in the section on standards. Some products may not be sold if they have not passed the tests of an appropriate testing laboratory.
10. Make sufficient information available to the production engineers to eliminate hazards. This information may be transmitted in a variety of ways, examples being notes on drawings, component specifications, and so forth. The information may specify that there be no burrs on certain faces of a part, that hardness must not exceed a certain value (or that it must have at least a certain value), that plating in certain areas must be done in such a way that it will not chip, that torques on bolts shall be at least X lb-ft and not more than Y lb-ft, and so on. This type of information is necessary because it may, for example, dictate that a die is made in one way rather than another to avoid burrs. If rolled copper is to be bent, it may crack if the bend is parallel to the grain and the hardness too high, whereas it could bend very well if the bend is perpendicular to the grain. Plating that is intended to provide a low-friction surface for a sliding part can easily seize the sliding part if the plating chips. The goal is to avoid making products which are unsafe or which could become unsafe.
11. Make a permanent record of the history of the product development giving sufficient information on all activities from the needs analysis to the pilot product run. This record should be so complete that others can understand

the reasons for each design decision. In addition, a complete record will buttress one's own memory if necessary, and may be a determining factor in the case of a product liability suit.

12. Wherever there is a question regarding safety of a product, document the risk/utility considerations made during the design phase. Also, document possible safety improvements that could be made but that would probably result in the product becoming noncompetitive economically or functionally, or becoming unmarketable.
13. Use warning labels on the product when this is appropriate. (See Section 2.3.4 for further discussion.)
14. Supply unambiguous instructions for the proper installation or use of the product. Write the instructions for the least qualified installer or user. Test the instructions on someone no more skilled that the person for whom you wrote the instructions. If that person has difficulty in interpreting the instructions, you can be sure that someone in the field will also have difficulty.
15. Determine any service or maintenance necessary to keep the product in a safe operating condition and in a condition to perform the function for which it was designed. If appropriate, provide a maintenance record form to encourage the user to comply with the suggested maintenance program. For example, every manufacturer of electrical ground fault interrupters for personal protection insists that the device be checked monthly by pushing a test button. A form provided with each unit can be used by the homeowner to record test dates.
16. Where feasible, have all products inspected after manufacture (100% inspection). If possible, use functional rather than visual testing as a final check. It is important that the design engineer be at least a participant in deciding the verification tests to which each device produced will be subjected. It is also important to keep in mind that tests at elevated stress levels may be destructive. For example, gloves used by those working on high-voltage lines, if tested at a higher voltage than that for which they were designed, may become defective because of the test intended to verify their quality.
17. The quality control supervisor must be informed of manufacturing errors that may result in an unsafe product. Tool engineering and production personnel may allow wear of dies or molds that they believe to be tolerable because they see only minor variations, but those variations may result in unsafe operation of the product. Quality control must be aware of these possibilities. Jigs and gages must also be kept in good condition.
18. Test the effects of mass production on the product by having a pilot run made by production personnel using production tools. Randomly selected units from this run should be subjected to the same accelerated life tests and safety analyses as were the experimental samples made during product development. The results of these tests should then be compared to the

PRODUCT LIABILITY 45

results of tests made previously on the engineering samples. If performance of the production devices is deficient, remedial action must be taken.

19. Work with the advertising/marketing department to guard against overstatements of product performance. This point is discussed below in the section on warranties, Section 2.3.3.
20. Encourage sales and service personnel and dealers to report all complaints, especially those that have to do with injury or economic loss. The report should include:
    a) Whether or not a defect in the product is alleged to exist.
    b) If so, a clear description of the alleged defect.
    c) How the defect is said to have caused the problem leading to the complaint.
    d) The nature and extent of any injury or economic loss.

The engineer is, of course, only one of the people responsible for product safety. This list reflects what would be done by the engineer, but it does not address the relevant activities of the many other employees of the manufacturing organization that are just as necessary for production of safe products.

Most of the guidelines above are adequately explained in the brief statement on each point. However, hazard and failure analysis, warranties, warning labels, and instruction manuals are important enough to require the elaboration of the following subsections. A section on standards is included in the next chapter.

2.3.2 Hazards and Failure Analysis

A *hazard* can be defined as a condition having the potential to cause harm or injury to people, animals, or property. A *risk* is the probability of an accident occurring when a hazard is present, and *danger* is a simultaneous occurrence of a hazard and a risk with the likelihood of serious consequences.

Products can possess both inherent hazards and the potential for contributing to or initiating a sequence of events that results in a hazard. The existence of hazards is determined from experience, analysis, and careful study. Perhaps the most common type of hazard is that which is an inherent property or characteristic of a product. In a belt and pulley drive, the point at which the belt reaches the pulley—the "pinch point"—is an inherent hazard; the use of guards at such points is routine. In an electrical device, electrical shock may be an inherent hazard; proper grounding, use of the correct insulation, and correct use of clamps to secure line cords at entrance points will reduce the risk.

The interaction of people with products introduces hazards associated with human performance and behavior. For example, human error was blamed for the extensive damage to the Three Mile Island nuclear reactor after what should have been a minor incident. The fiasco at the Chernobyl nuclear plant, resulting in widespread distribution of radioactive materials, was strictly due to improper actions on the part of the operators, although it is generally recognized that the plant design contained a large number of unsafe features. Guidelines for reducing the probability of human error are discussed in Chapter 15.

Evaluation of danger from human error or product failure must be carried out by the designer in an organized and methodical manner. The evaluation procedure and the results should be recorded and made a part of the development record of the device or system. An evaluation done early in the design cycle is not sufficient; it must be repeated several times throughout the cycle, especially when design modifications occur. Changes that improve one characteristic of a product may have negative effects on others.

One method of exploring the dangers associated with a given product is called "failure modes and effects analysis" (FMEA) [2]. FMEA is a method in which the possible failure modes of individual parts of a product or system are considered and the consequences of each failure mode are noted. Because the method begins with the individual parts or components, it is a bottom-up approach. Another method of failure analysis is the "fault tree analysis" (FTA) [3], which considers the relationships between functional effects and product elements (subsystems, components, and so forth), looking from the top down. Both methods, if properly carried out, will lead to the same results.

The format for FMEA varies somewhat depending on the objectives, but generally the following steps are involved:

1. Describe the component, assembly, subsystem, or system whose failure modes are sought.
2. Identify and describe the ways in which the unit can realistically fail. These are the failure modes.
3. Determine the symptoms of each failure mode to aid in early detection of failure.
4. Determine the effects of the occurrence of each failure mode.
5. Determine the probability of occurrence of each failure mode using statistical data where feasible. If statistical data are not available, a qualitative ranking must be used, even though this will inherently be subjective in nature. The values used should be numerical, such as 1 = very low, 2 = low, 3 = intermediate, and 4 = high.
6. Assess the injury potential (i.e., the hazard) associated with each failure mode. Again, a qualitative ranking may be used, such as 1 = minor property damage and/or no injury, 2 = medium property damage and/or minor injury, 3 = substantial property damage and/or major injury, and 4 = major property damage and/or death.
7. Determine the danger index based upon the combination of factors in steps 5 and 6. The danger index must also include additional considerations, such as the lack of time to take corrective action after the first symptoms appear and the effects of failure on the environment.

PRODUCT LIABILITY

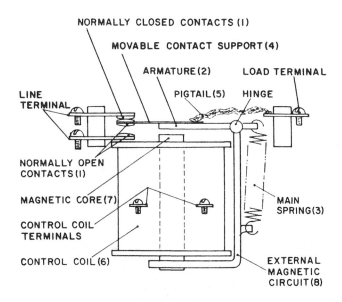

Figure 2.2 Basic components of an electric relay.

Developing a failure modes and effects analysis, as noted above, is a bottom-up process. If one starts at the lowest level, the designer may use the parts list to ensure that all components are included in the analysis. However, it is not necessary to start at the lowest level. Analysis could begin at the device level, such as a speed reducer, amplifier, or relay, proceed to the simple subsystems which are combinations of devices, and then to more complex subsystems and finally to the complete product.

All failure modes of each device or system must be analyzed in a way that expresses the effect on the next higher level in order to determine what will happen if a postulated failure occurs. FMEA is not concerned with determining the cause of the failure, but simply the consequences thereof. For example, the relay shown in Figure 2.2 can fail by loss of contact, welding of contacts, an open circuit in the coil, overheating of the coil, a short circuit in the coil, failure of the pigtail by fatigue, by breakage of the main spring, and so forth. However, as far as the effect of failures is concerned, only two conditions need be considered: (1) Contacts closed when they should be open, or (2) contacts open when they should be closed. If either of these conditions will create a hazard, a design modification should be made if at all possible.

The electrical distribution switchboard shown in Figure 2.3 is an assembly of busbars, insulators, switches, control devices, meters, and transformers, all inside a metal enclosure. Failure can occur because of inadequate design of the structure, failure to take proper account of the heat transfer necessary because of the internal losses and the ambient temperatures, improper installation, or other

48 **CHAPTER 2**

Figure 2.3 Switchboard serving several laboratories at the University of Cincinnati. (Courtesy of the Department of Electrical and Computer Engineering, University of Cincinnati.)

mistakes in manufacture or design. The busbar subsystem is composed of rectangular aluminum or copper bars running both horizontally and vertically within the enclosure, connecting the switches (whose handles can be seen) to the transformer and to the external circuits. These bars must be insulated from each other and from the enclosure. Physical separation of the busbars provides some of the insulation needed, but the busbars must be supported at various points. At these points, the insulation used in the support must have sufficient structural strength to support the bars and must also have the dielectric characteristics necessary to withstand the voltage differences that can reasonably be expected during both normal operation and when voltage surges occur.

Table 2.1 shows a failure modes and effects analysis of a subsystem of the switchboard. A similar analysis should be made for every device and subsystem in the board. The ranking of the probability of occurrence of a failure and of the hazard of the effect thereof are multiplied to give a qualitative measure of danger.

Fault tree analysis (FTA) [2,3] begins with the identification of an event (or events) that can cause injury or property damage. This is called the "top event." Then the logical combination of faults occurring at the next lower level of the product (subsystems or components) that could cause the top event are determined. This top-down process is repeated at all levels until the most primitive faults that could lead to the top event by a sequence of faults have been identified.

As noted earlier, if both a fault tree analysis (FTA) and a failure modes and effects analysis (FMEA) are done, the results should be equivalent. The experienced designer may find the fault tree more advantageous than FMEA, and this is also true for products that are minor modifications of existing products because the fault tree need be developed only to those levels at which reliability data and consequences of failure are already known.

A fault tree for the switchboard of Figure 2.3 is shown in Figure 2.4, assuming that the top event is destruction of the switchboard by a short circuit. The symbols in the diagram include rectangles, which represent events resulting from a fault; OR and AND gates; diamonds for fault events that need not be further developed to identify the basic fault; and circles, which show the basic (or initial) fault events. Qualitatively, the diagram demonstrates that a phase-to-enclosure (that is, busbar to ground) short circuit can be caused by failure of door hinges, an access door that does not have an adequate stop on closure, bridging of the gap between busbar and enclosure by an animal or an electrician's tool, loose structural elements, especially when combined with an inadequate busbar-enclosure clearance, and so forth. Some obvious design changes will reduce the possibility of a fault. Examples of such changes are the use of piano hinges and improved stops on access doors, avoidance of any horizontal surfaces inside the enclosure and above the busbars where electricians might be tempted to place tools, and warning labels that no work should be done inside the enclosure until the electrical supply to the switchboard is locked out.

Table 2.1 Failure Mode and Effects Analysis

Description	Failure Mode	Symptom	Effect	Prob. of Failure	Hazard	Danger
1.1 Bus bar assembly	Loose connection	Local temp. increase	Insulation damage	1	2	2
	Bar-to-bar contact	Arcing	Massive internal destruction	2	3	6
	Bar-to-enclosure	Short-term arc [a]	Minor damage [a]	3	1	3
			Major damage [b]	2	3	6
1.2 Support	Insulation tracking	Leakage current to enclosure	Area near insulator needs repair	1	1	1
	Insulation rupture	Large current to enclosure	Area near insulator needs repair	1	2	2

[a] Designed with ground fault interrupter which disconnects power in seconds after fault occurs.
[b] Major damage occurs only if arc caused by momentary ground fault results in uncontrolled phase-to-phase fault.

PRODUCT LIABILITY 51

The possibility of loose structural components is an element of the fault tree whose importance must be stressed to quality control personnel. Ventilation of the switchboard is important for thermal reasons, but screening adequate to bar rodents from entrance must be included in the design.

When sufficient data are available, the fault tree diagram may be used quantitatively to determine the probability of occurrence of the top event. The first step is to determine the minimum number of cut sets. A cut set is a set of basic events whose occurrence will cause the top event to occur, and it is minimal if nonoccurrence will prohibit occurrence of the top event. Identification of cut sets is best shown by example. Referring to Figure 2.4, note that all gates are OR gates except the gate leading to event G, that being an AND gate. OR gates increase the number of cut sets whereas AND gates increase the size of cut sets. An array of events is formed in which inputs to OR gates are put in separate rows and inputs to AND gates are placed in separate columns.

Begin at the OR gate just below the top event and place events B and C in separate rows, as follows:

    B
    C

Since each of these events can cause the top event to occur, each is a cut set. Now B can be caused by either S *or* T. Hence B can be replaced by S and T in two separate rows. Similarly, C can be caused by D, E, *or* F, and thus may be replaced by D, E, and F in three rows. Similarly, F may be replaced by G, H, U, *or* V. G can be replaced by W *and* X (in two columns because they come through an AND gate), and H by Y *or* Z in two rows. Diagrammatically,

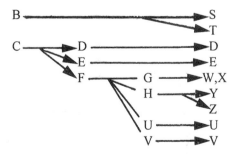

The right column shows that there are nine cut sets or paths, which in this simple example can be identified by inspection and followed from the primitive events to the top event. In a more complex example, a large cut set may be dominated by a smaller set containing some of the same events. For example, if the right-hand column in the array above contained another row with only W because W was also part of a path through an OR gate, the W,X row would be

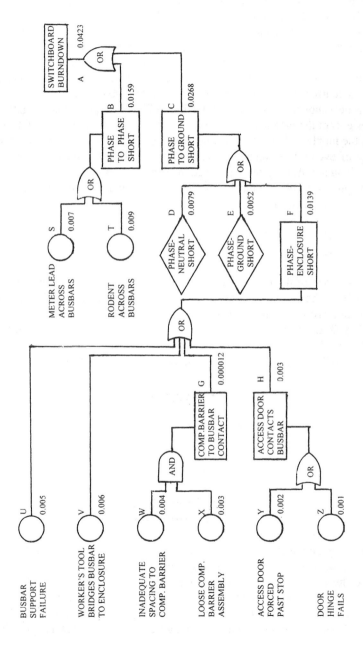

Figure 2.4 Fault tree analysis of one possible major failure of a switchboard.

PRODUCT LIABILITY 53

removed as unnecessary. Event X alone could not lead to the top event and W, if it occurred, would be sufficient.

Assuming that the input events to an AND or an OR gate are statistically independent, the probability of occurrence of the output event is given by

$$P(E_A) = P(E_1)P(E_2)\cdots P(E_n) \qquad (1)$$

for the AND gate, and by

$$P(E_0) = 1 - [1 - P(E_1)][1 - P(E_2)]\cdots[1 - P(E_n)] \qquad (2)$$

for the OR gate.

In order to use Equations (1) and (2) to determine the probability of failure of the switchboard, the probabilities of the primitive factors must be known. These are best obtained from accumulated experience, from data provided by suppliers, or from accelerated life testing (Chapter 14). However, if verifiable data from these sources is not available, use of reasonable values of probability will at least alert the designer to any unusually high level of probability of failure as the analysis proceeds. The probability of the top event, that is, event A, switchboard burndown, is of course the value the designer want to know. For Figure 2.4, assume that

$P(S) = 7 \times 10^{-3}$ $\qquad\qquad$ $P(W) = 4 \times 10^{-3}$

$P(T) = 9 \times 10^{-3}$ $\qquad\qquad$ $P(X) = 3 \times 10^{-3}$

$P(U) = 5 \times 10^{-3}$ $\qquad\qquad$ $P(Y) = 2 \times 10^{-3}$

$P(V) = 6 \times 10^{-3}$ $\qquad\qquad$ $P(Z) = 1 \times 10^{-3}$

We also assume that the undeveloped sections of the complete diagram have been analyzed elsewhere, and that

$P(D) = 7.8653 \times 10^{-3}$ $\qquad\qquad$ $P(E) = 5.2321 \times 10^{-3}$

Applying Equations (1) and (2) yields $P(A) = 42.333 \times 10^{-3}$, or the probability of 42 failures per 1000 identical switchboards. The manufacturer, of course, will not make 1000 identical switchboards, because this is a product made to the customer's specifications. These numbers can, however, be looked on as odds: that is, 1000/42, or 24 to 1 that the switchboard will not be destroyed.

This simple diagram can be analyzed by inspection. For example, if one looks at G, H, U, and V, it is immediately obvious that the compartment barrier-to-busbar contact (Event G) is much less likely to cause a problem than the event that a worker's tool will bridge from a busbar to the enclosure (Event V). Real-life products are likely to be much more complicated, and to require a computer program to avoid errors.

The diagram may be analyzed to determine the effect of taking various steps to reduce the probability of a fault. For example, if the designer takes steps to reduce the probability of a meter lead falling across busbars (Event S) to 1×10^{-3}, the probability of event A falls to 36.5×10^{-3}, a reduction of only 13.7%. If, in addition, better screening of ventilation ports reduces the probability of a rodent getting across busbars to 2×10^{-3}, the probability of burndown falls to 29.7×10^{-3}, a reduction of 29.7%. An examination of Figure 2.4 along the row of events comprised by S, T, D, E, and F (all of which may be assumed to go through one OR gate to event A) shows quite clearly that the major contributor to the probability of event A is event F. Hence, the designer needs to concentrate more effort on those parts of the design represented by the events that lead to F.

In analyzing hazards associated with any product, it is beneficial to look at the problem from as many different aspects as possible. Hazards should be considered during special situations, such as maintenance and transportation, as well as during normal operation. Another way to look at hazards is in terms of the potential severity of injury, loss of property, or loss of production while a key unit is out of service.

Since not all accident situations can be avoided, steps should always be taken to minimize the consequences of failure by insisting on increased reliability of components, by adoption of adequate safety factors and safety margins, through the use of warning labels, and by recommending procedures for monitoring of the product while in use. The efforts of the design engineer to eliminate or reduce hazards will help not only to avoid costly lawsuits but, more important, to reduce the human misery resulting from injury or death.

2.3.3 Technical Literature and Warranties

A person who has sustained injury or financial loss can hope to receive compensation under the principle of breach of warranty. A warranty is the representation of the character or quality of a product. In order to pursue this legal strategy successfully, it is necessary to prove the existence of warranty, that it was breached, and that the breach was the proximate cause of the injury or loss. The misrepresentation must be known to the purchaser (or to the lessor or renter) and must have influenced the purchase. This restriction, however, has probably had very little effect on the outcome of any case. With the passage of time and the prolonged discovery process, purchasers may be hard-pressed to remember just when they heard or read the laudatory comments about the product on which the breach of warranty suit is based.

Perhaps the most important thing to remember about warranties is that they are comprised of every bit of information given to the purchaser by the manufacturer, the seller, and their representatives. The courts do not consider merely the formal written warranty statement. In fact, the very act of offering the product for sale represents it as having at least the minimum qualities of similar products and thus constitutes a warranty.

It is convenient to divide warranties into two major classes, depending on how the product representation is communicated. These classes are *express* and *implied*. *Express* warranties are defined by the Uniform Commercial Code to be:
1. Any affirmation of fact or promise made by the seller to the buyer that relates to the goods and that influenced the sale.
2. Any description of the goods that influenced the sale.
3. Any model or sample that influenced the sale.

Catalog descriptions, pictures, advertisements, and sales brochures all affirm facts, make promises of quality, or describe advantages, and thus become part of the warranty. Furthermore, there is no limitation that affirms that the statements or descriptions on which the purchaser relied must be in print. A statement by an engineer or sales person about a product can also be held to be part of an express warranty.

Such statements, however, do require individual evaluation as to their seriousness. The courts have recognized that statements are sometimes made during the interchange before a sale that are laudatory without constituting a warranty. For example, a salesperson selling an electric razor might very honestly say, "I've used this model for years and have had no problems with it." This is merely a statement of fact about one unit out of the entire population of the product, and it is not reasonable to assume that all others will have the same performance record. Or a car dealer's representative might say, "I believe this model is the most reliable of its class on the market." This is obviously an opinion, one that must be accepted or rejected by the buyer as simply that, an opinion. These presale statements are clearly "sales talk" or "puffing," and the courts treat them as such. However, if a buyer is told by the manufacturer's engineer that a fuse has a time delay feature that will withstand the inrush current of a motor while still protecting the motor against overload conditions, that statement becomes part of an express warranty.

In 1975 Congress passed the Magnuson-Moss Warranty Act in an attempt to clarify for the consumer the extent of any express warranty given by a manufacturer. Basically, the act requires that a manufacturer must indicate in a conspicuous manner that a written warranty for any product costing more that $10 is a *limited* warranty (as opposed to a full warranty) if:

1. Remedy of the customer's problem with the product causes the customer to bear any expense.
2. The duration of the warranty is limited.
3. The compensation to the customer is limited.
4. The customer cannot choose from among the options of replacement, repair, or a refund if a defect exists.

5. The customer must follow a fixed procedure to receive remedy for a defect.
6. An unreasonable depreciation is used in fixing the value of a product that becomes defective after some fraction of its expected life.
7. Everyone claiming breach of warranty is not given prompt repair, replacement, or refund without cost.

However, the Act states that the Federal Trade Commission is not authorized to require a written warranty nor to require that the express warranty have a certain duration.

Implied warranties are divided into two subclasses, merchantability and fitness. Merchantability may be understood by noting that the courts have held that the mere act of offering a product for sale communicates in a subtle way, and hence *implies*, that the product is safe and that it has at least the minimum qualities one would expect of such a product. Most cases relating to breach of implied warranty of merchantability involve allegations that a specific device that caused loss or injury was not suitable for the ordinary use of that type of product. This can happen, for example, if the tolerances on parts of a device all cumulate "in the wrong direction," resulting in a "worst-case" unit. The plaintiff need not try to prove that the entire population of these devices is not worthy of being sold but only that the particular one purchased is unworthy.

The second type of implied warranty is called fitness. This form of warranty generally comes into existence when a specific recommendation is made by a representative of the manufacturer or seller in response to a situation described by the purchaser. An example of a product liability case based on a breach of fitness warranty is that of an industrial circuit breaker sold for use in air conditioners made to military specifications. The particular breakers being used had been recommended by the engineers of a circuit breaker manufacturer. In the field, however, many of the air conditioners experienced such frequent tripping of the breakers that the air conditioner manufacturer had to institute a worldwide maintenance program. During the discovery process that accompanied the suit to recover costs of this remedial program, it was learned that the recommended circuit breakers were designed with a tripping characteristic suitable for protection of electronic equipment, and *not* for the starting of electric motors. The case was settled favorably for the plaintiff shortly before the scheduled start of the trial.

As in so many product liability situations, there are interpretations that must be considered. As one example, the Uniform Commercial Code states that, if the circumstances are such that the seller has reason to realize what purpose is intended or that the purchaser is relying on the seller's skill and judgment, the buyer need not state to the seller full details of the particular purpose for which the goods are intended or of his or her reliance on the seller's skill and judgment. In other words, the seller's recommendation of the product need not be precisely verbalized in terms of the buyer's particular purpose for the implied warranty of fitness to apply.

PRODUCT LIABILITY

As always, product liability risk is best reduced by good design. However, the warranty of any product can be escalated by overstatements to the point that even a well-designed product cannot meet the promises. Every person involved with the sale or promotion of products must accept the fact that warranty will be breached if the product (the single unit involved in a failure or accident) cannot do what has been promised either by expression or by implication. To avoid such a situation, engineers should review all promotional material and set guidelines for statements made by salespeople. All statements must be technically correct and there must be no overstatements of fact.

There is some relief from possible breach of warranty by the use of disclaimers. However, the courts have tended to subject disclaimers to very strict interpretation in cases of economic loss and to hold them to be of no value if personal injury has resulted from a product defect. If disclaimers are used, they must be written using words chosen so carefully that they are not only technically correct, but they must be legally correct as well. Although a first draft should be written by an engineer, the final version should be written by a lawyer skilled in the disclaimer area to insure legal correctness, and "signed off" by an engineer to insure technical correctness.

The most important points to remember about warranties are: (1) That they exist, either in express or implied forms, for all products, and (2) that the best defense against breach of warranty is to be certain that all communication to prospective customers is devoid of overstatements. This requires avoidance of such phrases as "assures positive contact," "provides perfect circuit protection," "precise positioning," "no danger of overheating," and any other phrases of a similar nature that one is tempted to use to effect a sale.

As an example of the increasing precision being used in product description, it may be noted that for many years the National Electrical Manufacturers Association (NEMA) used descriptive terms in their Standard MG 1 for the enclosures of electric motors, each of these being based on the enclosure having passed certain prescribed tests. Some of the terms used were "splash-proof," "guarded," and "totally enclosed fan-cooled," but the terms were easily misinterpreted. For example, water could get into the splash-proof enclosure under certain circumstances, a guarded enclosure had openings which were protected by screens or expanded metal grilles which would prevent the entry of probes of a designated size or shape but into which smaller probes could be inserted, and totally enclosed fan-cooled motors were not required to be air-tight, but only to prevent the free exchange of air between the inside and the outside of the enclosure. NEMA has now adopted the International Electrotechnical Commission (IEC) system of designations, the descriptive terms being replaced by an alphanumeric code, which precisely define the tests used. The "splash-proof" motor is now designated IP03, which means that there is no special protection for persons against access to dangerous parts or against the ingress of solid foreign objects, but that there is protection against water sprayed upward at any angle from vertical to 60 degrees from the vertical.

2.3.4 Warning Labels [2,4,5]

The manufacturer and/or seller of a product assumes the duty to warn of any dangerous aspects of the product. This duty requires that the warning have a degree of clarity, intensity, and intelligibility that will cause a reasonable person to exercise caution commensurate with the potential danger.

Although warnings are used in an attempt to protect the manufacturer from liability as well as the user from injury, it must be understood that warnings are not to be viewed as a simple and inexpensive way of dealing with hazards and risks that should be taken care of by redesign of the product, especially when redesign can be accomplished without substantial alteration of the cost/utility factors. Many court decisions in favor of plaintiffs on failure-to-warn grounds are really directions to the defendant to redesign the product in order to avoid future liability risks. That is, warnings are related to design and vice versa.

There is no duty to warn if the danger is a matter of common knowledge. For example, the fact that a sharp knife is capable of cutting a careless user is a matter of common knowledge. Similar remarks can be made for guns and propane torches. However, manufacturers are increasingly apt to add warnings even when the danger should be common knowledge. For example, some containers for microwaveable foods display the warning "Carefully remove film [after the product has been microwaved for the period of time for cooking as given in the instructions] in order to avoid steam burns" in the instructions for preparation.

The hazards to be warned against are those that are less obvious in nature, those where the manufacturer has knowledge of inherent, latent, or concealed dangers that the user cannot foresee. For example, a sheet metal feeder for a punch press used compressed air for the feeding mechanism. Shop air was supplied to the machine for this purpose, and an air reservoir was built into the machine (in a concealed position) so that pressure drop in the supply line to the machine would not cause improper operation. When the machine was shut off electrically, a solenoid-operated valve at the inlet to the reservoir closed, but pressure remained in the reservoir. On one occasion, the sheet metal was feeding on an angle and jamming, and the machine was shut off so that the sheet metal jam could be cleared. As soon as it was cleared, the pressure in the reservoir drove the feeder forward, injuring the worker, who had assumed that the machine was in a safe condition when it was shut down. There were no labels on the machine warning that the feeder could operate under stored air pressure when electrical power was shut off. (It should be noted, however, that reliance on a warning label in this case is the wrong approach. The control circuitry should be redesigned to avoid the possibility of unanticipated operation of the feeder. For example, a second solenoid valve could be placed between the reservoir and the feeder, arranged to shut off the air from the reservoir and to vent the air in the line running downstream to the feeder.)

PRODUCT LIABILITY

Although the duty to warn of obvious danger is not required as a matter of law, it should be remembered that the law changes rapidly and, more important, the interpretation of this duty has been shifting in favor of the plaintiff. Interpretations may also vary from one jurisdiction to another.

If it is not feasible to design or guard against product hazards, then warnings must be included in labels on the machine or product, in instruction manuals, and in other printed material associated with the product. Warnings must be communicated in a clear, complete, unambiguous, and conspicuous manner. Labels must be carefully designed and instruction manuals for installation, operation, and maintenance should be carefully drafted, stressing safe practices and methods. Labels are probably more important than the manuals because many operators learn by "doing the thing," and refer to an operating manual seldom or never, but the label is in front of them when they are doing their work. The importance of labels becomes even more evident when one considers that instruction manuals are frequently kept in the engineering files, and that with the passage of time or in the event of resale of a machine they tend to become lost. However, remember that a poorly written label can itself become part of the evidence on which litigation is based.

Warning communications, whether in the form of labels on the product or machine or in the manuals for the product, should be drafted with the following general considerations in mind:

1. *Clarity and intelligibility.* Warnings must be written in such a way that they are easily understood, even by those having a below-average level of education. In the case of the sheet metal feeder mentioned above, a warning label having several lines of text that describe the operation of the pneumatic system would be useless. The very wordiness would discourage one from reading the entire label. Far better would be a label saying simply: WARNING! STORED AIR PRESSURE MAY CAUSE OPERATION AND INJURY EVEN WHEN ELECTRICAL POWER IS OFF.

2. *Adequacy.* Warning labels should not only warn of the danger but also tell what can happen. The proposed warning in (1) does both. If preventive measures are possible, they should be so stated, choosing words with great care. For example, a label reading "Do not heat or use near fire" was held to be inadequate when a bottle of nail polish exploded from contact with a user's lighted cigarette, a lighted cigarette not being considered as fire by most people. The warning label should also state the severity of harm that is possible if the warning is not heeded. A label that says WARNING! DO NOT DRINK is far less apt to receive the proper attention than one that says WARNING! DRINKING THE CONTENTS MAY BE FATAL.

3. *Completeness.* Warning labels should be complete. The presence of a warning label on a product implies that every nonobvious danger is explained. If drinking the contents of a bottle may cause death, but simply spilling it on one's skin may cause a severe chemical burn, the label is in-

complete if it refers only to the possibility of death but omits the burn possibility. If first aid treatment will reduce the severity of injury, that information should also be included.

4. *Warning label placement and durability.* The manufacturer has a duty to make sure that warning labels are placed so that the user of the product will see them, and has the additional duty to make certain that the warning labels are not easily removed. They should be durable, attached to the device itself in a permanent manner, and placed in a position where they are easily seen during normal operation. If there are multiple operating stations, the labels should be at each station. They should not be rendered illegible by normal wear and should be protected from obliteration by repainting.

5. *Language.* Many products and devices are shipped to other countries, and even those which are not may be used by individuals whose first language is something other than English. As noted below in the discussion of instruction manuals, many such manuals now appear in English, German, French, Spanish, and occasionally in other languages as well. Warning labels in the language of those who might reasonably be expected to use the product must be provided.

Warning labels must not be used in an attempt to escape litigation, but as part of an honest attempt to prevent accidents. Their message must be given, if possible, to all who might be injured by the product. This is difficult because injury may occur to infants, to innocent bystanders, and to the property of absent owners. One such case will illustrate the difficulty inherent in trying to convey the necessary information to everyone who may be concerned. An apartment owner brought successful suit against the manufacturer and retailer of hair rollers that had been used by one of his tenants. The hair rollers were to be heated in water when used. The warning with the product stated, "Rollers may be inflammable only if left over the flame in a pan without water. Otherwise they are perfectly safe." The tenant fell asleep after putting the rollers on the stove in a pan of water, the water boiled away, a fire ensued, and there was extensive damage to the building. The court held that the manufacturer should have anticipated that a user might fall asleep and thus the caution on the rollers was inadequate. As a matter of fact, the warning was worded in such a way by use of the phrase "may be inflammable" and the sentence "Otherwise they are perfectly safe" that the impact of the warning was considerably diminished, almost to the point of failure to be a warning.

This case also shows that the warning to the purchaser or user about a danger does not always shift the liability to them if a third party is injured or suffers loss of property. The pull cords on window coverings, cited above, are an example of a product for which a warning label could easily be ineffective because the victims, infants and young children, would be unable to read a label and unable to comprehend if anyone read the label to them.

2.3.5 Instruction Manuals

Many products cannot be assembled or used correctly without specific instructions. Furthermore, use of the product without instruction may expose the user or bystanders to unreasonable risk of harm. It may, therefore, be essential to provide instructions or instruction manuals with the product.

Instruction manuals have the potential for describing the product more completely than any other form of communication from the manufacturer to the user. Advertising brochures are not designed to convey detailed information, and verbal communication from a manufacturer's representative to the purchaser or an employee of the purchaser will generally be incomplete. Even if complete, the hearer will forget much of what was communicated. Instruction manuals, if properly prepared, provide a complete set of information on how to install the product, how to use, and how to maintain it. These manuals are frequently prepared by a technical writer, but the ultimate responsibility for their content resides with the product engineer, because the designer knows most about the strengths, weaknesses, and hazards of the product.

If there are hazards associated with the product or with its installation, warnings need to be incorporated into the instructions. These may extend to auxiliary equipment which might be used with the product. For example, one TV manufacturer's sheet of instructions (in English, French, and Spanish) includes a warning about the dangers of installation of an outdoor antenna in the vicinity of power lines, and also includes instructions on grounding of the antenna, citing the relevant sections of both the Canadian Electrical Code and the National Electrical Code. (These codes are standards; standards are discussed in Chapter 3.)

International trade barriers are becoming lower and in some areas no longer exist. As a result, products once thought of as being for use only in the United States are being sold throughout the world. The manufacturer of such products now has an obligation to provide instruction manuals or instruction sheets in many languages other than the English, French, and Spanish mentioned in the preceding paragraph. An inexpensive digital multimeter, purchased in 1997, came with instruction sheets in fourteen languages and a list of telephone contacts for service in sixty-three countries.

Depending on the complexity of the product, there will be from one to three manuals. If there are three manuals, they will be an installation manual, an operator's manual, and a maintenance or service manual, each of which may contain several sections. The installation manual will contain a variety of information, including a description of the product, specifications of the electrical and/or pneumatic supply required, foundation requirements, and detailed instructions on how to carry out installation correctly. An operator's manual will contain complete instructions as to the operation of the product or device. It will also contain adequate precautionary statements intended to preclude foreseeable misuse or mishandling that could result in injury to the operator of the product or in damage to the device itself. These warnings may also appear as labels on the device itself. A maintenance or service manual will contain drawings, circuit diagrams, parts lists, and similar information needed for troubleshooting or repair. It will also contain information as to proper maintenance procedures and schedules, and should identify potential hazards the maintenance personnel should be aware of, such as danger of electrical shock, pressure in pneumatic or hydraulic systems that may be present even after removal of power, toxic materials which may be present, and so forth. Adequate information on dealing with these potential hazards must be included.

For simple devices, there may be only one or two manuals. For example, a furnace for domestic use may have only two manuals, one for installation and the other for both operation and maintenance; it may have only one. An automobile will have two, one being the owner's (operator's) manual (which will contain information on maintenance schedules, but may have no information on maintenance procedures) and the other the maintenance manual, which the owner usually never sees because it is produced in much smaller numbers and is intended for use in the dealer's shop. Many household appliances have only a single manual. Simple devices, such as window coverings, may have only a single sheet of instructions.

However few or many instruction manuals there are, all of the factors discussed previously must be considered in addition to accuracy, validity, complexity, clarity, and suitable illustrations. Since manuals serve as documented, prima facie evidence of warnings and instructions concerning the avoidance of hazards, they should be reviewed by the manufacturer's legal staff after the design engineer finds that they are complete from a technical viewpoint.

Every product liability action based on failure-to-warn grounds is essentially a negligence action because it involves the alleged failure to carry out the duty of the manufacturer to warn of either reasonably foreseeable or intended uses of a product that may result in harm. The challenge to the designer is to determine all of the possible ways the user or an innocent bystander can be injured or property damage be caused and then to warn in a clear manner of those product hazards that creative design cannot remove. However, no matter how well they are prepared, instructions, warnings, and labels cannot by themselves guarantee consumer safety or freedom from product liability suits. The best way to be sure of maximum reduction of product liability is good design.

2.4 THE IMPORTANCE OF DESIGN REVIEW AND CHECKING

At 9:07 p.m. on April 13, 1970, an "event" occurred aboard Apollo 13 that came close to taking the lives of the three-man crew. At the time, Apollo 13 was outbound to the moon under the command of astronaut Jim Lovell, with the

PRODUCT LIABILITY 63

mission of landing in the Fra Mauro area. There was a sudden jolt to the entire craft, followed by loss of electrical power and oxygen from both of the tanks in the service module. The event was first thought to be the result of a collision with a micrometeorite. The remainder of the harrowing journey required heroic efforts on the part of the crew and very careful technical analysis on the part of the ground backup personnel in order to conserve the limited resources left in the service, command, and lunar excursion modules.

Because they were outward bound at the time, the only return possibility that was at all feasible was to continue the flight, pass around the back side of the moon, and then follow a trajectory which would bring the command module back to earth on a reentry path that had to be carefully controlled in order to avoid burning up in the atmosphere (if the path were too steep) or skipping out into space (if the path were too shallow).

The lunar excursion module, designed to be used by only two of the astronauts, had to support all three men and for a longer period than intended for only two. Internal temperatures in the spacecraft went so low that the astronauts had trouble sleeping, and there was constant concern that the limited oxygen and electrical energy supplies would not last for the duration of the journey. The full account of the entire sequence of events will be found in Reference 6.

When the service module was jettisoned late in the flight, good photographs were made by the crew as it moved away from the command module. These photographs indicated that the original damage was not due to a micrometeorite, but resulted from an internal explosion, apparently in the oxygen tanks. After the landing, a thorough review of the design of the tanks and of their accessories uncovered the real cause of the explosion.

Normal power on the Apollo was at 28 volts, provided by fuel cells. The oxygen tanks required an internal fan and a heater in order to keep the oxygen in such a state that it could be easily bled off as needed. In order to keep the tank's internal temperature (and hence pressure) from rising too far, the heater was controlled by a thermostatic switch that opened when temperature rose to a preset value. This switch was designed to operate properly on a 28-volt supply. On the launch pad, however, the engineers wanted to be able to vent the oxygen tanks quite rapidly if a launch was aborted. Early on, they had used 28 volts, but to empty the tanks more rapidly it was desirable to use a higher voltage on the heaters. Sixty-five volts was suggested.

This change was reviewed by a design team consisting of representatives of the manufacturer and of NASA, including as it happened Jim Lovell. After extensive discussion, the change was approved. *No one, either then or later, thought to question the capability of the switch to function properly on a 65-volt supply.* As it turned out, the switch contacts could easily weld closed when the 65-volt supply was used. If the heater was later turned on while in flight—as it was early in Apollo 13's flight—the thermostatic function was inoperative, and the temperature and pressure could rise to the point of rupture of the tank. NASA's investigation team concluded that this was exactly what had happened, and the sequence of events the team mapped out was shown to be valid by experiment.

Did the design review team fail? Whose responsibility was it to say "Will every component affected by this design change be able to handle the higher voltage?" It is not our intent to try to answer these questions—the important point is that members of a design team should not be reluctant to ask pointed and even embarrassing questions, nor should any member of the team be reluctant to ask what may appear to be a stupid question, and the engineer responsible for a design must beware of the assumption that a given design change is so minor that it is not necessary to recheck details. The complex of hardware in this case was far beyond that normally encountered and the possible consequences of a failure were far more serious than in all but a very few instances, but the design engineer must cultivate the habit for every design of checking, checking, checking! In hindsight, it was obvious that the capability of the thermostatic switch contacts when using 65 volts should have been determined. The failure to do so almost cost three astronauts their lives, it aborted the planned lunar landing, and it put the entire space program in jeopardy. No one asked the right question beforehand!

2.5 THE ENGINEER AS EXPERT WITNESS

Product liability cases usually require an individual having specialized scientific or engineering knowledge and the ability to draw accurate and technically sound inferences from the available evidence. The task of presenting this information often falls to those engineers who did the design, testing, or quality control of the product. The role of the expert witness is substantially different from that of other witnesses. The purpose of this section is to explain those differences.

Federal Rule of Evidence 701 states:

> If the witness is not testifying as an expert, the witness' testimony in the form of opinions or inferences is limited to those opinions or inferences which are (a) rationally based on the perception of the witness and (b) helpful to a clear understanding of the witness' testimony or the determination of a fact in issue.

These restrictions allow the lay witness to testify only as to facts that he or she has perceived. The witness is not permitted to draw conclusions or inferences from the facts. To do so would invade the prerogative of the jury, which has the duty of drawing its own conclusions.

In contrast, Federal Rule of Evidence 702 states:

> If scientific, technical, or other specialized knowledge will assist the trier of fact to understand the evidence or to determine the fact in issue, a

witness qualified as an expert by knowledge, skill, experience, training, or education may testify thereto in the form of opinion or otherwise.

That is, an expert is allowed to explain relevant principles to the jury, such as the relationship between force and acceleration or the separation between two conductive bodies at which a given voltage will allow the initiation of an electrical arc, and to give his or her opinion as to the most likely sequence of events.

Federal Rule of Evidence 703, which defines the sources from which the expert may draw facts, states:

> The facts or data in the particular case upon which an expert bases an opinion or inference may be those perceived by or made known to the expert at or before the hearing. If of a type reasonably relied upon experts in the particular field in forming opinions or inferences upon the subject, the facts or data need not be admissible in evidence.

There is no particular limitation on how these facts are made known to the expert, although this is usually done by depositions of persons having knowledge of facts of the case or by examination of physical evidence by the expert. This examination may involve tests of the physical evidence in the case, but care must be taken not to cause any changes in the evidence during this process. The physical evidence is usually retained by the attorneys for the plaintiff (or the defendant) in product liability cases. However, in cases in which a felony might have been committed, such as suspected cases of arson or energy theft, civil authorities frequently retain possession of the evidence and allow examination by the attorneys and experts only under their direct surveillance, and this is also beginning to be the practice for some product liability cases as well. Unfortunately, relevant evidence is not always found or retained.

In one case in which arson was initially suspected, but which on close examination turned out to be a product liability case, an important piece of evidence was swept out with the other debris. The component that was lost in the general cleanup after the fire was a float switch used to control a heater in a bath in which small parts were rinsed as part of a deburring operation. The switch itself was what is known as a reed switch, which is controlled by use of a magnet. The switch was housed in a stainless steel cylinder, and the magnet, annular in shape, was housed in a spherical float, also made of stainless steel, pierced by a tube so that the float and the magnet encircled the cylinder with the reed switch. The configuration is shown in cross section in Figure 2.5. When the float was at the bottom of its travel, the switch contacts were open. As the water level in the tank rose, the float moved up and the switch contacts would close. The submersible heater could then be energized.

In order to rinse the parts, they were placed in a basket, immersed in the bath, and the basket was then subjected to an up-and-down movement that set up

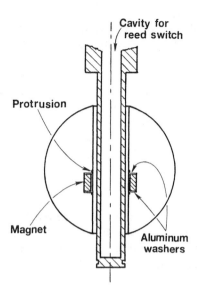

Figure 2.5 Cross section of a float switch. A clip snaps into the groove at the bottom of the cylinder in order to retain the float on the cylinder.

waves in the rinse water in the tank. These waves caused the float to slam against the top of the cylinder and then fall to the retaining clip (which snapped into the groove at the bottom of the cylinder). Investigation of other samples of the float switch revealed that the aluminum washers were so soft that the magnet could easily drive a washer over the internal protrusions intended to hold the washer-magnet-washer assembly in place. Because the float was driven against its upper stop but fell only a short distance against the lower stop, the tendency was to drive the magnet up. Its internal diameter was large enough in some samples that it could pass over the internal protrusions once it had driven the upper washer past them, and if it was driven past those protrusions it could easily "hang up" on them above its normal position. The dimensions of the magnet and the float travel normally required to close the switch were such that, with the magnet hanging on the upper side of the protrusions, the switch was closed when the float was at the bottom of its travel. An empty tank would appear to the control circuitry as a tank filled to its normal level. This was apparently the sequence of events that led to the fire, which occurred when the heater in an empty fiberglass tank was switched on by a timer sometime before the first shift was due to arrive.

The point of this discussion is that the lack of a piece of physical evidence, regardless of whether it was lost or of who took possession of those materials thought to be relevant immediately after the incident, does not mean that a defect in the lost evidence cannot be introduced into the record of the case and used to prove that the product was defective [7]. Experiments or tests on

PRODUCT LIABILITY

other units of the population of the device or devices in question may be used to show that the product was improperly manufactured or was inherently poor in design. In this case, it was determined that the float on the switch could fail in an unsafe manner, and that the float switch manufacturer was at fault because of the poor design of the float.

There are limitations on the testimony of the expert. First, he or she cannot invade the field of common knowledge, that is, knowledge that the jury is expected to possess. For example, an expert who has been qualified as an expert because of education and experience as a chemical engineer would not be allowed to advance the opinion that a device is too heavy or cumbersome for one person to carry because the jury is able to decide that as well as the chemical engineering expert. That is, the chemical engineer is a lay person in this respect, and such an opinion would then not be allowed under Rule of Evidence 701. On the other hand, if the expert had been qualified as an ergonomics specialist, the opinion could be allowed.

A second limitation is that the expert's testimony may not invade the province of the jury; that is, the expert cannot advance an opinion on the precise or ultimate fact at issue before the jury. However, Federal Rule of Evidence 704 states:

> (a) Except as provided in subdivision (b), testimony in the form of an opinion or inference otherwise admissible is not objectionable because it embraces an ultimate issue to be decided by the trier of fact.
> (b) [This subdivision relates to testimony concerning the mental state or condition of a defendant in a criminal case.]

The current trend seems to be not to exclude an expert's opinion merely because it amounts to an opinion of the ultimate fact. This is usually stated in terms of the probability or actuality of the fact.

The testimony of an expert witness is usually taken before trial by way of a deposition for discovery purposes and may be taken more exhaustively during trial. Depositions are taken under oath, there will always be a court reporter present to record the testimony, it may (and probably will) be recorded on audio tape, and it may be videotaped. A transcript is always prepared, and the witness may ask to review the transcript after it is prepared to make sure that no substantive errors have been made in the transcription. These transcripts are used by the lawyers for both plaintiff and defendant, and may be provided to other experts who are brought into the case. This practice is part of the discovery phase of the litigation.

The testimony at trial typically proceeds as follows. The direct examination begins with questions regarding the witness's education and experience so as to establish his or her level of expertise. This is followed by questions about the facts on which the expert's opinion is based. Hypothetical questions are especially useful when the facts of the case are so voluminous that it is difficult for the jurors to make a judgment about the basis of the opinion.

An issue that is likely to arise during direct examination is the degree of certainty in the opinion of the expert. The opinion will not be inadmissible because it is not stated with absolute certainty. A statement such as "Based upon the evidence and my education and experience, it is my opinion that so-and-so is true to a reasonable degree of engineering certainty" is acceptable. On the other hand, the expert is decisively barred from expressing an opinion only in terms of possibility or based merely on guess or conjecture.

Opposing counsel may elicit still more information under Federal Rule of Evidence 705, which states:

> The expert may testify in terms of opinion or inference and give reasons therefor without first testifying as to the underlying facts or data, unless the court requires otherwise. The expert may, in any event, be required to disclose the underlying facts or data on cross-examination.

That is, cross-examination of an expert witness is generally given more latitude than that of a lay witness. The opposing counsel can test the qualifications, skill, and knowledge of the expert. The basis on which the expert has formed an opinion can be explored to a greater depth under this rule than the lawyer who retained the expert might like, including questions designed to elicit the underlying data. The general rule is that the cross-examination extends to anything that might show that the opinion given during direct examination is unreliable or invalid. Even demonstrations to substantiate an opinion involving the exhibits can be requested, and matters not touched on in direct examination can be included. Hypothetical questions can be used and the answers compared to that of another expert in the field if the facts in the comparison are identical. The expert witness may be impeached by showing that his or her testimony is inconsistent with his or her previous testimony in the case. The reasoning, basis, data, and principles on which the expert relied are also likely to be explored by the opposing counsel, as are bias and prejudice. To avoid the taint of bias, engineers employed by the defendant manufacturer are usually used to explain technical details of the product alleged to be at fault, its operation, results of tests, the quality control procedures used, and so forth. Independent experts of broader experience are usually retained for matters that relate more directly to the cause of the incident. It might also be pointed out that the court, acting under Rule 706 [not quoted here], may appoint its own expert, although we know of no instance of this occurring.

Testifying under oath is an unusual experience for most engineers and, as such, may be somewhat frightening, especially on the first occasion. The outcome of the litigation can very easily be determined by the level of expertise of the expert witness, but his or her demeanor on the stand may have a greater impact on a jury than his or her level of expertise. Keep in mind that you and the

PRODUCT LIABILITY

attorney who retained you will have gone over the details of the case very thoroughly before a deposition and again before trial. You may even have had a "run-through" of the questions he or she intends to ask and of the answers you intend to give. The opposing attorney will have considerable knowledge of your position on various points as a result of having deposed you and from other information he or she has gathered. Hence you can have considerable confidence that no questions will be asked that you and your attorney have not already anticipated, at least in general terms.

Some guidelines that are appropriate to keep in mind are:

1. Before agreeing to act as an independent expert, make an honest assessment as to your level of expertise in the subject areas of the proposed testimony. The lawyer retaining you must be fully informed as to your limitations. Do not hesitate to inform the attorney of your lack of expertise in certain areas. You may wish to recommend other experts.
2. Study the material provided to you as soon as you get it, and form your opinion as to the validity of the view of the case held by the side wishing to retain you. You may find that you need to refer to other sources for information before forming your opinion. These sources may include texts, reference books, and standards covering the point at issue. If you consult standards, be certain that you use the standard in effect at the time the product was manufactured or the installation was made. Standards do not have a retroactive effect. (See Chapter 3 for discussion on standards.) If you require more information, ask for it at once. If your conclusion is that you disagree with the attorney's view, inform him or her at once, and withdraw from the case.
3. If you have agreed to serve as an expert witness, review all of the material before each deposition and before trial. Make a written summary of key points. Have available copies of relevant documents, such as pertinent codes and standards. Provide this information to the attorney who retained you, and discuss it thoroughly with the attorney so that he or she does not inadvertently ask a foolish question. You *must* be well prepared.
4. During the questioning, make certain you understand what is being asked. Lawyers, being generally lay persons as far as engineering or science is concerned, will sometimes ask questions that are based on an incorrect understanding of scientific or engineering principles. Do not hesitate to respond to such a question by simply saying, "I'm sorry, but I cannot answer that question because it is based on a physical impossibility."
5. Attorneys have a tendency to ask questions that are carefully prepared in such a way as to elicit an answer that tends to support their case unless responded to with a complete answer. For example, suppose that a test was performed in a way that imposed conditions somewhat more severe than prescribed by Underwriters Laboratories (UL). On cross-examination, an attorney may ask "Is it not true that UL does not approve the test as

performed?" The double negative in the question and the fact that UL is silent on the method used make a simple "Yes" or "No" answer misleading, although the attorney may try to insist on such a one word answer. The answer should be that "UL does not require the test to have conditions as severe as those imposed by this test." In our experience, whenever an attorney tried to force a "Yes" or "No" answer to a question that required a more complete answer, an appeal to the judge always allowed the complete answer to be given.

6. *Never* guess and *never* give a precise answer if the answer you know is only approximate. If you guess or if you give a precise value, the opposing attorney need show only that the value is incorrect.

7. Do not embellish your answers with information not required by the question. If done during a deposition, you may lead the opposing attorney to valuable discovery. If done during a trial, you may confuse the jury.

8. The opposing attorney may become contentious in an attempt to have you accept his or her point of view, or to "rattle" you. The proper strategy is to remain calm and polite, but firm. By exercising these qualities, the expert in one case put the opposing attorney somewhat on the defensive when, after the attorney had sketched on an easel-mounted pad a diagram intended to represent an electrical distribution line and asked a question of the expert based on his diagram, the expert asked whether the diagram represented a plan view or an elevation. Upon being told it was an elevation, the expert calmly pointed out that the diagram did not show sag of the conductors, a key point in the case.

9. Sometimes an attorney is reluctant to explore a facet of the case in direct examination. For example, in one case the simple question "Where do you believe the fire started?" asked on direct examination would have opened up a line of questioning on cross-examination that would have required some speculation on the part of the expert, something to be avoided. In this case, the opposing attorney, after a long series of questions closely related to the testimony on direct examination, asked what he thought would be the final question. "You did not visit the scene of the fire and therefore you do not have an opinion as to where the fire started. Is that correct?" Answer: "No. I do have an opinion." The attorney had no choice but to pursue the questioning, eliciting an opinion from the expert as to the origin of the fire that was damaging to his client. You must be prepared for some surprises on the stand.

10. Keep in mind that you have two roles. One is to provide the jury with technical information relative to the point at issue. The second derives from the first: When you provide information, you are in a teaching role.

PRODUCT LIABILITY

2.6 SUMMARY

The changes that brought about the set of principles on which product liability suits are based are summarized in Figure 2.1. Although legal pleadings usually recite all of these principles, most cases are tried on the theory that the product is defective in design or manufacture. Insight into the variety of situations that can result is illuminated by the examples of foreseeable use and misuse. These examples include all phases of the product life cycle after manufacture. The examples in Section 2.2 should make the designer aware of the many ways in which human beings can cause injury to themselves or to others or to cause property damage, and the ever-present need to be careful in carrying out the design process.

Ways to reduce product liability risk include, in the order in which they normally occur:

- Making safety an irrevocable specification.
- Using nationally and internationally recognized standards.
- Selecting materials and components of high quality.
- Using mathematical analysis to evaluate stress levels.
- Testing models using accelerated life tests.
- Conducting design reviews to obtain input from others.
- Making a failure and hazard analysis of the product.
- Making a worst-case analysis of the product.
- Obtaining an independent testing laboratory evaluation.
- Providing the production department with information that is complete in every detail.
- Making a history of the product development for future reference.
- Documenting all decisions made that impact on product safety.
- Writing correctly formulated warning labels, and positioning them correctly.
- Writing clear instructions for installation, use, and maintenance.
- Determining an appropriate maintenance procedure and schedule.
- Proposing a comprehensive product inspection program.
- Working closely with quality control personnel.
- Beginning manufacture with a test run using production tools for the verification of performance of the product to be shipped.
- Guarding against overly laudatory statements of product characteristics.
- Insisting on receiving copies of customer complaints.

In spite of all that is done, products as simple as an extension cord or a hammer are occasionally objects of product liability suits. Engineers who are called upon to be expert witnesses need to know that their task is to help the jury understand the technical information necessary to evaluate the evidence presented to them. Although the expert witness may be called upon to express an opinion based on his or her knowledge, it is the jurors who, in their deliberations, decide the importance of that opinion. Their evaluation is undoubtedly influenced by the witness's responses and demeanor under the stress of cross-examination. Expert witnesses must prepare for trial in the same way that students prepare for an examination in a difficult course.

REFERENCES

1. Thorpe, J. F., and W. H. Middendorf. *What Every Engineer Should Know About Product Liability,* Marcel Dekker, New York, 1979, pp. 38-41.
2. Kolb, John, and Steven S. Ross. *Product Safety and Liability,* McGraw-Hill, New York, 1980.
3. Fuqua, N. B. *Reliability Engineering for Electronic Design,* Marcel Dekker, New York, 1987, pp. 176-195.
4. Peters, George A. *Product Liability and Safety,* Coiner Publications, Washington, DC, 1971.
5. Freeman, Stanley H. *Injury and Litigation Prevention,* Van Nostrand Reinhold, New York, 1991.
6. Lovell, Jim, and Jeffrey Kluger. *Lost Moon: The Perilous Voyage of Apollo 13,* Houghton Mifflin, Boston, 1991.
7. Durst, John E., Jr., Proving a defect without the product, *Trial,* January 1989, pp. 102-108.

REVIEW AND DISCUSSION

1. "Strict liability in tort" is an important term in the product liability area. What is the full meaning of this phrase?
2. The decisions in the *MacPherson v. Buick Motor Co.* case of 1916 and that in *Henningsen v. Bloomfield Motors, Inc.,* in 1960 differ in what major respect?
3. Why is the *res ipsa loquitur* principle generally not an effective basis for a plaintiff's suit?
4. Why is the word "foreseeable" so important in product liability?
5. The strict liability doctrine has different effects, depending on whether one is the plaintiff or the defendant, as well as on others who may not even be a party to a given suit. Compare the effects of this doctrine on an injured party or someone who has suffered a financial loss, or on the manufacturer. What effect does this doctrine have on the public at large?
6. Every product deteriorates with time and eventually wears out. How do the liability risks change during this process? What goal should the designer have with respect to product liability during the life cycle of the product?

PRODUCT LIABILITY

7. Can a manufacturer be held liable for injury to someone who did not purchase the offending product?
8. Earlier in this chapter, there is a list of 20 ways to reduce product liability risk. Look for the key word in each one, and make a list of these words.
9. What are the factors that determine the extent of the express and the implied warranty? Which is likely to hold the product to a more severe accounting?
10. In some countries electrical outlets in bathrooms have limited current capability (about that required for the charging of an electric razor) and in other countries there are no outlets at all in bathrooms. The reason advanced is that electric shock is avoided or at least minimized. In the United States, however, we have electric outlets in bathrooms with sufficient current capacity to power hand-held hair dryers, and there is an electric shock hazard as a result. This hazard is avoided in newer construction by the incorporation of a ground fault circuit interrupter (GFCI), and a few older homes and apartments have been retrofitted with GFCI's. Recognizing the possibility of shock to the user that is present when using a hair dryer connected to an unprotected outlet, write a warning label that gives, in your opinion, the best liability protection possible.
11. Consider the float part of the float switch shown diagramatically in Figure 2.5. Suggest design modifications that would avoid the failure mechanism described in the discussion of that figure.

PRACTICE PROJECTS

Note: Some of the exercises that follow are based on actual cases, but they are not necessarily complete or correct in every detail, having been modified to be appropriate as exercises.

1. Make a liability analysis of the product you have chosen to develop. List all possible ways it could cause injury or property damage; suggest remedies.
2. The XYZ Corporation manufactures hair dryers for beauty salons. The unit consists of a base with an attached telescoping column and a dryer head. The dryer head is set on the tube at the top of the column. To partially offset the weight of the head, the tube is supported by a compression spring inside the column. Once the head is at the desired height, a set screw on the side of the column is tightened against the tube.

    The owner of several shops wanted to move one dryer from one shop to another. She hired a youth from the neighborhood to transport the dryer in his car. To do so, he had to remove the head. At the destination, he put the base with the attached column in the shop. Before he reinstalled the head, he noticed that the tube was somewhat higher than it had been when he removed the head. To get it back to its previous position, he loosened the set screw on the column. When he did so, the spring expelled the tube from the column and the tube struck him in the eye.

The young man's parents sued the manufacturer on the basis of inadequate design and the lack of a warning of a nonobvious danger. The manufacturer showed that a warning had been supplied on the back of the warranty card that accompanied each new dryer. The warning read: "Warning. Do not compress the chrome tube, or leave it compressed, unless the dryer head is attached to it." The boy violated this warning, but the manufacturer lost the case. Why?

3. Which of the following three statements would be the best warning for a stepladder? Why?
   a) This ladder is hazardous for use by a person weighing over 300 pounds.
   b) This ladder is not designed for persons weighing over 300 pounds.
   c) This ladder is not to be used by persons weighing over 300 pounds; it may collapse.

4. A woman stepped on a clear plastic clothes hanger lying in the aisle of a department store. As she stepped on it, the hanger slid on the floor, causing her to fall and be injured. The defendant (the store) claimed that the hanger could be seen and was therefore not unsafe. The plaintiff won the case on the basis that a hanger lying where people walk is a foreseeable misuse of the product. How could the likelihood of such an accident be reduced?

5. A manufacturer of electric toothbrushes included the usual 90-day warranty. A local distributor noticed that an unusually large number of orders were going to a certain drugstore and, upon inquiry, was told that they were being bought by one individual. After two months, returns began to flood in from this person, all the returns having burned-out motors. An investigation revealed that the brushes were being used to polish jewelry, being used for 8 hours each day. This is obviously an intensity of use for which they were not designed. What should the manufacturer do to avoid a similar problem in the future?

6. A scaffolding plank, purchased from a retail lumber company, had a defect and broke. The company was sued by the injured worker. If you owned the lumber company, how would you attempt to avoid similar suits?

7. The manufacturer of a commercial meat grinder was held liable when a worker was injured after he removed the guard that was designed to prevent the hand of the operator from entering the grinder. What design feature would you try to incorporate to avoid such an accident? Would you place a warning label on the grinder? If so, what wording would you use?

8. A New Jersey housing development was heated by hot water. The architect designed a system of supply and return pipes in which the pipes ran horizontally and parallel to each other along the wall. A child playing in her bedroom climbed on the piping and slipped, forcing her leg into contact with the supply pipe. By the time her mother was able to free her, her left leg was severely burned. Although skin grafts were made, her leg was left in a scarred condition. The New Jersey Supreme Court held that the

architect, the builder, and the owner of the development could all be sued. What was the basis for such a finding?

9. A car manufacturer used a plastic knob on a floor-mounted gearshift lever. The plastic deteriorated due to exposure to sunlight and developed hairline cracks. During a collision, a passenger in the rear seat was thrown against the lever. The knob broke and the lever penetrated the passenger's chest. Of the cases cited in this chapter, which is most like this case?

10. An electrical testing device composed of three neon lamps, a male plug, and an insulated wire with a bare metal tip was advertised in trade journals as follows:

"Model XXX ground tester is a simple, foolproof, totally reliable device that can check the ground on 3-wire outlets, 2- or 3-wire equipment and tools, 2/3-wire adapters, and even 2-wire outlets. Use requires nothing more than plugging the device into an outlet, using the ground probe, and noticing which of the three indicator lamps become lighted. A total of ten tests can be performed with 29 different indications for precise trouble shooting."

A little thought will show that the capabilities of the product are probably overstated. Suggest wording for a new ad to include the facts but eliminate the overstatements. Note: Although it is not needed for this project, it would be expected that the designer would have referred to the relevant standard, UL1436-93, Outlet Circuit Testers and Similar Indicating Devices.

11. Select a product and construct a failure modes and effects analysis chart. Household appliances and tools that you have available for close inspection, such as toasters, dishwashers or clothes washers, clothes dryers, rotary and reciprocating power saws, routers, and lawn mowers are candidates, but others, such as chain saws, power tillers, and brush chippers are also challenging candidates.

3

NEED ANALYSIS AND SPECIFICATIONS

Every human activity that involves the construction or manufacture of something new requires that there be an evaluation of the need or needs to be satisfied. This is true for such varied activities as those of a husband and wife who wish to install shelves in a child's room (how many, on which wall are they to be located, what will be kept on the shelves, how they are to be supported), the retail store that must expand its showroom (on the same site or at a new location, how large, display window considerations, and so forth), the company manu-facturing television receivers (size, stereo or monaural, remote control or not, on-screen menus), and the manufacturer of automobiles (family car, minivan, two-door or four-door, trunk space, accessories such as a radio or provision for a cellular phone). The items listed in parentheses in the previous sentence are needs or possible needs, even though very ill defined, with some rudiments of specifications.

    The design engineer is usually faced with a problem that can often be said to have been described only in a fuzzy or ill-defined manner, similar to the examples above. On occasion, he or she must respond to the mere suggestion that there is a need for a product to perform a certain function. One of the most important tasks in design is to determine as precisely as possible what really needs to be done. What does the prospective customer actually want? How does the designer or the design team ascertain which of the customer's desires are requirements, and which are only whims or preferences? If the product being considered is unlike any presently on the market, is it safe to assume that there will be customers for the product? The first major step in the design procedure

NEED ANALYSIS AND SPECIFICATIONS

is to obtain answers to questions such as these. The level of reliability of these answers should be as high as possible, although there will always be uncertainty.

Once the customer's needs have been determined, with whatever degree of uncertainty, the product will have been described as fully as possible in terms of functional needs. In addition to functional needs, one must also consider physical limitations as well as those limitations that are imposed by company policy or outside authority. Having functional needs and known limitations, one has a set of product specifications. At this stage, however, the specifications are still subject to alteration as new information becomes available. Moreover, the specifications themselves have different levels of importance; they may be divided in descending levels of importance into requirements, goals, and preferences.

There are three main sources of information concerning the product specification's requirements, goals, and preferences. These are the prospective customer, groups having authority to impose restrictions, and the company for which the product design is being made. However, anyone and any organization that may be affected by the product during any part of its life cycle should be considered.

3.1 NEED ANALYSIS FOR TOTAL PRODUCT LIFE

As discussed earlier, the life-cycle stages of a product are production, distribution, consumption, and retirement. Production consists of transformation of raw materials into specific shapes, adding value by assembling components into useful devices or systems, or both. As the product is designed, a constant interchange of ideas between the designer and high-level technical factory personnel is mandatory. This liaison should include at least the factory's tool engineer, quality control supervisor, and production methods engineer. A cost-effective, high-quality product cannot be developed if consultation with knowledgeable factory personnel is deferred until after all design decisions have been made. Exchange of information between the designer and those who will actually manufacture the product must occur concurrently with the development of the design. The production methods available, the tolerances needed or allowable, and the skill of those on the production line are important determinants of how a product may be designed, and the designer will never have as much knowledge in these areas as those charged with production.

In addition to the production personnel mentioned above, the designer needs to maintain close contact with the purchasing department. Since raw materials and components typically account for one-third or more of the manufacturing cost of a product, their characteristics are just as important as production methods in determining appropriate design. Prices depend heavily on the raw material to be used, the quality of those raw materials, the volume or quantity to be purchased, and so forth. Substitution of one raw material for another may result in a substantial decrease in price, but it may also require some design changes. One does not want to get into a situation where, the design having been completed, the purchasing agent tells you that "We can cut 50% of the cost of the raw material for this part if you will substitute X for Y," especially if the substitution mandates extensive redesign.

Distribution consists of the steps necessary to move the product from the manufacturer to the consumer. It may involve shipment to the manufacturer's warehouse, then to a wholesale distributor, thence to retailers or contractors, and finally to the consumer. The packaging must be designed so that the product is not damaged or parts lost during shipment or in warehousing operations, or by customer handling. Packaging specialists should be part of the design team, although they are probably of most value late in the design cycle. Their contributions may not be restricted solely to designing proper packaging for distribution, but they may also be a source of ideas that can promote sales, such as the packaging of a carpenter's steel tape that allows the tape to be pulled out of its housing for inspection without invasion of the clear plastic bubble in which the assembly has been packaged. Distribution may also involve technical personnel who advise the user on appropriate applications. This is especially true in the case of products intended to be sold to original equipment manufacturers (the OEM market).

Since every product is intended for use by a consumer, the needs of the consumer are of greatest importance. Sources of information for these needs are discussed in the next two sections.

Retirement is the final stage in the product life cycle, being that part of the cycle in which the product is withdrawn from use. How this occurs, of course, depends on the product. The major concern of the designer with regard to the retirement phase is that eventual disposal is as safe as possible for those who may come into contact with the product after it has been discarded, and that it has as little adverse effect on the environment as possible. This is the reason for the warnings on certain dry cell batteries that they not be incinerated and for the proliferation of stations to which one may take used motor oil and leftover paint. A classic example of a simple change of design having beneficial effects is that of the aluminum beverage can, which has evolved from being opened by use of a pull ring that separated a pear-shaped section of the top of the can from the rest of the can to today's design which keeps all parts captive. This change eliminated the casual discarding of the ring and attached part of the top (although not eliminating the casual disposal of the entire can), as well as the hazard of injury to small children, who found the pull rings to be attractive, and would wear them as rings or put them in their mouths.

It cannot be stressed too highly that design is not an isolated activity. It is part of a concurrent process in which those having special knowledge relating to manufacture, inspection, purchasing, and applications are consulted throughout the design cycle.

NEED ANALYSIS AND SPECIFICATIONS

3.2 DETERMINING WHAT THE CUSTOMER WANTS

The methods by which the customer's desires are determined depends to some extent on who the customer is. Some products are intended for direct sale to the general public. Some products are intended for sale to original equipment manufacturers, the so-called OEM's. Other products are sold to governmental agencies or to the military, and still others are sold to contractors, such as building contractors.

Probably the most useful approach to identification of customer needs and wants is through your sales department, which is typically the major link between the manufacturing company and the consumer. A good sales department will have an accurate analysis of what is being sold and a fair idea of trends in the case of products being sold to the general public. If the product is intended for OEM use, the sales engineer should have a very good understanding of what the customer wants, what improvements the customer would like to see, and of products that are not now available but that the customer would like to have.

Companies having a repair department for their products are in a favorable position to ascertain the customer's desires because they are in contact with the customer at precisely the time when the customer is most likely to be vocal about the shortcomings of a product. In addition, the repair department should have a record of what was needed to repair a product; that is, the weak points of the product should be readily identifiable. This is especially important to the designer if the product entering the design cycle is a redesign of an existing product or is similar to an existing product.

Customer needs for products that reach the general public through others, such as building contractors, are more difficult to determine because the manufacturer is to some extent insulated from the user by the contractor or distributor. This is not always the case, however. Some products, such as kitchen or bathroom sinks and faucets or kitchen dishwashers, are frequently selected by the ultimate user and the information as to make, model, and color or other possible choices are communicated to the contractor. Other products, such as circuit breaker panels and heating, ventilating, and air conditioning equipment, are most often selected by the contractor, with little or no input by the ultimate purchaser.

Surveys of prospective customers are another source of determining customer needs. This approach must be done with care, however. All too often the customer will identify a particular characteristic of a proposed product as being very attractive. However, if it turns out that provision of that characteristic requires a substantial increase in price, the customer may suddenly lose all interest. Another difficulty with conducting surveys of prospective customers is that of identifying the actual customer, that is, the person or group that makes the decision to purchase or not. This may be the ultimate user, but it may also be a supervisor in consultation with a purchasing agent.

When conducting surveys, it is important that the customer evaluate the actual product, and in the environment in which it will be used, if this is possible. For example, the home products industry frequently enlists homemakers to use soaps, detergents, cooking oils, and similar products (using containers which are identified only by numbers or letters). Following use, the homemaker fills out a questionnaire designed to elicit information as to product performance and desirability. If the manufacturer is sufficiently confident of the marketability of the product, a test of acceptance can be conducted using distribution to selected target areas in the country before introduction to the market as a whole. The reader may wish to consult Chapter 3 of Ulrich [1] or Chapter 6 of Salamone [2] for detailed methods of carrying out customer surveys.

If the designer does not already know what the manufacturers of competitive products are offering for sale, those products should be investigated. This should not be done with the idea of copying those designs, but rather with the idea of better understanding how others are attempting to satisfy the customer's needs or desires. Much can be learned from trying out the competitor's product, studying the way in which it is styled, how it is assembled, and so forth. In addition to studying the actual products, a great deal of information can be gained from review of advertisements for the products of others. They presumably are trying to appeal to prospective customers by touting the advantages of their products; one needs to consider whether the product entering your design cycle will exhibit at least these advantages, or whether a superior product can be achieved within a reasonable cost.

Finally, ideas for design need not come only from competitive products. The "Channel up" and "Channel down" controls on television receivers can give the designer an idea for a "Temperature up (or down)" setting of the oven of a kitchen range having electronic controls or for the temperature setting of the heating/cooling system for the passenger compartment of an automobile. The successful designer is often a practitioner of the art of coalescing ideas and features from products having vastly different uses into a feature that results in a product being considerably more attractive than that of the competition. The technical term "bisociation" has been coined for this process; it is discussed more fully in Section 6.3.

3.3 DETERMINING THE RESTRICTIONS OF GROUPS WITH AUTHORITY

Many products must meet the requirements of standards or codes imposed by organizations outside the company that employs the designer. These standards or codes may have the authority of law, they may result from the mutual consent of a group of manufacturers, or they may derive from respect for the customer. Because of their differing sources, some of these will have more influence on the design than others. Those organizations that derive their authority from law can easily demand that their standards be met, or the product may not be offered for sale. For example, the National Electrical Code (NEC) [3] has been adopted by most states and municipalities as the code governing electrical installations. Article 550-5(c) is on attachment plug caps for the electrical supply for mobile homes. That article in turn refers to the National Electrical Manufacturers Association (NEMA) *Standard for Dimensions of Attachment Plugs and Receptacles,* ANSI/NEMA WD6-1989. [The significance of the code designation is explained below in Section 3.3.1.] If you are producing attachment plug caps that fail to meet the NEMA standard, you may not sell them anywhere that the NEC is in effect. You effectively have no market.

As discussed in the next paragraph, some standards result from mutual consent of a group of manufacturers, and therefore do not *per se* have the force of law. However, as pointed out above, even these standards or portions of them may have the force of law because they have been referenced by standards which do have the force of law.

Standards that have resulted from the mutual consent of a group of manufacturers, as long as they have not been incorporated by reference into law, may legally be ignored. For example, NEMA Standard MG 1, *Motors and Generators,* contains tables of the dimensions that have been agreed on for various motor frame sizes. A motor manufacturer may use other dimensions, but may not then designate the motor as having a certain NEMA frame size when placing the motor on the market. Although the manufacturer is perfectly within its rights in manufacturing motors that fail to meet the NEMA dimensional standards, it will also find that the market is severely limited as a result.

Organizations that derive their authority from consumer respect can limit but cannot destroy the market for a product. Such an organization is Consumers Union, which publishes the results of its testing and evaluation in *Consumer Reports.* Products that receive high ratings may enjoy a favorable position in the market, although the reports themselves may not be used in advertising without the threat of legal action by Consumers Union. On the other hand, products receiving a "Not acceptable" rating because of a dangerous aspect of their design or construction may become the subject of recalls by the Consumer Products Safety Commission.

Standards may be classified in a number of nonexclusive ways. There are workplace standards, product standards, mandatory standards, and voluntary standards. Standards dealing with devices and systems (product standards) are usually voluntary. Workplace standards dealing with safety are usually mandatory and as such may affect products manufactured for use in the workplace.

A workplace standard sets the rules and regulations concerning the total environment of an employee when actively working on the job, while a product standard is one designed to encompass the requirements for a specific category of products. Standards developed by the Occupational Safety and Health Administration (OSHA) are workplace standards, but a product manufacturer cannot obtain approval of a product design from OSHA. In some cases, however, products can be submitted to independent testing laboratories to determine compliance with workplace standards.

A mandatory standard is one having the force of law. The various standards adopted by OSHA and published in the *Federal Register* are mandatory standards. A voluntary standard, on the other hand, is usually developed and promulgated by a trade organization or a similar group. Compliance with a voluntary standard is not a requirement of law, even though noncompliance can often compound legal problems. For example, a manufacturer involved in a product liability suit is in a weaker position if the product in question was not made in compliance with a voluntary standard than if it had been. Moreover, the manufacturer may be forgoing some of the potential market share that might otherwise be enjoyed. The hypothetical motor manufacturer mentioned above who fails to comply with a NEMA standard on dimensions will find that purchasers are far less likely to purchase the product.

Requirements imposed on a product by a standard or a code can be written in only one of two ways: As prescriptive specifications or as performance specifications. A *prescriptive specification* deals with materials and dimensions. For example, the materials used in communications wires and cables [Article 800-51 of the National Electrical Code] must have fire-resistant and low smoke-producing characteristics that are defined for various types of cables by three different Underwriters Laboratories standards or by a Canadian Standards Association standard. The Underwriters Laboratories Standard for Cabinets and Boxes (UL 50) covers steel enclosures for electrical equipment used outdoors. Among other requirements, the steel is required to be protected by a zinc coating defined as G90 by the American Society for Testing and Materials (ASTM) or by a G60 coating and certain types of enamels. These are prescriptive specifications because they prescribe certain design decisions, and do not permit the engineer to decide what is adequate. The designer must be sure that he or she has followed the trail of such specifications to the end to insure that the end product meets legal requirements.

A *performance specification,* on the other hand, contains the details of a performance test that the product must complete successfully to comply with the standard. Typically, these tests involve increasing the electrical, mechanical, or thermal stress on the product and then repeatedly operating it. This combination of increased stress and number of operations is designed to represent worst-case use or to accelerate wearout. For example, small motors are frequently protected by automatic reset thermal cutoff protectors. The test is described in Section 10.11.3 of Reference 4. One part of the test begins with a motor at room temperature with its rotor locked so that it cannot rotate and with test voltage, as measured at the terminals, applied to the motor. The test is run for 72 hours, during which time the protector must operate successfully (opening and resetting as temperatures rise and fall) and there must be no permanent damage to the motor or creation of a fire hazard. Permanent damage includes deterioration of the insulation systems, such as charring, brittleness, melting or destruction of

NEED ANALYSIS AND SPECIFICATIONS 83

insulation barriers, grounding of windings, flashover to the frame, severe or prolonged smoking or flaming, and electrical failure of components such as relays and capacitors.

Performance specifications allow the engineer to exercise the full latitude of his or her abilities in finding ways to comply with the specifications. They thus promote creative design and continual product improvement, and are therefore preferable from the designer's point of view to prescriptive specifications. It may be noted that most standards contain both prescriptive and performance specifications.

3.3.1 Standards and Other Important Sources of Information

As a designer, you must locate those standards that might apply to your product. The discussion that follows is intended to give you enough information to begin the search, but it is not possible to give a complete and definitive list, nor to point out all of the ways in which standards of various organizations may be cross-referenced from one standard to another. No publication is able to provide all of this information. The number of standards is large and new standards are issued as the need becomes evident. New standards may be issued by organizations not previously engaged in standards work. Once published, they cannot be ignored.

Acquisition of the necessary standards is not an overwhelming task. Information Handling Services (IHS) (Address: P.O. Box 1154, Englewood CO 80150) publishes a number of indexes that enable the designer to find the necessary standards rather quickly. One of their publications is *Industry Standards and Engineering Data.* This is published in two parts, a *Subject Index* and a *Numeric Index.* The latter is especially useful if someone has given you a number for a standard, and you wish to determine what it covers. The *Numeric Index* lists standards issued by various organizations that are categorized by the issuing organization; under each organization, the standards are listed sequentially by the number of the standard. A wide range of organizations are included in this publication, such as AISI, the American Iron and Steel Institute, ASA, the Acoustical Society of America, NEMA, the National Electrical Manufacturers Association, and PFI, the Pipe Fabrication Institute. The organizations listed are only those in the United States. International standards may be accessed by a process described below.

The *Subject Index* has entries by subject. For each subject entry, the relevant U.S. standards are listed. Suppose, for example, that your company is a manufacturer of electrical receptacles and switches for domestic use, and your management is interested in adding a line of ground fault circuit interrupters (GFCI's). Entering the *Subject Index,* one finds the following list of standards:

CHAPTER 3

NEMA 280-90, Application Guide for Ground Fault Circuit Interrupters
UL943-93, Ground-Fault Circuit-Interrupters
NEMA PP1-86, Procedure for Evaluating Ground Fault Circuit
        Interrupters for Response to Conducted Radio Frequency
        Energy

You now believe that you have the United States standards. However, the entry for ground fault circuit interrupters lists three other categories you should consult: Circuit Breakers, Electrical Grounding, and Ground Fault Protection. Reference to those entries in the *Subject Index* uncovers one more standard on GFCI's.

NEMA PB2.2-88, Application Guide for Ground Fault Protective
        Devices for Equipment

One's first reaction is that this standard is probably not relevant. However, reference to the NEMA standards catalog describes the content with the sentence: "Provides practical information containing instructions for the safe and proper application of ground fault protective (GFP) devices." This description still leaves one in doubt as to the relevance of the standard. When in doubt, acquire the standard.

U.S. companies have easily accessible markets in Canada and Mexico, and aggressive manufacturers are always looking for markets overseas. Are there any non-U.S. standards that may apply? To determine whether this is the case or not, make use of IHS's *Product/Subject Index*. This index will provide one with a locator code that can be used to access information on a particular subject in federal construction regulations, vendor catalogs, and military specifications, in addition to entries in another IHS publication, *Non-U.S. National, International, and U.S. Industry Standards*. Only the last of these is of interest here.

In the *Product/Subject Index,* one finds the following entry:

Ground
    Fault
        Circuit Interrupter   A-11-27
        Circuit Interrupting Distribution Panel   A-31-27
        Portable Power Service Center   A-31-11
        Relay   A-11-27
        Tester, Electrical Outlet   F-61-23

The alphanumeric designations are the locator codes. With these codes, enter the *Non-U.S. National, International, and U.S. Industry Standards*. Only the locator code A-11-27 yields any standards that apply to GFCI's. In addition

NEED ANALYSIS AND SPECIFICATIONS

to the standards already located, there are two more U.S. standards that may apply. These are:

UL943A-93, Leakage Current Protective Devices, and
UL1053-94, Ground Fault Sensing and Relaying Equipment.

Under this locator code, there are 38 non-U.S. standards listed, including one Canadian standard, CSA C22.2 No. 144-M91, Ground Fault Circuit Interrupters. Because sales in Canada are a definite possibility, this standard should be acquired and studied also, although in all probability there will be no surprises in it.

The publications referred to above are available in engineering college libraries, and may be available in large public libraries as well. The search described above took about one-half hour, including the time to become acquainted with the system. Information Handling Services also has available on CD-ROM an index that contains listings for over 275,000 standards from over 400 organizations, worldwide. In addition to the index, complete collections of standards from more than 65 organizations are available. These are available in the same format as hard-copy originals, including figures. The system includes several options for conducting a search. If your company is in frequent need of standards, this source should be explored.

There are relatively few organizations that account for the majority of standards. These are discussed below. Active participation in professional society activities and those trade organizations relating to the kinds of products you are designing will enable you to have an influence on content of standards by participation in the committees that draft new standards or revise existing ones. You will also be able to learn about changes in standards that are being contemplated. Another source of information is the Index of Federal Specifications and Standards, which is available from the Superintendent of Documents, U.S. Government Printing Office, North Capitol and H Streets NW, Washington DC 20401.

The following organizations are the major nongovernmental sources of standards:

1. The American National Standards Institute (ANSI) does not write standards. However, it is the clearinghouse for standards. To gain ANSI recognition, a standard must be written with full participation of all parties concerned with the subject matter. Many of the standards of other U.S. organizations, as well as foreign standards, have been designated as ANSI standards, frequently with such compound designations as ANSI/NEMA. This organization is therefore the first place to look for standards that might apply to a given product. Address: 11 West 42nd Street, New York NY 10036.
2. The Underwriters Laboratories (UL) is probably the most widely known standards-writing and testing organization in the United States. UL listing is

recognized as being important even by nontechnical users. Their standards apply to materials and devices and are intended to prevent loss of life or property from fire, crime, or casualty. There are over 350 UL standards, four major testing laboratories (New York, Chicago, Raleigh, and Santa Clara), and offices of local inspectors throughout the U.S. to provide surveillance of products before they leave the manufacturer. Products that comply with UL standards bear a label as evidence of compliance with relevant UL standard(s), usually "UL" circumscribed by a circle and with a numbered code. Beginning in 1997, products from China must have a holographic UL label [5]. The UL label is sufficiently well recognized that many products have virtually no market without it because inspectors will not approve their use without the label. An example that will be found in every residence with electrical service is the circuit-breaker or fuse enclosure. Main office address: 333 Pfingsten Road, Northbrook IL 60062.

3. The American Society for Testing and Materials (ASTM) develops standards on the characteristics and performance of materials. It publishes over 4000 standards in over 30 volumes, each standard being available as a separate publication. A typical ASTM standard identifies a certain property of a material that is of interest, describes the equipment needed to carry out the test that quantifies the property, and gives a careful description of the test procedure. Address: 1916 Race Street, Philadelphia PA 19103.

4. The National Fire Protection Association (NFPA) promotes and improves methods of fire prevention and protection. The NFPA publication list is extensive and touches any subject that conceivably could be connected with fire. One of its publications is the National Electrical Code (NEC) [3], referred to earlier. It is accepted by most political subdivisions of the U. S. as the standard for electrical installations within buildings and service to buildings. It is one of the sources of changes eventually made in UL standards. Updating of the NEC is done on a continuous basis by 20 panels of knowledgeable individuals. Anyone may make suggestions for changes. The NEC is issued every third year. Other publications of the NFPA of special interest to engineers involved with product improvement are its *Fire Protection Handbook* and its *Fire Protection Guide on Hazardous Materials.* Address: Batterymarch Park, Quincy MA 02269.

5. The National Safety Council (NSC) devotes its entire effort to the prevention of accidents. Many of its publications relate to the design of chemical plants. Address: 1121 Spring Lake Drive, Itasca IL 60143-3201.

There are certain nonmilitary government agencies that write standards or solicit other organizations to write standards in which they have an interest. The activities of these agencies should be monitored if the products of your company are likely to come under their scrutiny, because these agencies are empowered by Congress to impose their standards on the marketplace.

NEED ANALYSIS AND SPECIFICATIONS

6. The Food and Drug Administration (FDA) is the oldest consumer safety and protection agency. Although it began as a consequence of the Food and Drug Act of 1906, its activities have been broadened to include cosmetics and the agency is currently attempting to bring tobacco products under its control. The agency generally relies on voluntary compliance, although it has used court orders to seize dangerous products. Those designers who are engaged in the design of food or drug processing equipment or of containers for food or drugs should be cognizant of agency requirements. Address: 5600 Fishers Lane, #1490, Rockville MD 20857.
7. The Federal Trade Commission (FTC) has a relationship to restrictions on products mainly as a result of the Magnuson-Moss Warranty Act (see Section 2.3.3). It is most familiar to the public from its warnings about products that it believes to be unsafe or dangerous. Address: Pennsylvania Avenue at 6th Street NW, Washington DC 20580.
8. The Occupational Safety and Health Act of 1971, the act that established the Occupational Safety and Health Administration (OSHA), adopts consensus standards such as the NEC or sets its own standards. Its principal charge is to determine whether workplaces comply with these standards. Its inspectors can levy penalties if violations are found, and repetition of violations may lead to heavy fines or even prison sentences. A citation by OSHA implying that a device or system is unsafe will have obvious negative effects on the salability of the product. Hence anyone involved with the design of products that are likely to be used in the workplace should determine what standards OSHA relies on in the relevant product area and be certain to comply as fully as possible. Address: 1120 20th Street NW, Washington DC 20036.
9. The Consumer Product Safety Commission (CPSC) came into existence in 1972 with the mission of protecting the public from unreasonable risk of injury from consumer products, that is, products whose principal use is in the home. One of its first actions was to determine the product categories most likely to cause injury. It found sixteen categories that "subject the consuming public to unreasonable hazard." Among these were color TV sets, high-rise handlebar bicycles with elongated seats, vaporizers, infant furniture, rotary lawn mowers, toys, and unvented gas heaters.

Other categories have appeared since the original list was formulated. For example, in the mid-1980's the all-terrain vehicle (ATV) became very popular, but at the same time there were many deaths from these vehicles, over 900 in a five-year period, many of which were children, and injuries were being reported at the rate of 7000 per month. The CPSC negotiated a consent decree with the manufacturers of the ATV's that resulted in the end of production of the three-wheel models and a requirement that free, hands-on training be given with the purchase of the four-wheel models.

Just what is a consumer product? That is one of CPSC's problems. In general, a product can be classified as a consumer product if it can provide

its function when standing along. Carbon monoxide detectors can detect carbon monoxide whether mounted in a home or not, and are therefore consumer products even though a contractor may install one in a home. On the other hand, vinyl siding is functional only when installed, and is thus not a consumer product. A more recent product reported as causing large numbers of injuries is the grocery cart, the claim being made that there are 25,000 injuries per year due to children falling out of the carts. It will be informative to follow developments. CPSC Address: East West Towers, 4340 East West Highway, Bethesda MD 20814.
10. The U.S. Environmental Protection Agency (EPA) came into existence in 1970 to coordinate federal environmental activities. The agency's authority derives from the Clean Air Act, Water Pollution Control Act, Safe Drinking Water Act, Solid Waste Disposal Act, Federal Insecticide, Fungicide, and Rodenticide Act, Toxic Substances Control Act, and Noise Control Act. The EPA sets standards for quality of the environment, monitors pollution levels, and does research related to environmental pollution. Most of the research is contracted out. One of its major efforts has related to pollution from automobile exhaust gasses. As a result of those efforts, there have been major changes in automobile engines, in their control systems, and in the formulation of gasolines, including the use of ethanol. Most companies, however, come into contact with the EPA because of the EPA's surveillance of effluents from factories. Chemical or environmental engineers are most likely to be in contact with this agency. Address: 401 M Street SW, Washington DC 20460.

Two consumer research groups should also be mentioned for the benefit of those engaged in design of products for the consumer market, even though they have no regulatory authority.

11. One of these groups, Consumers Union (101 Truman Avenue, Yonkers NY 10703-1057), has been referred to above. It is an independent organization that maintains a sizable laboratory for testing a wide variety of products. These products are purchased on the open market, not directly from the manufacturers. Consumers Union is supported largely by subscriptions to *Consumer Reports* and by sale of reference books. Another independent research organization is Consumers' Research, Inc. (800 Maryland Avenue, NE, Washington DC 20002). Although it does no testing, it publishes informative articles on subjects of interest to consumers in the periodical *Consumers' Research*.

In addition to these publications, the designer should follow popular periodicals in the area of interest. Many of these will rate the products in their field (e.g., automobiles and computers), listing advantages and disadvantages to buttress their ratings. Because these ratings are made by individuals having

NEED ANALYSIS AND SPECIFICATIONS

considerable expertise in their fields, the designer will do well to pay attention to the reasons for the ratings assigned.

Standards such as those issued by UL undergo changes on a more or less continual basis. Others, such as the NEC, are updated periodically. UL and other organizations issue replacement pages to those who subscribe to the service as standards are modified. The NFPA publishes a revised NEC every three years; because the complexity of the technical world is growing, each revision includes more regulations. For example, the 1987 edition discusses fire resistance of communication cables as part of Article 800-3, which is on installation of conductors. Three years later, fire resistance of communication cables had been moved to a new article, 800-51, with considerably more detail on cable types and sources of standards for determining fire resistance. Similar changes occur in other standards, and the designer must insure that he or she is using the most recent standard. Service on standards panels, if company policy permits, will alert the designer to changes that may be anticipated.

Standards for military equipment are a separate area. When proposals are requested for military hardware, the relevant standards and specifications (MILSPEC's) are referenced in the proposal. The designer must work within very tight boundaries, and the end product will be subjected to very close inspection and testing.

Although the method described above is a rapid method of finding standards, you will probably want to obtain a catalog of the standards issued by the organization most relevant to your work (e.g., electrical, mechanical). These catalogs generally list standards by area of interest (product category). Each standard is described briefly so that the prospective purchaser can usually determine if the standard is relevant to the design being initiated. If in doubt, purchase the standard. The catalogs and standards should become a part of your company's library or of your own personal library.

3.4 DETERMINING COMPANY RESTRICTIONS

Every manufacturer imposes certain constraints on its designers. These constraints are intended, for example, to reduce inventory needs or to limit the variety of manufacturing operations to those of which the company is capable. Suppose the company uses a large number of 1/4 in. diameter bolts. There are two standard threads per inch, 24 and 28, but the second of these is seldom used. Unless there is a compelling reason to do otherwise, the designer is expected to specify 1/4-24 bolts. A manufacturer of controls that incorporate integrated circuits (IC's) will probably have a short list of IC's that are preferred, and the designer is expected to use one of those on the list. A casting operation using a die already in stock is to be preferred to one requiring a new die. Don't design a mechanical part that requires a milling operation to manufacture if the company does not have a milling machine.

These restrictions are obviously intended to reduce expense by avoiding costs associated with inventory or purchase of new machines. They may also relate to the abilities of those in the sales department and idiosyncrasies of prospective purchasers. "Design a product that we can manufacture and that our salespeople know well enough to be able to sell." The restrictions may also derive from limitations of those making the parts. Someone accustomed to holding a part to a 0.005 inch tolerance will have difficulty in holding a 0.001 inch tolerance.

One of the functions of a designer is to be a leader in a company. A capable designer can frequently demonstrate to management that, in order to remain competitive, some of the restrictions set by company policy must be lifted, that new production machines are needed, and that skills of the manufacturing personnel must be improved.

3.5 NEEDS ANALYSIS

The best method of showing an effective way of determining the requirements, goals, and preferences for a given device or system is by means of an example. In order to be sure that there are no hidden surprises later on, the needs of all affected persons and organizations throughout the four stages of the life cycle must be factored into the analysis.

The example chosen is that of an electrical distribution panel for mobile homes. This panel will be similar to that shown in Figure 3.1 on page 93, which shows a distribution panel for conventional construction. One of the first steps to take is to refer to relevant standards. One standard is that of the NEC, Article 550, Mobile Homes, Manufactured Homes, and Mobile Home Parks [3]. (The title should alert the manufacturer to the idea that there is a market in manufactured homes as well as in mobile homes, and that the proposed product may also be sold in that market, possibly without modification.)

Most of the sixteen pages of this standard are relevant to the design of the electrical system of the mobile home, but the entire standard needs to be understood by the panel designer because specifications on power supply connections, neutrals and grounding, disconnecting means, lighting and appliance loads, numbers of receptacle outlets, and so forth will determine some of the features needed to be incorporated into the panel. Table 3.1, spread across the next two pages, is the needs analysis. There are several things to be noticed in this table. The first is that the consumption portion of the life cycle is divided into two phases. The first phase is the installation phase. The needs in this phase have to do with having a product that will certainly meet UL specifications and which will also be easy to install and which has at least adequate flexibility as viewed by the mobile home manufacturer as to the combinations of circuit breakers that may be installed. The second phase is that which begins after the installation is completed. These needs relate to long-term service requirements

NEED ANALYSIS AND SPECIFICATIONS

Table 3.1 Distribution Panel for Mobile Home Needs Analysis

Life cycle stage	Requirement, goal, or preference	Needs
Production		
From incoming material to finished product	G	Use standard parts
	G	Use readily available raw materials
	R	Device must be insensitive to normal manufacturing variations
	G	Parts to be designed for minimum scrap
	G	Minimum material within present standards
	G	Minimize labor requirements
	G	Design for dedicated tooling if cost analysis shows financial benefits
	G	Design for dedicated tooling if necessary to maintain quality product
	G	Minimize the number of parts
	G	Design parts so that jigs may be used
	G	Design for robotic assembly
Distribution		
In transit from manufacturer to mobile home factory	R	Egg crate packaging; 12 panels to container
	R	Quick open; readily accessible panels
	R	Shock-absorbent packaging to avoid damage
	R	Adequate labeling to insure prompt delivery
	P	Caution shipper not to drop, warning on stack height, etc.
Consumption		
Phase I:	R	Comply with UL67 so that Underwriters Installation listing can be obtained (this standard contains both prescriptive and performance requirements)
	R	Design suitable for either flush or surface mounting (avoids double stock)
	G	Convenient to install; use keyhole slots or equivalent for mounting
	R	Adequate number of well-placed knockouts, easily removed
	R	Machine screw slots of proper size and configuration for installer's tools

(Continued)

Table 3.1 *(Continued)*

Life cycle stage	Requirement, goal, or preference	Needs
	R	Main circuit breaker position to be capable of accomodating a 100 ampere breaker
	R	Busbars arranged to accomodate all possible combinations of branch circuit breakers
	R	All bus bar positions capable of accepting 15 to 70 ampere breakers
	R	Provision for installation of two ground fault circuit interrupters
	R	Terminal bar for connection of all grounds
	R	Neutral to be isolated from the enclosure
	R	Connectors approved for use of either copper or aluminum wire
	R	All parts of enclosure mounted before shipping; no loose parts
	G	Complete and understandable instruction sheet and warning label for installer
	R	Instruction sheet to emphasize torque to be used in tightening connectors
	R	Circuit breaker ratings selected and breakers installed during mobile home manufacture
Phase II: After installation	R	Highly reliable metal to metal electrical contact where necessary
	R	Adequate heat dissipation
	R	Durable paint finish
	R	All metal exposed after installation to be free of burrs
	R	Enclosure to be reasonably dust-tight
	G	Cover that is attractive enough to be exposed
	R	Company name or logo on cover for easy identification by customer
	R	External instruction and warning plate with regard to restoration of service
Retirement		
From end of consumer use until scrapped or otherwise disposed of		Product will be discarded at same time as mobile home. No foreseeable safety or environmental problems

NEED ANALYSIS AND SPECIFICATIONS 93

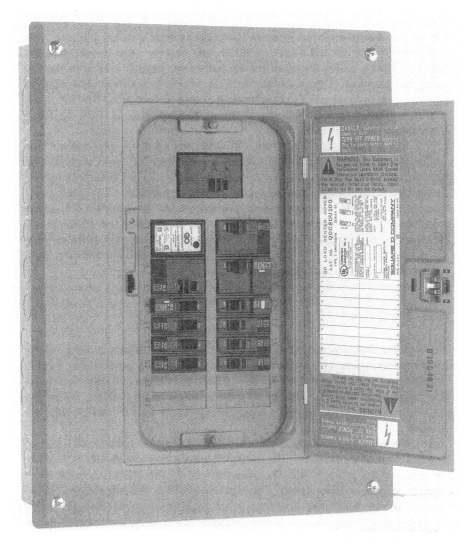

Figure 3.1 Small residential circuit breaker service entrance panel. (Courtesy of Square D Company.)

(e.g., heat dissipation capability and dust-tightness) and to the ultimate consumer's satisfaction (attractive cover and lack of burrs).

Note that this is an instance in which the ultimate consumer (the mobile home owner) was never consulted as to the distribution panel selection. The panel was selected by the mobile home manufacturer for reasons such as low cost, easy installation, past history of few or no problems in the field, and so forth.

The entries in Table 3.1 were compiled from information gathered from the kinds of sources described earlier in this chapter. You should note the detail necessary before design of the hardware even begins. Note also that the needs are not listed in the order in which the designer must eventually consider them; the list is developed by considering the needs in each part of the life cycle. Those in "Consumption, Phase I" are the needs most likely to drive the major activities in the design cycle.

The needs in this listing are obviously not of equal importance. The center column headed "requirement, goal, or preference" [R, G, or P] is the last part of the table to be completed, and this is not done until the needs list has been completed. Note that the needs in the production stage are largely designated as goals, whereas those in distribution and consumption are largely requirements. This simply reflects the fact that the manufacturer of the panels has more latitude as to the way in which the panel is produced than the user has as to what is required by standards, even though the panel manufacturer must insure that those standards are met.

3.6 SUMMARY

Early in the design procedure, the designer must spend time searching for the requirements, goals, and preferences that will result in a product that is safe, reliable, and cost-effective. It may later be decided that the resulting wish list is too demanding and that compromises must be made where possible. Decisions as to compromises will be made more intelligently if all the relevant facts are known.

The emphasis in this chapter has been on the need to consider all stages of the life cycle of the product. The customer, groups having authority to evaluate or approve the product, and the manufacturer of the product are the three primary sources of information. They are not, however, the only sources. Consumer groups that have the confidence of the general public, especially those that do testing and evaluation, may have a strong influence on the eventual success or failure of the product in the market.

Any attempt to describe in greater detail how designers go about gathering information other than that given here would probably be misleading. Although there are many ways to approach the task, a needs analysis similar to that of Table 3.1 is a good start, but only if thoughtfully and patiently completed. Perhaps the best advice that can be given is to gather enough information so that each design decision can be made with confidence.

NEED ANALYSIS AND SPECIFICATIONS

REFERENCES

1. Ulrich, Karl T., and Steven D. Eppinger. *Product Design and Development,* McGraw-Hill, New York, 1995, Chapters 3 and 4.
2. Salamone, Thomas A. *What Every Engineer Should Know About Concurrent Engineering,* Marcel Dekker, New York, 1995, Chapter 6.
3. *National Electrical Code,* National Fire Protection Association, Quincy, MA.
4. Engelmann, Richard H., and William H. Middendorf. *Handbook of Electric Motors,* Marcel Dekker, New York, 1995.
5. *U. S. News & World Report,* March 3, 1997, pp. 69-70.

REVIEW AND DISCUSSION

1. Devise a chart that shows which sources of requirements, goals, and preferences are most important for determining the needs of each of the four product life stages.
2. Why are the decisions as to purchase of materials and components so important in the design phase?
3. What needs are associated with the retirement of a product from service?
4. Which department of a manufacturing company is likely to be the primary source from which to learn about consumer needs?
5. What can you learn from competitive products about specifications?
6. Name three organizations you would look to for standards on equipment.
7. Which of the following excerpts from a standard on semiconductor power converters are prescriptive specifications and which are performance specifications?
    a. Except for potential circuits below 50 V, fuse holders shall be either of the cartridge or fuse type. Above 250 V, they shall be of the cartridge type.
    b. When a semiconductor power converter is tested in accordance with its rating, the total temperature rise of buses, connecting straps, or terminals shall not exceed the values given in paragraph X.
    c. All splices and connections shall be mechanically secure and shall have adequate current-carrying capacity.
8. In what ways do specifications reduce the options available to the designer?
9. Give several reasons why a designer may not ignore voluntary standards.
10. What type of standard is enforced by OSHA and what effect does this have on the designer of industrial products?
11. List some products that would be considered by the Consumer Products Safety Commission as consumer products and make another list of products that would not be considered as consumer products. Can you suggest some products for which you can make no clear-cut decision?

12. Enumerate some reasons why a manufacturing company will impose its own specifications.

PRACTICE PROJECTS

1. Develop a needs analysis similar to that of Table 3.1 for the product you have selected to design. The needs in that table may or may not be relevant to your product. Be prepared to spend some time and considerable thought on this analysis.
2. Make a list of products that are chosen by someone other than the consumer. The tires on a new car and the circuit breaker panels used as an example above are obvious candidates for this list, but there are many more.
3. Select a product advertisement from a trade magazine (available in any library) and list the qualities it claims that you would endeavor to include in a competitive product. If you can select an advertisement for the product you have chosen for design or a similar product, you will be able to improve your needs analysis in (1) and/or decrease the amount of time to complete that project.
4. Locate a product standard (available in most engineering libraries) and find examples of prescriptive and performance specifications.

4

THE FEASIBILITY STUDY

You have now been able to determine a tentative set of specifications, goals, and preferences, and have taken into account the limitations imposed by standards and by company policy. Although you will probably make changes in the specifications as the design proceeds, you must now take the specifications as they stand and, together with others in your company, answer the question: Can and should the product, as described by the specifications, be made?

As you try to answer this question, you soon discover that the question must be broken into three parts:

1. Is the product, as described by the specifications, physically realizable?
2. Is the product, as described by the specifications, compatible with the engineering, production, and sales capabilities of your company?
3. Can the product, as described by the specifications, be produced at a cost that provides an acceptable profit to the manufacturer and distribution chain and still be within the price range that customers are accustomed to?

The gathering of the information needed to answer these questions is called the feasibility study. All three questions must be answered affirmatively before design proceeds.

Feasibility studies have not received a great deal of attention in the literature, possibly because such studies are difficult to describe in general terms. Moreover, the tasks involved in answering the three questions that lead to a final GO/NO GO decision are not trivial. Even after completing the study, there is usually some doubt about the validity of the results, regardless of the skillfulness, experience, and knowledge of those doing the work.

If doubt actually exists, why do a feasibility study at all? The reason for doing such a study is that it is better to have an answer to the original question, even though there may not be a high level of confidence in the answer, rather than no answer at all. Even if there is doubt, the study gives the best available indication of probable success or failure. It can also bring to light possible difficulties that may have to be faced as the project moves along. It may show that the project is physically impossible because it would require violation of the laws of physics. If the study brings out reasons why the project cannot succeed, this is the time to cancel it. The designer will have become sufficiently familiar with the proposed product to have a good understanding of it, but the investment of time and other resources has been modest.

The lack of an absolute answer to the feasibility question immediately after completing the feasibility study should alert the designer to the fact that, as the project proceeds, the feasibility question can be answered with greater and greater precision, but the designer must remember to ask the question anew at various stages of the project. You will never have spent less money on a project than you have spent at the time that feasibility is assessed or reassessed. If it becomes clear at the time of any of the reassessments that the project is not feasible, cancellation should occur at once. The cost of a development project rises rapidly toward the end. It is at that point that materials begin to be purchased, jigs and fixtures built, dies cut, any machines needed for production that are not already in the plant are purchased, and the production personnel trained for the new product. If it becomes obvious at this point that costs of materials and capital equipment will make the project unprofitable, the project should be canceled.

4.1 PHYSICAL REALIZABILITY

The first of the three requirements for feasibility, physical realizability, may be determined very easily in some cases, but will require considerable effort in others. The easiest situation of all is that of an updating of a product you are already producing. If the changes proposed are minor in nature, it may never be necessary to address the question of physical realizability at all. If, on the other hand, the product is to be something that has never been made before, either in your company or, to your knowledge, elsewhere, physical realizability may be very difficult to determine, and may require considerable time and effort. Most product designs lie between these extremes. The designer must, however, be wary of pitfalls when considering realizability.

As one example of a pitfall, suppose that your company has a group of products of various ratings that has been successful in the marketplace. Can the line be extended by increases or decreases in some of the specifications? One's first reaction is to answer with a "Yes," and this will be true in many cases. However, suppose that the "product" is a paper airplane, made out of a sheet of paper 8-1/2 x 11 inches in size with a thickness of 0.005 in. We have all made these toys, and we know that the resulting "airplane" when hand-launched will glide for several feet. It would certainly be possible to make modest changes in the size of the original sheet of paper and not have any feasibility problems.

However, is it possible to increase the size by a factor of ten and have an "airplane" that can be hand-launched and that will glide for a comparable distance? The first thing to note is that the material of the large "airplane" will have to have more strength than the original sheet of paper. Drawing on your knowledge of strength of materials, you conclude that it will be necessary that the thickness of the material to be used, assuming it has the same inherent bulk properties as the original paper, will have to be increased by a factor of ten to at least the second power. If this is so, the material will have to be 0.5 in. thick. Because the length and width of the "airplane" are each being increased by a factor of ten and the thickness of the material is being increased by a factor of ten to at least the second power, the weight of the product will have increased by a factor of at least 10,000. The weight of one sheet of the original 8-1/2 x 11 inch paper is about 1/6 ounce; the scaled up version will weigh at least 105 pounds. Without further consideration, you can easily draw the conclusion that increasing the size by a factor of ten will result in an "airplane" that cannot be hand-launched. The weight by itself would prove to be a substantial obstacle, and the launch speed would no doubt have to be increased to a value far beyond what a human being could achieve. It should be noted, however, that an entirely different answer might be obtained if a much stiffer material were to be used.

A different set of problems arises if the size is to be reduced by a substantial factor, not the least of which is the difficulty of handling small sheets of paper. Between these two extremes, there is of course a range of size increase or decrease that is feasible without considering the use of a material other than paper or something similar to it. This example simply points out that having a product similar to the proposed product does not guarantee success, although it should certainly be used as a starting point. Later, a technique will be described by which you can analyze a product to determine the range for which the product is feasible.

Before leaving the topic of scaling, it may be pointed out that problems arising from attempts to scale up in physical size are probably seen most often in civil engineering structures. For an interesting discussion of the failures of bridges, as larger and larger spans were attempted by using extensions of a previously successful design, the reader may wish to refer to Petroski [1].

In many cases, the design task is to develop a product similar to one already being produced and sold by a competitor. Many design projects are of this nature because many companies want to have a line of products comparable to the line available from competitors. Suppose that you are employed by an appliance manufacturer, one of the lines being domestic refrigerator/freezers. For many years, these appliances made by your company were single-door models, having a separate door inside that provided access to the freezer compartment, which was always located at the top. When the company went to two-door models, the freezer remained on the top. A new line of refrigerators is now in the discussion stage, and the advantages and disadvantages of a freezer-at-the-bottom design, used by some of your competitors, are being considered.

The advantages stem from the fact that the freezer is accessed far less often than the remainder of the unit. Hence, the items that one usually wants, being in the refrigerator section, are more nearly at eye level, and the user does not have to bend or stoop to get to the lower shelves or slide-out drawers. Moreover, the freezer door in the units now being produced is somewhat of an obstacle when accessing the lower compartment because the door projects out over the area in which one is retrieving or storing various food items; occasionally the user inadvertently strikes the back of the head on the freezer door. On the other hand, the freezer-at-the-top design is slightly better when considering the ratio of the useful internal volume to the volume of the entire unit. That is, for the same external dimensions of the unit, the freezer-at-the-top design can be advertised as having X cubic feet of useful space, whereas the freezer-at-the-bottom design has somewhat less useful space. It is obvious that either design is physically realizable. The decision as to which of the two to pursue will be reached after weighing other considerations, such as any image the company may have built by advertisements meant to convince the public that the freezer-at-the-top design is superior, or as a result of an economic analysis that demonstrates smaller profits if the change in design philosophy is pursued.

The two examples above are of products similar to existing products. Suppose, however, that a basically new product is being considered. An advantage enjoyed by the designer in this case is that there is no previous history or company tradition or constraint to consider. Because it is a product not produced before, new tooling, materials not previously used by the company, and different methods of assembly can all be considered. As an example of the freedom one has in such a situation, consider the first manned space vehicles. There were no precedents to be considered, and the designers had free rein as to selection of the methods to explore. The difficulties that were encountered stemmed from the fact that there was very little relevant information on which to base the initial work. The designers had to consider the set of basic phenomena likely to be involved and to estimate the range of values of the important variables. For example, the space vehicle, when in orbit, has a certain amount of kinetic energy, and virtually all of that energy has to be dissipated as heat during reentry. The minimum vehicle size and weight suitable for the mission and the maximum weight that could be put into orbit at the time established a range of vehicle mass. The design of the heat shield and the materials available for such shields became prime factors in the studies, as did the aerodynamic properties of the possible shapes of reentry vehicles that would be stable (that is, not tumble during reentry through the atmosphere).

An example of the kind of study needed for determination of physical realizability is instructive. The following section gives a simple example.

THE FEASIBILITY STUDY 101

4.2 A PHYSICAL REALIZABILITY STUDY

Human beings have had an age-old desire to be able to fly, as can be seen by the tale of Icarus and Daedalus in Greek mythology. Although people learned early that they could not become airborne by means of wings strapped to their arms, various other schemes by which an individual could rise from the ground and propel oneself through the air have been proposed and tried, often at considerable expense, and most probably without benefit of a feasibility study.

In 1959, Henry Kremer, a British businessman, announced a prize of £5,000 for the first human-powered flight that met the following conditions:

1. The aircraft must be a heavier-than-air machine, powered and controlled entirely by its pilot.
2. The aircraft must take off from level ground in still air entirely by human power.
3. The aircraft must fly a figure-eight course with two turning points not less than one-half mile apart.
4. The aircraft must fly over a 10-foot altitude marker at the starting line and cross the same marker again at the finish line.

Although there were numerous attempts to construct an aircraft with which the prize could be won, the best effort was still far short of the goal in 1973. In that year, Mr. Kremer raised the award to £50,000—about $129,000.

An August 14, 1967, article in *Product Engineering* about the prize had caught the attention of one of the authors (WHM) at the time it was published. Believing that engineers should be able to venture outside their own avowed field of expertise, he assigned to a group of electrical engineering students the project of doing a study to determine whether it was feasible to design and build a vehicle capable of meeting the requirements. The requirements were made somewhat simpler for this study, the length of the flight being reduced to 1,000 feet without the necessity of making any turns, but retaining the 10 ft altitude requirement.

The result of this study by the students (developed, no doubt, with the help of students from aerospace engineering) was the analysis that follows:

> A search through *Jane's All the World's Aircraft* [2] revealed that the Swiss Neukom AN-66 sail plane has the lowest minimum sinking speed of any aircraft and the highest glide ratio. (Glide ratio is the ratio of horizontal distance covered in unit time to the altitude lost in the same time in still air.) Since this seems to be the optimum for the state of the art in 1967, this aircraft was used as the basis for the study. Pertinent data for the aircraft are:

Wing aspect ratio (defined below) = AR = 23.2
Glide ratio = $-\gamma$ = 45:1 @ 60 mph (88 ft/sec) — See Fig. 4.1(a).
Minimum sinking speed = V_s = 1.64 ft/sec
Empty weight = W_s = 635 lb
Gross wing area = S = 150.7 ft^2
Maximum wing loading = P = 5.86 lb/ft^2

The wing aspect ratio, AR, is the ratio of the length of the wing to the average chord length. That is, it is essentially the ratio of the square of the length of the wing to the wing area.

Since aircraft of this type are built to withstand 5-6 g forces, a reasonable assumption is that a low-altitude, manpowered aircraft could be made with lighter structural members. On this basis, the total weight of plane and pilot, for which the symbol W will be used, for manpowered flight could certainly be well below the empty weight of the Neukom.

In level flight, the only retarding force on the aircraft is drag, D, as shown in Figure 4.1(b). The value of drag is given by Perkins and Hage [3] as:

$$D = C_D \left(\frac{\rho}{2}\right) S v^2 \tag{1}$$

where

D = drag force
C_D = drag coefficient
ρ = density of air = 0.00238 slug/ft^3
v = velocity of the aircraft

The drag coefficient C_D is given by Perkins and Hage [4] as:

$$C_D = C_{DO} + C_{Di} = C_{DO} + C_L^2 / \pi AR \tag{2}$$

where

C_{DO} = profile drag coefficient
C_{Di} = induced drag coefficient
C_L = lift coefficient

In level flight the lift equals the total weight and the lift coefficient is given by [5]:

$$C_L = \frac{W}{(\rho/2) S v^2} \tag{3}$$

THE FEASIBILITY STUDY

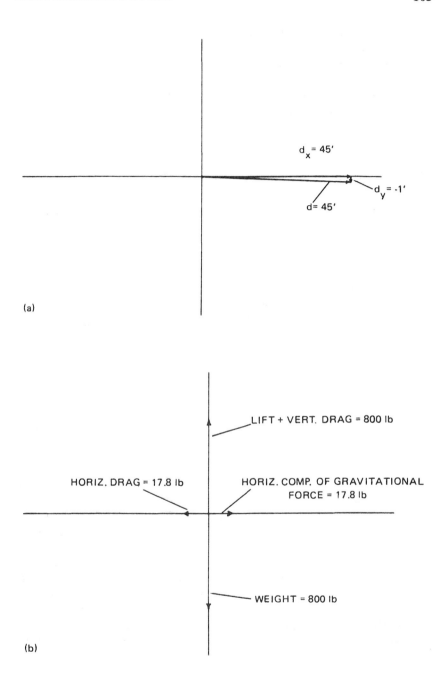

Figure 4.1 Displacement (a) and force (b) diagrams for a Neukom AN-66 sailplane at 60 mph.

Profile drag can be found by considering that the drag is equal to the change in potential energy per unit distance necessary to keep the airplane moving at constant speed. That is, the potential energy lost as the airplane sinks is all used up in overcoming the drag. Assuming that the glide ratio is given for a pilot (with gear) weighing 165 lb, the total weight W is 635 + 165 = 800 lb.

The change in potential energy per unit vertical distance is

$$Wd_y = 800 \times 1.0 = 800 \text{ ft lb}$$

This energy is equal to that required to overcome the drag in the same interval of time, Dd_x. During the time required to lose one foot of altitude, the glide ratio tells us that there will be 45 ft of horizontal travel. Hence, 45D = 800, and

$$D = 17.8 \text{ lb at a speed of 88 ft/sec.}$$

The solution of Equation (1) for C_D followed by substitution of numerical values yields a value of 0.0128 for C_D.

From Equations (1), (2), and (3), D may also be written as

$$D = \left(\frac{\rho}{2}\right) S v^2 \left(C_{DO} + C_{Di}\right)$$

$$= \left(\frac{\rho}{2}\right) S v^2 \left[C_{DO} + \frac{1}{\pi AR}\left(\frac{W}{(\rho/2)Sv^2}\right)^2\right] \quad (4)$$

In the second form of Equation (4), all terms are known except C_{DO}. Solving, $C_{DO} = 0.00827$. From Equation (2), one can then find $C_{Di} = 0.00455$. These are values for the Neukom aircraft as built.

Assume that 50 ft/sec is a reasonable minimum flying speed for this aircraft. The 1000 ft specified in the project assigned to the students will require 20 seconds. To this must be added the time to accelerate the aircraft from standstill to flying speed and the time to climb to 10 ft. Allowing 45 seconds for the acceleration phase and 5 seconds for climbing yields a total mission time of 70 seconds. The power for this mission will be supplied by a well-trained man, very likely using both pedaling and hand cranking. The available output power in horsepower for a 70-second period may be calculated [6] from:

$$hp = 4.4 t^{-0.4}, \text{ t being in seconds} \quad (5)$$

For 70 seconds, this is 0.804 horsepower on average. This is shown in Figure 4.2, as are horsepower capabilities for shorter times.

The 45-second takeoff time was found by iteration. The horsepower was calculated for a given takeoff time. The result was then compared to the horsepower available as calculated from Equation (5). The power required for acceleration was based on the assumption that, since the aircraft did not need to withstand the g forces of the aircraft as built, the weight can be reduced so that the total weight of aircraft and pilot will be no more than 500 lb. The power required from the pilot for constant acceleration of the aircraft to flying speed (50 ft/sec) in 45 seconds was calculated as follows:

$$f = ma = \left(\frac{500}{32.2}\right)\left(\frac{50}{45}\right) = 17.3 \text{ lb}$$

$$s = \frac{at^2}{2} = \frac{50}{45}\frac{(45)^2}{2} = 1125 \text{ ft}$$

$$P = \frac{fs}{t} = \frac{17.3(1125)}{45} = 432 \text{ ft lb/sec or } 0.786 \text{ hp}$$

This calculation accounts only for the power required to accelerate the mass of the aircraft and pilot, and allows a very slight excess of power available over that needed. However, the power required to overcome the drag as the speed increases was not taken into account. Referring back to Equation (1), it is seen that the drag increases as the square of the speed assuming that the coefficient of drag remains constant; hence the power required to take care of the drag will increase as the cube of the speed.

Since we were engaged in a feasibility study, the power required to overcome the drag was not calculated as a function of either the speed or the time. Rather, the drag force was calculated from Equation (4) for the proposed plane in level flight at 50 ft/sec, yielding D = 11.4 lb. At this velocity, the required power is 11.4 x 50 = 568 ft lb/sec, or 1.033 hp. That is, the horsepower required by the aircraft at the point at which the aircraft reached minimum flying speed was 1.033 + 0.786 = 1.819 hp. This level of effort is achievable for short periods, provided that the effort required is smaller both before and after the burst of demand toward the end of the acceleration period. As noted below, however, the power demanded from the pilot will be larger because propellers are not 100% efficient.

Having reached flying speed at the end of 45 seconds, the airplane must climb to 10 ft in 5 seconds. For the 500 lb assumed weight, the airplane must then gain an additional 500 x 10 = 5000 ft lb of energy in 5 seconds, a rate of 1000 ft lb/sec, or 1.818 hp.

The average thrust required from the propeller in order to sustain level flight was calculated above as 11.4 lb. The pertinent relationship relating efficiency for the ideal propeller and the other variables [7] is given by

$$\frac{1-\eta}{\eta^3} = \frac{2P_i}{\pi\rho v^3 L^2}$$

where

L = propeller length
v = speed of the aircraft, and
$\eta = \dfrac{P_O}{P_i}$ = propeller efficiency

Examination of this equation shows that, the larger L, the higher the efficiency. Keeping in mind that we are considering a design similar to an existing sailplane and that sailplanes have a small cross-section fuselage, it seems that we cannot choose L greater than 4 ft. Using this value,

$$\frac{1-\eta}{\eta^3} = \frac{2(11.4)(50)}{\pi(0.00238)(50)^3(4)^2}$$

and $\quad \eta = 0.937$

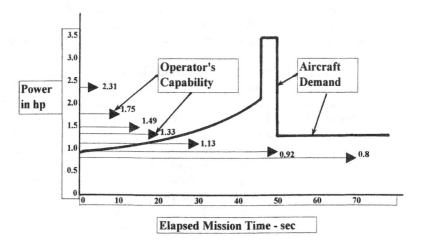

Figure 4.2 Aircraft power demand and operator power capability versus time.

THE FEASIBILITY STUDY

However, actual propeller efficiency is smaller than this theoretical value [8]. Using 85% of theoretical, $\eta_{act} = 0.796$, and

$$P_i = \frac{11.4 \times 50}{0.796} = 716 \text{ ft lb/sec} = 1.30 \text{ hp while in flight.}$$

Summary: Based on the assumptions of 500 lb total weight and 50 ft/sec minimum flying speed and taking into account the propeller efficiency, the aircraft will require the pilot to provide 0.787/0.85 = 0.926 hp at standstill. This will increase to 1.819/0.85 = 2.14 hp at the end of the acceleration phase, the point of takeoff (50 ft/sec). The power to maintain minimum flying speed then falls to 1.30 hp. However, during the 5-second climb, the pilot will have to provide an additional 1.818/0.85 = 2.14 hp, for a total of 3.44 hp. The major results of this study are shown in Figure 4.2.

Conclusion: Manpowered flight to meet the specified mission would be possible if an aircraft were designed to be substantially lighter than the Neukom AN-66. This would permit a much lower flying speed and result in less drag. Also, such flight will be more likely if a pilot can be found who is capable of truly exceptional power output and endurance.

The Kremer prize was won on August 23, 1977, by the *Gossamer Condor*, which was built by a team led by Dr. Paul MacCready. The wingspan was 96 feet, wing area was 760 square feet, weight without the pilot was 70 pounds, and the propeller was 12 feet long, driven through a plastic chain drive from a bicycle-type pedal and crank assembly. The specified flight path was completed in 6 minutes, 22.5 seconds, at an average speed of 10.8 mph, with the expenditure of 0.33 hp (i.e., 25% of the value obtained in the feasibility study for level flight) by the pilot during the straight-line portions of the flight. Laboratory tests had established the power capabilities of the pilot, who regularly trained as a bicycle racer, as 0.35 hp for 30 minutes, 0.45 hp for 7 minutes, and 1.2 hp for short bursts. These figures cast doubt on the validity of Equation (4).

Less than two years later, on June 12, 1979, the team won a second Kremer prize—£100,000—with the *Gossamer Albatross's* flight across the English Channel at an average speed of about 8 mph. That plane had a slightly shorter wing span and a design weight of 55 lb. Even that achievement was far outdone on April 23, 1988, although by a different team, when the *Daedalus 88* was flown from Crete to the island of Santorini, a distance of 72.4 miles, only to break up in flight 30 yards short of the beach. (The reader interested in reading more about these flights should consult References 9, 10, and 11.)

There are three lessons to be learned from this feasibility study. The first of these is that you start with state-of-the-art designs and determine if reasonable improvements will meet the objectives. This example showed quite clearly that incremental improvements in technology would not suffice. It made very clear the fact that major changes in the technology, different materials, and radically different construction concepts would be necessary.

The second lesson is that, if the study is done correctly, it will indicate the changes that must be made. In this case, the power requirement at 50 ft/sec exceeded what the pilot could theoretically provide even for the limited mission of the study. Lower flying speed, a larger propeller, and a reduction of weight were all indicated by the results of the study. The specifications of the successful aircraft showed that all three were necessary.

The third lesson (one that becomes obvious if one looks at References 9 through 11) is that the students did not do an adequate search of the technical literature. There had been a number of straight-line human-powered flights prior to 1967, some of which had covered distances sufficient to win the Kremer Prize. This information was in the technical literature. The major problem still remaining was that of control, especially in turns. The MacCready team achieved what no one had been able to do previously: They built an airplane that could be flown for the requisite distance by a human being, *and* they built an airplane that could be steered around the course.

Before leaving the topic of physical realizability, it should be noted that there are differences in feasibility studies when applied to a device from studies applied to a system. A system may be defined as a group of activities or components that can be bounded in such a way that all relevant interdependencies and interactions are enclosed [12]. This definition allows one to class groups of roads, communication networks (e.g., Internet), and individuals with common goals as systems. The definition also includes most devices.

A definition which is more to the point for our purposes is that a system is an operating unit in which some of the major components were not designed specifically for that unit; a device is limited to those operating units in which all major parts were manufactured specifically for the unit. The difference is that, in a device, the designer theoretically has complete control over the design. (As noted earlier, company policies may dictate certain aspects of the design.) In system design, the designer must select components whose external characteristics come closest to filling the needs of the system. As a matter of fact, the major challenge in system design is to determine the order in which components will be sought and to search for the best of those that are available.

Frequently, the best component or subsystem available will dictate the direction taken by the designer of the system. As an example, suppose that a generator is needed to supply power to test electronic components for use in aircraft. The drive motor must have a speed control which will keep the generator at 15,000 rpm with a speed deviation of less than 0.05%. Shunt motors are not available for this speed, but series motors are, and an ingenious speed control [13] for series motors will meet the specification. However, with modern electronic controls, there are many other drives which do not use motors with mechanical commutation that will also meet the specification. The designer must maintain an open mind in the search for the best drive, weighing all of the advantages and disadvantages (mechanical commutators are a distinct disadvantage) and not become committed to any particular one until all reasonable drives have been assessed.

It cannot be stressed too strongly that feasibility studies, properly done, are worth far more than an offhand opinion from an "expert." During World War II, Admiral E. J. King, who was an expert on conventional explosives, told President Roosevelt that expenditures on the atom bomb were a waste of time, money, and manpower: "It will not explode." Lord Kelvin, noted physicist and president of the Royal Society, stated in 1895 that "Heavier-than-air flying machines are impossible." After the Wright brothers had proved Lord Kelvin wrong, Marshal Ferdinand Foch, while professor of strategy at the Ecole Superieure de Guerre, said that "Airplanes are interesting toys, but of no military value." Possibly the most sweeping statement of all in the category of wrong predictions was that of Charles Duell, commissioner of the U.S. Office of Patents, who proposed in 1899 that the office be closed because "Everything that can be invented has been invented." All of these statements, which we now see as ludicrous, were made by people who were really not experts in the field in which they were making their pronouncements.

The last point to be made concerning physical realizability is that the feasibility study should be a "pencil-and-paper" study that is completed long before a model is built. The term "pencil-and-paper" as used here is intended to include computer analysis and simulation, in addition to more traditional methods.

4.3 COMPANY COMPATIBILITY

An engineering department may have the capability to design and develop many different products. The production department, however, has a more limited capability, whether because the skills of production personnel do not match those needed to manufacture a certain kind of device or because the necessary production machinery is not owned by the plant. Moreover, the sales department may not have the capability to sell the product without extensive instruction in the applications of the product.

To evaluate whether the production department can make the product, the designer must look for requirements in the skill of personnel or in the available production machinery that are not present in the company. A lack in either area may be compensated for by purchasing parts or services. In fact, every company depends on others to supply some parts that cannot be made in the plant or that the company chooses not to manufacture. Specific information as to what can be made with present facilities and personnel and what will require additional investment or will need to be purchased will assist the company management in deciding whether the company has the financial capability to produce the product. This part of the feasibility study should be made very carefully. It can be done correctly only with the cooperation of the production department personnel.

Engineering and production are not the only departments involved in the feasibility study. Every department that will be affected must be kept apprised of what is being developed. One department requiring special consideration is the marketing or sales department. Without attempting to enumerate all of the considerations in determining whether a given sales force can effectively market the proposed product, it may be pointed out that there are two important constraints to be evaluated:

1. The product must be sold to the same class of customer (original equipment manufacturers, distributors, wholesalers, or retailers) as are the other items made by the manufacturer.
2. Sale of the product must require technical expertise of the sales force at about the same level as the remainder of the manufacturer's line.

If either of these constraints is not met, the company will have to build a substantially new sales network or recruit a new sales force, or both. Table 4.1 shows the technical knowledge, skills, and capabilities that the engineering, manufacturing, and marketing departments must have to design, make, and distribute the mobile home panel discussed in Chapter 3. A table such as this is an effective way to evaluate the degree of company compatibility with a proposed system or device. Consultation with the sales force is important during this phase of the study.

4.4 ECONOMIC DECISION ANALYSIS

The final requirement that a proposed system or device must meet is that of having an acceptable manufacturing cost relative to the price the manufacturer can charge. That is, the engineer is expected to design products that will sell with a fair return on the investment. There are several different kinds of problems that the engineer may encounter that have little or nothing to do with the technical feasibility of the design. Among these are:

1. Evaluation of two or more manufacturing techniques. That is, the designer always has alternatives which must be weighed. When considering these possibilities, one should focus on those aspects of the alternatives that are different, because it is only the differences that are relevant. If, for example, a large switchboard such as that described in Chapter 2 is to be built, there are alternative methods of connecting sections of busbars. One method may be a welding process; the other may be drilling, tapping, and fastening

Table 4.1 Technology Needs Analysis for Mobile Home Electrical Distribution Panels

---

Engineering needs to know:
    Heat transfer
    Surface-to-surface contact theory
    Wire termination technology
    Electrical insulation characteristics
    Human factors engineering as applied to electrician's work
    National Electrical Code (Article 550)
    Underwriters Laboratories Standard UL67 (panelboards)
    Underwriters Laboratories Standard UL50 (cabinets and boxes)

Production needs to know:
    Punch-press operations applied to steel and copper
    Tool and die design
    Assembly jig design
    Aluminum extruding
    Insulation fabrication
    Tapping
    Plating copper and aluminum
    Painting steel enclosures
    Spot welding
    Riveting

Sales needs to know:
    Methods used in electrical installations in mobile homes
    Functional advantages of product compared to competitor's
    Data on reliability, avoidance of callbacks, etc.
    Pertinent articles of National Electrical Code
    Overview of Underwriters Laboratories standards

---

together with machine screws. The circuit breakers to be used will be the same in either case, as will the enclosure with its ventilation ports, doors, and all of the other hardware. If each of the two methods is acceptable from a technical point of view, then the decision as to which to choose becomes one to be based on the economics of the situation.

2. Make-buy decisions. Any given part may be made by the manufacturer, or it may be purchased. Anything made in-house will require equipment, personnel, floor-space, and so on. Without long experience, it may not be possible to make the part at a lower cost than the identical product commands on the market. Moreover, a part made in-house may need to

have certain features that differ from a purchased part. One must avoid any patent infringement, there are limitations on the in-house manufacturing equipment available, and obvious imitation is to be avoided.
3. Evaluation of a design modification. Companies are constantly searching for ways to reduce manufacturing costs. Manufacturers frequently assign personnel—value engineers or cost engineers—to pursue cost reduction as their major or only task. Cost estimation is discussed in more detail below.
4. Evaluation of a new product development or of a major redesign. The manufacture of new products or redesigned products can be justified only if their production earns enough to recover the total development and tooling costs with sufficient additional financial return to meet the manufacturer's minimum attractive (or acceptable) rate of return (MARR). (This concept is discussed below.)

In every one of these cases, financial or economic analysis is needed to make the necessary decision. One might think that the necessary analysis is the province of the cost accountant. However, as in many other aspects of the design process, the cost accountant is interested only in working on the "final" design, but the decisions as to alternatives to select, make-buy decisions, and so forth must frequently be made far earlier in the design cycle. In any case, the design engineer must know enough about the ways in which economic data are viewed by others that he or she can discuss financial analyses in a knowledgeable way. The remainder of this section is intended to provide the basics, but those who wish to delve further into the subject should consult suitable references, such as 14, 15, and 16.

4.4.1 Some Basic Concepts

The common denominator in engineering economic analysis is *money*. If alternatives are to be compared quantitatively, there is no other available measure by which they can be placed on an equal footing. It is an objective measure; others that might be used involve subjective aspects that will contribute to uncertainty. This is not to say that all uncertainty is removed by comparisons based on dollars, but it does say that there is a higher level of objectivity using dollars as a measure than otherwise.

The principles to be used in engineering economic analyses are few in number and are relatively straightforward. They are:

1. Define the viewpoint or system to be used.
2. Identify cash flows into and out of the system.
3. Focus on the differences in alternatives.
4. Compare the cash flows over equal time periods through interest calculations.

THE FEASIBILITY STUDY

If these principles are properly applied, decisions as to which of two or more alternatives to choose frequently become obvious.

4.4.2 Time Value of Money

Money received or money paid out at different times does not have the same value. The earlier that money is received, the sooner it can be reinvested. The later that money is paid out, the longer the time available to earn a return on the funds. The words "reinvested" and "earn a return" immediately bring up the additional concepts of *interest* and of *discounting*. A simple example will illustrate the difference between a loan at simple interest and a discounted loan, and will reinforce the concept that money received or paid out at different times has different values, depending on the viewpoint.

Suppose that you have a need for $1,000 for one year. Your bank agrees to a loan at a rate of 10%. You are also given two options. One option is to agree to borrow the $1,000 and to repay it a year later with interest of $100, a single payment of $1,100. The bank receives its interest at the end of the loan period. The other option is that you agree to borrow $1,000, but that you pay the 10% interest at the time the loan is made. Assuming that you have $100 in hand, this appears to be a tempting offer. You accept only $900 from the bank and sign a note to repay the bank $1,000 at the end of the year. This is known as a discounted loan. You don't have to repay $1,100 at the end of the year as in the first option, but only $1,000.

In the first case you had the use of $1,000 of other people's money for one year, and you repaid that amount plus 10% more. In the second case, you again had the use of $1,000 for one year (although $100 of that amount was already yours), but the bank earned interest on the $900 it gave you at the rate of $100/\$900 = 0.1111$, or 11.11% for the year. From the viewpoint of the bank, this is a better deal. It receives the $100 in interest under either option, but under the second option it "receives" it a whole year earlier (that is, it retained $100 of its resources at the beginning of the year), and that money can be invested elsewhere. From your viewpoint also, the discounted loan is probably the better of the two options. If the rate of return on the $1000 you have at your disposal is above 0%, you will have made money by the end of the year. If you do not have the $100 initially, you will of course have to take the first option.

Before looking at capital investments and repayment plans, it must be pointed out that other elements of the financial analysis necessary to reach a final decision on whether to proceed with a project under consideration will eventually have to be included. These include such items as wages or salaries of those using the equipment, training costs, material costs, utility costs, and taxes. For the moment, we will consider only the capital investment

Most loans extend for several years, and the loan is seldom discounted. In the following, it will be assumed that cash receipts and disbursements (cash flows) occur only at the end of an interest period. This convention requires that there be a zero (0) period to allow for the initiating transaction to occur at the end of a period. Interest payments and reduction-in-principal payments are made at the end of period 1, period 2, and so forth until the end of the agreed-upon number of periods, which is designated by the symbol n. The interest rate is expressed in percent, but for calculation purposes, it must be expressed in decimal form for the interest period. For example, if the annual rate is 18% and the interest period is one month, the interest rate i is 0.18/12 = 0.015.

A number of different repayment plans may be devised. For our examples, we will consider four different plans. We will assume that the loan required (referred to as the *principal*) is $100,000, that the annual interest rate is 8%, and that the payoff time is at the end of the fifth year.

Plan I. The principal is repaid at the end of the fifth year. Interest is paid at the end of each year.
Plan II. There is a uniform repayment of principal at the end of each year, with interest paid only on the balance of the principal.
Plan III. There is a uniform payment each year, part of which reduces the principal, the remainder paying the interest on the balance of the principal.
Plan IV. No payments are made until the end of the fifth year, at which time the principal and the accrued interest (which has compounded) are paid as a lump sum.

Table 4.2 shows these repayment plans, with all dollar amounts rounded to the nearest whole dollar. The total cost for the use of the $100,000 for the five years is not shown, but it is instructive to compare those costs for the various plans.

$$\begin{aligned}
&\text{Plan I} &&- \$40{,}000 \\
&\text{Plan II} &&- \$24{,}000 \\
&\text{Plan III} &&- \$25{,}227 \\
&\text{Plan IV} &&- \$46{,}933
\end{aligned}$$

Before looking at the calculations behind the numbers in the table, it should be pointed out that, despite the differences in cost for the use of the $100,000 over the life of the loan, the four payment plans are said to be *equivalent*. Their equivalence derives from the fact that, given an interest rate, any payment or series of payments that repays a present sum of money with interest at that rate is equivalent to that present sum.

The calculation for the payments in Plan I is very straightforward, the payment at the end of each year except the last being simply $P \times i$, where

P = present value (in this case, $100,000), and
i = interest for the period in decimal form (in this case, 0.08)

THE FEASIBILITY STUDY

Table 4.2 Four Plans for Repayment of $100,000 in 5 Years with Interest at 8%

	End of year	Interest due	Total owed before payment	Year-end payment	Money owed after payment
	0				$100,000
	1	$8,000	$108,000	$8,000	$100,000
Plan I	2	$8,000	$108,000	$8,000	$100,000
	3	$8,000	$108,000	$8,000	$100,000
	4	$8,000	$108,000	$8,000	$100,000
	5	$8,000	$108,000	$108,000	0
	0				$100,000
	1	$8,000	$108,000	$28,000	$80,000
Plan II	2	$6,400	$86,400	$26,400	$60,000
	3	$4,800	$64,800	$24,800	$40,000
	4	$3,200	$43,200	$23,200	$20,000
	5	$1,600	$21,600	$21,600	0
	0				$100,000
	1	$8,000	$108,000	$25,046	$82,954
Plan III	2	$6,636	$89,591	$25,046	$64,545
	3	$5,163	$69,708	$25,046	$44,663
	4	$3,573	$48,236	$25,046	$23,190
	5	$1,855	$25,046	$25,046	0
	0				$100,000
	1	$8,000	$108,000	0	$108,000
Plan IV	2	$8,640	$116,640	0	$116,640
	3	$9,331	$125,971	0	$125,971
	4	$10,078	$136,049	0	$136,049
	5	$10,884	$146,933	$146,933	0

The final payment must, of course, include both the principal and the interest for the last year.

The calculations for Plan II are also straightforward. Interest is calculated using $P \times i$ as for Plan I, but the value of P is decreased by $20,000 at the end of each year. Hence the repayments become smaller as the repayment plan is executed.

The calculations for Plan III are based on the *capital recovery equation*, which yields the installment amount A required to completely pay off a present debt P in n periods at a given rate of interest. This equation is

$$A = P \frac{i(1+i)^n}{(1+i)^n - 1} \tag{6}$$

It may be rewritten as

$$A = P \frac{i}{1-(1+i)^{-n}} \tag{7}$$

Substituting values,

$$A = \$100,000 \frac{0.08}{1-(1+0.08)^{-5}} = \$25,045.65$$

or, to the nearest whole dollar, $25,046.

The calculation for Plan IV is made using the *single payment present worth equation*,

$$P = F \frac{1}{(1+i)^n} = F(1+i)^{-n} \tag{8}$$

where F = the future amount, and the other terms have been defined earlier.

Because F is the value we wish to find, the equation may be rearranged and numerical values substituted therein to yield

$$F = \$100,000(1+0.08)^5 = \$146,932.81$$

Although not relevant to the present example, one other equation is frequently of value. That is the *equal payment compound amount equation*, also known as the *uniform series compound amount equation*, which is used to calculate the future value F that A equal installments will accumulate in n periods at interest rate i. The equation is

$$F = A \left[\frac{(1+i)^n - 1}{i} \right] \tag{9}$$

This equation would be especially useful when computing the future value of funds set aside in nontaxable individual retirement accounts (IRA's). If, for example, you could set aside $2,000 per year at a growth rate of 10% per year for thirty years, the equation tells us that the value of the fund at the end of that time would be

$$F = \$2,000 \left[\frac{(1+0.10)^{30} - 1}{0.10} \right] = \$328,988$$

THE FEASIBILITY STUDY 117

even though the actual amount of cash invested was only $60,000.

Equation (9), when solved for A, is known as the *sinking fund equation,* perhaps because it tells you the amount that you must *sink* into investments in order to yield a certain amount in the future. If, for example, you wished to accumulate $1,000,000 at the end of thirty years and the interest rate was still assumed to be 10%, the equation yields an annual investment amount of $6,079.25 to achieve this result.

4.4.3 Cash Flow Diagrams

A cash flow diagram is a pictorial way of showing the effects of various economic plans. In these diagrams, time is shown on the horizontal axis, increasing to the right. The convention previously mentioned as to the time of transactions (that is, the end of year) is used, again requiring the use of a zero period to identify the start of the process. Cash flow in is shown above the axis, using arrows pointing upward in the usual convention; cash flow out is shown below the axis, using arrows pointing away from the axis. Other conventions may be used for the arrow directions. The arrow lengths are not to scale. Figure 4.3 shows cash flow diagrams for the four plans of Table 4.2.

Cash flow diagrams are not restricted to use in comparing repayment plans. They can be and are used for any situation in which cash flows in and out. Salary costs, personnel benefit charges, cost of new machinery, material costs, maintenance costs, income from sales, and so forth may be included in such analyses in the same way as principal repayments and interest payments have been shown.

4.4.4 An Important Tax Consideration: Depreciation

Every tangible item used in a business has a limited useful life. The tangible item may be a building, a lathe, a computer, office furniture, and so forth. The reduction in the value of these assets is a legitimate item of expense, and is used as an accounting charge against before-tax income; it is an accounting expense as real as the operating expenses, such as salaries, office supplies, and gasoline to run the delivery truck. This reduction in value is known as depreciation. The accounting procedure used in calculating the dollar figures for depreciation generally spreads the original capital cost over a number of years, although under certain circumstances the Internal Revenue Service (IRS) rules allow deduction in the year the property is acquired.

Because depreciation is treated as an accounting expense and thus is a means of reducing the before-tax income, it is advantageous to the company to be able to make the depreciation figure as large as possible, even though the

118   CHAPTER 4

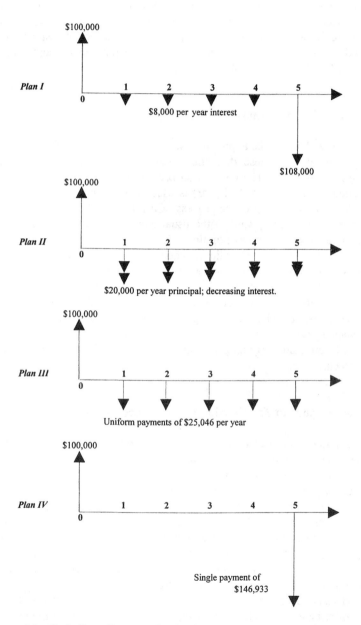

Figure 4.3 Cash flow diagrams for the four repayment plans of Table 4.2.

entire amount of the original capital expenditure less salvage cannot be exceeded during the depreciation period. The reason this is advantageous is that, as has been pointed out earlier, money recovered (received) in the future is of less value than money recovered now. The company, therefore, may want to have as short a period as possible for depreciation, and to accelerate the depreciation if it can. However, the IRS wants to get as large a tax payment as possible from the company, and hence wants long depreciation periods and does not like accelerated depreciation methods.

The statements of the preceding paragraph are not always true, however. Suppose that a small company, just starting up, expects little or no taxable income in the early years. That company will want to defer depreciation until later years, those which are more likely to be profitable.

The tug-of-war between business on the one side and the IRS on the other has resulted in numerous court cases concerning reasonable equipment lifetimes and acceptable methods of calculating depreciation. The IRS has tables in its Publication 946, "How to Depreciate Property," showing the so-called "class life" of equipment for tax purposes for various classes of equipment. These tables show allowable class lives for tax purposes that range from as little as three years for race horses and automobiles to as much as fifty years for railroad hydroelectric generating stations. Even this is made more complex in certain cases. The class life of an automobile, for example, is three years, but the IRS specifies a depreciation recovery period of five years under either of the two depreciation methods in its Modified Accelerated Cost Recovery System (MACRS). To make matters still more complicated, changes in the tax code at various points of time have placed equipment purchased before the effective date of a change in a category subject to different depreciation rules than equipment purchased after the date of the change. It is obviously important to use the correct class of equipment and an allowable method of calculating depreciation. Publication 946 must be consulted.

There are several methods used to calculate depreciation, including:

1. Straight line.
2. Double-declining balance.
3. A combination of (1) and (2).

A fourth method is referred to as the "sum-of-the-years digits." This method, although obsolete, is still referred to occasionally, and is explained later.

The basic idea behind each of these methods is discussed below, using a simple example for comparison purposes. It must be noted, however, that these examples are simplified by assuming that the equipment being depreciated was placed in service on January 1 of a given year and that the full depreciation for the first year can be claimed. IRS rules, however, are formulated so that equipment placed in service at any time during the year does not qualify for the full depreciation calculated by these methods during the first year, although the depreciation in subsequent years will be larger, eventually offsetting the effect of the first year's shortfall.

Straight line. The straight line method is the easiest to use. An asset is expected to decline in value at a constant rate from the date of acquisition to the day of salvage. Using D = depreciation, C = original cost, S = expected salvage value, and L = number of years of write-off life,

$$D = \frac{C-S}{L} \tag{10}$$

The depreciated value (the book value, B) at the end of the first year is simply B = C − D. At the end of subsequent years, the book value at the end of the previous year is again decreased by D. The process ends when B = S. Figure 4.4 shows the book value as a function of time for an asset with an original cost of $100, salvage value of $8, and expected life of 10 years.

Double-declining balance. This method results in a higher rate of depreciation initially, and is therefore a method that the company will usually prefer to the straight line method. In this method,

$$D_t = (2/L)B_{t-1} \tag{11}$$

That is, the depreciation taken in year t is twice the book value at the end of the preceding year divided by the number of years of write-off life. If L = 5, the straight line method yields 20% of original cost less salvage in the first year and every year thereafter until the salvage value is reached. The double-declining balance method yields 40% of original cost in the first year and 24% in the second year. Salvage is ignored in this method because the book value rarely declines to the salvage value during the asset's life, as can be seen in Figure 4.4.

Combination of double-declining balance and straight line. If the salvage value will be less than the book value reached by the double-declining balance method at the anticipated end of the asset's life, the IRS will permit the calculation to be switched from the double-declining balance method to the straight line method. The changeover from one method to the other is made when the straight line depreciation of the remaining book value for the remaining estimated life of the asset exceeds the depreciation calculated by the double-declining balance method. From the company's point of view, this is usually desirable because the tax savings are recovered more quickly toward the end of the expected life. Figure 4.4 shows this changeover occurring at the end of the eighth year, the final part of the double-declining balance curve being completed by a dashed line running down to the $8 salvage value.

THE FEASIBILITY STUDY

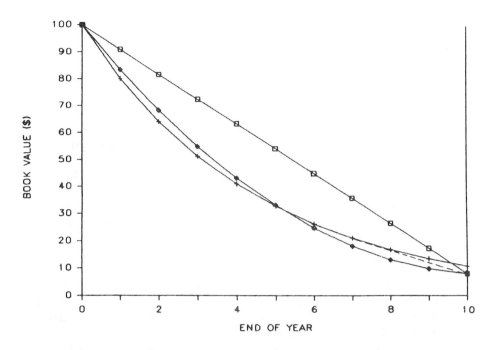

Figure 4.4 Comparison of depreciation methods: □, straight line; +, double-declining balance; ◊, sum-of-the-years digits.

Sum-of-the-years digits. This method, like the double-declining balance method, accelerates write-off in the early years. The amount charged off in any year is calculated by multiplying the original cost less the anticipated salvage by a fraction determined as follows. The denominator is the sum of the digits of the expected life. For a five-year life, this sum is $5 + 4 + 3 + 2 + 1 = 15$. The numerator is the years of life remaining at the beginning of the year for which the depreciation is being calculated. Hence for the first year, the fraction is $5/15 = 1/3$, or 33.3%, and for the second year it is $4/15 = 27\%$. Because the numerators of the fractions are the same numbers used in computing the denominator, their sum will be the same as the denominator, and the entire original cost less salvage is charged off over the estimated life of the asset.

Salvage value. Suppose an asset is sold for more than its book value. The difference between the sale price and the book value represents income to the company for tax purposes because the depreciation has previously been used to reduce before-tax income. Keep in mind, however, that any costs of removal are part of the cost of doing business.

It is important to realize also that assets are usually not totally useless at the end of their estimated life for tax purposes. In fact, there are many buildings and many pieces of equipment having zero book value (or salvage value) being used in industry.

4.4.5 Other Tax Considerations

Depreciation is obviously a major consideration for a company with regard to its tax situation. In general, the company wants to write off as much as it can as early as it can. There are, however, other tax considerations to be kept in mind. In what follows, most discussion relates to federal income tax considerations, but state and local tax codes must also be consulted. Discussion of these codes is obviously outside the scope of this work. Only a cursory discussion of each of these tax considerations follows.

Depletion allowance. This is a tax deduction for exhaustible natural resources and for timber. The natural resources category includes oil and gas, minerals in mines, and minerals recoverable by means other than excavation. There are two methods of computing the depletion allowance; if the taxpayer is given a choice, the larger of the two results should be used.

One of the depletion allowance methods is known as *cost depletion*. The cost depletion allowance is equal to the cost per unit multiplied by the number of units sold during the year. The cost per unit is determined by dividing the value of the property by the total number of recoverable units, where the value of the property is the original cost less all depletion allowances previously taken. Determination of the number of recoverable units is an engineering problem, and there will be some uncertainty about the result of that determination.

The second depletion allowance method is known as *percentage depletion*. This is a certain percentage of the gross income from the property during a given year. The percentage used in the calculation depends on the mineral being extracted, and varies from a value as low as 5% to a value as high as 22%. The IRS publishes a table that lists the allowable percentage by type of mineral and whether it is located in the United States or not. In no case may the depletion allowance taken exceed 50% of the taxable income computed without the allowance. Under this method, it is possible that the total depletion allowance over the entire life of the extraction process will exceed the cost. This provision in the tax code exists because of recognition of the fact that there are unusual risks associated with this kind of activity, but that it is in the public interest that someone take such risks.

Investment tax credits. This tax credit was originally introduced in 1962 in order to stimulate business. Since that time, it has had a checkered history, being suspended, reinstated, repealed, reinstated, and again repealed.

State and local taxes. In general, companies pay property taxes to state and local governments on the basis of value of property. The valuation will certainly include tangible items such as buildings, capital equipment, and inventory, but may also include other property, such as patents. In some cases, calculation of the property value may not be possible for the plant as a whole. The General Electric Aircraft Engine plant, just outside Cincinnati, lies in two different local tax districts. The company must, on the same date each year, determine the property value for each district, including the value of engines under construction on that date, some of which will be in one tax district and some in the other.

4.4.6 Comparison of Alternatives

The decision to put a new product into production can be justified on one of two bases. One of these is the need to transact business effectively; the other is that there is an economic gain to be realized. The first basis leads a manufacturer to produce some products whose purpose is "to fill out the line," to keep distributors from having a reason to take on a competitive line for the needed item (with the unpleasant possibility that it may then drop some other products you are producing), to provide customers with a product that will move rapidly, or simply because management wants to be able to advertise that the company is a "full line" supplier. If any of these is the reason for a project, the best strategy is to design the product to be as inexpensive as possible, consistent with retention of the same level of quality as your other products. Even with this reason, there will still be alternatives to evaluate.

In addition to a management-driven decision to introduce a new product, it should be pointed out that there are other instances in which a new product must be designed and introduced to the market. When the U.S. Congress enacted legislation requiring the reduction of emissions from automobile engines, the engineers employed by automobile manufacturers had to design engines and/or engine controls that would meet the legal limitations.

Whatever the origin of the decision to produce a new product, the decision as to which of two or more alternatives to select requires that there be some economic basis on which to compare the alternatives. Each alternative will include present expenditures which are expected to be offset by future benefits. We have already seen that money received in the future is of less value than money received now because of the interest that may be earned. Hence the comparisons must factor in an interest rate. Management will have in mind a minimum attractive (or acceptable) return rate (MARR). An alternative that fails to reach this value will probably be rejected. For convenience in the following, the symbol i^* will be used instead of MARR.

When decisions are to be made concerning alternatives, it is important to realize that alternatives may be placed into two classes. One class is referred to by the term *independent alternatives,* the other by the term *dependent alternatives* or *mutually exclusive alternatives.* A simple example will illustrate the differences between these classes.

Suppose that a homeowner has a seventeen-year-old central air conditioner and a four-year-old gas furnace. The air conditioner has been in frequent need of servicing and repair, and must be replaced before the next cooling season. The homeowner decides to investigate a number of possibilities, including replacement of the air conditioner, replacement of the air conditioner by a heat pump while retaining the existing gas furnace as backup, replacement of the air conditioner and replacement of the gas furnace by a high-efficiency furnace, and replacement of the air conditioner by a heat pump and replacement of the existing furnace by a high-efficiency furnace. These alternatives will be *mutually exclusive* or *dependent* alternatives, because the homeowner can select only one of them to implement.

Suppose that the same homeowner would also like to remodel the kitchen and the master bathroom at the same time that the heating/air conditioning system is modernized. Once the decision has been made as to which of the alternatives should be selected for the heating/air conditioning system, the homeowner still has three alternatives from which to select. But these are now *independent* alternatives, it being possible to do any one of them or, if finances permit, all of them.

There are four methods generally used for comparing alternatives. These are:

1. *Present worth method (P/W).* Using i^*, convert the cash flow for each alternative to its equivalent present value and compare the alternatives on the basis of these present values.
2. *Annual worth method (AW).* Using i^*, convert the cash flow for each alternative to its equivalent uniform annual value and compare the alternatives on the basis of these annual values.
3. *Benefit/cost method (B/C).* Using i^* and separating costs from benefits, convert the cash flows to their equivalent annual (or present) values and compare the equivalent benefits to the equivalent costs for each alternative. Benefits must obviously exceed costs.
4. *Return-on-investment method (ROI).* Find the interest rate i that balances present and future cash flows, and compare the result to i^*.

There is a fifth method, *years-to-payback,* that can also be used. It is theoretically deficient, but it has the advantage that it is direct and simple. This method is discussed briefly toward the end of this section.

Before the comparison methods can be discussed, a subsidiary problem must be mentioned. This is the problem that arises when the alternatives being proposed for consideration have unequal lives. Suppose that alternative A has a longer theoretical life than alternative B, and suppose furthermore that B is less

THE FEASIBILITY STUDY 125

expensive. In order to compare the two, one must also ask what happens when B expires. If it must then be replaced, the follow-on project and its costs must be included in the comparison. On the other hand, if the company can foresee that the product to be manufactured will no longer be marketable at the end of B's life, then alternative A should be evaluated on the basis of the shorter lifetime. For example, the present system of transmission of television signals is expected to be replaced within a few years with a high-definition television system. Both alternatives A and B are for production of television sets that will use the present system of transmission. If alternative A envisions production using this line that would extend for, say, five years after the introduction of high-definition television and alternative B is planned to have a life that terminates concurrently with the end of the present system, there is no economic justification for using the longer life span of A in evaluating these two alternatives.

4.4.6.1 Present Worth Method

The present worth method (PW) allows one to compare alternatives by converting the cash flows for each alternative into an equivalent present amount. The conversion is done using the minimum attractive rate of return, i^*. If dependent (mutually exclusive) alternatives are being considered, the alternatives may then be ranked according to their present worth. Independent alternatives having positive present worths have met the criterion for minimum attractive rate of return (MARR).

When determining cash flows, a reduction in a cash flow out from the present value is counted as a cash flow in. For example, in the case of the alternative heating/air conditioning systems being considered by the homeowner above, a reduction in heating or cooling cost is taken as a cash flow in. Repair costs on the existing air conditioner will also be avoided, and could be counted as cash flows in. If this is done, a more favorable PW will result for all four alternatives, but the ranking of the alternatives will not change.

Example 4.1 Heating and cooling alternatives. Let us look in some detail at the question the homeowner has to resolve as to which of four alternatives to select. The alternatives are:

Plan A: Replace the air conditioner with a new air conditioner; continue to use the present gas furnace.
Plan B: Replace the air conditioner with a heat pump; use the present gas furnace for heating backup.
Plan C: Replace the air conditioner with a new air conditioner; replace the furnace with one of the new, high-efficiency, furnaces.
Plan D: Replace the air conditioner with a heat pump; replace the furnace with a high-efficiency furnace.

The homeowner gathers the following data:

1. The annual cost of cooling has averaged $500.
2. The annual cost of heating has averaged $800.
3. The cost of repairs to the air conditioner for the past few years has averaged $100 per year.

Contact with several vendors of heating and air conditioning equipment results in bids (including installation) and other relevant data as follows:

1. A replacement air conditioner will cost $2,000; energy use will be about 30% smaller than with the old unit.
2. A heat pump will cost $3,000; energy use as an air conditioner will be 30% less than with the existing unit, and in addition there will be a 30% saving in heating cost.
3. A high-efficiency furnace will cost $3,000; heating costs will be reduced by 35%.
4. The combination of heat pump and high-efficiency furnace will yield a further reduction of 10% in heating costs.
5. Because of newer technology and better materials, the life of the new equipment can be expected to be 20 years. The homeowner has a dilemma at this point, because the existing furnace is already four years old, and will theoretically need replacement under either Plan A or Plan B before the end of the 20 years. He decides to ignore this difficulty.

The homeowner decides that i^* should be 0.10, that is, 10%. The capital recovery equation (Equation (7)), rearranged to solve for P, can then be used for each of the plans to obtain the present worth P of future cash flows (that is, of the expected savings). This value will, of course, be offset by the initial expenditure for equipment and installation. For Plan A, we find

$$PW_A = -\$2,000 + (\$500 \times 0.30 + \$100) \frac{1-(1+0.10)^{-20}}{0.10} = \$128.39$$

The following tabulation shows the relevant data and the results for the other three plans. It is assumed that there will be no salvage value, and the effect of the avoidance of the repair costs has been taken into account.

	Plan A	Plan B	Plan C	Plan D
Initial Cost	$2,000	$3,000	$5,000	$6,000
Estimated Life	20 yrs.	20 yrs.	20 yrs.	20 yrs.
Savings	30% cooling	30% cooling	30% cooling	30% cooling
	0% heating	30% heating	35% heating	45% heating
Present Worth	$128.39	$1,171.65	−$487.81	−$806.73

THE FEASIBILITY STUDY

It is obvious that Plan B is the best, followed by A. Both C and D are unacceptable.

Example 4.2. Wechsel Motor Company: PW method. The Wechsel Motor Company is an old-line manufacturer of small definite-purpose electric motors. It has just been bought out by a group of entrepreneurs at a low price because the company has not kept its production equipment up to date. The new management intends to modernize by installing computer drafting equipment, rebuilding the production line, upgrading the rotor die casting equipment, and replacing the coil winding machines. Note that these are independent alternatives. There is therefore no point in comparing them against each other, but each should be compared against the often forgotten alternative: Do nothing. The table on the next page shows the relevant data for these projects, with all dollar values in thousands of dollars.

The present worth of each project may be calculated in the same manner as the calculations for present worth in Example 4.1, except that the present value of the salvage is calculated using Equation (8). The investors have agreed that they want a minimum acceptable rate of return of 20%. For Project A,

$$PW_A = -\$300 + \$60 \frac{1-(1+0.20)^{-5}}{0.20} + \$100(1+0.20)^{-5}$$
$$= -\$300 + \$179 + \$40 = -\$81$$

Project:	A	B	C	D
	Computer drafting equipment	Rebuild production line	Upgrade rotor casting	Replace coil winders
First cost	$300	$75	$150	$100
Life	5 yrs.	10 yrs.	10 yrs.	15 yrs.
Salvage	$100	$5	$7	$20
Annual savings	$60	$20	$40	$25

That is, with round-off the present worth of Project A is –$81,000.

The present worth of the other projects is calculated in the same way. The results rounded off to the nearest dollar may be tabulated as follows:

Project A: –$80,375
Project B: $9,657
Project C: $18,830
Project D: $18,510

Since Project A has a sizable negative present worth, it is obvious that this project should not go forward. The other three are acceptable from an economic point of view. Which of these to proceed with depends upon the total capital available to the entrepreneurs.

4.4.6.2 Annual Worth Method

The annual worth (AW) method resembles the present worth (PW) method. In this method, however, the alternatives are made comparable by converting their cash flows into equivalent uniform *annual* amounts. The initial cost is converted into a (negative) annual amount by use of Equation (7). Salvage value is converted into an annual positive amount by use of Equation (9). Annual savings are, of course, positive amounts in this summation. In Equations (7) and (9), we use $i = i*$. If the alternatives are mutually exclusive, the alternative with the greatest equivalent annual worth is the best choice. With independent alternatives, all alternatives with positive annual worths are acceptable.

Example 4.3. Wechsel Motor Company: AW method. The relevant data are shown in Example 4.2. For Project A, the annual worth in thousands of dollars is:

$$AW_A = -\$300\frac{0.20}{1-(1+0.20)^{-5}} + \$60 + \$100\frac{0.20}{(1+0.20)^5 - 1}$$
$$= -\$100.31 + \$60 + \$13.44 = -\$26.87$$

That is, the annual worth of Project A is −$26,870. The calculations for the other projects are made in the same way, so that a tabulation for the four projects is as follows:

$$AW_A = -\$26,870$$
$$AW_B = \$2,300$$
$$AW_C = \$4,490$$
$$AW_D = \$3,890$$

The rank ordering of the four projects is the same as with the results of the present worth method, showing very clearly that Project A should not be undertaken.

4.4.6.3 Benefit/Cost Method

The benefit/cost ratio method has been used for many years by governmental entities to evaluate proposed projects or the effect of regulations. A frequently cited Act of Congress is the Flood Control Act of June 22, 1936, which put in

THE FEASIBILITY STUDY 129

place the concept that certain public flood control projects should be undertaken "if benefits to whomsoever they may accrue are in excess of the estimated costs" [15, page 134]. Although this concept was originally applied only to flood control, over the years it came to be applied to all kinds of other public works, including bridges, highways, air and water pollution control, toxic and radioactive dumps, improvement of waterways, and so forth. One of the problems with application of the concept has, of course, been the identification of the benefits, especially in placing a dollar value on them.

In the case of a governmental agency, the costs are the cash outflows of the agency, whereas the benefits usually accrue to others, and are far more difficult to quantify. For example, if a new highway is to be constructed or an existing highway is to be rerouted, the costs to the governmental entity responsible for construction can be assessed with considerable accuracy. The benefits, however, accrue to the motorist as a result of reduced operating costs, fewer accidents, shorter commuting time, and so on. They may also accrue to those living in the area because of reduced air pollution or an abatement of noise levels. In a sense, however, some of these same benefits may be looked on as negative benefits (or disbenefits) for others: The companies supplying gasoline because their sales are reduced, the automobile repair garages because they have fewer damaged automobiles to repair, the medical community because they have fewer accident cases to treat, and—one is reluctant to mention it—the legal community because there are fewer suits to file.

In the case of a business, the benefits are far more easily identified and quantified, as are the costs. If the benefits are symbolized by B and the costs by C, then the ratio B/C gives a numerical measure that tells one whether the benefits exceed the costs (B/C>1) or whether the benefits will not cover the costs (B/C<1). The analyst may have to develop a new set of conventions when doing the computations. For example, in the PW and AW methods, the money obtained from salvage is counted as income, which one might consider to be a benefit. However, it can just as well be counted as a negative cost. Similarly, maintenance expenditures are not costs, but negative benefits. Before embarking on B/C analysis, what is counted as a benefit and what is counted as a cost (either positive or negative) must be defined.

Benefits accrue throughout the life of the project, but capital expenditures usually occur only at the beginning and salvage, of course, occurs only at the end. It is therefore necessary to annualize both benefits and costs. Initial costs are annualized by the use of the capital recovery equation (Equation (7)); salvage is annualized by use of the annuity compound amount equation (Equation (9)).

Example 4.4. Wechsel Motor Company: B/C method. The data for the four proposed projects appear in Example 4.2. The benefit/cost ratio for Project A is calculated as follows, with all amounts in thousands of dollars:

$$B = \$60$$

$$C = \$300 \frac{0.20}{1-(1+0.20)^{-5}} - \$100 \frac{0.20}{(1+0.20)^5 - 1} = \$86.9$$

Hence B/C = 0.69<<1.

The benefit/cost ratio is calculated for the remaining projects in the same way. The final results show that

$$(B/C)_A = 0.69$$
$$(B/C)_B = 1.13$$
$$(B/C)_C = 1.13$$
$$(B/C)_D = 1.18$$

4.4.6.4 Return on Investment (ROI) Method

The return on investment method, like the B/C method, yields a dimensionless number that may be compared to some standard. In this method, the present worth equation is set to zero, and the value of the interest i that will satisfy the equation is determined. This is a trial and error solution. The equation, after rearrangement, is:

$$\text{Initial cost} = \text{Annual savings} \times \frac{1-(1+i)^{-n}}{i} + \text{Salvage value} \times (1+i)^{-n}$$

Having determined i, the value is compared against i*. If the interest rate i obtained from this equation is less than i*, the project is obviously not a satisfactory use of capital. If the inequality is reversed, then the project can go forward, subject to availability of capital.

Example 4.5. Wechsel Motor Company: ROI method. Because Project A has shown up so poorly in each of the previous evaluations, it is clear that the rate of interest that will satisfy the equation above is probably considerably less than 20%. A first trial is made using 10%. The right side of the equation is then, in thousands of dollars, $60 x 3.7910 + $100 x 0.6209 = $289.54.

This is smaller than the initial cost of $300(000), and indicates that a still smaller interest rate should be tried. Trial and error results in an interest rate of 8.8% to satisfy the equation. In all likelihood, the interest rate calculation would not be pursued far enough to obtain this degree of precision. The first result (that for 10%) would have made it immediately apparent that the project came nowhere close to satisfying the entrepreneurs' i*.

For projects B, C, and D, the calculations yield interest rates of 23.6%, 23.6%, and 24.2% respectively.

THE FEASIBILITY STUDY

4.4.6.5 Years-to-Payback

"There is always an easy answer to every problem—neat, plausible, and wrong." The easy answer to the problem of which criterion to use when comparing alternatives is to use the years-to-payback method. As is shown below, it is also wrong.

The years-to-payback method is simple, it is easy to understand, and it is easy to make the calculations. In fact, the calculation can frequently be done in one's head with fair accuracy.

The number of years-to-payback is simply the initial cost divided by the positive cash flows that result. In the examples above of the Wechsel Motor Company, the positive cash flows are the annual savings. *The method assumes that money is available at zero interest.* If one wants to make the method a bit more elaborate, the salvage value can be subtracted from the initial cost before doing the division on the ground that one will eventually recover that value. Another variation is to calculate what is referred to as the discounted payback period [16]. If the calculations are done for the Wechsel Motor Company, the following table results.

Project	Without Salvage	With Salvage
A	5.00 yrs.	3.33 yrs.
B	3.75 yrs.	3.50 yrs.
C	3.75 yrs.	3.58 yrs.
D	4.00 yrs.	3.20 yrs.

With this simple-minded approach, Project A looks as if it is comparable to the other projects, even without considering salvage. If salvage is included, Project A looks better than all of the others except D. Yet we have already seen that A is distinctly poorer than the others by any of the other measures: Present worth, annual worth, benefit/cost ratio, and return on investment. The fundamental misapprehension that creates this incorrect result is that there is no cost for the use of money.

4.4.6.6 Other Economic Decision Analysis Topics

The foregoing discussion has deliberately been kept rather simple so that the basic concepts would not be lost in a forest of details. A number of important topics have therefore been omitted. Among these are leveraging, inflation, cash flows that are expected to vary with time because of anticipated sales increases (or decreases), and so forth. The references [14-16] mentioned above contain discussions of all of these topics, as well as many others. Shupe [14] is the easiest of these to read and to understand, but both Grant [15] and Thuesen [16] are more complete.

4.5 COST ESTIMATING

The purpose of cost estimating by design engineers is to arrive at data sufficiently accurate that they may be used with confidence when choosing between two or more alternative designs and to determine whether the one chosen is likely to meet competition. It is, therefore, one of the planning tools the company uses. The importance of cost estimating by the design engineer is made evident by the fact that it is now generally accepted that over 70% of the cost of the final product is determined during the design phase [17, page 2]. This is true even though only about 5% of that cost can be directly assigned to the design activity. Earlier comments regarding the importance of extensive interaction among all parts of the company—design, manufacturing, purchasing—during the design phase are reinforced by this knowledge.

Cost estimates are forecasts of what a product will cost. The data used and the method of calculating the cost are dependent on the type of product, the trade-offs between manufacturing in house versus buying the component from another source, and the manufacturing techniques used, as well as other factors. Various methods have been used to make these estimates, but all of them incorporate past experience as a necessary ingredient in the process. A new product under development will probably be similar to other products your company makes. The actual costs of making those products will provide a large amount of information that can be utilized when estimating the cost of a new product. The experience factor makes cost estimating difficult to teach, although much progress has been made in recent years [17, 19].

Regardless of the product, there are a number of ingredients that must be included when making the estimate. These include the obvious direct costs, such as material costs, parts purchased from others, labor costs, and travel costs, as well as other direct costs that may tend to be overlooked, such as computer services, reproduction services, and training of production and inspection personnel. To these direct costs must be added indirect costs, administrative costs, and profits. Indirect costs and administrative costs are often lumped together under the rubric "overhead costs."

Material costs and the cost of parts purchased from others typically make up anywhere from 30% to 50% of the total cost. This is true in spite of the fact that manufacturers, because they purchase in large quantities directly from other manufacturers, pay much less than the price that would be charged at the retail level. These prices are known as OEM (original equipment manufacturer) prices.

The most obvious labor cost is the hourly rate of pay to the worker. This differs from one worker to another, depending on the work being done and in some cases on the longevity of the worker's service with the company. Knowing these rates and the length of time to do a particular task required in production enables one to get a first estimate of labor costs, but there are other factors that must be taken into account. The cost of fringe benefits, such as contributions to pension plans, the employer's share of the contribution to Social Security, sick leave, vacation pay, workmen's compensation premiums, and medical insurance premiums, must be included when calculating labor costs. These fringe benefits can easily add 25% or more to the amount of direct compensation to the worker. Labor costs for a product, including fringe benefits, will range from as low as 15% to as much as 30% of total cost. The low end of the range is more likely if the design was made with an eye toward ease of manufacturing and assembly, but some products are inherently more labor-intensive than others, resulting in labor costs tending toward the high end of the range.

Indirect costs and management costs, that is, overhead, are not directly related to each device produced or even the current volume of production. These are costs for activities required for the company to stay in business regardless of the level of production. Even a partial list of these activities will demonstrate that these costs are a sizable part of the total cost of production. Such a list will include:

Salaries and fringe benefits of foremen and supervisors.
Material handling costs.
Packaging.
Custodial services.
Amortization of facilities and equipment.
Industrial relations.
Electrical power and lighting.
Water and sewage charges.
Licenses.
Maintenance costs of buildings and equipment.
Insurance.
Communications charges.
Research and development in anticipation of new or improved products.
Purchasing department costs.
Sales department costs.

These costs are usually assigned in proportion to the direct labor required, but other algorithms may be used in deciding how much of the overhead costs to assign to a product. Moreover, different assignment of costs to the various cost categories may be made, depending on the accounting practices of the company. Fringe benefits, for example, which are assigned above to labor costs, may be included instead with overhead. Whatever the division of costs used within a company, a very rough rule is that one-third of the cost is due to each of the three main categories, materials, labor, and overhead.

There are three degrees of precision that can be used in estimating cost. They are: (1) Quick approximations, (2) semidetailed estimates, and (3) detailed estimates. Quick approximations are made using rules of thumb. For example, a given type of product made by a given manufacturer will have a nearly constant production cost per pound. This rule is based on the fact that a major element in the cost of a product is material and that similar products tend to have like proportions of various materials and labor content. Another way of estimating cost is to use the 1-3-9 rule. This rule, proposed by Rondeau [18], seems to hold well for mechanical and electromechanical products for which the manufacturer purchases most of the materials in basic sheet, bar, and coil forms and uses them to product a standard, off-the-shelf product. If this rule is used, it is assumed that the manufacturer's cost will be three times the material cost, and the retail price will be nine times the material cost. It is easily seen that these rules of thumb must be used with extreme caution, especially if substantial changes have been made in a particular design. Nevertheless, such rules can provide a quick estimate to see whether the product is likely to have a reasonable cost or not. Because of the high degree of uncertainty attached to such estimates, they would be made only very early in the design phase.

Semidetailed estimates can be made by working from the actual cost of parts of products already in production in your company. Increasing or decreasing the cost of each part by considering differences in material and labor to make it comparable to the similar one already being made can produce a fairly accurate estimate. To do this takes time, the availability of cost records, and enough experience to make accurate judgments as to the necessary adjustments in labor and material costs. Semidetailed estimates may be of value around the midpoint of the design process.

Detailed estimates are made when the design has essentially been completed. These estimates are considerably easier to make today than even a few years ago. To make these estimates as precise as possible, it is necessary that five basic tools be available [19, pages 16-18]. These are a qualified estimator or a team of estimators, a methodical approach, in-depth knowledge and data concerning the product or device, computation capability, and a publication capability.

The estimator(s) must be able to devote undivided attention to estimate preparation. The qualifications of an estimator should include experience in estimating techniques. He or she should have a good general knowledge of the product for which a cost estimate is needed and must also have the ability to develop a detailed understanding of the product and the way in which it is produced as the work progresses.

A methodical approach will be applicable to a wide variety of products, and its use will insure that two or more alternative designs will be evaluated on the same basis. Fortunately, there are now a number of software programs available for estimating costs. Stewart [19, pages 252-254] lists a number of companies engaged in developing such programs. Contact with those companies will enable the estimator to locate a company having programs applicable to the kind of business in which you are engaged.

The third basic tool listed above is in-depth knowledge and data concerning the product or device. Some of this knowledge can be acquired from appropriate handbooks, but most of it should be available in-house. A large amount should be available in computer-accessible databases. In addition, anyone in the company (for example, tool engineers and industrial engineers) with the ability to visualize any part of the task and with the ability to convert this knowledge into a statement as to what machines are needed, what kind of worker would do the task, and how long it would take to do it, is part of this tool. When in doubt, consult those who have done similar tasks.

The fourth tool is computation capability. Those methodical approaches available in software form and databases containing large amounts of information require the use of a modern computer. The result will be much more rapid results with more accuracy, as well as the ability to explore the effect of possible variations in the design. This may be especially important if one is designing with a specific cost limitation.

Finally, the results must be communicated to others. This will require a publication capability, such as a desktop publishing system, if the printout from the computer is not in a form readily understandable by anyone other than the estimator(s). If the project is a rather simple one, the computer printout may suffice, but for a complex product the final document may include a number of cost estimates for subassemblies or components as well as a summary for the project as a whole. It may also include graphs illustrating trends. This is an important document; a decision to proceed or not will be made on the strength of the results presented. A well-organized document will enable those making the decision to do so with a clear understanding of the ramifications. A poorly organized presentation of a project that should go forward too often leads to rejection.

As an example of detailed cost analysis, consider the manufacture of an electrical connector of the type shown in Figure 4.5. This connector is used to connect copper or aluminum cable to electrical equipment such as panelboards, large switches, motor controllers, and so forth. The connector is mounted on the busbars of that equipment by a single machine screw into a tapped hole in the bottom. The cable is placed into the chamfered hole after the insulation has been removed from the end. The screw shown at the top of the connector will then be turned down tight to clamp the cable against the bottom surface of the cable hole.

The material is purchased in 10-foot extruded aluminum bars, the cable hole being formed during the extrusion process. Table 4.3 lists the operations necessary to manufacture the finished connector in the order in which the operations are carried out. The time is in minutes per 100 connectors as estimated by an industrial engineer. Once the cost estimate is completed, the engineer is in a position to obtain quotations based on the anticipated volume from outside vendors so that a comparison can be made against the in-house cost and a make/buy decision made.

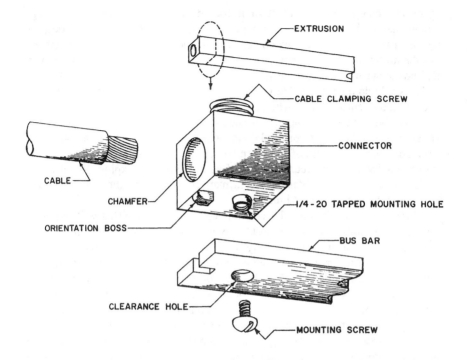

Figure 4.5 Illustration of electrical connector used in detailed cost analysis.

This cost estimate may not be complete. For example, if it is necessary to purchase equipment to do some of the operations on the connectors, the cost of that equipment must be prorated as an additional cost.

4.6 WHAT DETERMINES SELLING PRICE?

The manufacturer can attempt to set a selling price for its product by taking the cost and adding a reasonable profit, perhaps based on the i* desired. As long as that price comes in near the price being charged by other manufacturers, the product will obtain a share of the market. However, in the long run, the selling price is related to the price of the most efficient producer. Others must meet or come close to that price. If they cannot, they will ultimately go out of business. Schumpeter [20] described this as the creative destruction of capitalism, in which true competition in the long term results in the survival of the fittest.

The concept of competition deserves comment. All products are meant to satisfy human needs and as such must compete with the way in which those needs were met before a new product appeared on the market. As an example, a product that appeared on the market a few years ago was a small refrigerator/

THE FEASIBILITY STUDY

Table 4.3 Manufacturing Cost of an Electrical Connector

Operation number	Description of operation	Time (min/100)
1	Cut bar into nine 12-11/16" pieces	2.3
2	Drill 5/16" holes, 12 simultaneously	10.0
3	Pierce and extrude mounting holes	10.0
4	Tap 1/4-20 bottom holes, 12 simultaneously	12.0
5	Tap 5/8-16 top holes, 12 simultaneously	13.0
6	Cut into 1" lengths	25.0
7	Form orientation boss	10.0
8	Chamfer boss end of wire hole	8.0
9	Tumble to remove cutoff burrs	1.5
10	Screw wire holding screw into connector	10.0
	Total	101.8

At $0.167/min wages and 30% fringe benefits, direct labor is $0.221 each.
Material costs per connector:

Aluminum extrusion	$0.125
Wire-clamping screw	$0.105
Plating to meet UL requirement	$0.013
Total	$0.243

The total cost per connector is:

Labor	$0.221
Material	$0.243
Overhead (at 100% of labor)	$0.221
Total cost	$0.685

heater based on the Peltier effect. The product was designed to be operated from the 12-volt dc system of automobiles and recreational vehicles, drawing about 4 amperes. It comes equipped with a line cord having at one end a plug that fits into the lighter socket. The other end of the line cord is designed so that, when inserted into the receptacle on the unit one way, the inside of the unit is cooled, and when inverted (so that positive and negative sides of the line are reversed) the unit will heat the contents. One embodiment of the unit is 16" x 11" x 11" externally, having internal dimensions of 11" x 8" x 8". That is, only about 36% of the volume is usable.

This device, although unique at the time, had competition in the form of the simple picnic cooler (having roughly the same external dimensions) for cooling and one-burner butane cooking units for heating. A higher percentage of the external volume of the picnic cooler is available internally, but since some of that volume must be used for ice the net usable volume is about equal. The disadvantage of the Peltier unit used as a cooler is that it has a continuous draw of 4 amperes from the automobile or recreation vehicle supply. If one leaves the vehicle for an excursion or overnight when in a motel or camping, there is a distinct possibility that the battery charge remaining when the driver wants to start the engine will be so small that the engine cannot be cranked. The disadvantage of the unit when used as a heater is that it cannot raise the temperature of the contents nearly as far as the butane cooking unit. It cannot, for example, boil water. Although these units are still on the market, sales are small because of the combination of high price and known disadvantages. The competition is not with other Peltier units, but with the products produced all along that performed the same functions just as well or possibly better, and at substantially lower cost.

As another example, consider the residential-type ground fault circuit interrupter (GFCI), introduced about 1972. These devices are designed to interrupt electrical service if the difference of current in the two conductors exceeds 6 mA. (If there is a difference in these currents, it is because some current is going to ground through a fault of some sort.) They have prevented many residential electrocutions. When they were introduced, the retail price was about $50 each. Was there competition? Yes, the acceptance of the risk that one might be electrocuted when, for example, a small radio fell into a bathtub. The price was so high that, even when the manufacturers reduced the price to levels well below their cost, the public still resisted. The market, however, was created when the National Electric Code began requiring about 1975 that the GFCI be used in hazardous locations, such as bathrooms. The price at retail has since fallen to $10 to $12, and the practice of daisy-chaining several outlets behind a single GFCI has made the additional cost negligible when compared to the improved safety of the individual.

Suppose you have a way of producing an existing product at a price substantially lower than the usual market price, and you can do this without any sacrifice of quality. Should you set the price in accordance with your costs and a reasonable markup, and expect to get a large share of the market? When ball-point pens were in their infancy and were selling for a few dollars each, one manufacturer learned how to make a pen with the same or better quality so inexpensively that one could be sold at retail for 19 cents. With that price, the pen did not sell. Upon investigation, it was discovered that the public believed that the price was so low compared to the price to which they had become accustomed that they could not believe that the pen could possibly write at all well. When the company discovered this, it promptly raised the price to 79 cents each, and the pens began to sell in ever increasing numbers. As other manufacturers became competitive, the price was of course driven down from that level. The problem with the 19-cent price was public perception.

What determines selling price?

- The prices of competitive products.
- The increment, if any, the consumer is willing to pay to satisfy the needs that have previously been taken care of in another way.
- The perception of the consumer as to the value of the product. A high price with no perceived exceptional value will make the product very difficult to sell. On the other hand, a very low price when compared to similar or nearly identical products will lead to a quick judgment that the product is not well made.

4.7 SUMMARY

Feasibility studies are a difficult part of the design process. Nevertheless, they play an important part. Properly done, the designer and the company management will know whether goals of the project will be achieved easily, whether the probability of achieving the goals is marginal, or whether the project is not feasible or desirable. In any case, the reasons that success or failure may eventuate will be known and understood. While there will always be some unresolved problems to nag the designer, a well-done feasibility study will reduce their number and insure that none of them will have a major effect. In other words, there will be no surprises later.

For a specific design project, questions will arise as the project develops that were not envisioned at the beginning. As they arise, the designer must be certain to revisit the feasibility question. Should the answer to any of these new questions show that the project will not be feasible, a halt should be called at once. As noted earlier, you will never have spent less time and money than at the point at which a project is terminated.

REFERENCES

1. Petroski, Henry. *Design Paradigms: Case Histories of Error and Judgment in Engineering,* Cambridge University Press, Cambridge, 1994.
2. Taylor, J. W. R., Editor. *Jane's All the World's Aircraft,* Sampson Low Marston & Company, London, 1972-1973, p. 541.
3. Perkins, C. D., and R. E. Hage. *Airplane Performance Stability and Control,* John Wiley & Sons, New York, 1949, p. 17.
4. Perkins, C. D., and R. E. Hage. *Airplane Performance Stability and Control,* John Wiley & Sons, New York, 1949, p. 26.
5. Warner, E. P. *Airplane Design,* McGraw-Hill, New York, 1936, p. 64.

6. Baumeister, T., and L. S. Marks. *Standard Handbook for Mechanical Engineers,* 7th Edition, McGraw-Hill, New York, 1967, p. 9-210.
7. Glauert, H. *The Elements of Aerofoil and Airscrew Theory,* 2nd Edition, University Press, Cambridge, 1948, p. 204.
8. Glauert, H. *The Elements of Aerofoil and Airscrew Theory,* 2nd Edition, University Press, Cambridge, 1948, p. 203.
9. Dwiggins, Don. *Man-Powered Aircraft,* TAB Books, Blue Ridge Summit, PA, 1979.
10. Grosser, Martin. *Gossamer Odyssey,* Houghton Mifflin, Boston, 1981.
11. Dorsey, Gary. *The Fullness of Wings,* Viking Penguin, New York, 1990.
12. Starr, M. K. *Product Design and Decision Theory,* Prentice-Hall, Englewood Cliffs, NJ, p. 10, 1963.
13. Howell, E. K. Solid-state control for DC motors provides variable speed and synchronous-motor performance, *IEEE Transactions,* 1966, IGA-2(2): 132-136.
14. Shupe, Dean. *What Every Engineer Should Know About Economic Decision Analysis,* Marcel Dekker, New York, 1980.
15. Grant, Eugene L., W. Grant Ireson, and Richard S. Leavenworth. *Principles of Engineering Economy,* 8th Edition, John Wiley & Sons, New York, 1990.
16. Thuesen, G. J., and W. J. Fabrycky. *Engineering Economy,* 6th Edition, Prentice-Hall, Inc., Englewood Cliffs, NJ, 1984.
17. Boothroyd, Geoffrey, Peter Dewhurst, and Winston Knight. *Product Design for Manufacture and Assembly,* Marcel Dekker, New York, 1994.
18. Rondeau, H. F. The 1-3-9 rule for product cost estimation, *Machine Design,* August 21, 1975.
19. Stewart, Rodney D. *Cost Estimating,* 2nd Edition, John Wiley, New York, 1991.
20. Schumpeter, J. A. *Capitalism, Socialism, and Democracy,* Harper and Brothers, New York, 1942.

REVIEW AND DISCUSSION

1. What three criteria must a product meet to be considered feasible for development and manufacture?
2. Describe how you would determine whether a proposed product is physically realizable.
3. If a feasibility study indicates that a proposed product is beyond the state of the art, what other equally valuable information will it give?
4. Give as many reasons as you can for a product not to be company compatible.
5. What number is determined by the capital recovery equation?

THE FEASIBILITY STUDY

6. Which method of calculating depreciation ignores salvage value? Why is it ignored?
7. What is the principal advantage of each method of computing depreciation? What is the principal disadvantage?
8. Discuss the ways in which the cost of a product would be ascertained at the beginning of a design project, at the middle, and at the end. How does the methodology for computing cost change as the project develops?
9. Four methods of comparing alternatives have been presented. These are: The present worth method, the annual worth method, the benefit/cost method, and the return-on-investment method. In addition, the years-to-payback method was shown, and reasons why it is not valid were discussed. What are the advantages and disadvantages of each method? Can your choice of method be a major factor in reaching a conclusion as to which of several alternatives to select? Ignoring the years-to-payback method, is there one of the other four that seems to you to be superior? From what does that superiority stem?
10. What determines selling price? Give some examples of products for which price seems to be a good indicator of quality, or lack thereof.
11. Domestic garbage disposal units sell for $50 to $100. A door-to-door salesman tries to sell you a unit for $300, and explains that the additional cost is because the unit is built with all of the best components. Do you believe him, or do you show him the door?
12. Two projects are essentially the same in all details except that one will require the addition of temporary personnel costing $50,000 and the other will require acquisition of a special machine for $50,000 to be used exclusively for that project. If only one of these projects can be done, which would you choose? Why? If the second project required the acquisition of a standard machine (say, a milling machine) for $50,000 instead of a special machine, would your choice be different?

PRACTICE PROJECTS

1. Do the necessary study on the product you have selected to design to determine whether it is physically realizable or not. Put your analysis in a form suitable for presentation to the lead engineer in your group.
2. A visitor to the Washington Monument decides to climb the stairs to the 500-foot level, rather than to take the elevator. Is it reasonable to believe that this can be done in 10 minutes?
3. A 15-lb bicycle with a 135-lb rider aboard coasts down a smooth straight road having a constant slope of 10%. The only restraining force is the profile wind drag. Bearing friction and the retarding force of the tires on the road surface are negligible. The gross area of the bicycle and rider is 24 square feet. With these conditions, the bicycle attains a steady-state speed

of 25 mph. The cyclist now enters a 1-mile race on a level portion of the road. Ignoring the time to accelerate and assuming that the rider travels at a constant speed, what is the best time possible if the rider's output power capability in horsepower is given by $2.8t^{-0.4}$?

4. Estimate the cost of manufacturing the product you have selected, and estimate the retail selling price. Is your product financially feasible? Try to be as realistic as possible in your estimates.

5. Many states now conduct lotteries in order to increase their revenues. Suppose that the winner can take the prize in twenty annual payments, receiving an equal amount (less taxes) each time. If the prize for a particular drawing is $5,000,000, how much must the state invest at 9% so that the annual installments can be paid without having to add funds from later receipts? What is the total cash disbursement by the state? (Be careful. These are payments made at the *beginning* of each period, not at the end.)

6. An inventor sold a patent to a corporation and was given her choice of three offers: (a) $100,000 in cash upon signing the contract; (b) an annual royalty of 7% of sales revenue (estimated at $15,000 per year) for the remaining 10 years of the patent's life; or (c) a 20-year annuity of $7,500 per year. Under the second and third options, she will receive the income at the end of each year. Neglecting the effects of income taxes and assuming that she can obtain an 8% rate of return, which offer is the best?

7. Equipment purchased for $55,000 has an estimated 8-year life, and no salvage value. Compute the book value at the end of 3 years using the double-declining balance method and the sum-of-digits method.

8. A grading contractor owns earthmoving equipment that cost $659,000. At the end of its anticipated 7-year life, the salvage value is expected to be $45,000. Calculate the depreciation used for tax purposes for each of the first 2 years and the book value at the end of 5 years using each of the following methods: (a) Straight line, (b) double-declining balance, and (c) sum of digits. (Ignore the fact that the IRS may not agree to the anticipated life or to any of the depreciation methods.)

9. In the previous problem, determine in which year straight-line depreciation will exceed that computed from the double-declining balance method.

10. The homeowner in Example 4.1 is not very pleased by the results of the present worth method for comparing the four alternatives. What are the results using the annual worth method, the benefit/cost method, and the return on investment method? Does the years-to-payback method yield comparable results?

11. A company has a contract with an independent inventor to make and sell a product covered by a patent held by the inventor. The royalty is 2% of their net sales revenue. The product has been on the market for 4 years, and the sales in those 4 years have been $400,000, $800,000, $1,000,000, and $1,000,000 respectively. The inventor projects sales of $1,000,000 for each

THE FEASIBILITY STUDY

of the next 2 years, followed by annual decreases of $100,000. The patent expires in 7 more years, and the inventor will of course no longer receive royalties from the company. If the company offers to buy the rights for the next 7 years for a lump sum, how will the inventor decide whether its offer is a good one? What additional information does the inventor need before making a decision? Make any necessary assumptions, and determine for those assumptions the minimum lump sum that the inventor should agree to.

12. Copper weighs 0.321 pounds per cubic inch and costs $1.25 per pound. Aluminum alloys suitable for electrical conductors weigh 0.097 pounds per cubic inch and cost $0.80 per pound. Their conductivity is 65% of that of copper. If you wish to conduct a given current from point A to point B using either copper or aluminum subject to the constraint that the power loss is to be the same in either case, what is the ratio of the cost of making the conductor of copper to the cost of making it of aluminum?

13. A sheet steel box is made as shown by the exploded sketch in Figure 4.6, shown on the following page. Each end is spot welded at the positions shown by the x's (four at each end). The lid, which has a 1/4" lip all around, is simply placed on top of the completed box. The work standards show that the notching for the corners is done at 400 n/hr, the bending operation is done at the rate of 240 b/hr, and the welding operation runs at 320 w/hr. Painting after assembly requires 30 seconds and uses $0.042 of paint. Labor, fringe benefits, and indirect labor amount to $11 per hour. Steel weighs 0.281 lb per cubic inch, and costs $0.31 per pound. Scrap steel can be sold for $0.025 per pound. The flanges on the end pieces are 1/2" wide. Other dimensions are shown on the sketch.

Determine: a) Net material cost.
 b) Cost of material plus labor.
 c) Estimated total manufacturing cost.
 d) Estimated retail price.

14. A computer can be purchased by a consulting engineer for $46,000. She estimates a useful life of 6 years for the computer with a salvage value of $4,000. A maintenance contract is available at a rate of $2,400 per year, and an operator will cost $15 per hour when the computer is in use. As an alternative, a computer service is available to the engineer to do all her computing at a rate of $24 per hour. Assuming interest at 8% compounded annually and insurance at 6% of book value, determine the annual usage in hours per year for which the two alternatives are equal. If annual usage is expected to be greater than this result, which of the alternatives is more favorable?

15. A present investment of $600,000 is necessary to provide the facilities and equipment required for the production of Product A. The estimated annual receipts will be $110,000 and the estimated annual expenses will be $50,000. An alternative to Product A is Product B, which will require an

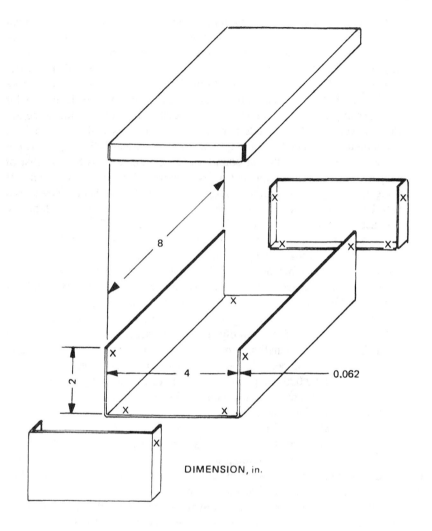

Figure 4.6 Metal enclosure for detailed cost analysis.

$800,000, but is estimated to produce annual receipts of $150,000 with annual expenses of $70,000. Over the 30-year projected life, it is assumed that income and expenses will remain unchanged. Salvage for each is estimated to be $100,000. Do you recommend that the company proceed with Product A, with Product B, or with neither? Be prepared to give detailed reasons for your answer.

16. A manufacturer is considering purchasing a personal computer for each engineer in order to increase his or her productivity. The model being considered costs $11,000 with software, and maintenance contracts will add another $1,000 per year of cost. The expected increase in productivity is estimated to be worth $3,000 each year. Salvage is estimated to be $1,000 at the end of a 5-year life. The company's minimum attractive rate of return, i*, is 24%. Should the manufacturer proceed with the purchase? Regardless of the answer to this question, assume that the purchase is made. What is the actual rate of return on the investment?

5

PATENTS

The Congress shall have Power . . . To promote the Progress of Science and useful Arts, by securing for limited Times to Authors and Inventors the exclusive Right to their respective Writings and Discoveries; . . .
                                      Constitution of the United States, Article I, Section 8

The first U.S. Patent Act came into effect on April 10, 1790. At that time, the responsibility for granting patents was placed in the Department of State. Until 1836, however, the system of granting patents did not even approximate the system in effect today. From 1790 until 1793 there was an examining board that consisted of the Secretary of State, the Secretary of War, and the Attorney General, and patents were granted if a majority of the board approved. Beginning in 1793, and lasting until the establishment of a formal review procedure in 1836, patents were merely registered. Two or more individuals could patent the same device, and the issue of priority could only be settled in the courts.

The Patent Act of July 4, 1836, established an Examination Corps and the office of Commissioner of Patents. This act required that copies of issued patents be available and that each patent contain a written description of the invention in such terms as to enable any person skilled in the art to which it pertained to make and use the device patented. The principal test of whether an invention had been made or not was the *flash of genius* test. This test was so difficult to apply that in the 1952 Patent Act this standard was replaced by the *novelty* and *nonobviousness* tests that are used today.

A patent is *intellectual* property, but there are other kinds of intellectual property that are of importance to the engineer. These are *trade secrets, copyrights, semiconductor maskwork,* and *trademarks.* These other kinds of intellectual property are discussed briefly in a later section.

5.1 WHY THE DESIGNER IS INTERESTED IN PATENTS

One of the reasons that a designer should be interested in patents stems from the fact that many designs have some feature or features that are new to the designer and possibly to everyone else in the company. Are these features already covered by a patent issued to someone else? That is, will you infringe on someone else's patent rights, and open your company to possible litigation? If so, the feature may not be used unless suitable arrangements (licensing) can be made with the owner of the patent, or the patent in question can be "designed around."

One of the authors (RHE), early in his career as a faculty member, was interested in audio oscillators. He thought of a way to build such an oscillator that was substantially different from anything he had seen described in the technical literature. Three other faculty members in the department had never seen the concept either, and the chief engineer of a company building electronic organs—which used audio oscillators by the hundreds—expressed the opinion that the idea was novel and nonobvious, and might be of interest to his company. Before proceeding farther with the idea, a patent search was made. This turned up a patent, thirteen years old at the time, that covered the whole field of audio oscillators "like a tent." The dreams of royalties evaporated on the spot.

This anecdote illustrates one of the reasons the designer should be interested in patents. That is, you do not want to unknowingly design and build something that is covered by someone else's patent, thus subjecting your company to the possibility of having legal action brought against it. To avoid such unhappy and expensive events, a patent search is conducted.

It should be pointed out that searches of United States patents will cover only issued patents. *Pending* United States patent applications are still secret by law and so cannot be located in a search. However, there is a trade agreement with Japan that may require that the United States begin to publish patent applications (as most of the rest of the world does) and legislation is being considered in this regard.

There are actually several different varieties of patent searches, each having a specific purpose. Moreover, although the engineer may conduct the search, as the author did in the case of the audio oscillators, it is far more common and generally much more cost-effective for patent attorneys to conduct these searches. The skills required to do high-quality searches are not easily developed. Since the designer will rarely have these skills, it is better to rely on patent attorneys, who exercise these skills on a daily basis, rather than run the risk that something of importance will be overlooked. (Even with these skills, relevant patents will still be overlooked from time to time, as will be evident below.) Moreover, patents are legal documents, and the interpretations to be placed on certain words or phrases will come easily to the attorney, whereas the engineer may misunderstand their import. Nonetheless, the engineer must work closely with the attorney, because the attorney lacks the in-depth technical knowledge of the engineer. Moreover, the engineer is the person who needs to know, in the first instance, the outcome and the significance of the results of the various patent searches, even though the patent attorney will, at the completion of any of these searches, submit a written report either to the designer or to the company management setting forth in detail what his or her conclusions are.

It is important to understand the difference between an expired patent and an unexpired patent, and to know when a patent expires. Patent protection used to run for 17 years from *the date on which the patent issued*. After that time, the patent rights have expired and the patent is in the public domain; anyone may feel free to use its teachings. With special exceptions, patents issuing or applications filed on or after June 8, 1995, however, will expire 20 years after *the earliest effective date of filing*. Why the change? There has been a perception that various ploys have been used to delay issuance of a patent so as to gain a longer effective period of protection; the new rule obviates that possibility. It would seem that it is best to defer submitting an application in order to extend the life of the patent as far as possible. There are other reasons for filing as early as possible, as will be seen in Section 5.4. Congress may change these rules yet again, thus making it important to consult with a patent attorney.

State-of-the-art searches. The reason alluded to above for making a patent search was to avoid infringing on an unexpired patent. This is only one of the reasons for doing patent searches. Return to the ground fault circuit interrupter (GFCI) used in Chapter 3 as an illustration of searching for standards. If your company has never been in the business of making GFCI's and now wishes to add them to its line of electrical products, you need to know what has been done previously. A major source of information is in the patents that have been issued in the field, and both expired and unexpired patents are of value. Hence a *state-of-the-art* or *collection* search should be one of the first tasks in the design process. The objective of this search is to sweep up all patents that relate to the subject matter of concern so as to get an overview of the technology. Such a search may focus on only a few of the subclasses, those that appear to be the more critical ones.

One of the advantages of doing such a search very early is that you thereby avoid the reinventing-the-wheel syndrome. You will not spend time and money developing an approach that others have followed before you and that may be well protected by unexpired patents. However, you will gain considerable knowledge of the art, and that is of benefit of itself.

Patentability searches. Suppose now that you have avoided the approaches shown in the patents turned up in the state-of-the-art search and have developed what you believe to be a new approach to making a GFCI. Is it really new? You have the state-of-the-art collection and have become very familiar with what it contains. However, the question of whether you now have a concept that is potentially patentable is not as yet answered. The concept is presumably not in the collection you have assembled, but is the concept one that was possibly incorporated into a patent that covered a different type of device or was the concept disclosed in a technical article or in a paper presented at an engineering meeting? Is the concept nonobvious? It is at this point that a *patentability* or *novelty* search should be made.

The collection of patents already in hand is a starting point for such a search. The patentability search requires more focus than a collection search because it is now concerned with specific concepts, and not just a general area or category. It will, therefore, focus on the most pertinent subclasses. Also, closer attention will be paid to the references for the details of their disclosures, and the number of subclasses to be considered will usually be greater than in a collection search. The scope of such a search must necessarily include some subclasses that are not directly "on the money" because related areas may also contain useful disclosures. Obviously, the broader the scope of the search, the more expensive. If a cost estimate at any point in time suggests that it will cost more for the remaining work to complete the patentability search than it costs to file a patent application, the application path becomes advantageous in the economic sense. Once a patent application is filed, the Patent Office Examiner must do the search for the "needle in the haystack." The negative aspect of this approach is that there will be a delay between the time of application and any response from the Patent Office. The length of delay may or may not be acceptable to management.

The purpose of a patentability search is to find patents or other publications in which the concept has been disclosed. The difficulty is that one is trying to prove a negative—that is, one hopes to find nothing! It is not surprising that these searches sometimes fail to lead to prior art.

For example, early in the development of digital computers the AND circuit was developed, and a patent application was filed covering the concept. The United States Patent and Trademark Office (hereinafter referred to as the Patent Office) disallowed the application on the ground that the concept had been disclosed in Patent #613,809, issued on November 8, 1898, to Nikola Tesla. That patent was titled "Method of and Apparatus for Controlling Mechanism of Moving Vessels or Vehicles." The patent had long since expired, of course, and the company could use the circuit without fear of infringement, but it could not protect itself from the use of the same circuit by anyone else. This was an example of a search in which an important patent was overlooked, but, in defense of the patent attorney doing the search, it must be pointed out that it was not at all obvious that one should search the group of patents in which this unhappy surprise was found.

Infringement searches. Suppose that the patentability search has led to the conclusion that the new concept for GFCI's is patentable. Before proceeding with production, one should make a still more detailed search, known as an *infringement* search. This is generally more time-consuming and more expensive than the previous searches because a wider swath must be cut in an infringement search than in the patentability search. During this search, the patent attorney studies the patents in fields pertinent to the presumed invention. For example, if you are developing a new tuner for an FM receiver, the attorney may find two or three tuner patents of concern, but may also turn up a patent on a voltage-controlled oscillator and a relevant TV patent as well.

If patents are found that appear to be relevant, the patent attorney will also study the patents cited by the examiner during the proceedings that led to issuance of those patents. If any patent appears to have claims broad enough to cover the subject matter, the Patent Office's file on that matter will also be studied. (The Patent Office file was formerly called a "file wrapper", it is now referred to as a "prosecution history.") The attorney is looking for any grounds that may lead him or her to conclude that the patent's claims are limited in such a way that they do not cover your concept.

Validity searches. Now let us suppose that in the course of developing your GFCI, you decide that you want to use a certain principle that has already appeared in one of the unexpired patents you have at hand, and that is covered by one or more of the claims in that patent. Or perhaps the infringement search has led to the conclusion that the concept you wish to use is covered by one or more claims in a patent turned up during that search. However, when you look at the description in the patent of this feature (that is, in the specification, which is discussed later), you conclude that the feature in question would have been obvious at the time the patent was filed, or that the way in which the device was described as operating was not physically realizable. That is, you believe that the patent—at least insofar as that feature is concerned—is defective, and you can advance a sound set of reasons for your conclusion. If you and your management wish to continue on the same course, a *validity* search is needed. This search expands into all sorts of published literature.

The patent attorney will attack this search by a still more thorough study of the patent at issue, its file in the Patent Office, the patents cited during the prosecution of the patent application, and whether the specification and claims are appropriate according to patent laws. The attorney will also review foreign patent publications, which may include published patent applications. The engineer can assist in this effort by researching old sales catalogs, trade journals, and the technical literature (including foreign language publications, M.S. theses, and Ph.D. dissertations) in an attempt to find evidence that the claimed subject matter in the patent at issue was in the public domain—either by reason of having been published in one form or another or having been sold—more than a year prior to the patent's filing date.

Another avenue which may be followed is to consider the possibility that the description of the way in which the previously patented device operates is not physically possible, or was not at the time of filing. For example, a patent

PATENTS

showing direct coupling of several transistors to form an amplifier was shown to describe a mode of operation that was not possible at the time of filing because only germanium transistors had been available at the time, and the circuit shown would have put at least the last transistor in the chain into saturation. While the circuit would probably have operated as described using silicon transistors, those were not available at the time of filing.

To sum up this section, there are four different kinds of patent searches that can be performed. These are:

1. A state-of-the-art search, which gathers those patents relevant to the product to be designed and developed.
2. A patentability search, intended to determine whether the principle on which your design is based can be patented or not.
3. An infringement search, intended to determine whether an unexpired patent is likely to be infringed.
4. A validity search, intended to prove that a previous patent (or patents) is invalid in one or more respects.

As noted above, these searches are best performed by a patent attorney, although a first cut at a state-of-the-art search can be done easily and quickly by the engineer, as described below. Moreover, the designer plays an important role in these searches by supplying the technical information that the attorney may lack. This is a team effort, and it requires a considerable amount of interchange of information between the designer and the patent attorney.

5.2 STRUCTURE OF A PATENT

If you are going to make use of patents or if you have an idea that may be patentable, you need to know how patents are structured. Every patent application consists of the *specification* and an *oath* or *declaration*. The original of issued patents has a cover (transmittal) page which formerly named the inventor and the date of issuance of the patent; presently the cover page is generic in nature, with no information as to inventor or date of issuance. Copies of patents began for many years with the first page of the drawings, such as that shown in Figure 5.1. Starting with patents issued in 1970, copies begin with a page that includes a heading with the last name of the inventor, the patent number, and the date of issue. The remainder of the page contains more detailed information, including the title of the patent, the full name(s) of the inventor(s) and place(s) of residence, the assignee if any, the date of filing, the fields of search by the Patent Office Examiner, the references cited, an abstract, and a drawing. This page therefore provides a wealth of information for both the patent attorney and the designer. An example is shown in Figure 5.2. The first page of the drawings follows this page.

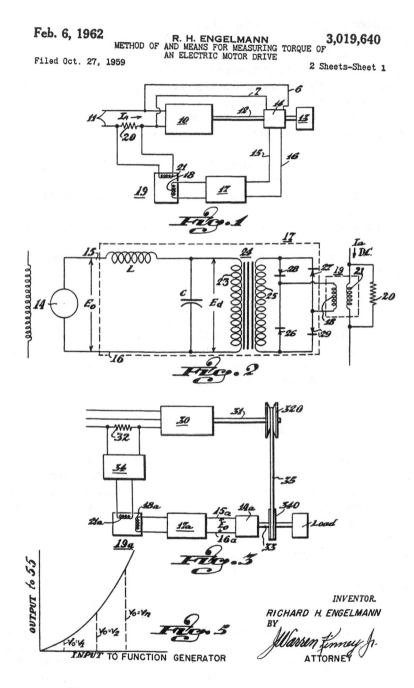

Figure 5.1 First page of drawings of a typical patent.

PATENTS

United States Patent [19]
Martin, Jr. et al.

[11] Patent Number: **5,398,277**
[45] Date of Patent: **Mar. 14, 1995**

[54] FLEXIBLE MULTIPROCESSOR ALARM DATA PROCESSING SYSTEM

[75] Inventors: Edgar C. Martin, Jr.; Kevin P. Duffied, both of St. Louis, Mo.

[73] Assignee: Security Information Network, Inc., St. Louis, Mo.

[21] Appl. No.: **831,978**

[22] Filed: Feb. 6, 1992

[51] Int. Cl.^{6} H04M 11/00
[52] U.S. Cl. 379/39; 379/45; 379/201; 379/207; 379/211; 379/265; 379/266; 379/309
[58] Field of Search 379/39, 40, 41, 42, 379/43, 44, 45, 46, 47, 49, 50, 51, 201, 207, 265, 266, 309, 211; 395/775, 700, 200

[56] **References Cited**

U.S. PATENT DOCUMENTS

Re. 32,468	8/1987	Le Nay et al. 340/506
3,626,098	12/1971	Lee .
3,694,579	9/1972	McMurray .
3,700,823	10/1972	Chulak .
3,881,060	4/1975	Connell et al. .
3,914,692	10/1975	Seaborn .
3,922,498	11/1975	Aul et al. .
3,937,889	2/1976	Bell et al. .
4,023,139	5/1977	Semburg .
4,141,006	2/1979	Braxton .
4,228,424	10/1980	LeNay et al. .
4,257,038	3/1981	Rounds et al. .
4,259,548	3/1981	Fahey et al. .
4,262,283	4/1981	Chamberlain et al. .
4,310,726	1/1982	Asmuth .
4,319,337	3/1982	Sander et al. .
4,489,220	12/1984	Oliver .
4,493,948	1/1985	Sues et al. .
4,540,849	9/1985	Oliver .
4,555,594	11/1985	Friedes et al. .
4,577,182	3/1986	Millsap et al. .
4,578,536	3/1986	Oliver et al. .
4,622,538	11/1986	Whynacht et al. .
4,623,988	11/1986	Paulson et al. .
4,688,183	8/1987	Carll et al. .
4,710,919	12/1987	Oliver et al. 370/96
4,774,658	9/1988	Lewin 364/200
4,800,583	1/1989	Theis 379/67
4,839,892	6/1989	Sasaki 379/45
4,839,917	6/1989	Oliver 379/45
4,893,825	1/1990	Pankonen et al. 379/45
4,922,514	5/1990	Bergeron et al. 379/49
4,924,491	5/1990	Compton et al. 379/37
5,022,067	6/1991	Hughes 379/95
5,046,088	9/1991	Margulies 379/211
5,048,075	9/1991	Katz 379/97
5,061,916	10/1991	French et al. 379/106
5,077,788	12/1991	Cook et al. 379/45
5,164,983	11/1992	Brown et al. 379/207
5,249,223	9/1993	Vanacore 379/266
5,278,898	1/1994	Cambray et al. 379/266

OTHER PUBLICATIONS

P. Ruggieri, "Dial 911 for Profits", *Telecommunications*, May, 1984 (3 pages).
E. Delong, Jr., "Making 911 even better", *Telephony*, Dec. 14, 1987 (pp. 60–63).
E. DeNigris, J. Shanley, P. Weisman and R. Keltgen, "Enhanced 911: emergency call with a plus," *Bell Laboratories Record*, Mar., 1980, (pp. 74–79).

Primary Examiner—Curtis Kuntz
Assistant Examiner—Stella L. Woo
Attorney, Agent, or Firm—Wood, Herron & Evans

[57] **ABSTRACT**

A multiprocessor alarm data processing system includes a database with identifying information for all the central stations and their subscribers who use the system. Several input processors receive alarm data from the central stations to identify subscribers incurring alarm events. The input processors generate event records for alarm data validated against the database. Several output processors are provided to process the event records to select the appropriate municipal authority and transmit thereto identifying information for a sucscriber incurring an alarm event. The processors may be switched between input and output processing for dynamic internal load shifting and redundant processing systems are provided for inter-system load sharing. A busy-out capable modem is provided to facilitate such inter-system load sharing.

90 Claims, 5 Drawing Sheets

Microfiche Appendix Included
(350 Microfiche, 5 Pages)

Figure 5.2 Typical first sheet of a recent patent.

The present patent laws provide in Section 112 that:

> The specification shall contain a written description of the invention, and of the manner and process of making and using it, in such full, clear, concise, and exact terms as to enable any person skilled in the art to which it pertains, or with which it is most nearly connected, to make and use the same, and shall set forth the best mode contemplated by the inventor of carrying out his invention.
>
> The specification shall conclude with one or more claims particularly pointing out and distinctly claiming the subject matter which the applicant regards as his invention . . .

The main body of the patent is the specification. It has two main functions. The first of these is to describe the invention as completely, clearly, and exactly as possible, while at the same time being concise. The second function is to point out with particularity the subject matter the inventor claims as his or her invention. Both functions are of value to the designer examining the patents of others for any of the purposes mentioned in the previous section.

The specification contains the following:

1. Title of the invention. This is selected by the inventor and the patent attorney, and is intended to be a concise description of the invention.
2. Abstract. The abstract is intended to provide information useful for those searching patents, and should therefore pinpoint the principal technical advance that is the real subject of the patent. When reading the abstract, one should understand that it is not allowed to be used in interpretation of the claims.
3. Background of the invention. This portion of the specification sets the stage by providing a brief statement describing the general field to which the invention applies. It then goes on to describe the prior state of the art relative to the invention's scope. In this description, the inventor may refer to relevant prior patents and published papers or articles in the technical literature. In the course of this description, previously unsolved problems that are solved or at least alleviated by the patented invention are discussed and the fact that the invention provides remedies is pointed out.
4. Summary of the invention. The summary is intended to be a concise statement of the nature of the invention, its operation, and its purpose. It is a brief synopsis of the claims.
5. Brief description of the drawing or drawings. This portion of the specification enables the reader to understand the broad purpose of each of the figures.
6. Detailed description of the invention. This is usually the longest portion of the specification. The invention is described in great detail, and includes a

description of how to make and use the invention. This portion of the specification refers to various portions of the drawings using numbers placed on the elements of the drawings. These numbers are referred to in the description to insure that there is no misinterpretation of the description by inadvertently confusing one element with another. The same element appearing on two or more drawings will be identified by the same number. There is more discussion of drawings below.
7. Claims. Because the detailed description does not precisely delineate the scope of the invention, this section is the most important part of the patent. It is in the claims that the coverage of the invention is established, just like a deed to property.

The patent whose first page is shown in Figure 5.2 has a number of pages of drawings. One of those drawings is shown in Figure 5.3. Following the drawings, the disclosure begins as shown in Figure 5.4. The first few of the ninety allowed claims in this case are shown in Figure 5.5.

Although our present interest is in the patents of others, a few comments are in order on some of the ramifications of applying for a patent. The work required to write a patent application will be largely that of the patent attorney, but the engineer/inventor must provide a large amount of information. One of the authors has found that an efficient method of providing the necessary information is to write the first (and very likely the second) draft of the detailed description himself, making the necessary sketches with numerical identification of the elements as he develops the description. The patent attorney reviews this material to determine whether the engineer has omitted steps that are obvious to the engineer but not necessarily to others, whether extraneous material has been introduced, and so forth. Following this review and discussion between the engineer and the patent attorney, the patent attorney is ready to write a first draft of the application in the form required by the Patent Office, and to hand the sketches on to the draftsman. There will, of course, be further interchanges to clarify sundry points before the application is in final form because, once filed, no new matter may be added to that application. A word of caution, however: Not all patent attorneys are willing to work in the manner just described; be prepared to adapt to various styles and personalities.

It is important to understand that patent drawings are not the same as engineering drawings. Their purpose is to illustrate the patentable features of the invention. That is, they are intended to be descriptive, not, as with engineering drawings, prescriptive. Hence these drawings are not scaled (and many are not to scale). They are frequently drawn in perspective with shading and other artistic devices so that they present a view such as would be seen by an observer, but they may also include cross sections, exploded views, and breakaways. The Patent Office has a number of rules specifying how these drawings must be made, including a specification on the material on which the drawings are made. The techniques required are not those ordinarily used by engineers or draftsmen

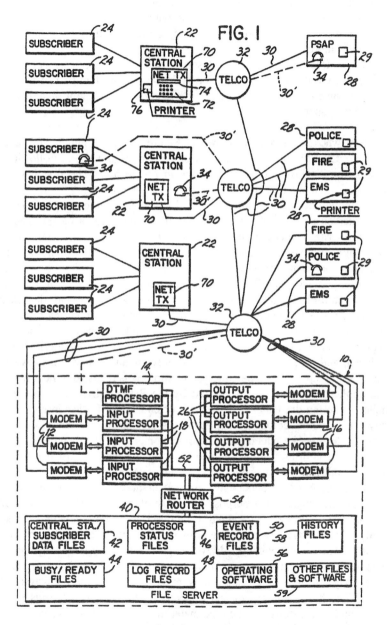

Figure 5.3 Drawing from the patent of Figure 5.2.

5,398,277

FLEXIBLE MULTIPROCESSOR ALARM DATA PROCESSING SYSTEM

APPENDIX

Attached hereto as Appendices A and B which is on microfiche having 350 microfiche and 5 pages which show source code listings of software for use with the present invention. The contents of Appendices A and B are incorporated herein by reference. Further, Appendices A and B contain material which is subject to copyright protection. The owner has no objection to facsimile or microfiche reproduction of the appendices, as they appear in the Patent and Trademark Office patent file or records, but otherwise reserves all rights whatsoever.

BACKGROUND OF THE INVENTION

I. Field of the Invention

The present invention relates to flexible multiprocessor data processing systems and more particularly to such systems adapted to provide computerized management of alarm data from a plurality of geographically dispersed sources and for advising municipal authorities such as public service answering points (PSAPs), police, fire or EMS departments of the particulars of those alarms.

II. Description of the Prior Art

A typical alarm reporting system monitors a variety of subscribing residences or businesses connected, such as by telephone lines, to a security company's central station ("CS") whereat electronic signals are received from the monitored location(s) in the event an alarm is triggered such as due to fire or break-in or the like. Upon receipt of that signal, in accordance with the central station's protocol, an operator at the central station telephones the appropriate municipal authority or authorities (such as the police or fire department or both) and orally advises the emergency services operator of the appropriate alarm information. The police or fire services or other necessary emergency services would then be dispatched by the operator. At the same time, other security company central stations as well as various members of the public may be trying to call the municipal authorities for emergency assistance. With such a typical alarm reporting system, multiple, simultaneous calls may thus tie up the emergency services operators causing sometimes tragic delays in response times. Indeed, in the event of a disaster, such as a hurricane, tornado or large scale fire, for example, the volume of calls generated by a single emergency occurrence may be so overwhelming that the operators simply can not process them all. The volume of calls might even be so excessive that the equipment is shut down. In that event, should the municipal authority otherwise encounter difficulty, the caller may not know that alternative action must be taken to summon assistance. Also, false alarms cannot be easily cancelled in which event limited emergency service resources are caused to respond while another, real emergency, possibly goes unanswered.

It has been proposed to computerize the interconnection between security company central stations and the municipal authority, thus reducing some of the sources of tie ups and the possibility of human error. But such proposed computerized systems have not otherwise overcome many of the problems with currently employed methods. Moreover, the proposed computerized systems have generally required that each security company central station have a computer that is not only equipped with a database containing all the necessary information (such as address, contact persons, special warnings or cautions, nearest police departments, etc.) for each of its subscribers, but which is also powerful enough to process alarm data for each subscriber encountering an alarm condition and compile the necessary reports for electronic transmission to the municipal authorities, all without significant delays. Such hardware is costly and places control of the various computers in the hands of central station operators where error may occur and be difficult to locate. Similarly, standardization is not established as a consequence of which each and every police, fire or EMS department or every PSAP or other municipal authority may be required to have a plurality of terminals adapted to communicate with the respective security company computers making the operation of those municipal authorities not only costly but prone to error and, again, leading to significant delays while the various terminals are monitored by the operators.

SUMMARY OF THE INVENTION

The present invention provides a flexible data processing system such as for a centralized alarm data processing system and which provides the advantages of computerized interconnection between central stations and municipal authorities without requiring added complex computer equipment at the central stations or the municipal authorities and without other drawbacks of previous proposed computerized interconnections. To this end, and in accordance with the broadest principles of the present invention, a multiprocessor data processing system is provided in which information such as for the subscribers of a plurality of central stations throughout the country is maintained in a centralized database coupled to a plurality of input processors and a plurality of output processors within the multiprocessor system. The input processors communicate with the various geographically dispersed data sources such as central stations to receive alarm data therefrom while the output processors concurrently analyze the received data and communicate necessary information to appropriate reception locations such as police, fire, EMS, PSAPs or other responding municipal authorities throughout the country.

Separate input and output processors allow a multiplicity of alarm calls from numerous, different central stations to be handled without delay or tieing up the communication or alarm processing capability of the data processing system such that multiple alarm events may be processed and the municipal authorities provided timely and thorough visual (e.g., printed on a printer at the municipal authority) alarm reports all in virtually real time and without clogging the emergency services telephone lines with numerous alarm calls. Moreover, the data processing system of the present invention may communicate with central stations and municipal authorities throughout the country thus eliminating the possibility of multiple communication standards from the various central stations. More specifically, because the database of the data processing system is compiled from data for all subscribers nationwide, standardization is readily accomplished. Thus, only one data processing system standard is necessary for all central stations and municipal authorities. The

Figure 5.4 First page of detailed description of the invention of Figure 5.2.

5,398,277

quently, if a subscriber 24 is deactivated, for example, the systems 120A and 120B need not check each and every subscriber record 62 but need only check and update that group of records in the affected file.

By virtue of the foregoing, there is thus provided a flexible multiprocessor data processing system such as for processing alarm data with only minimal delay, i.e., in virtually real-time and with the advantage of computerized interconnection between central stations and municipal authorities but with few or none of the drawbacks of prior computerized interconnection proposals.

While the present invention has been illustrated by the description of embodiments thereof, and while the embodiments have been described in considerable detail, it is not the intention of applicants to restrict or in any way limit the scope of the appended claims to such detail. By way of example, and not by way of limitation, more than two redundant systems 120 could be involved. Further, data other than alarm data may be processed by the flexible multiprocessor data system of the present invention. Still further, E-mail messages may be communicated with the system of the present invention by building the message into an event record for processing as if it were an alarm event. Additional advantages and modifications will readily appear to those skilled in the art. The invention in its broader aspects is therefore not limited to the specific details, representative apparatus and method, and illustrative examples shown and described. Accordingly, departures may be made from such details without departing from the spirit or scope of applicants' general inventive concept.

Having described the invention, what is claimed is:

1. An alarm data processing system for reporting to municipal authorities alarm events incurred by subscribers of central stations comprising:
   database means for storing identifying information for each of the subscribers;
   record means for storing event records;
   several input processor means each for (i) receiving, from at least one of the central stations, alarm data identifying at least one subscriber incurring an alarm event, (ii) validating the received alarm data based upon the identifying information in the database means, and (iii) generating and storing in the record means an event record having data based upon the validated alarm data;
   several output processor means each for (i) processing one of the event records from the record means so as to select a municipal authority and generate an alarm information report including at least part of the identifying information from the database means for the subscriber incurring the alarm event based upon the data in the event record being processed, and (ii) transmitting the alarm information report to the selected municipal authority.

2. The alarm data processing system of claim 1 further comprising:
   network transmitter means associated with at least one of the central stations for selectively transmitting alarm data from the central station to one of the input processing means.

3. The alarm data processing system of claim 1 further comprising means to temporarily establish a communication link between one of the central stations and one of the input processor means, whereby after the alarm data is received from the one central station, the one input processor means may thereafter receive alarm data from another central station.

4. The alarm data processing system of claim 1 further comprising:
   means for temporarily establishing a communication link between one of the output processor means and the municipal authority selected thereby, whereby after the output processor means transmits the alarm information report to the selected municipal authority, it may thereafter transmit an alarm information report to another municipal authority.

5. The alarm data processing system of claim 1 further comprising:
   busy/ready means associated with a first of the output processor means for indicating that the first output processor means has already selected the municipal authority being selected by a second of the output processor means;
   means associated with the second output processor means for generating a busy/ready record with data correlated to at least the event record being processed by the second output processor means; and
   means associated with the first output processor means for generating a second alarm information report based at least in part upon the data in the ready/busy record, whereby both the first-mentioned alarm information report and the second alarm information report are transmitted to the municipal authority by the first output processor means so the second output processor means may be freed up to process another one of the event records in the meantime.

6. The alarm data processing system of claim 1 further comprising:
   DTMF processing means for (i) receiving cancellation alarm data, the cancellation alarm data including data identifying one of the subscribers, (ii) validating the received cancellation alarm data against the identifying information in the database means, and (iii) generating and storing an event record having data based upon the validated cancellation alarm data;
   wherein each output processor means includes means operable when processing an event record based upon validated cancellation alarm data ("cancel event record") for (i) examining a group of other event records for an event record based upon validated alarm data ("alarm event record") identifying the same subscriber, (ii) if such an alarm event record is found, processing the cancel event record as if it were an alarm event record and including in the alarm information report that the report is of a cancellation, and (iii) if no such alarm record is found, discontinuing processing of the cancel event record, whereby a municipal authority need not be selected and an alarm information report need not be generated.

7. The alarm data processing system of claim 6 wherein the DTMF processing means includes means for receiving the cancellation alarm data from a telephone.

8. The alarm data processing system of claim 1 further comprising:
   DTMF processing means for (i) receiving cancellation alarm data, the cancellation alarm data including data identifying one of the subscribers, (ii) vali-

Figure 5.5 End of description and beginning of claims for the patent of Figure 5.2.

in an engineering office, and hence these drawings are made by a specially trained class, the patent draftsmen.

Finally, a word about the oath or declaration. This is a sworn statement by the inventor(s) that he or she knows of no impediment that would make the patent improper or invalid. It should not be signed until after the application has been read very carefully and no errors have been found, the drawings have been studied to make certain that they are complete, accurate, and that all elements have been identified, and the claims have been studied to make certain that they provide for every possible variation that the inventor (or the patent attorney) can visualize. (The patent attorney will probably devise many more variations than the inventor believes to be possible.)

5.3 PATENT SEARCHES: THE CLASSIFICATION SYSTEM

In the preceding section, patent searches were described in general terms, but nothing was said about how to conduct such searches. As pointed out also, these searches are best done by the patent attorney, but the engineer should understand how they are done and may even do a preliminary search, as described later. Since there are over five million U.S. patents now in existence, this appears to be a formidable task. As will be seen, however, there is a straightforward process to be followed, and the task has been facilitated as on-line searching techniques have become more readily available.

The U.S. Patent Office has developed the U.S. Patent Classification System (USPCS) so that patents may be found relatively easily from their subject matter. The USPCS is an arrangement of classes and subclasses of apparatus, articles of manufacture, processes, and compositions. These are found in the "Manual of Classification," which contains over 400 classes. At the beginning of the manual is a list titled "Classes Within the U.S. Classification System Arranged by Related Subjects." The classes are arranged in this list in three major groups: Chemical, electrical, and mechanical/miscellaneous. This is really a table of contents, and is used to direct one to the lists of subclasses. The definition of each class and subclass is available as a separate document.

To begin the search, one simply runs down the list under the appropriate group until the class is found that is applicable to the subject matter (known in patent parlance as the *object of the invention*) being sought. A typical page from these lists is shown in Figure 5.6. Having the class, the schedule of subclasses for that class is consulted, scanning the left-most entries in the schedule, that is, those having only one dot (.) before the entry. A typical example of such a schedule is shown in Figure 5.7. This should lead one to an entry that is relevant to the object, but it may not be the most definitive one. If there are indented entries below the one selected (that is, entries preceded by two dots (..)), these should be scanned in order to find one which includes a further feature of the

CLASSES ARRANGED BY ART UNIT

	CLASS TITLE	SUBCLASS RANGES FROM	TO
ART UNIT 2101			
CLASS 984	MUSICAL INSTRUMENTS	All	
ART UNIT 2102			
CLASS 310	ELECTRICAL GENERATOR OR MOTOR STRUCTURE	1 306 DIG 2	300 800 DIG 6
CLASS 318	ELECTRICITY: MOTIVE POWER SYSTEMS	114 35	135 38
CLASS 322	ELECTRICITY: SINGLE GENERATOR SYSTEMS	All	
ART UNIT 2103			
CLASS 200	ELECTRICITY: CIRCUIT MAKERS AND BREAKERS	80 A	86.5
CLASS 335	ELECTRICITY: MAGNETICALLY OPERATED SWITCHES, MAGNETS, AND ELECTROMAGNETS	All	
CLASS 336	INDUCTOR DEVICES	All	
CLASS 337	ELECTRICITY: ELECTROTHERMALLY OR THERMALLY ACTUATED SWITCHES	All	
CLASS 361	ELECTRICITY: ELECTRICAL SYSTEMS AND DEVICES	600	837
ART UNIT 2104			
CLASS 218	HIGH-VOLTAGE SWITCHES WITH ARC PREVENTING OR EXTINGUISHING DEVICES	All	
CLASS 290	PRIME-MOVER DYNAMO PLANTS	All	
CLASS 314	ELECTRIC LAMP AND DISCHARGE DEVICES: CONSUMABLE ELECTRODES	All	
CLASS 341	CODED DATA GENERATION OR CONVERSION	1 200 50	17 899 172

Figure 5.6 Typical page from the Manual of Classification showing classes.

CLASS 361 ELECTRICITY: ELECTRICAL SYSTEMS AND DEVICES

1	SAFETY AND PROTECTION OF SYSTEMS AND DEVICES	46	...With more than two wires
2	.Arc suppression at switching point (i.e., includes solid-state switch)	47	..In a polyphase system
		48	...With more than three wires
3	..Synchronized or sequential opening or closing	49	..In a single phase system
		50	...With more than two wires
4	...Counter electromotive force	51	.Overspeed responsive
5	...With current sensitive control circuit	52	.By regulating source or load (e.g., generator field killed)
6	...With voltage sensitive control circuit	53	..Prime mover control
		54	.Load shunting by fault responsive means (e.g., crowbar circuit)
7	...With combined voltage and current sensitive control circuit	55	..Disconnect after shunting
		56	..Voltage responsive
8	...Shunt bypass	57	..Current responsive
9	....With sequentially inserted impedance	58	.Impedance insertion
10	..By inserting series impedance	59	.Circuit automatically reconnected only after the fault is cleared
11	...Nonlinear impedance		
12	...By arc stretching (e.g., horn gap)	60	..With differential voltage comparison across the circuit interrupting means
13	..Shunt bypass of main switch		
14	..Arc blowout for main breaker contact (e.g., electromagnet, gas, fluid, etc.)	61	..Reclosing of the nonfaulty phases of a polyphase system
15	.Capacitor protection	62	.Feeder protection in distribution networks
16	..Series connected capacitors		
17	..Shunt connected capacitors	63	..With current responsive fault sensor
18	.Voltage regulator protective circuits	64	...With communication between feeder disconnect points
19	.Superconductor protective circuits		
20	.Generator protective circuits	65	..With current and voltage responsive fault sensors
21	..Voltage responsive		
22	.Compressor protective circuits	66	...With communication between feeder disconnect points
23	.Motor protective condition responsive circuits		
		67	.Series connected sections with faulty section disconnect
24	..Current and temperature		
25	..Motor temperature	68	..With communication between disconnect points
26	...With bimetallic sensor		
27	...With thermistor sensor	69	...Pilot wire communication
28	..With time delay	70	..Constant current system
29	...During energization of motor	71	.Automatic reclosing
30	..Current and voltage	72	..With lockout means
31	..Current	73	...Including timer reset before lockout
32	...Bimetallic element	74	..Continuous
33	..Voltage	75	...With time delay before reclosing
34	...Bimetallic element	76	.With phase sequence network analyzer
35	.Transformer protection	77	.Reverse phase responsive
36	..With differential sensing means	78	.With specific quantity comparison means
37	..With temperature or pressure sensing means	79	..Voltage and current
		80	...Distance relaying
38	..Transformer with structurally combined protective device	81	....With communication means between disconnect points
39	...With lightning arrester and fuse	82	...Reverse energy responsive (e.g., directional)
40	...With lightning arrester (e.g., spark gap)		
		83	...With time delay protective means
41	...With fuse	84	..Reverse energy responsive (e.g., directional)
42	.Ground fault protection		
43	..Fault suppression (e.g., Petersen coil)	85	..Phase
		86	..Voltage
44	..With differential sensing in a polyphase system	87	..Current
		88	.With specific voltage responsive fault sensor
45	..With differential sensing in a single phase system		
		89	..With time delay protective means

Figure 5.7 Schedule of subclasses for Class 361.

Prop			INDEX TO CLASSIFICATION			Protection		
	Class	Subclass		Class	Subclass		Class	Subclass
Bicycle	280	293*	seismometers)			Fire escapes with fire protection	182	47
Clothesline	248	353	Chemical analysis with			Fluid	182	51
Draft pole	278	87	Apparatus	422	50*	Fly nets for animals	54	80.4
Flower	47	44*	Methods	436	25	Garment and body part		
Ladders	182	165*	Earth boring combined	175	58*	Apparel type	2	2*
Collapsible	182	156*	Electrical	324	323*	Aprons bibs shields etc	2	46*
Hinged extension	182	163	Radar	342	191*	Design	D 2	860*
Separable extension	182	22*	Neutron using	250	253*	Arm guard	D29	120
Wheeled	182	16	Radiant energy	250	253*	Back and chest safety	2	92
Scaffolds building supported	182	87	Radioactive using	250	253*	Design	D29	100*
Shocker	56	431	Sound using	181		Baseball gloves	2	19
Thill support	278	85	X ray using	250	253	Design	D29	115
Tree	47	43	Prosthetic Resector (see Resector)			Boxing gloves	2	18
Trunk lid	190	34	Prosthetic Article, Design	D24	155*	Buttonhole	24	659*
Propeller (see Impeller)	416		Prosthetic Article, Utility	623		Eye and face	D29	108*
Aircraft combined	244	62*	Coating	427	2.1*	Garment protectors	D 2	860
Stabilizing	244	92	Prostoglandin Acids	562	503	Goggles	2	426*
Design	D12	214	Esters	560	121	Design	D16	303*
Locomotive combined	105	66	Protection (see Cover; Fender;			Hand	D29	113*
Making, by milling	409	120	Prevention)			Head protectors for permanent		
Power plant X-art	60	904*	Against earth currents for electric			waving	132	243*
Screws	416	176	railways	204		Headwear	D 2	891*
Ship combined	440		Against radiation damage	376	277	Helmets	2	410*
Buoyant	440	98*	Antiabrasion or protective layer	430	961*	Design	D29	102*
Chain	440	95*	Arm guard	D29	120	Leg guard	D29	120*
Oscillating	440	13*	Baking food	126	22	Leggings and gaiters	36	2R
Portable	440	53*	Barrel bung	217	114	Design	D 2	901
Reciprocating	440	13*	Bonnets for animals	54	80.1	Leggings stocking type	2	242
Screw	440	49*	Book	281	20	Overshoes	36	7.1 A*
Testing	73	147	Building	52		Design	D 2	909*
Toy	446	36*	Animal blocker	52	101	Shirt bosom	2	120
Airplane combined	446	57*	Coating	52	515*	Shoe, boot, and legging	36	72R*
Propicillin	540	341	Earth quake	52	167R	Skirt edge	2	222
Propioic Acid	562	598	Earth supported coping	52	102	Slippers inside shoe	36	10
Propionic Acid	562	606	Exterior flashing	52	58*	Spats	36	2R
Proportional			Marker or monument	52	103	Stocking heel from shoe	36	55
Counter	250	374*	Metal monument markers and			Trouser	2	231*
Geiger muller tube	313	93*	guards	52	102*	Trouser guards and straps	24	72
Radioactivity	250	336.1	Pole shell	52	727*	Wringer release	68	256*
Scintillation	250	361R*	Safes and banks	109		Hand guard or bonds	D29	113*
Tube structure	376	153*	Snow stop	52	24*	Headware & helmets	D29	120*
Feeding			Tubular edger	52	244	Heat exchanger	165	134.1
Automatic	431	90	Window	52	171.3	Heating system parts freezing	237	80
Dispenser automatic control	222	57	Burner heat protector	431	350	High voltage for electric apparatus		
Dispensing from nonserial traps	222	426*	Automatic	431	23	and conductors	361	1*
Dispensing from plural containers	222	134	Casings			Horse		
Dispensing from plural outlets	222	482*	For cigars	206	242*	Boots	54	82
Fluid distribution	137	87*	For watch movements	206	18	Shoes	168	
Gas & air mixer for burner	431	354	Clock key with dust guard	81	123	Hose against abrasion or bending	138	110
Automatic	401	90	Closure edge	49	462	Jacket engine freezing	123	41.1*
Mortar mixer	366	16*	Collapsible steering post	280	777	Leg guard	D29	120
Mortar mixing process	366	8	Drill rod			Light modifiers and combined	362	317*
Water purification	210	101	Applying well protector to	29	236	Lightning	361	117*
Measuring			Guide or slide for	175	325.1*	Marine structure	405	211*
Flow meter	73	202*	Electric devices			Metals	422	7
Gauge point markers	33	663*	Anti-inductive structures	174	32*	Casting device safety	164	152
Proportional dividers	33	358.2	Arc lamps liquid electrode	313	164	Molten metal oxidation prevention	75	709
Proportioner, Rectangular	33	DIG. 9	Circuit breaker arc	200	144R*	Object against corrosion by electrical		
Propositional Logic	395	67	Conductors and insulators	174		neutralization	204	147*
Propulsion Devices and Systems			Connector cord terminal	439	135*	Electric current apparatus	204	196*
Aircraft	244	62*	Demagnetizing	361	267*	Magnetic apparatus	422	186.2
Ammunition and explosive devices			Discharging static charges	361	212*	Oxen		
Design	D22	112	Implosion of tubes	445	8	Shoes	168	5
Land mines	102	362	Lightning arrester	361	117*	Padlock	70	54*
Propelling charge combined	102	374*	Lightning arrester and thermal			Pencil sharpener and point		
Pyrotechnic rocket	102	347*	current switch	337	28*	Piano case	84	183
Electric of railway cars	104	288*	Lightning conductors	174	2*	Piles	405	211*
Land vehicles			Lightning rods	174	3	End caps	405	255
Convertible	280	7.15*	Meter circuit	324	110	Pipe ends	138	96R
Runner	280	12.1*	Plural terminal insulator rod type	174	179	Pipe pressure compensator with		
Simulations	280	828*	Railway right of way	246	121	freeze protector	138	27
Skates	280	11.115	Ray energy shields	250	515.1	Pipe thawing and freeze protector	138	32*
Wheel propelled	280	3	Safety series	361	1*	Protective coating compositions	252	381*
Wheeled	280	200*	Shock hazard	174	5R	Removable	427	154*
Marine	440		Signal box key	340	302	Protectors	30	460
Manual power	440	21*	Signal system circuit maintenance	340	292	Railway car heating stove	126	57
Motor vehicles	180		Telegraph system circuit			Roadway		
Propellers	416		maintenance	178	69R	Curbs or corners	404	7*
Railway			Telephone	379	412*	Railway automatic switch traffic		
Cable	104	173.1*	Anti-inductive devices and			protected	246	330
Car	104	287*	systems	379	416*	Railway crossings signals and		
Locomotives	105	26.5*	Thermal current switch arc	337	273*	gates	246	293*
Switches and signals	246		Third rail	191	30*	Railway drawbridge	246	118*
Reaction motors	60	200.1*	Track connections	238	14.5*	Railway grade crossing	246	111*
Store service			Trolley rope	191	71	Railway right of way defect	246	120*
Cable propulsion systems	186	14*	Voltage distribution regulator	191	2*	Railway switch snow protected	246	428
Self propelled car systems			Erosion	405	15*	Railway track fastener boxes etc.	238	312
Single impulse systems	186	8*	Eye and face guard	D29	108*	Third rail	191	30*
Prospecting (see Measuring; Testing)	73	151*	Fiber protecting during fluid			Traffic barrier	404	9*
Acoustic, vibratory or seismic wave			treatment	8	133	Traffic guides	404	6*
(see seismographs and			Color	8	685	Vault cover	404	25*

Figure 5.8 Page from the Index to the U.S. Patent Classification System.

object. Should there be another level below this, it should be scanned also. Once the final level is reached, the Patent Office files may be searched by class and subclass to obtain patent numbers. For more information on this and some of the other search methods discussed below, the reader may wish to consult Reference 1.

A second entry method (and one that is preferred by some patent attorneys) that is available through the Patent Office is the "Index to the U.S. Patent Classification System." There are approximately 60,000 terms in this index. Using this approach gives the user the advantage of having only one entry point into the system, but it has the deficiency that related classifications can and probably will be missed. It is intended only for a first step.

Another available entry method is by use of the Patent Office's Cassis CD-ROM System. This system is updated periodically, and contains titles of about thirty years worth of patents, and the abstracts for the last three years. (This is a rolling window with each new disk update.) Terminals are available to the public at the Patent Office's building in Arlington, Virginia. They are also available at some of the Patent and Trademark Depository Libraries. These are university, municipal, county, and state libraries, as well as libraries at some research institutions, that receive and maintain collections of patents and trademarks for the benefit of the public.

Patent searches may also be conducted using commercial databases. Reference 1 lists 49 on-line databases, 10 of which are classified therein as major. Of the 10 major databases, five include not only the United States but as many as 56 other countries. Four of the databases cover the United States only; one is for Japan. At least one of the United States databases now has the capability of displaying patent drawings with their Windows software. All of these databases have limitations of one kind or another, and one should determine what those limitations are before making use of them (and expending funds). The remaining 39 databases listed in the reference are narrower in scope, being tailored to meet the needs of particular clienteles.

As an example of searching, return to the GFCI. Reference to Figure 5.6 enables one to find that GFCI's are either in Class 200 or in Class 361. A search of Class 200 came up empty-handed. However, a search of Class 361, shown in Figure 5.7, shows that there are a number of subclasses that may be relevant, running from 42 through 50. If you are interested only in residential applications, subclasses 45, 46, 49, and 50 appear to be the ones to search.

If the "Index to the U. S. Patent Classification System" is consulted, it would seem obvious to search under the heading "Electrical." If this is done, there seems to be no entry relating to GFCI's. Since they are protection devices, however, the entry "Electrical, protection" was tried. This entry refers one to "Protection, electric." Figure 5.8 shows the relevant page from the "Index." Under "Protection, Electric devices," there are no entries for ground fault circuit interrupters, ground fault interrupters, interrupters, or any other entries that appear relevant until one reaches "Safety systems." This gives Class 361 as the appropriate class to search. Referring back to Figure 5.7, it becomes very obvious that one must still know the relevant subclasses before starting a search. Beginning with the "Manual of Classification" seems to be preferable in this case, although it may not always be so.

Many databases are readily available to the engineer through one of the commonly available networks, such as CompuServe®. There are, of course, charges for use of the system and there are additional charges for searching the databases. However, one can quickly find information on patents that can then lead to a more detailed and sophisticated approach. The cost for such a quick look is small compared to the rates of a patent attorney. However, it cannot be emphasized too strongly that anything more than a first cut at a search should be done by someone skilled in the art, and this is rarely the engineer.

One of the authors carried out a quick search using one of the networks, using the words GROUND FAULT INTERRUPTER. In a few minutes, this search turned up a considerable amount of information with the expenditure of only $2.00. Several databases were searched, one of which was U.S. Patents Fulltext. This database extends back only to January 1971; it is updated on a weekly basis. In this database, there were 126 U.S. patents on this subject for the years 1971-1979, another 111 for 1980-1989, and 151 for 1990 to early 1996. A further—and admittedly casual and inexpert—search retrieved seven patents by title, inventor, assignee, date of issue, and U.S. patent number, for four of which the abstracts were called up also. One of those four was then printed out in its entirety, except for the drawings, which were not available in the database used.

Regardless of how a search is to be carried out, knowledge of the class number is of importance; it is the key to all of the searches. One of the important aspects of the computer search approach is the ease with which the class number and some of the subclass numbers can be found. Note that the class number was not known and was not even needed for the search just described, although it and the subclasses were acquired during the search because of the information provided in the full text printout. It was not certain when the search was initiated whether the words used for access corresponded to an entry in the "Index." Reference to Figure 5.8 shows that it is not.

If your company is contemplating entering the GFCI market, there are two other interesting bits of information to be gleaned from the data in the previous paragraph. The first of these is that, although the GFCI has been on the market since about 1970, there have been 388 patents issued since that time (and there were undoubtedly some issued before that time). The second bit of information is that there seems to be a high and apparently increasing level of activity in the field, with almost 40% of the 388 patents having issued since 1990, only a little more than a five year period, while about 60% were issued in the preceding twenty years.

This information should be of interest to your company management, because it indicates that this may be an exceedingly difficult field to enter both because of the intense current activity and because of the large amount of prior

PATENTS

art that must be studied and digested. However, no immediate action should be taken on this information. The patents need to be searched more carefully to see exactly what is covered. For example, if management intends to enter the residential GFCI market and most of the patents in the 388 are directed solely toward three-phase use, the situation may not be as gloomy as the first glance would indicate. They may have claims that are so narrow that they are not applicable to the product you wish to market. In addition, many of these patents are no doubt "improvement" patents, being based on an earlier patent that has either expired or may have only a few years of life remaining. Patents such as these are still based on the principle covered by the original patent. If your design begins from a different point, improvement patents are less likely to be relevant. However, this is the point at which your patent attorney should be put to work!

Progressive companies make it a point to keep up with patent activity that may affect their business. This is especially important if the technology in your field is in a period of rapid development. As noted above, databases are frequently updated on a weekly basis. A source of information that is readily available through many libraries is the *Official Gazette of the United States Patent Office*. This is published weekly, and groups patents using the major divisions of the Manual of Classification. Furthermore, patents are usually assigned patent numbers in the order of the class number, making it possible to enter the *Gazette* with the class number of interest and quickly find all patents in that class that issued during the previous week. The complete patent is not shown; each entry shows a drawing and one claim or the abstract of the patent.

5.4 BASICS OF PATENT LAW

Although the treatment of patent law that follows is only a cursory treatment of the field, it will aid in understanding the patent system. The vehicle used to aid in this understanding is a description of the process followed in prosecuting a patent application, and covers a wider variety of events than is likely to take place in the prosecution of any one patent. For those interested in more detail, References 1 and 2 should be consulted. It should be pointed out also that the following discussion relates only to U.S. patent law; other countries have laws that differ significantly from those in effect in the United States. Reference 1 provides a considerable amount of information on this subject, with discussions of different ways of filing for patents abroad and comparing the United States patent system to those of several other countries.

As the attorney develops the application, he or she may decide that more than one application is needed. The inventor may not have thought of his or her work as constituting more than one invention, but multiple applications are not an infrequent occurrence. For example, if in designing an electric switch a significant improvement was made on the operating mechanism and another on the method of making the electrical connections and an application is submitted trying to claim improvements in both areas, the Patent Office will recognize that there are two separate inventions present in an application that attempts to cover both improvements, and will insist that one invention be claimed in the original application and that a divisional application be filed for the other. (The patent attorney should recognize this situation and consult with you about your options.)

A major reason for requiring two applications is that each patent must be identified by a classification number as we saw above. Part of the work of the Patent Office is a search of that class, and those numbers would be different for the two patents. The drawings and disclosure—that is, the specification up to the point of the claims—can be exactly the same, but the claims will be directed only to the invention being claimed in each application. One of the authors (RHE) was once active in a project in which exactly this situation arose. There were two applications, one for improvements made by the author, the other for improvements made by a second inventor, a mechanical engineer. The applications, except for the claims, were exactly the same. Two different examiners were involved, and the references cited in the two issued patents were completely different.

The inventor or inventors must be correctly identified. An issued patent that can be shown to list other than the true inventor(s) can be declared invalid. Hence supervisors, executives employing consultants, or technicians who contributed some excellent work to the development are not to be named unless they actually contributed to the insight on which the invention is based. You are *not* submitting a technical paper, in which recognition as a co-author of someone whose contribution has been minimal has often been done as a courtesy. There is further discussion of this point in Chapter 6.

You should keep an accurate record of when the idea of the invention was conceived and when it was reduced to practice. The date of conception is not when you had some nebulous idea about how to solve a particular problem, but is the date on which your fuzzy notions became sufficiently crystallized that you were able to describe the solution in such a way that others could understand it. The best way to establish the date of conception is to write the description and have others who are skilled in the art read the description, agree that they understand it, and sign and date a statement to that effect. Most companies want their engineers who are engaged in design work to maintain a log of their activities. This is the place for the descriptions and for the signed statements of those who have read it. The log also serves other purposes, as will become evident in the next paragraphs.

The date of reduction to practice is the date when you actually tested a model under typical operating conditions or, if you do not do that, the date your application is received in the Patent Office. Between the time of conception and reduction to practice, you must practice due diligence; that is, you must be able to show that you pursued the development with reasonable attention. This becomes important if another inventor submits an application that is similar to yours.

When two or more applications covering essentially the same invention are being processed simultaneously by the Patent Office, an interference is declared. That is, it becomes necessary to determine which of the inventors should be awarded the patent. The Patent Office has a Board of Patent Appeals and Interferences to handle such cases. If one person is first to conceive the invention and also first to reduce it to practice, he or she will receive the patent. If the second person to conceive the invention is the first to reduce it to practice, the first to conceive must show diligence by records (the log, for example) and witnesses. If this cannot be shown, the patent is awarded to the second inventor on the basis of having first reduced the invention to practice. The first to conceive is not always the prevailing party if an interference is declared. However, patent attorneys look on this as a murky area; the discussion above has been simplified in order to give the reader some idea of the possible outcomes and of possible reasons for those outcomes.

Since reduction to practice is so important in interference proceedings, it would seem prudent to make a patent application as early as possible so that the date of reduction to practice becomes the date of receipt by the Patent Office. However, continued work may show further improvements or possibly confirm some lingering doubts about the earlier work. You and your management have two options. One is to delay the application so as to be able to put the work on a firmer basis or to have time to make improvements that can be used to broaden the claims, recognizing that one is simultaneously running the risks of others filing first and that questions may be raised as to exercise of due diligence if an interference is declared. The other option is to make an early application, even though you may not be satisfied with the total content, in order to gain the advantage of the earliest reduction to practice date. Under this option, if it is necessary to change an application because you have additional information resulting from pursuit of the development, there is a way to submit a new application using what is known as a continuation-in-part. The continuation-in-part will contain the still-valid information of the original application as well as the new information developed since the original application was submitted. The date of submission of the original application will remain the date of reduction to practice for that part.

With the change in law such that the term of a patent is now twenty years from the filing date, a new element has to be considered in making the decision as to when to file. Obviously, this additional element comes down on the side of delaying filing. A new procedure, available for the first time in the United States, allows filing of a provisional application. That filing provides an extra year before filing of a regular application without starting the 20-year clock.

After the application has been filed and after the examiner has made a search to determine patentability, negotiations begin between the attorney and the examiner. If the examiner has agreed that the invention is patentable, these negotiations will mostly concern the claims. The attorney, the inventor, and company management want the broadest claims possible; the examiner wants the claims to be no broader than he or she believes to be deserved. These negotiations are usually carried on only by correspondence. That emanating from the Patent Office is called *office action*. Two office actions are typical, with the second being "Final." Each must be responded to within a specified time with arguments for the validity of the claims as written or with proposed changes in the claims to meet the examiner's objections. Agreement is generally reached, but occasionally the examiner will find it necessary to issue a final rejection. If that occurs, the claims must be put into language the examiner will accept, an appeal process begun, or the patent abandoned. Although done infrequently, it may be advisable to ask for an interview with the examiner to resolve differences. If an interview is to be requested, this should be done after the first office action, because they are rarely allowed after the final action. The inventor may participate with the attorney in that interview. (One of the authors once went to an interview with an examiner accompanied, not by the attorney, but by the inventor under a patent being cited by the examiner against the author's application. The patent was granted.) How long does this process take? Typically, a patent issues about two to three years after filing.

Until the patent issues, information about the application should be held in confidence. However, after the patent issues the Patent Office will sell copies of the entire application file (file wrapper) to anyone wishing to buy it. File wrappers are especially valuable in defending allegations of infringement because they might indicate limitations of interpretation to which the inventor agreed during the prosecution of the application.

While the patent application is being prosecuted, "Patent Pending" may be put on the product. This has the effect of warning others of the risk they would take in copying your design, although they have no way of knowing exactly which features are subsumed under this warning. The inventor actually has no right to object to use of the invention before the patent issues because no legal right has as yet been created. Indeed, were he or she to do so, the confidentiality of the application would have to be breached. "Patent Pending" must not be used unless a patent application has been made and a patent is pending. Violation can result in legal action.

After the patent has issued, the patent number can be placed on the product. Doing so has the advantage that anyone thinking of copying the product is immediately liable for any loss of profit if he or she infringes. If it is not placed on the product and you or your management becomes aware of another manufacturer using the invention for which you now hold a patent, their liability begins only when the other party is notified of their infringement.

5.5 COMPUTER SOFTWARE

If you are interested in patenting computer software, the first thing to do is to consult a good patent attorney in the field. The reason for this statement is simply that the question of computer software has been one of the most hotly debated issues of intellectual property law for the past thirty years, and we are still at a point where the courts seem to be pulling in one direction—toward patentability—although slowly and under great pressure, and the Patent Office is continuing its pattern of digging in its heels. In the following discussion, there is no attempt to say which is the "correct" point of view; the intent is simply to try to set some boundaries on what may and what may not be patentable. If the reader is in a state of confusion at the end of this section, he or she is in no poorer condition than the patent law on this subject seems to be in at the moment.

To begin, what is *not* patentable? Some subject matter is clearly not patentable. An example of unpatentable subject matter (known as *nonstatutory* subject matter) is an "invention" that merely presents and solves a mathematical formula or algorithm. One of the early cases (Benson) in which an attempt was made to obtain a patent in the software area was finally rejected by the Supreme Court in 1972. The invention was a mathematical algorithm that converted the binary coded decimal (BCD) notation for a number into the pure binary notation for the same number. The applicant claimed that the method was to be used only on signals in telephone switchboards. The Court decided that the algorithm was a procedure for solving a given type of mathematical problem, and that despite the applicant's assertion that it was to be used only in the telephone field the claim was so broad that it included all possible applications. Award of a patent would have given Benson a monopoly on the algorithm itself.

Six years later, the Court heard another case (Flook) in which the application dealt with a method of updating alarm limits during catalytic hydrocarbon conversion. The catalytic conversion process was old, the monitoring techniques were old, the concept of computing alarm limits was old, and so was the concept of using computers for monitoring, computing, and triggering. All of this was freely admitted by the applicant. The novelty claimed for this process was the algorithm used to determine the updated alarm limit. The Court held that this application merely presented a new and presumably better method of calculating alarm limit values, the algorithm that Flook was using. Because that was a mathematical formula, the patent was not allowable.

In the Benson case, the process was to be performed entirely within a computer, and the claims were directed to an algorithm that would have wide applicability. In the Flook case the algorithm was only a part of a larger chemical process, and the claims were limited to that single process, yet a patent was denied. In both cases the Court seems to have been focusing on the use of an algorithm, and relying on the unpatentablity of algorithms. The possibility of the award of a patent involving software appeared to be very small.

In 1987, the case of Diehr and Lutton was decided by the Supreme Court. The application for patent in this case was for a process using a computer to regulate the curing time of raw rubber in a mold (that is, a press). Rubber is cured under heat and pressure for a specific time, the optimum time for a good cure depending in part on the temperature inside the press during the curing cycle. Uncontrolled or uncontrollable variables during the curing process make it difficult to arrive at the exact temperature in the press to use in the calculation for the optimum time. Industry practice was to calculate a curing time which was the shortest time in which all parts would definitely be cured. Obviously, the manufacturer knows that this method will lead to overcuring of some parts, and it can also lead to undercuring of other parts at times because of the uncontrolled variables. Dierh and Lutton solved the problem by measuring the temperature of the mold repetitively and using those values in the computer to recalculate the curing time by use of the Arrhenius equation. When the actual elapsed time equaled the calculated curing time, the process was terminated. The result was a more uniform product with less waste. The Court found for the applicants, stating (in part):

> The respondents [the applicants] here do not seek to patent a mathematical formula and although in the process a well-known mathematical equation has been employed, the respondents do not seek to pre-empt the use of that equation. Rather they seek only to foreclose from others the use of that equation in conjunction with all of the other steps in their process.

Patents may be granted on combinations of devices or processes that include an algorithm if it can be shown that the combination interacts to produce a functioning product or process. In this case, all of the steps in the process were old individually and the Arrhenius equation has been known for about a century, but the combination was such that the Court did not preclude a patent. The ruling in this case seems to make it "clear that an invention *involving software* will fall within a statutory category, *i.e.*, can be regarded as patentable subject matter, if an arithmetic expression is implemented in a specific manner to define structural relationships between physical (hardware) elements of a claim or to refine or limit process steps" [Reference 1, page 130].

In 1981 the Supreme Court upheld a decision in favor of Bradley and Franklin for their software-oriented invention. They invented a computer data structure that allowed multiprogramming and system modification. The decision was apparently based on many of the same reasons used in the Dierh and Lutton case, with emphasis on the fact that Bradley and Franklin's invention involved hardware as well as mathematical algorithms.

It is noteworthy that these cases were eventually decided by the Supreme Court. The Patent Office and/or its Board of Patent Appeals had ruled against patentability, and the cases were then argued through the courts, eventually

PATENTS 171

reaching the Supreme Court. Reference 3 cites two cases (Alappat and Warmerdam) which reached apparently opposite conclusions at the Court of Appeals level. In addition to references previously cited, Reference 4 may also be consulted. The interested reader will from time to time find relevant articles in trade journals and similar publications.

The best advice to the prospective applicant, however, remains that in the first sentence of this section: "If you are interested in patenting computer software, the first thing to do is to consult a good patent attorney in the field."

5.6 OTHER INTELLECTUAL PROPERTY

The patents we have been discussing so far are known as *utility* patents. These patents relate to processes or methods, machines, improvements in manufacturing techniques, or composition of matter. There are two other kinds of patents that you should know about. The first of these is the *design* patent. Despite its name, it has nothing to do with the kind of design we are concerned with, but rather with the appearance of a product. A design patent bars others from making a product that will look the same as yours. The other kind of patent is the *plant* patent, which relates to distinct and new varieties of plants.

Other than patents, you may need to know about other kinds of intellectual property. These are copyrights, trademarks and other marks, maskwork for semiconductor applications, and trade secrets, each of which is treated separately, but briefly, in the remainder of this section. There are legal mechanisms by which each of these may be protected.

Copyrights. A copyright is a grant of certain rights by the U.S. Government to the originator of a copywritable work, or, if the work was by an employee, to the employer. Works that may be copyrighted include all sorts of literary works, art, motion pictures, music and the lyrics, engineering drawings, and so forth. One of the items included in the "and so forth" catch-all is computer software. (An interesting discussion of this aspect of the copyright law may be found in Section 14.3 of Reference 1, including discussion of the kinds of cases that have been filed.) The rights granted to the copyright holder include the right to reproduce and distribute the work, prepare derivative works such as abridgments and translations, perform copyrighted musical or dramatic works in a public setting, and hold public displays of paintings, graphics, and statuary. Moreover, these rights can be licensed or transferred to others.

Acquisition of a copyright is a fairly simple process. It can be done by placing on each copy in a reasonably prominent place (for example, on the page after the title page of a book) a "copyright notice." For a book, this notice must include the symbol ©, the word "Copyright," or the abbreviation "Copr." The notice also includes the year of first publication of the work, and the name of the copyright owner. This is usually followed by a statement such as "All rights reserved. No part of this book can be reproduced ...without written permission from the publisher." Although it is not necessary to do so, the copyright may be registered with the U.S. Copyright Office; the usual fee for this service is $20. It is wise to do this for anything other than an ephemeral work, because infringers usually may not be sued if the work has not been registered.

Although the copyright owner has the exclusive right to make and distribute reproductions of the copyrighted work, others can do so also to a limited extent and not be infringers of the owner's copyright under what is known as the "doctrine of fair use." This allows reproduction for purposes such as news reporting, provision of copies of technical papers for the use of students in a class, and use in current research. In determining whether any of these uses is fair, several factors need to be taken into account. One of these is the purpose of such use. Use for commercial purposes might not be fair, whereas use in an instructional setting would probably be construed as being fair, unless one were reproducing entire volumes. Another factor would be the effect on the potential market for the copyrighted work. One way of looking at the issue of fair use is to consider whether the contemplated use will deprive the copyright owner of profits that might otherwise be realized; if it does not, the general trend in court decisions seems to be that such use is allowable.

Under the present copyright law, which became effective in January 1978, copyrights begin when the work is created and extend for 50 years after the death of the author. Works copyrighted before the present law became effective were under a different set of rules. If a copyright was still in effect on January 1, 1978, that copyright was automatically extended for a period of 75 years from the date of first publication, and the date of the author's death does not affect the date of termination.

More detailed information on copyrights for various materials will be found in Reference 2. If it appears that a copyright may be of importance, consult an attorney skilled in the copyright law.

Trademarks, service marks, certification marks, and collective marks. Trademarks are words, names, symbols, devices (logos), or a combination of these that has been adopted by a manufacturer or other organization and which is intended to identify in a unique manner the source or the origin of a product or a publication. For example, the spine of this book carries a logo that is unique to the publisher. The Institute of Electrical and Electronics Engineers (IEEE) has a logo utilizing the right-hand rule, on which electrical engineering is based. Trademarks should be registered with the U.S. Patent Office. Once registered, the symbol ® should follow the mark. The cost of registration is nominal, and the user is then protected in a legal sense against the use of the trademark by others. Moreover, one may find when seeking registration that some other organization has rights to the trademark, and that a different trademark must be devised. To avoid this problem, an attorney can perform an availability search first.

Service marks are the same as trademarks except that they are used in connection with the sale or provision of services rather than products. These *may* be the same as a trademark if a company, such as General Electric, is engaged both in manufacture and service. Examples of companies that are engaged only in service activities and that are easily identified by the public by their service marks are Anderson Consulting, Delta, and TWA.

Certification marks are used on products to indicate that they meet certain standards or were produced in a certain way. The mark that is probably most important to the engineering designer is the UL designation for those products that have been listed by the Underwriters Laboratories. Other marks, such as the "Seal of Approval of *Good Housekeeping*," will also be of importance for certain products.

Collective marks, which can appear on similar products originating with different manufacturers, indicate membership or sponsorship by a particular organization. For example, clothing produced by a company having employees belonging to the International Ladies Garment Workers Union often bear the mark ILGWU on the label. Labels on industrial electric motors frequently refer to a NEMA frame size, NEMA being the collective mark of the National Electrical Manufacturers Association. The mark (and logo) in this case have some of the characteristics of a certification mark as well as a collective mark. Reference 2 contains much more detailed discussion of the various kinds of marks.

Maskwork for semiconductors. This is a relatively new kind of intellectual property. The masks for early semiconductors were relatively simple. As the industry developed, the devices became smaller, it became possible to increase the number of devices placed on a chip of any given size, and the size of the chip increased. The complexity of the interconnections between devices on a chip as well as the number of interconnections increased in a quasi-exponential fashion, as did the complexity of the masks required for fabrication. Protection for the companies which had invested large sums of money in these masks was afforded by the Semiconductor Chip Protection Act of 1984 [5]. This act provides for a 10-year term of protection for the designer of the masks, thus effectively precluding the use of reproductions of mask works in the manufacture of competing chips. However, competitors are not prevented from reverse engineering the chip in order to analyze its functions.

Trade secrets. A trade secret may take on many different forms. It may be a chemical formula, a recipe (for example, what is in Coca-Cola?), a pattern for a machine, customer lists, a method of doing machining in order to be able to reduce tolerances, or a prototype for a new product. It is any confidential information that the company believes gives it a competitive edge over others. The key point, of course, is that, whatever it is, it must be something that can be kept secret. If someone else independently discovers how to do the machining process referred to above, that person is free to use the process. The only legal protection one has against loss of the secrecy is against someone obtaining or trying to obtain the information by stealth. Bribery of someone in your own organization who has been entrusted with the information as a necessary part of his or her work is an obvious example of illegal tactics.

At the same time, the company has certain responsibilities to preserve the secrecy and, if legal action becomes necessary, to demonstrate that appropriate steps were taken to try to do this. For example, the machining process referred to above should be placed in a portion of the plant that is off limits to visitors and to those employees who have no business to conduct in that area. The company should have employment contracts which require the employees trusted with trade secrets not to divulge any of those secrets if they leave the company's employ.

If the trade secret is so important, why not patent it and thus foreclose its use by others? To begin with, it may not be patentable at all, and there is no choice but to try to maintain it as a trade secret. Moreover, if it is patentable and does confer an economic advantage, as soon as the patent issues your competitors will be looking for ways to achieve the same advantage without infringing on your patent. Some years ago, a manufacturer patented a spot welding process for copper that avoided the annealing process by doing the spot welding under running water. Before the patent issued, the competition did not know how the spot welding was being done because there was no way of detecting the use of running water from examination of the final product. After issuance, the process was known in detail. Would the manufacturer have been better off to maintain the process as a trade secret rather than to patent it?

5.7 SUMMARY

Patents are an important source of information for the designer. In order to access this pool of information, it is necessary to make a search of issued patents. With a little training, an engineer, a technician, or other company personnel can make a preliminary search using any one of several methods. On-line searching is becoming a primary tool. With the patents in hand, the designer can determine the state of the art as it existed possibly two or three years earlier and will be able to understand the way the technology developed. The designer can also determine whether an approach that is being considered or is in development appears to infringe on an existing patent, and if it does not he or she has some knowledge about how to protect the new approaches being developed. However, it must be kept in mind that this is an extremely complex area of the law. Although the discussion in this chapter should be of help, only a patent attorney can hope to be able to do thorough patent searches and advise as to the consequences.

In addition to the various references cited in this chapter, there are several useful brochures that are available at no cost. These include *Basic Facts About Patents; Copyright Basics (Circular 1),* and *Basic Facts about Registering a Trademark.* The copyright brochure is available from the Copyright Office, Library of Congress, Washington DC 20559-6000. The second brochure, as well as others, is available from the U.S. Department of Commerce, Patent and Trademark Office, Washington DC 20231.

REFERENCES

1. Lechter, Michael A., Editor. *Successful Patents and Patenting for Engineers and Scientists,* IEEE Press, The Institute of Electrical and Electronics Engineers, New York, 1995.
2. Konold, William G., et al. *What Every Engineer Should Know About Patents,* Second Edition, Marcel Dekker, New York, 1989.
3. Harrington, Curtis L. Computer program patentability update, *Motion,* March/April, 1995.
4. Galler, Bernard A. *Software and Intellectual Property Protection,* Quorum Books, Peoria AZ, 1995.
5. 17 U.S.C. § 901 *et seq.*

REVIEW AND DISCUSSION

1. What is a patent?
2. What does the inventor exchange for the limited monopoly granted by a patent?
3. Make a list of exclusions from patent coverage.
4. Suppose you have received a patent on an improvement to a device already covered by a patent issued to someone else. Can you make use of your patent?
5. What are the three main parts of a patent application? Can any of them be changed after the application is filed?
6. Why do patents typically have some very broad claims while other similar claims have limitations?
7. How many inventions can be included in a single patent?
8. What is the most important reason for listing only the true inventor(s) in a patent application?
9. What is meant by reduction to practice? How can the date of reduction to practice be established?
10. Why are the dates of conception and of reduction to practice important?
11. "Patent Pending" may be placed on a product after an application has been submitted. Discuss whether this is a wise move to make.

12. The patent number(s) may be placed on a product after the patent has issued. What are the advantages of doing so? Can you think of any disadvantages?
13. Suggest some possible trade secrets other than those enumerated earlier. What precautions to maintain secrecy should be taken in each of these cases?
14. A product made by your company requires the spot welding of copper, and you have just learned of the patent disclosing the idea of welding under running water so that annealing may be avoided. Can you suggest some approaches that may be just as successful and that you think will avoid infringement?

PRACTICE PROJECT

Using any means of searching that is available to you, locate a patent that is related to the product you have chosen to design. Identify the improvement(s) in the state of the art that the patent claims has been made by the inventor.

6
ALTERNATIVE DESIGNS AND INVENTIONS

Don't keep forever on the public road, going only where others have gone. Leave the beaten track occasionally and dive into the woods. You will be certain to find something you have never seen before. . . . All really big discoveries are the results of thought.

Alexander Graham Bell

As you continue to follow the design procedures discussed previously, you will reach a point where you have a firm set of specifications, you are confident that the product is feasible, and you have some insight into what you must avoid so as not to infringe the patents owned by others. Having reached that point, you must now generate some specific ideas on how to accomplish your objective. The purpose of this chapter is to develop some of the concepts of creativity and to discuss some strategies the individual can adopt to make himself or herself more creative. One of the best ways to gain a better understanding of these concepts and strategies is to examine great inventions of the past, and how the inventors proceeded to reach their solutions. We begin, however, with some discussion of alternative designs.

6.1 ALTERNATIVE DESIGNS

A capable designer strives to produce the best design to meet the specifications. As discussed earlier, "best" is not quantifiable; that accolade is frequently conferred by the users of the product by the act of buying that product in preference to others of similar design. An example is the large share of the minivan market held by the Chrysler products. The favorable evaluation by the public that is evident from the sales data may, however, differ from that of automotive engineers. Regardless of the product under consideration, it is only rarely that the designer happens upon the best solution to a design problem when he or she sketches out the first system or embodiment of a device that seems to be workable. The first solution should be evaluated, but then set aside while the designer develops alternative solutions. As this is done, one of the solutions will be found to be best from the viewpoint of the designer, the production personnel, the sales department, and management.

How does one develop alternatives? Patents are valuable sources of ideas, as was discussed in the preceding chapter. Certainly no one should feel defensive about basing a design on an expired patent. The patent system was designed so that, although the inventor or the assignee retains monopoly rights for 17 or 20 years, the public is subsequently free to use the patented invention. If you base a design on a patent that has expired or will have expired before your product reaches the market, you have the assurance that you need not worry about infringing an unexpired patent that has been overlooked. One must, of course, be careful about follow-on patents, the "improvement" patents, and be sure to avoid infringing on any of those that are still in effect. Some very successful products have been developed using expired patents as the base. The principal disadvantage of following this course is that others can copy your product without fear of infringement, just as you have done.

Other sources for ideas for alternative designs are the publications of professional societies, trade journals, and manufacturer's catalogs and application brochures. Other engineers have probably been confronted with the same design problem as the one on which you are working, or one that is similar. Because one way of acquiring status in the profession is publication of new ideas, engineers are usually eager to publicize their work and their results. However, one must use such ideas with caution. Patent applications must not be preceded by publication of the idea by more than a year; some of this work may be the basis for a patent application that is close to filing. (But note some of the *caveats* in Chapter 5, and keep in mind that patent law is in a state of flux.)

After you have searched all of the sources you can find, you may conclude that a basically new approach is desirable. It may be necessary to do this to avoid infringement of an unexpired patent, it may be that you believe that a superior design is possible, or it may simply be that patent protection is a high priority of your management. Any of these reasons is proper motivation to try to invent a new device or system to meet the specifications. If something is to be invented, there must be a creative act on the part of the inventor. Much of the remainder of this chapter is directed toward the creative process as it is applied to engineering design and invention.

Whether creativity can be taught has been a matter of some controversy. Kivenson in the introduction to Reference 1 claims that

> Inventing, like art, can be taught. The mental mechanisms for increasing personal creativity can be stimulated. Undisciplined but highly innovative

thought processes can be systematically channeled to produce streams of problem-solving concepts. Skills useful to the invention developing process can be learned.

Many great inventors have claimed that creativity is an inherited trait, possibly because they have not tried to analyze the process they follow in inventing. The authors believe that there are certain ways of looking at problems that stimulate the creative process, and that every engineer benefits by understanding these methods. Before proceeding to a discussion of these methods, we will look at the way some of the great inventors of the past worked. As we will see, certain keys to the inventive process will become evident from their stories.

6.2 WHAT CAN BE LEARNED FROM SOME GREAT INVENTIONS OF THE PAST

Most inventions are not great, complicated discoveries that appear full-blown in the mind of the inventor. Even those that one ordinarily thinks of as great inventions are usually the culmination of a series of small steps. Moreover, the new or improved product is usually the result of the integration of many of these small steps, some of which (and possibly all of which) were made by others. To set the stage for our study of creativity, a few landmark inventions will be discussed.

"The steam engine was invented by James Watt in 1765." This sentence is generally accepted as an accurate statement of the origin of reciprocating steam engines. As a matter of fact, however, a steam engine was built as far back as the first century A.D., by Hero of Alexandria. This first engine was a rotary engine—a primitive turbine. It is not certain whether it was merely a novelty, or whether he used it to pump water for a fountain. Whichever it was, it did not become a useful device, and was probably never the basis for the work done by later inventors.

The first reciprocating steam engine was built in 1690 by Denis Papin in France. The piston in this engine was raised by steam produced in a vertical cylinder by heat from a fire under the bottom of the cylinder. After the piston reached the top of its stroke, the fire was removed, the cylinder gradually cooled and the steam condensed so that atmospheric pressure on the top of the piston would return the piston to the bottom of the stroke. Many of the first engines were used to pump water, and this engine and many that followed it could do so by use of a lever with one end attached to the piston in the engine, the other end being attached to the piston in a reciprocating pump. The next improvement was made by Thomas Savery in 1698. The steam was now produced in a separate boiler, but was still allowed to condense in the cylinder after the piston had reached the top of its stroke. The next step was to hasten the condensation of the steam by spraying cold water into the cylinder at the top of the stroke. This was done by Thomas Newcomen in 1705. Beginning in 1712 Newcomen engines were used to pump water from mines, and some of those engines were still in use in 1800. None of these engines was very efficient, but each was better than its predecessor as a result of a small step that eliminated one deficiency of its predecessor.

The reason that James Watt is remembered is that he made a major step forward. He had been retained to recondition a Newcomen engine at the University of Glasgow, and it was obvious that it was not a very efficient machine. His great contribution was that he exhausted the steam from the cylinder into an added auxiliary chamber, the condenser, for cooling. The cylinder could then be insulated and remain hot, rather than being required to go through a cycle of heating and cooling for each stroke. Watt himself recounted how he had been on a walk on a Sunday in Scotland, and that "I had entered the green and had passed the old washing house, I was thinking of the engine at the time. I had gone as far as the herd's house when the idea came into my mind . . . ." So successful was this step that the fuel consumption for steam engines was reduced by 75%. He and his partner in the steam engine business, Matthew Boulton, were able to obtain from the purchasers of their engines the payment of one-third of the fuel savings as compensation. The addition of a crank and crankshaft allowed the steam engine to be used in many applications other than pumping water out of mines. By 1787 John Fitch was using a steam engine in boats and by 1814 George Stephenson had built a steam locomotive. We shall return below to some aspects of Watt's contribution.

The process now known as xerography was the brainchild of Chester Carlson [2]. Carlson's first job, while still a boy, was with a printer. During the Great Depression years of the 1930's, he graduated from college with a degree in physics. He then worked briefly as a research engineer at Bell Telephone Laboratories in New York City, then for a patent attorney, and after that for P. R. Mallory & Co., a manufacturer of electronic components. While he was employed by Mallory, he earned a law degree in an evening program, and eventually became manager of Mallory's patent department. This was a fortuitous combination of experiences, because he learned that it was very difficult to get words into clear, hard copy and that there was a need for a convenient process to duplicate printed documents such as patents.

Using his physics background, he began to do library research on imaging processes and found accounts of work done by a Hungarian physicist, Paul Selenyi, on producing images by electrostatic processes. This phenomenon was new and was unexplored, and there was no single idea that came to him that directed him onto a clear road to success. Instead, there was a long period of careful research that taught him what he needed to know about the process and how to begin to make improvements. Although he filed his first patent application in October 1937, more years of research were needed to improve the process. In this work he was aided first by another young physicist and later by Battelle Memorial Institute, a research institute in Columbus, Ohio. It was not

ALTERNATIVE DESIGNS AND INVENTIONS

until 1957, however, 21 years after the first patent application, that the first useful office copier was available. The entire development required dedication to and faith in the eventual successful outcome of his invention far in excess of what most individuals are willing to invest. However, the honors he received and his financial reward were spectacular. At age 14 he had been the sole support of his parents; at his death at age 62, he had given away an estimated $100 million to foundations and charities.

Another landmark invention was that of the negative-feedback amplifier, invented by Harold S. Black on August 6, 1927. Black began work with Bell Laboratories in 1921, immediately after graduating with a BSEE degree. He was not the typical employee. To learn about the company and the telephone business, he began coming in on Sundays to read through a collection of important memoranda the company kept on file. The file began with 1898; by the time he reached the 1921 file he knew the technical problems facing the company. An immediate problem was to reduce distortion in push-pull vacuum-tube amplifiers carrying three channels over 1,000-mile lines. However, Black's studies had convinced him that telephone traffic would grow so rapidly that amplifiers handling 3,000 channels over 4,000 miles would soon be necessary, and the requirements for such amplifiers were far beyond the state of the art.

The research people at Western Electric (Bell's manufacturing arm) were also aware of the need to reduce amplifier distortion. An obvious approach to the problem of distortion was to improve the linearity of the vacuum tube, but this soon proved to be a dead end. However, Black took the important step of directing his attention to the amplifier as a whole, because the objective of the work was really to remove all distortion from the amplifier output. He later [3] wrote, "In doing this, I was accepting an imperfect amplifier and regarding its output as composed of what was wanted plus what was not wanted. I considered what was not wanted to be distortion and I asked myself how to isolate and eliminate this distortion. I immediately observed that by reducing the output to the same amplitude as the input and subtracting one from the other only the distortion would remain. This distortion could then be amplified in a separate amplifier and used to cancel out the distortion in the original amplifier output." The block diagram of Figure 6.1(a) shows the concept.

Black's line of reasoning occurred about 2:00 a.m. on March 16, 1923, after returning home from a meeting of the American Institute of Electrical Engineers at which Charles Steinmetz gave a lecture that impressed and evidently inspired Black by its clarity and logic. The next day Black sketched two embodiments of the scheme and set them up in the laboratory, thereby inventing the feed-forward amplifier.

This invention reduced the distortion by 40 dB. However, it required such precise balance and subtraction that it was difficult to maintain the advantage that was theoretically possible. Black continued his work, but every circuit he devised turned out to be far too complicated. What he was seeking was simplicity and perfection.

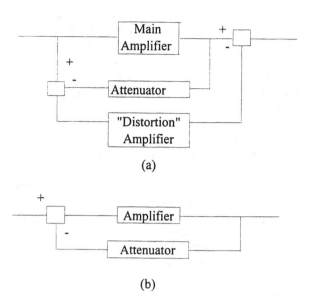

Figure 6.1 Amplifier block diagrams. (a) Feed-forward amplifier. (b) Feedback amplifier.

On the morning of August 2, 1927, the concept of the negative feedback amplifier came to him in a flash while he was crossing the Hudson River on the Lackawanna ferry on the way to work. After years of intense effort, he had suddenly realized that if the amplifier output was fed back to the input in reverse phase the means of canceling out the distortion in the output would be realized. He sketched a simple diagram (such as that of Figure 6.1(b)) of a negative-feedback amplifier on the copy of the *New York Times* he was carrying and derived the equation for amplification with feedback. On December 27, 1927, with typical input signals covering a frequency band from 4 to 45 kHz, he achieved a reduction of distortion of 100,000 to 1 (50 dB) in a single amplifier. Although Black's work was directed toward a specific problem, the result of his work was quickly found to be of value in the field called at that time servomechanisms, more commonly referred to today as feedback control.

There is a common thread that runs through all three of these examples: The dedication and intensity with which Watt, Carlson, and Black pursued their goals. Beyond that, and except for a common characteristic to be discussed below, many differences exist. Watt, as well as many others, was fully aware of the dreadful inefficiency of the steam engine as it existed when he began his work, and he had worked at it for at least two years before he saw a clear solution. He addressed his efforts toward improvements in an existing device. Carlson and Black were looking forward to ways of satisfying needs that had previously been satisfied only in a cumbersome way (or not at all) by the use of

ALTERNATIVE DESIGNS AND INVENTIONS

carbon paper or by photography in one case, or to tremendous growth in the use of the telephone in the other. Each of them realized that existing technology was at a dead end as far as achieving their objectives was concerned, and that new concepts were necessary. As to the manner of approach to solution of their problems, Carlson and Black were methodical and scientific. Carlson's biographies tell of the exciting moment when he first achieved success in making a copy, but there was no dramatic moment of insight—the *flash of genius*—when the process flashed into his mind's eye, as was true of Watt and Black. Carlson himself is quoted as saying that ideas came slowly, but he was deeply immersed in his investigation; Black had prepared himself by a long period of study of the telephone business and by detailed investigation of distortion, but it was four years from the feed-forward amplifier to the feedback amplifier; Watt had been working on the reconditioning of the Newcomen engine for a two-year period. The common thread of dedication and intensity of each man is quite evident when one studies the events that culminated in a major invention.

The other common characteristic was stated by Black when he said he sought a solution that was "simple and perfect." This characteristic is difficult to define, but it is easy to recognize. It includes the idea of simplicity, in the sense of not being complex, as well as being clever, reliable, and cost-efficient. It should be conservative of material and energy. It must of course be novel and nonobvious, although almost without exception one is inclined to say in retrospect, "That's so obvious." These ideas are frequently summed up by use of the one word "elegant."

A number of years ago, one of the authors (WHM) attended a talk given by a manufacturer of a line of heat pumps. During the presentation, the speaker mentioned that twice the rate of fluid flow was required when the pump was cooling as when it was in the heating mode. It was also obvious that the flow

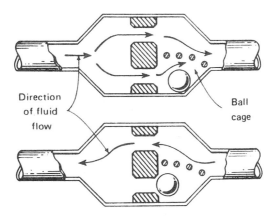

Figure 6.2 Valve having a two-to-one flow rate depending on the flow.

direction is reversed for the heating and cooling modes. Thoughts began to be generated almost immediately about a monitoring unit and transducers to sense the direction of flow and a controller to open or close a valve. Such a system can certainly be built. However, later on in the talk, the speaker showed a slide illustrating how the rate of fluid flow was changed when switching from one mode to the other. The system appears in Figure 6.2. This is an excellent example of an *elegant* solution. "It's so obvious!"

6.3 A THEORY OF INVENTION

What is a *flash of genius?* We saw above that both Watt and Black experienced those moments of insight, and many other inventors have had similar experiences. One theory [4] seems plausible and conforms to personal experiences of the authors. The value in presenting it here is that knowing it will give direction to what you can do to develop your creative ability.

The basic tenet of the theory is that all creativity (including art, poetry, and humor) has the common characteristic that a relationship is seen to exist between two entities not previously recognized as being capable of connection. Figure 6.3 is an attempt to show such a relationship diagrammatically. In this figure, the vertical plane is intended to represent a specific area of thought; all of the ideas normally associated with that area of thought are in that plane. Hence, as our mind scans the plane there are no surprises; we might even say that any train of thought we follow is "common sense" and is familiar to those "skilled in the art." If we have a new problem to solve, we may wander in a tortuous path in this plane during this frustrating search for a solution. Suppose, however, that there is another plane of thought that intersects the first in some subtle way that our mind might jump to and in so doing reach a solution. This new area of thought is represented by the horizontal plane. Koestler [4] calls this jump "bisociation."

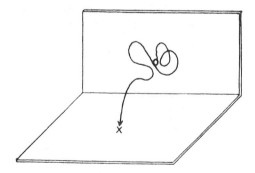

Figure 6.3 Two intersecting planes of thought found to have an unexpected problem/solution relationship.

ALTERNATIVE DESIGNS AND INVENTIONS

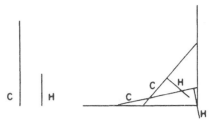

Figure 6.4 Problem: Draw a right triangle with hypotenuse C and distance H from the hypotenuse to the apex of the right triangle.

Kestin [5] gives a simple but excellent example of this moment of insight. As a boy he was challenged by the problem of drawing a right triangle when given two lines of fixed length, one being the hypotenuse C and the other, H, being the distance from the hypotenuse to the right angle. H is, of course, perpendicular to C. His first approach is shown in Figure 6.4. He drew a right angle with sides of indefinite length. Then he visualized the end points of the hypotenuse sliding along these two sides while H was slid along C, always keeping it pointed at the apex of the right angle. It is obvious that a solution can be obtained by trial and error, at least to a close approximation, but a precise solution was wanted. The next morning the problem appeared on a quiz. He visualized it in a different orientation, that shown in Figure 6.5.

Kestin had seen this orientation before, related to a theorem of geometry that states that the angle subtended on the diameter of a circle by a point on the circle is a right angle. The reorientation resulted in the jump to recall a theorem that did not occur to him during his study the previous evening. However, once the connection was made, the exact solution was easily obtained. For the young Kestin, this was an invention.

In his extensive treatment of this theory of creativity, Koestler cites the invention of the printing press as an example of this moment of insight, that is, of bisociation. Gutenberg had tried many different ways of improving the old art of printing. The method used was to engrave a picture on a block of wood, stone, or metal, sometimes with a few words of text. The block was then inked, and

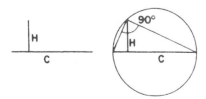

Figure 6.5 The perfect solution to the problem of Figure 6.4.

paper was placed on it and rubbed to transfer the picture to the paper. This technique had been used for many years to produce playing cards and pictures of saints, and a variation still survives today in the practice of making rubbings of headstones in cemeteries. But consider the enormous amount of time that would have been required to produce even one page of text in woodblock form, and the virtual impossibility of correction of errors. Even if that task were completed successfully, uneven inking of the woodblock or the paper not being rubbed uniformly over the entire surface would result in the production of a defective copy.

Gutenberg made two major contributions to the art of printing. The first of these was the idea of movable type. (Although he had been anticipated in this by 400 years by the Chinese, the knowledge of that advance had not reached Europe.) The second was the idea of the application of pressure to the paper. That idea came to him when he was participating in a wine harvest. He wrote, "I watched the wine flowing and going back from the effect to the cause I studied the power of this press which nothing could resist . . . ." At this moment it occurred to him that the same steady pressure might be used to press type to paper and also provide a means by which the type could be lifted off the paper without smudging the page. One of the most important inventions of all time came about because a person who understood printing, was skilled in it, and who recognized the need to improve the process happened to witness an operation that seems totally dissociated, the pressing of grapes necessary in winemaking, and made a connection between the two. For Gutenberg this was a completely creative act; for civilization it was the beginning of an era of information distribution and storage.

The intersecting planes of thought shown in Figure 6.3 take on clearer meaning if the concepts involved in Gutenberg's invention are identified on each plane as in Figure 6.6. The requirements or characteristics that are common to winemaking and printing by a press define the line of intersection of the two planes. As shown in the figure, these common elements are the movement along

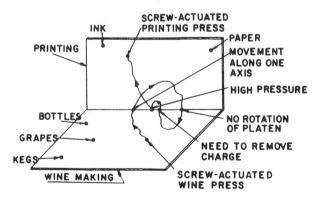

Figure 6.6 Gutenberg's invention of the printing press.

only one axis, the fact that the platen (the plate that presses the paper against the type or the pressure plate acting on the grapes) did not rotate, the high pressure, and the need to remove the charge (the paper in printing, the crushed grapes in winemaking). If such mutual requirements are identified as the inventor seeks a solution, then bisociation usually follows. In both planes in this figure, there are other factors that are pertinent only to one activity or the other, and these lie far from the line of intersection. Gutenberg, watching the action of the wine press, had the insight to realize that this same mechanism could be used in the printing process. We are so accustomed today to the fact that printed matter is readily available that we may have difficulty in appreciating how large a leap Gutenberg made. The books available before the invention of the printing press were all hand-lettered, and many were quite beautifully done. However, they were also available only to the very few who had plentiful resources at their command. There were no newspapers or magazines, and most people other than the clergy and a fraction of the nobility were illiterate. After Gutenberg, broadsides and pamphlets became commonplace, and books were more readily obtainable. People at all levels of society became literate, ideas were disseminated, and the world changed in ways that Gutenberg and his contemporaries never dreamed of.

A more modern example of bisociation resulted in Patent Number 2,292,387, issued to H. K. Markey and George Antheil on August 11, 1942. (Hedy Markey was the legal name of the actress Hedy Lamarr.) This was the basic patent covering what is now known as frequency hopping, a system designed to make communications by radio as secure as possible by changing the carrier frequency in a random pattern from one frequency to others. This is easily achieved in the transmitter, but the receiver has to follow the same pattern and make the local oscillator frequency change to the correct frequency simultaneously with that of the transmitter. The idea for frequency hopping was apparently Lamarr's, but the synchronization aspect was the contribution of Antheil, who was a musician. He had successfully synchronized sixteen player pianos for his *Ballet Mécanique.* The patent calls for slotted paper rolls similar to player-piano rolls to achieve the necessary synchronization. The system was too complex for easy use, and practical implementation was not achieved until transistor circuitry became very common. An interesting article on the collaboration of the two inventors appears in Reference 6.

6.4 BLOCKS TO CREATIVITY

It is not at all unusual for one to wander more or less aimlessly in the kind of looping mental activity indicated in the vertical plane of Figure 6.3 without making a leap to another plane, the step referred to above as bisociation. These blocks are entirely normal. During his evening study of the triangle problem, Kestin's creativity was blocked by the partial success of a poor approach to the solution. Once one starts down the wrong road, it is frequently difficult to discard an approach that is leading nowhere and that we know is leading nowhere. It is best at this point to put the problem aside for a time before looking at the problem again, and then to do so without even looking at the previous attempt. Kestin's fresh start the next morning showed that his block had somehow been removed during his night's sleep. (The advice to "Sleep on it" should not be taken lightly.) However, under some circumstances such blocks can permanently prevent the formation of the necessary connection between the two planes of thought.

How do such blocks arise? Kubie [7] argues that there is no single cause, but that all possibilities can be lumped under the term "neurotic." He cites examples of individuals whose research was unsuccessful because of deep-seated emotional problems that caused certain obvious solutions to be overlooked, among them being the compulsion to expend mental energy on criticizing associates or on a desire to prove a preconceived notion. On the other hand, the person who is at peace because he or she understands such conflicts and can put them aside is free of this unwanted mental burden. His or her brain is free to act as a communication center to process bits of information on the three levels that Kubie calls the conscious, the preconscious, and the unconscious levels.

On the conscious level, a person deals with subjects in terms of communicable literal ideas and realities. On the preconscious level, he or she processes data at an extraordinarily rapid rate and with great freedom, assembling and disassembling many diverse patterns. At the unconscious level, a person uses his or her special competence and knowledge to express those needs dictated by innermost concerns and emotions. Preconscious processes operate best when they are neither restricted by the conscious level nor interfered with by the unconscious level. Preconscious thought activity can and does proceed when we are asleep, when we are engaged in some other activity, or even if we are not especially alert. The "sleeping on" a problem of deep concern and finding an obvious solution the following morning is not an unusual event for many people.

In other research giving insight into causes of mental blocks, Hyman and Anderson [8] report tests where slides in color of familiar objects—for example, a fire hydrant—were projected on a screen and people tried to identify the object while the picture was out of focus. As the test proceeded, the focus was improved in discrete steps. The striking finding was this: If an individual identified an object incorrectly while it was far out of focus, it had to be brought to a significantly better state of focus for him or her to correctly identify it than was the case for others who had not made an appraisal at all. The general conclusion from their study was that it takes more evidence to overcome an incorrect hypothesis than to establish a correct one. In plainer words, a false start can create a mental block.

Mental blocks can occur because of the limits of one's education. The electrical engineer tends to find solutions in electrical technology, the mechanical engineer in mechanical technology, and the chemical engineer in the

ALTERNATIVE DESIGNS AND INVENTIONS

technology most familiar to him or her. Recall the valve of Figure 6.2 for the heat pump. What were the thoughts that sprang into the mind of one of the authors? They revolved around elements most likely to be found in electrical devices, the field in which he worked, but the solution shown in that figure contains not a single electrical device, and the mechanical device that resulted in an elegant solution is one of the simplest kinds of valves, a check valve. Once a need is recognized, the would-be inventor must consider all possible ways of accomplishing the goal. A less familiar technology, after a reasonable learning period, may result in a superior solution. Chester Carlson's invention of xerography is a case in point, although one might be tempted to question the duration of the "reasonable learning period."

This discussion of blocks to creativity has been included to provide a positive basis for advice on improving creativity, not to discourage you. Keep in mind that much of the useful problem-solving activity goes on at the preconscious level. You must therefore recognize that you do not want to inhibit this activity by too rigid an approach to a new problem, and you must learn to keep personal biases and emotions as far away from your problem-solving activities as you can. Finally, the problem must be studied in depth before making a decision on the mode of attack. Once committed to the wrong course, it is very difficult to begin anew.

6.5 WHAT KIND OF PERSON INVENTS?

As with so many other aspects of the human experience, there is no complete agreement on the characteristics of creative people. There are, however, certain characteristics that seem to predominate, and there are characteristics that seem not to have great relevance. Relevant characteristics include the ability and the willingness to explore tenuous connections between only remotely connectable things. While the vast majority of such attempted connections lead to nothing useful, occasionally one will yield a novel and useful insight into the problem at hand. When this occurs, it is apt to lead to a very important result.

A very important characteristic for a creative person is the ability to visualize constructions in the medium in which he or she works. For a productive author, this is the ability to see sentences, paragraphs, and even outlines of a chapter or a book in the mind's eye before the first word is written. The benefit of possession of this ability is obvious. Where the less able must laboriously write, revise, rewrite, rearrange, delete, and augment, the writer with the ability to visualize more clearly the finished product can perform many of these steps with great speed. The correct word always seems to be available when needed, but the truth probably is that as the author approaches the point at which a word will be needed several possibilities will already have been considered subconsciously and the one the author believes to be best has been selected.

This same sort of thing occurs in all creative occupations. The artist "sees" the final picture, and knows what colors to use (and how to obtain them from the various paints at his or her disposal) and what kind of brush and stroke to use to achieve the end result being sought. The interior decorator must visualize how the room will appear when the selected upholstered furniture, the drapes, and the rugs are brought together. Inability to do this well results in extra expense for the homeowner and loss of business for the decorator.

Some engineering designers work with systems or devices that require the ability to visualize relationships in ways similar to that required of an author. For example, the electronic circuit designer needs to have a clear mental picture of component characteristics so that he or she can quickly consider the effect of choosing one transistor or integrated circuit over another, or of choosing one circuit connection as against another. The common-base, common-emitter, or common-collector transistor amplifier will be chosen without more than a moment's consideration because the properties of each are so well understood. Other designers work with systems or devices that require ability more like that of the artist. That is, they must evaluate the effects of spatial relationships among materials of various shapes and of forces or potentials.

Where does one put the transistor heat sinks to insure adequate ventilation, and where does one put the transformer for the power supply so as to minimize the distortion of the circuit board caused by its weight? Where does one put the hydraulic cylinder in order to obtain the most effective motion of the load? Although these appear to require different abilities, they are much alike and a person who can do one kind of visualization can usually learn without much delay to do the other. Having the basic ability is the key.

A creative person does not hesitate to think unconventionally. However, a truly creative person does not select the unusual approach just because it is different. It must also be elegant. One can easily invent new devices or systems if being unusual is all that is required. For many years, an engineer named Rube Goldberg supported himself by selling his cartoons showing ludicrously complex systems intended to achieve trivial or useless results. The cartoons were extremely creative, but the end results were certainly not to be classed as elegant. The term "Rube Goldberg" was used for a long time to characterize an overly complex design, and the term is still used on occasion today.

Creative people tend to be dissatisfied with the products within their field with which they come in contact. This is a natural consequence of being creative. So many alternatives are evident to creative individuals that they are apt to judge some other design as being more desirable, or at least worthy of being developed.

Lastly, creative persons maintain enthusiasm about their work, often in the face of disappointment. Creating something new requires full involvement of the skills of the inventor. Moreover, the inventor usually takes great pride in his or her own accomplishments. Participation in a half-hearted way will probably produce nothing of great value. If you lose interest quickly, you will not last

ALTERNATIVE DESIGNS AND INVENTIONS

long enough to invent. If you don't care whether you produce or not, you will not. Pride in accomplishment is a vital motivation.

Just as there are relevant characteristics, there are others that are irrelevant. One that seems to be irrelevant is the IQ of the individual, although there is probably a minimum IQ of about 110 below which it appears to be unlikely that a creative person will be found. We have, of course, no way of measuring the IQ of many individuals who are acknowledged to have been great inventors or outstanding scientists simply because they are no longer with us, but the characterization of some of these individuals by those who knew them intimately early in their lives is rather interesting, as the following list will show.

Thomas Edison	Bottom of the class.
Benjamin Franklin	Poor mathematician.
Wilhelm Roentgen	Expelled from school.
Albert Einstein	Mentally slow.

Another characteristic that seems to be irrelevant is age. Innovative individuals tend to exhibit creativity into their 60's and 70's. Press [9] has referred to "the many important contributions made by scientists over the age of 55" and Tsang [10] has pointed out that innovative older scientists and engineers are often unsung heroes, acting as "big brothers" by passing on their ideas to younger colleagues. Tsang cites Einstein's assistance to DeBroglie and to Bose, and Rutherford's to Bohr. A more recent example of a person who remained active and who contributed beyond the age of 80 is Grace Murray Hopper (1906-1992), who was a mathematician and a leader in the computer field.

6.6 IDENTIFYING THE STRATEGY

Almost as soon as you are aware of the need for an invention, you are a candidate for falling into a trap. As mentioned earlier when discussing the research of Hyman and Anderson, a false start can produce a mental block. An example of a false start that led to a solution for something that was not really a problem has been provided by Jacob Rabinow, one of the most prolific inventors in the United States. He used this story in after-dinner talks.

In 1945 he received a waterproof watch as a gift from his wife. That was at a time when good watches and many clocks kept time by the oscillation of a flywheel-spring combination. If the watch ran a bit fast or slow, one moved a rate control lever on the back of the movement to shorten or lengthen the spring in order to change the period of oscillation. Of course, there was no way for the owner to know how far to move the lever to compensate for a given error rate. In order to get the error rate to zero, a trial and error approach was used or one took the watch to a jeweler who had a device called a Watchmaster™, which picked up the sound from the escapement and used it to draw a graph of the actual oscillation period as a function of time

To make the adjustment it was necessary to remove the back of the watch, which required a special tool and which was frequently difficult to replace. Nonetheless, many people used whatever was handy and damaged the case as well as their composure. After repeated openings of the case, Rabinow had become aware of a need. The obvious approach would have been to use his inventive talents to devise an easy-to-open waterproof case. He had the good fortune not to pursue that course. Instead, he followed the strategy of looking for a way to eliminate the need to open the case at all. He recognized that whenever a person corrects the setting of a watch, he or she introduces motion into the mechanism by rotating the stem, and if done periodically, say, as the watch is wound each day, the extent of that motion is a measure of the error rate of the watch. Rabinow's invention was the addition of a simple gear mechanism to adjust the position of the rate control lever by the movement of the stem when setting the watch. There were, to be sure, some complications, such as the effect on the rate control lever when switching to or from daylight saving time or changing time zones. These two were solved in due time.

The invention never sold well on watches for marketing reasons which Rabinow explained in a humorous way, but one reason was that most watches did not need adjustment. However, there was another market in which it sold very well. Clocks used in automobiles were notorious for being poor time-keepers, and getting access to the rate control lever was impossible without removing the clock from the dashboard. Most of these clocks were spring-driven (the spring being wound by an electric actuator when it had run down a certain amount) and incorporated the same flywheel-spring mechanism. The addition of Rabinow's mechanism converted these clocks from useless ornaments to useful clocks after a series of resettings. Today, of course, most clocks and watches use quartz crystals and electronic circuits to establish their rate, and have high accuracy.

Rabinow, whose many inventions cut across a spectrum of disciplines, gave another example that is very instructive. Screws intended to be driven by a screwdriver or a similar device originally had a simple slot for the conventional screwdriver, one having a flat blade. Philips screw heads followed at a later time, having a head that could in some cases be driven by a conventional screwdriver, but which were intended to be driven by a screwdriver whose shaft terminated in a cross (+) shape. This had the advantage that greater torque could be applied to the screw. This has been followed by screw heads having an indentation shaped in the form of a square, a hexagon, or even a six-pointed star. The difficulty with all of these, as far as preventing someone not having the proper tool from driving or removing these screws, was that a conventional screwdriver having the proper width of blade could be used, even though it was not the proper tool. Another screwhead variation that is common is the flat head having two indentations diametrically opposite to each other. The tool to drive

ALTERNATIVE DESIGNS AND INVENTIONS

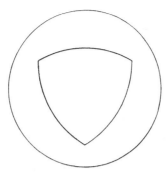

Figure 6.7 Rabinow's proposal for a screw head designed to defeat the possibility of being driven by a conventional screwdriver blade.

or remove such a screw has a U-shape at the end of the shaft, with the two ends of the U having the correct size and the proper spacing. The improvement in security is substantial, because few people have the proper tool, but a punch and hammer can be used to put impact torque on the screw head, and some Swiss army knives have a blade that can be used in the same way as the correct tool.

What other kind of screwhead indentation and driving tool can be devised? Can a new design simultaneously provide substantial torque capability and defeat the person who wants to use a conventional screwdriver blade?

Rabinow has proposed another variation that seems to provide an affirmative answer to these questions. It is shown diagrammatically in Figure 6.7. The screw head has a triangular depression, with the sides in the shape of three arcs, each point of the "triangle" being the center of the arc for the opposite side. A straight screwdriver blade, regardless of the width, cannot be used to move this screw because the blade will pivot at each corner and slide over the opposite curved surface. Rabinow admits that he does not know whether this idea is new, or whether it has been patented. (In any case, it is no longer patentable.)

The advice that needs to be given with more emphasis than anything else you have read in this chapter is to stop very early in your work and ask yourself how many different strategies are available to you. This ought to become an automatic step in analysis of problems, in design projects, and in the challenge of inventing. Perhaps one reason more inventions are not made is that most people concentrate on improving what is already there (the watch case, for example), only to find that not much can be done. They have gone so far down the wrong track that they have created a mental block. They are trying to devise better ways of opening the watch case than of adjusting the rate lever.

6.7 HOW TO IMPROVE YOUR ABILITY TO INVENT

What can you do to improve your ability to invent? The discussion earlier of theories of creativity, results of psychological research, and examples of inventions was presented to provide a background to make it a bit easier to answer this question. The suggestions given below should not be taken as a recipe to consult when you see that an invention is needed. Instead, they are intended to outline a way of organizing your life as an engineer, one that you will find useful even if you do not become directly involved in product design or invention. Many of these suggestions also require much more than a passing effort. Some of these suggestions will be seen to be formalizations of what has already been said, but others reflect personal experience on the part of both of the authors and have resulted from reflection on the question with which this paragraph began.

6.7.1 Suggestions for the Would-Be Inventor

1. Develop a desire to invent. This desire can be enhanced by reading about inventors and important engineering projects. Trade magazines and periodicals often have articles about inventors and inventions of note. Interesting accounts of the accomplishments of inventors can also be found in articles in business magazines and even newspapers. Finding such material does not mean that you must read these publications completely; they can be scanned rather quickly for items of interest. Once you have found them, make use of Chester Carlson's invention to make copies, and file them. No suggestions will be made as to how to organize those files because each person's needs are different, but being able to leaf through a file when you realize that you have seen something of interest in connection with a design project is far more efficient than trying to go back through all of the publications you scan on a regular basis.
2. Channel your mental energy into the creative process. One arena that many engineers ignore but in which their training in organized thinking can provide solutions is in the political and social spheres, but beware of allowing the rhetoric accompanying most discussions of social problems to divert you from the real problems and dissipate your energies uselessly. One area in which engineers can make substantial contributions because of their habit of orderly thinking is in civic and charitable activities. Engagement in these spheres can provide many opportunities to exercise their creative abilities, and those abilities are usually welcomed by others.
3. Become familiar with inventions by reviewing the *Patent Gazette* on a regular basis. If your employer maintains a file of current patents, study them. (Remember what Black did very early in his career.) Keeping current on patents will make you familiar with the state of the art and the kinds of

ALTERNATIVE DESIGNS AND INVENTIONS

problems others in the industry are addressing. You will thereby become more conscious of invention as an activity important to your work.

4. Be on the lookout for unusual physical phenomena or ones that you have never learned about. Carlson built his work in xerography on the earlier work done by Paul Selenyi on the use of electrical charges to produce images. When the first laser was built, it was described as "a solution looking for a problem," but the number of problems it has solved is still increasing. The first transistor came about because scientists at Bell Telephone Laboratories were investigating the surface states of germanium, and were probing the surface of a point-contact diode with a second point-contact electrode so as to better understand the surface state. On the other hand, some phenomena have been observed, but their significance disregarded. An example is Edison's failure to invent the vacuum tube diode in spite of having made observations as to the effect of a second electrode in his early lamp bulbs. (That disregard of an unexpected phenomenon may have been the result of a mental block. Edison was committed to direct current, and diodes are used to rectify alternating current.) You will find reports of unusual phenomena in trade and professional journals and in the popular press. There is a list of physical phenomena in a book by Alley and Hix [11]. The examples cited earlier of bisociation are illustrative of the desirability of knowing about and having some understanding of phenomena outside your own immediate field. When you learn about a physical phenomenon previously unknown to you, make notes or a copy of the relevant information, and place in one of your files.

5. You must seek to develop a true insight into any phenomenon that seems to be of value in your work. This does not mean that you understand the mathematical formulations with which the phenomenon may be described, but a true understanding of its operation. For most inventors, it is more important to be able to visualize how physical events occur than to be able to express them mathematically. Nikola Tesla, the inventor of the induction motor, apparently visualized the entire concept of the rotating magnetic field without having any recourse to even the simplest mathematics. Thomas Commerford Martin, while describing Tesla's contributions to rotating machines in Reference 12, shows only one set of equations, and a review of Tesla's patents on the induction motor shows none. Yet his contribution to the development of rotating machines was so great that virtually no new induction motors have been devised since the 1880's.

6. Don't be too proud to acknowledge that there are areas of engineering and science with which you should be familiar but which were not included in your formal training, and take steps to overcome the deficiencies. Moreover, if you do not make a planned effort to keep abreast of your own field, you will discover that you have deficiencies in areas in which you believe yourself to be competent. Short courses, seminars, and self-study

are all important. Read the technical journals in your own field, and be familiar with those in other fields!
7. Improve your hands-on working ability. Black and Carlson were actively engaged in experimentation, building models, and so forth. Such skills are valuable assets of the inventor. The excitement of trying a new idea is strong motivation for early pursuit of a new idea. If you have to wait while the model you need is scheduled through the company's model shop, you will be less likely to pursue it than if you can produce the model—or most of it—yourself. Moreover, the hands-on experimenter receives instant and unfiltered feedback from an experiment and can frequently see quickly how to make improvements; an experiment in which the model is built by others and then tested by a technician who records the results but does not really understand the purpose of the test may fail, not because the idea is basically defective, but because there are minor problems that need to be resolved. One of the colleagues of the authors was a very prolific inventor, and it was common knowledge that when he wanted something built to try out a new idea, he would not waste time; he would go and build it!
8. If you are going to do some of the experimental work yourself, you must take the necessary steps to acquire the facilities to try out your ideas. Tools, model-building materials, test equipment, and measuring instruments are needed to pursue an idea. You do not need research quality equipment, and it need not even be the easiest to use, because you are not going to build really complicated devices. Simple models are often sufficient to test an idea to see if you are on the right track.
9. If you want to be an inventor, you must immerse yourself in a climate of creativity. Work for a company where new ideas are welcomed and appreciated. It is difficult to become interested in self-development if is clear that the effort will not gain respect, status, and appropriate compensation in your company.
10. Recognize that opportunities vary depending on the maturity of the technology. Newer fields usually present the better opportunity to do something novel, but new approaches to old technology can present great opportunities. The field of electric machinery was for many years rather stagnant, but forward-looking engineers, beginning about 1970, began to develop solid-state circuits to control these machines. The field today is one of the most dynamic in engineering.
11. If you are good at design or have a desire to be an inventor, keep in mind that there are jobs in industry that are not conducive to those objectives. An engineering manager will direct his or her attention to budgetary matters, personnel problems, equipment needs, reports, and so forth. Although theoretically a person in that position can also be a designer or an inventor, the probability is very small, certainly far less than for the engineer primarily involved with design of new products, redesign of older ones, or having responsibility for manufacture. The offer of a management position

ALTERNATIVE DESIGNS AND INVENTIONS

to a younger engineer needs to be considered very carefully both by the engineer and by more senior management. The engineer should keep in mind that, even if a management position is an eventual goal, a delay until significant engineering accomplishments have been made may give one credentials for a much better management position. Moreover, if the engineer makes an early switch to management and later concludes that he or she is not really interested in that kind of a career, a switch back to engineering will be far more difficult because one cannot keep fully abreast of engineering developments while in management.

12. As soon as you identify a need important enough for you to initiate an invention effort, start a notebook in which to record every idea. If the idea is discarded, make an entry giving the reason. Whatever the reason, it may not be valid at some later date due to a change in technology or in the economic situation, or putting two of your previous ideas together may result in an elegant solution. This will not occur, of course, unless you review the notebook from time to time. There is another advantage: Unless the need is satisfied by some other means, you may find that a mental block that led you to discard an idea earlier may no longer exist, and you see a way to use the idea in a different way to reach the objective. Both Carlson and Black kept extensive records, and both worked for a long time in order to achieve their objectives. It would have been impossible to remember details of early experiments, and those results might be important later on.

13. Set aside a portion of each work day or work week—depending on your individual situation—to think about how you can improve your company's products. If you know in what way the products are less than ideal, you are in a better position to make suggestions for improvements. Black saw that it was necessary to address the problem of 3,000 channels with 4,000 miles of transmission while his superiors were still worrying about how to solve the problem of three channels over 1,000 miles. He clearly understood the problem of future telephone communication far better than those who were his superiors because he had studied the problem more carefully than they.

14. Above all, do not expect instant results. Black worked on the telephone amplifier problem for several years, Tesla was thinking about the problem of alternating current motors even while he was working for telephone companies and while he was working for Edison (who was committed to direct current), and Carlson's patience seemed inexhaustible. If the need is important, be prepared to expend a considerable amount of time and effort on a solution.

6.8 METHODS TO STIMULATE INVENTION

Is it possible to stimulate the creative process? Because the process is not and probably cannot be well defined, many people think that creativity is an innate characteristic of some people, and that others cannot develop it. One of our students said some years ago, "If there are those so creative that they do not need procedures to follow, fine—let them do it their way. For others, a way to guide one's thoughts might be necessary." That student's attitude is supported by an increasing number of investigators, as References 1, 4, 5, and 7 show.

Structured methods to aid creativity may help principally by identifying those intersecting planes of thought remotely associated with the plane of interest and thereby promote bisociation. It is also probable that such activity will promote one's interest and emotional involvement in the needed invention to the point that the preconscious activity will be urged in that direction. With these ideas in mind, we offer only a few possible procedures out of many that have been suggested.

6.8.1 A General Method

Consider an approach that is useful in itself but that may be readily changed to accommodate other methods. This is called the general method. The steps are:

1. Define the problem.
2. Gather information.
3. Repeatedly review the elements of the problem.
4. Try for a solution that works even if it is not as elegant as you hoped for.
5. Try for an unusual and elegant solution even if it does not satisfy all specifications or work perfectly.
6. Repeat steps 1 through 5 several times.
7. Direct your attention away from the problem for some time—weeks, if possible. This "vacation" is intended to provide time for your preconscious mental activity to organize some solutions or partial solutions to the problem.
8. Return to your problem when you feel enthusiastic about working on it. Having such a feeling is a strong indication that you are ready for some creative insight.

If this process seems vaguely familiar, it is because it tends to parallel that given in Chapter 1, Figure 1.2, for a design procedure.

6.8.2 Adaptation

A very important method that can be used in conjunction with the general procedure to develop the solutions required in steps 4 and 5 is called the adaptation method. In this method, a solution of a problem in one field is applied to a similar problem in another field. A modern example of adaptation is

ALTERNATIVE DESIGNS AND INVENTIONS 199

the CD-ROM. The compact disc was first thought of and developed as a superior way to provide recordings of voice and music. Because of the fact that the information on the disc is in digital form and because of the immense amount of data that can be placed on one disc, it was a natural step for someone to conceive the idea of using a compact disc to provide read-only storage for other kinds of information, such as that needed for computers. The technology that was already available to do one task was easily adapted to another, without the necessity of developing an entirely new functional relationship.

In most situations where adaptation is used, the significant step is to recognize the functional relationship. Consider how one might design a clip to be mounted in a hole in a piece of sheet metal, the clip being intended to hold small tubing or a cable for a thermostat, for example. The part of the clip that is intended to retain the tubing or cable in position is straightforward, but the part that holds the clip in the hole is not. Here an adaptation of the principle that keeps the human foot in a shoe (a loafer, for example) can be used to hold the clip in the hole in the sheet metal, as Figure 6.8 shows. This idea of adapting from the biological world for engineering purposes is important enough that it has spawned research into an area referred to as bionics, which is the study of living things as engineering prototypes. For example, intuition tells us that the bow of a ship should be a knife edge, or in the case of a submarine, be a cone. But the bow of submarines today is a rounded surface, a design stemming from investigations into the way dolphins and other sea creatures are able to swim at high speeds. The navigation system of bees has been found to depend on the ability of the bee to detect polarized light and on the fact that polarization of light is greater for light coming from a northerly direction than from other directions. The principle has been adapted for use in a nonmagnetic compass for use at high latitudes.

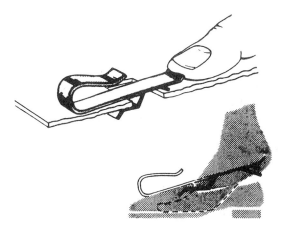

Figure 6.8 Mounting a clip on sheet metal.

6.8.3 Analogs and Duals

Two systems of different technologies that perform in a similar way under comparable circumstances are said to be analogs. For the systems to be analogs of one another, the defining system equations must be of the same form, with the variables of the two technologies positioned in a corresponding way. Any mechanical system consisting of a mass, spring, and damper can be represented by the equation

$$f = M\frac{dv}{dt} + Dv + K\int v\,dt$$

where f = the applied force,
M = the mass,
D = the damping constant,
K = the spring constant,
v = the velocity of the mass, and
t = time,

all in a consistent set of units. The analog in an electrical system consisting of a series combination of a resistance, inductance, and capacitance has the defining equation

$$v = L\frac{di}{dt} + Ri + \frac{1}{C}\int i\,dt$$

where v = the applied voltage,
L = the inductance,
R = the resistance,
C = the capacitance,
i = the current, and
t = time,

again all the parameters and variables being in a consistent set of units. That is, for a given force and voltage—say a ramp function that runs up from 0 to 100 newtons or volts—the velocity and the current will have exactly the same variations with respect to time. The analogous equations are said to be force-voltage analogs from the driving forces in each case. Analogs are easily constructed in other systems as well, such as fluid, thermal, acoustic, electric, and magnetic fields.

Two systems within the same technology are said to be duals of each other if a relationship similar to that of analogs can be established. For example, the equation

ALTERNATIVE DESIGNS AND INVENTIONS

$$i = C\frac{dv}{dt} + \frac{v}{R} + \frac{1}{L}\int vdt$$

has the same form as either of the equations above, but is the defining equation for a parallel circuit consisting of a capacitance, a resistance, and an inductance. The series circuit used originally and the parallel circuit of this equation are duals; if the current in this case follows the function that the voltage followed in the first case, then the voltage in the parallel circuit will have the same form as the current in the series circuit.

What can the concepts of analogs and duals do for the designer or the inventor? First of all, it provides a way to transform a problem into a form that may be more familiar. If I am an electrical engineer and am faced with a problem that includes mechanical components, I may gain a better insight if I convert the mechanical components into their electrical analogs because I am more comfortable and more knowledgeable thinking about electric circuits than I am about mechanical systems. On the other hand, the mechanical engineer who is looking at a series electric circuit having negligible resistance will more easily understand that the circuit will exhibit oscillatory behavior if he or she first converts the circuit into its mechanical analog, which would be a spring-mass system with virtually no damping.

Secondly, the concept of analogs can be used to generate alternatives to a proposed design by using a different technology in which the resulting device may be far more effective. For example, Jacob Rabinow's invention of the magnetic clutch came about because he understood the value of analogs. Early in Rabinow's career, while he was an employee of the National Bureau of Standards (now the National Institute for Science and Technology), he was working on a clutch that used electrostatic attraction. The clutch was built with two closely spaced plates with a suspension of dielectric material, such as starch or limestone particles in oil, between the plates. If a high direct voltage was applied between the plates, the particles suspended in the oil would chain up into a kind of conglomerate that would permit torque to be transmitted from one clutch plate to the other. The inventor, a man named Winslow, had submitted a patent application, and a patent had been issued after Winslow demonstrated to the examiner that the idea did indeed work. The problem with the clutch, however, was that only modest torques could be transmitted; as the torque was increased, the limit on shear force of the oil-dielectric suspension was exceeded and the clutch went into a slipping mode.

At the time, the National Bureau of Standards had some consultant firms, and one of these firms brought the clutch to Rabinow's group. While Rabinow was trying to improve the clutch, he suddenly realized that if electrostatic attraction could produce a clutch action, then its analog, electromagnetic attraction, should be able to do so also. Moreover, it should be much better, because the permittivity of most materials is very low, rarely exceeding a relative permittivity of 10. On the other hand, the relative permeability of iron is about 2000 and other magnetic materials have even higher values. Substitution of a magnetic field for an electric field with the change of particles in the oil from insulator-type materials to magnetic materials should therefore allow an increase in the shear force in the clutch of about three orders of magnitude. There were other advantages that accrued as a result of this change, among them being the fact that the electric supply to energize the clutch is low voltage and therefore much safer than the high voltage required for the electrostatic counterpart. The magnetic clutch was an instant success commercially.

6.8.4 Area Thinking

The objective of this approach is usually to improve an existing product by concentrating on one of its characteristics that is important to the consumer. Hence the areas of cost, performance, function, appearance, safety, reliability, repairability, and so on can be considered. One does not use this list in the manner of a laundry list, running down it and checking off each item in turn. Rather, the list should be used merely to remind the engineer of the various aspects of a product that may be of importance to the consumer, and attention should be concentrated on only one item in the list at a time, and then for an extended period. If the company has found that one of its products is being returned at an unusually high rate, one should obviously concentrate on the reliability question. On the other hand, if your competitor seems to be taking some of your market, you may find that your selling price is too high and the concentration should be on cost reduction factors.

This approach may also be used to invent new products. Most residences have running water available at pressures in the range from 0.10 to 0.17 MPa (15 to 25 psi). When domestic clothes washers with more sophistication than the scrubboard were first devised, they were driven by a variety of power sources, one of which was waterpower. This was obviously a system that was extremely wasteful of water, and was never very popular. However, a device that uses little water and that is used today is the water-powered toothbrush. Another device produced by a manufacturer in the Netherlands is a means for assisting disabled individuals in getting into and out of bathtubs. This consists of a chair that can be raised by waterpower, as shown in Figure 6.9. This does not require much water, and the spent water can easily be discharged into the tub. Such a system could also be built very easily using an electric motor, but the waterpower approach has the advantage that there is no possibility of electric shock. Although disadvantages to water-powered devices are readily apparent, such as the need to modify plumbing, there may be other devices that could be used in the home and in industry that might advantageously utilize water power with low flow rates.

ALTERNATIVE DESIGNS AND INVENTIONS

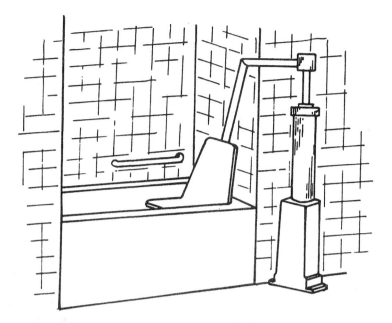

Figure 6.9 Water-powered chair for easy bathtub entry and exit.

6.8.5 Brainstorming

Brainstorming was a very popular activity shortly after World War II. The method requires a group of people who are familiar with the general area of a problem, but none of them need be an expert. The problem is stated by the leader of the group, and the participants say whatever comes to mind. The rules for a brainstorming session are very simple:

1. You are not allowed to judge the value of any idea that is presented. They are simply listed on a flipchart or a chalkboard, and evaluated later.
2. The more unconventional the idea the better.
3. The more ideas the better.

One advantage to this method is that many ideas are presented, and one or more of them may trigger a bisociation that would otherwise never occur. The disadvantage is that it seems to have a low success rate in the engineering setting, although it has shown more success in generating advertising ideas or possibilities in the sales or management areas. Kivenson (Reference 1, pages 93-95) has an interesting section on how to conduct such sessions.

6.8.6 Involvement

The method of involvement seems to be most useful with mechanical devices. It is assumed that the general method outlined earlier will be followed down to steps 4 and 5. At that point you must visualize yourself as being part of the mechanism. For example, suppose the problem is to develop a way of cleaning leaves out the gutters on a house. If you assume that you are a leaf, what mechanism can be used to get you out of the gutter?

An obvious one is for someone to climb a ladder and lift you out. Another possibility is for someone to bring buckets of water or a garden hose up and sluice you down the downspout. That will work only if there is no strainer at the downspout, and even if there is not, there is still the possibility of getting caught in the downspout and stopping it up. Another possibility is to use a strong stream of air to blow you out. The stream of air can be provided by the motor and blower on a shop vacuum sweeper, such as is found in many residential garages, and the idea has the further advantage that on one-story houses it may be possible to do the entire job from the ground using extension tubes and angle fittings to get the air stream into the gutter. (Tubes and fittings to let the homeowner do exactly this are now available.)

Maybe the statement of the problem was wrong to begin with. Still thinking like a leaf, "Why was I ever permitted into the gutter in the first place? Is it possible to devise some kind of guard that will keep me out but still permit the water to get in?" (Guards of this kind are also available.)

6.8.7 Functional Synthesis

This method was called "orderly creative inventing" when it first appeared in print in 1957 [13]. It was identified by introspection after the first invention by one of the authors (WHM). Discussions with G. T. Brown of National Cash Register Company of Dayton, Ohio, who had similar ideas, led to the development of the method. (National Cash Register has since undergone name changes, but is now back to the familiar trademark NCR.)

The method can best be applied by individual effort. The most important part of the procedure is the act of describing a sought-after device in terms of functional requirements rather than by descriptive adjectives. There are six steps in the method:

1. Define the problem.
2. Gather information.
3. Divide the system into subunits.
4. Describe each subunit by a complete list of its functional requirements.
5. List all the ways the functional requirements of each subunit can be realized. Each of these is a partial solution.
6. Study all combinations of partial solutions.

ALTERNATIVE DESIGNS AND INVENTIONS

Notice that the first two steps are identical to the general method with which this section began. If you have determined the specifications, made a feasibility study, and searched the patents, you have what is at least a tentative definition of the problem and you have gathered information.

The third step, division of the system into subunits, is a necessary step because in any device or system there is usually a key part or subunit whose characteristics influence the other parts to a major extent. This element should be separated from the others and treated first. Then proceed to another subunit. If you wish to develop a completely new method of transporting a small number of people and a few relatively small things, such as luggage, your subunits could be the source of motive power, a compartment in which the people will ride, and another compartment in which to place the luggage. Another subunit is the means of coupling the vehicle to whatever the source of power will react against. This all sounds like an automobile, but it could as well be a small airplane or a boat. In any case, which of the subunits is the most important?

The fourth step, describing the subunits by functional requirements, is the key to the success of the procedure. The subunit *must* be described by what it must accomplish and never in terms of a possible realization. If this is skillfully done, you free yourself from thinking in terms of how the problem has been solved in the past.

The fifth step is to list, as briefly as possible, all the ways in which the functional requirements of step 4 can be satisfied. If the subunit you are considering is the means of coupling the vehicle to whatever the source of power will react against and if the vehicle is to traverse water, one tends almost immediately to think of a propeller to push against the water and thrust the boat forward (or reverse, if necessary). But propellers are not the only way. You will probably want to disregard sternwheels or sidewheels, but what about jets of water from pumps inside the hull? What about the equivalent of an airplane propeller? Can you lift the vehicle on a pad of air and force it in a preferred direction by venting more air in one direction than in another? Is it possible to use magnetohydrodynamic effects instead of conventional pumps to provide jets of water? Could this effect be made to work in fresh water as well as in salt water? The more alternatives that can be listed, the greater the chance for success. It is at this point that brainstorming may be extremely useful.

In the sixth step one considers all the combinations of the ways that been listed to realize each subunit. This results in many solutions, some of which will be old, and some of which will be ridiculous. For example, one of the sources of motive power would probably be a gas turbine, and the combination of a gas turbine with sidewheels would evoke a laugh but nothing else. Out of the entire list of solutions, however, one hopes to find at least one that is novel, useful, nonobvious, and physically realizable. These are the main requirements of an invention.

This method is a very natural one to use and, in fact, it is often used but is not often identified as a structured method. Engineers are taught to think in terms of functions that must be accomplished, and this method reinforces that approach. Its main advantage is that, properly done, it strips away the embodiments of the past that provided the enumerated functions and challenges the engineer to find new ways to provide those functions. Another advantage of this method, especially in a large organization, is that it may be possible to simulate all of the combinations using computer programs and thus try all of them out very easily. Kivenson [1] refers to the last four steps in this method as the matrix method, and gives an example on pages 78 to 80 of its application to solar heaters.

A major advantage to this method is that you have the ability to use existing devices as a starting point. In fact, it is almost impossible not to start with something already in existence. However, there is a potential roadblock in that you may fail to provide the most general description possible of the device, that is, a description of the functions. Nevertheless, starting with an existing device frequently enables you to list the subunits to be considered. Figure 6.10 shows a simple embodiment of a can opener. Such can openers have been available for almost two hundred years, and they have only two essential parts, the handle and the cutter. The function of the cutter is solely to separate metal but the handle has two functions. One of these is to enable the user to position the opener and the other is to enable the user to apply power. If one begins with these functions for a can opener, step 5 in the procedure will yield the possible ways to provide these functions, as is shown in Table 6.1.

The lists in Table 6.1 are not exhaustive. Separate brainstorming sessions on how to realize each function would add to the lists of alternatives. Once this table is completed, all possible combinations are to be evaluated. There are 168 such combinations in this case. Some of these will probably be eliminated because of company policies. For example, your company may simply not wish to consider a hydraulic actuator because it has no experience with such devices and does not wish to develop the expertise. Other combinations in the table are definite possibilities. Combination 2-1-3 describes an opener that tears metal by manual power with the opener built on the can. Beverage cans are made in this

Figure 6.10 A simple manual can opener.

ALTERNATIVE DESIGNS AND INVENTIONS

Table 6.1 Functional requirements of a can opener.

Part	Function	Realization
Subunit 1	Separate metal	1. Shearing 2. Tearing 3. Fatigue 4. Melting 5. Drawing thin 6. Chemical erosion 7. Mechanical erosion
Subunit 2	1. Supply power	1. Hand 2. Electric motor 3. Hot wire 4. Hydraulic motor 5. Flame 6. Chemical reaction 7. Mechanical vibration 8. Laser
	2. Position	1. Bring can to opener 2. Bring opener to can 3. Have opener built on can

fashion, as are the cans for some meat spreads. Other combinations will describe openers that are already in use. Some will describe openers that are impossible or impractical. There may be other openers which are not only practical but may even be less expensive than openers now on the market. There are some on the list that are not practical today but which may be in the future as technology changes.

One of the frustrations of inventors is the lack of necessary supporting technology or the fact that the invention on which they have spent so much time and effort is not needed at the time. One such example is the entire area of facsimiles (FAX's). In the 1920's, George Finch, later a Captain in the U.S. Navy, demonstrated the necessary equipment to transmit FAX's, but there was no interest. The equipment was "reinvented" twenty or thirty years later, with the same result. The concept came to the surface again in the 1980's, and FAX machines are now part of everyday life in the business world. A similar example is that of remote-controlled vehicles, demonstrated by Nikola Tesla in 1898, but rejected by the military and unused for at least the next twenty years.

The method of functional synthesis or orderly creative thinking has been described as being followed in a sequential manner, but do not assume that once a step is completed you need no longer pay any attention to it. In the can opener example, combination 4-5-1 describes an opener that opens the can by use of heat, presumably a very high temperature tightly confined flame, that is, a small version of a plumber's propane torch. Is such a device available? If so, how would one incorporate it into an opener that would be inherently safe to operate?

6.9 SUMMARY

Does the list of activities in Section 6.7.1 overwhelm you? Why must one do all of that? Look at Carlson, he didn't seem to do anything other than stick to the single-minded idea of reproducing images by using electric charge transfer.

The suggestions offered throughout this chapter are meant to be extensive. While they do not guarantee invention, they certainly make it more likely. The suggestions, if followed, will assist you in self-development and help you in your career. However, do not become a fanatic about inventing or assume that there must be a better way to do everything. There are many products that are already elegant designs. One of the important reasons for following a program such as that suggested in this chapter is to prepare yourself for those times in your professional career where a truly significant opportunity arises where a need must be satisfied. No one can predict when this will happen, nor how important it will be. It may be something as important as inventing a device that will save many lives, or it may be something as important as helping children in their education. Invent to make contributions to society; the inventor who is inventing simply to invent will have so narrow a viewpoint that most of his or her work will not fill a need. Inventing should be a fun exercise.

If you are interested in reading about inventors, there are many books one can find in any library about the more famous inventors. Kivenson [1] gives a few case histories. Kock [14] has interesting tales about the development of the transistor, lasers, waveguides, and so forth, as well as some thoughts about the traits of creative individuals and how to foster creativity in oneself and in one's children.

REFERENCES

1. Kivenson, Gilbert. *The Art and Science of Inventing,* Second Edition, Van Nostrand Reinhold Company, New York, 1982.
2. Xerox Corporation, *Interim Report to Stockholders,* September 30, 1978.
3. Black, H. S. Inventing the negative feedback amplifier, *IEEE Spectrum,* 14 (12), 1977, pp. 54-60.
4. Koestler, A. *The Act of Creation,* Danube Edition, Macmillan, London, 1969.

5. Kestin, J. Creativity in teaching and learning, *American Scientist*, 58, 1970, pp. 250-257.
6. Braun, Hans-Joachim. Advanced weaponry of the stars, *Invention & Technology*, 12(4), Spring 1997.
7. Kubie, L. S. Blocks to creativity, *International Science and Technology*, 1965, 42: 69-78.
8. Hyman, R., and B. Anderson. Solving problems, *International Science and Technology*, 1965, 45: 36-41.
9. Press, F. Age and tenure, *Science (Letters)*, October 17, 1975, p. 219.
10. Tsang, T. Creativity vs. age, *Physics Today (Letters)*, August 1974, p. 9.
11. Alley, R. P., and C. F. Hix, Jr. *Physical Laws and Effects*, John Wiley and Sons, New York, 1958.
12. Martin, Thomas Commerford. *The Inventions, Researches and Writings of Nikola Tesla*, Barnes & Noble Books, New York, 1893, reprinted 1992.
13. Middendorf, W. H., and G. T. Brown. Orderly creative inventing, *Electrical Engineering*, 1957, 76(10): 866-869.
14. Kock, Winston E. *The Creative Engineer—The Art of Inventing*, Plenum Press, New York, 1978.

REVIEW AND DISCUSSION

1. What change of strategy did Harold Black make in his attempt to reduce amplifier distortion? Was it a key step in his later successful development of the feedback amplifier?
2. The "distortion" amplifier of Figure 6.1(a) must certainly have distortion of its own. Given this information and assuming that the amplifiers are absolutely identical, can you see any way by which the output of that circuit could be absolutely distortion-free? Give reasons for your answer. It may be helpful in thinking about this if you assume an internal (distortion) generator in each of the amplifiers. Would there still be an overall reduction in the distortion?
3. From the examples of inventions given in this chapter, select two or three and find other references describing those inventions and the trials and tribulations of the inventors. List for each of these as many lessons as you can deduce from those accounts.
4. a) Find three examples of elegant inventions or designs in your home, office, or plant and study them to determine why they are superior. These will probably not be complex devices, but more likely simple ones, such as airtight food containers that are easily opened and closed. b) Now find three examples of devices in which you see deficiencies. What could be done to eliminate each of those deficiencies?

5. Every day each of us is likely to say "Why doesn't someone make a _____ to do _____?" Can you identify three or four needs from your own personal experience at home, at school, or at work?
6. Make a list of the characteristics of creative people given in Section 6.5, running your list down one side of a sheet of paper. Opposite each entry indicate your perception of whether you have the characteristic or not.
7. Of the suggestions in Section 6.7.1 on improving creative ability, identify those which seem most important for you to pursue.
8. Make a first try at developing another method to stimulate invention that you believe would be beneficial to you.
9. The water-driven hoist for disabled persons described briefly in Section 6.8.4 appears to be a useful device. Without getting into a detailed design, determine reasonable dimensions for the cylinder needed to lift the user. That is, what stroke would be required, and what piston area would be needed? Do your results make it appear that this would be a device that might be readily accepted, or can you envision reasons that the average homeowner might have negative thoughts about having the device installed in the bathroom?
10. For many years, blowers in residential furnaces were driven by small general-purpose induction motors, sometimes with the capability of running at either of two speeds, using a V-belt to couple the motor and the squirrel-cage blower rotor. In recent years, a definite-purpose induction motor has been used, placed inside the squirrel-cage rotor so that the rotor is driven directly, and having two- and even three-speed capability. Make lists of all the advantages and disadvantages you can think of for each system. Which system is more advantageous for the manufacturer? Which is better for the homeowner?

PRACTICE PROJECTS

1. Set aside the design of the product that you have chosen as a project (as it stands at the moment) and try to develop at least three alternative designs.
2. Pick three products with which you are most familiar and list possible improvements for each. Try to develop designs that incorporate these improvements.
3. After doing suitable library research on the subject of creativity, write a term paper on the subject.
4. Find a large mail-order catalog and select two pages at random. Select one item from each page and try to combine those items into a new item that is both useful and novel.
5. Infusion of fluids into the human body is part of the treatment for various diseases or deficiencies. The infusion process is generally quite long, ranging up to several hours duration, and the infusion rate is very important

ALTERNATIVE DESIGNS AND INVENTIONS

and must be controllable. The source of the fluid is a plastic bag or bottle; the fluid flows through disposable and deformable plastic tubing, reaching a vein of the patient through a needle. To control the rate, a pump is usually used that can be set to deliver a fixed number of cubic centimeters per hour. The delivery system must be closed; that is, the pump mechanism must be made so that it is not possible for fluid contamination to occur. Devise one or two methods for effecting the pumping action.

7

MODELS

Experimenting with models seems to afford a ready means of investigating and determining beforehand the effects of any proposed estuary or harbor works; a means, after what I have seen, I should feel it madness to neglect before entering upon any costly undertaking.

                                                            Osborne Reynolds

Every design eventually reaches a point at which decisions must be made about the material to use, the dimensions of the parts, the components of a system, the spatial relationships among the parts and components, and their interconnections. Once this point has been reached, the designer begins to represent the product by a series of models. The first models are likely to be crude and incomplete, but as the designer learns from these models later versions will be more representative of the final product.

Engineers usually design by creating models that make it easy to visualize the product, to analyze it, to rearrange it, to exercise it mathematically, and finally, when a physical model is created, to test it under realistic operating conditions. In the past, many products were developed by trial and error using physical models. The present accelerated development of technology and the ever-present possibility of legal action relating to product liability make it mandatory that designers use techniques that enable them to design products that are energy-efficient, that conserve material, and that are safe when they are first placed on the market. Fortunately, better analytical methods and the availability of better means for doing the necessary calculations make it possible for the engineer to meet these challenges when executing a design, and more sophisticated test equipment and methods makes it possible to learn more from the physical models created in the design process. Even with the improved tools available today, however, recalls of new models still occur because of details that were overlooked or thought to be unimportant, or because testing procedures did not fully duplicate the way the products are used in the field.

MODELS

The term "model" frequently carries the connotation of a physical entity, such as a model of an airplane or a car. A broader definition is used in engineering, a model being *any* representation of the proposed system or device that contains enough information to be useful in making design decisions. That is, models are not restricted to being only physical replicas of the product, although such replicas can certainly be very useful. There are also sketches, drawings, block diagrams, electrical network diagrams, and mathematical models. In this chapter, we will look at these possibilities, and will also consider physical models in more detail.

7.1 SKETCHES AND DRAWINGS

In most of your courses in engineering, you have seen a sketch used in the explanation of some engineering principle, and you may have taken a course in sketching and drawing. In what follows, it is assumed that you understand orthographic projection drawings (plan, front elevation, side elevation) and that you have had some experience in making such drawings and have some skill in making two- or three-dimensional sketches.

Sketches are important to the creative process. Typically, the making of a freehand sketch is the first act after an engineer conceives a new design concept. [The ability to make sketches quickly and with reasonable accuracy is an ability that can be developed through practice.] When this is done, the idea is immediately tested for obvious flaws, it is easy to give it more thought, and it may confirm the idea as being adequate for the next step in the development process. Those who are skillful at sketching will find that this sequence can occur very rapidly. Since more ideas are likely to be rejected than to be found acceptable, valuable time will not be expended on ideas that are not suitable. Another effect can also occur as one becomes better at making sketches: The ability to visualize objects in finer detail improves. This in turn enhances the ability to detect flawed ideas without a sketch being made, thus improving the efficiency of the process. However, it is best to make a sketch in any case, even though you foresee that it will disclose some flaw in the idea, because the sketch itself may suggest still other ideas.

How does one use a drawing as a model during the design process? Because designing involves determination of spatial relationships, sketches or drawings that show these relationships in a clear way can certainly aid the designer in making the next decision. There is no better way to determine whether parts will fit, whether tolerances on dimensions will cause difficulty, or whether metal parts of opposite electric potential are dangerously close than by reviewing carefully made drawings of the proposed product. The principal advantages of the drawing as a model are that it is relatively inexpensive to make and it can easily be changed. These two characteristics are especially important in the early stages of product development. Drawings of large objects are gener-

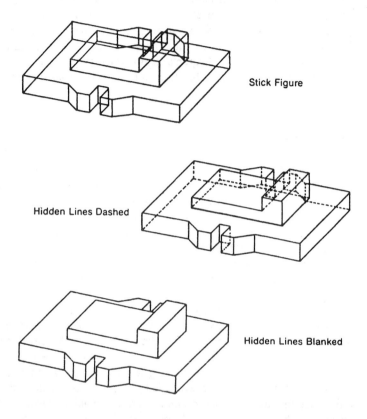

Figure 7.1 A mechanical part modeled with a wire-frame representation, and the same part shown with hidden lines dashed and completely removed.

ally made to a smaller scale, although they are occasionally made full size because the designer will then have a better feel for the appearance of the device. Small objects are usually drawn enlarged so that details are not lost in a plethora of lines that are close to one another, or when the line widths themselves are comparable to some of the dimensions. A prime example of drawings that are many times larger than the final product are the masks for semiconductor work.

The modern tools of computer-aided design (CAD) have converted the tedious task of engineering drawing, carried out largely by engineering draftsmen, into a relatively easy task that can and should be incorporated into the decision-making activities of the designer. Interactive graphics allows the

MODELS

engineer to explore various alternatives. Most graphics terminals can be used to display orthographic projections in three or more views, with changes in one view being reflected immediately in the other views. Isometric and perspective views can also be obtained. All of these capabilities aid the designer in visualizing and improving the design. For example, the wire-frame representation of a mechanical part shown in Figure 7.1 can be made more understandable by showing hidden lines in dashed form, and can be shown in isometric form by removing the hidden lines completely. Very early in the use of CAD an economic analysis [1] showed a payback time of about a year and a half because design and drafting time were reduced by two-thirds.

The use of CAD also improves design by facilitating rigorous analysis of alternatives. These analyses can range from simple calculations of volume and mass to the more complicated calculations of moments, as shown in the wire-frame drawing of Figure 7.2. In addition, still more complex calculations leading to information on stresses, deflections, and vibration of mechanical systems can be carried out. The analysis of electrical circuits and the effects of the tolerances of components can be carried out quickly. For example, a filter having a certain desirable frequency response may have been designed, and the values of the various components ascertained through a synthesis process. However, the component values in the filter circuit will probably not be readily

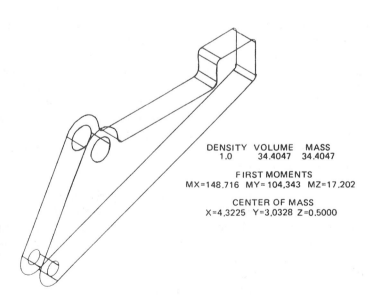

Figure 7.2 Wire-frame drawing of a simple mechanical part, showing typical data that may be calculated to assist the designer.

obtainable off the shelf, and off-the-shelf components will always have some tolerance. If allowable variations from the ideal response are established, the filter response using components having standard values and standard tolerances can be determined, and the designer can then specify a standard value for each component and the allowable tolerance for that component.

Automated drafting systems are most beneficial for products that use standard parts but that must be specifically designed for each application. Examples of such products are industrial conveyor systems that must meet specifications as to the path they are to follow, the unit loading, and the material to be handled; web-offset printing presses made from standard parts configured to meet specific customer requirements; and integrated circuit photolithography masks having many transistors, diodes, resistors, and capacitors with the necessary interconnections. For products such as these, it is easy to justify the most sophisticated interactive graphic systems available.

7.2 BLOCK DIAGRAMS

Sketches and drawings are of principal use when designing components or assemblages of components. Design of systems frequently begins with a model called a block diagram. Figure 6.1 shows two block diagrams, one for the feedforward amplifier, the other for the feedback amplifier, that were useful to Harold Black in his work on reduction of distortion in amplifiers. Black had decided to treat the amplifier as a unit (his "imperfect amplifier"). That is, he focused his attention on a major section of the system and on necessary interconnections to show the flow pattern of electrical signals. However, block diagrams are not restricted to electrical systems, but are used to show the flow of materials, energy, liquids, or any other variable with which the engineer is concerned. Designers use block diagrams to consider various alternatives as they strive for the best system to meet their specifications. For example, look at Figure 7.3, which shows an experimental arrangement for calibrating a linear variable differential transformer (LVDT), and compare the block diagram of the figure to the description of the system [2].

> Consider the experimental and equipment setup shown in [Figure 7.3]. The laser's retroreflector . . . was mounted to a plate that was itself mounted vertically to the forcer of the linear step motor. This vertical plate was the sensing surface for the LVDT. Straddling the linear motor's track was a stationary frame to which the LVDT was mounted. The frame, the track, and the interferometer's beam splitter were mounted to a common platform. The forcer's motion was controlled by a set of instructions that were transmitted over a three-wire RS 232 connection to the linear motor's controller and amplifier. The forcer's resolution was specified as 10,000 steps/inch (0.0001 inches) and the magnitude of the

MODELS

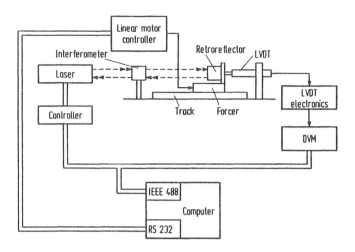

Figure 7.3 Experimental setup and instrumentation for the calibration of an LVDT using a laser interferometer. (Reproduced with permission from Edward B. Magrab, *Computer Integrated Experimentation,* Springer-Verlag, 1991.)

forcer's velocity and acceleration were programmable. The laser interferometer's measured and digitized displacements were transferred to the computer via an IEEE 488 interface. The output voltage from the LVDT's electronics was recorded by a 6-1/2 digit DVM [digital voltmeter], which also communicated with the computer via an IEEE 488 interface.

This figure shows blocks representing major functional parts of the system, such as the laser, linear motor controller, and so forth, but it also includes a sketch showing how certain parts of the system are related to each other spatially. Combinations of block diagrams and sketches of components are not at all unusual.

There is another kind of block diagram, sometimes referred to as a "functional block diagram," that is used to depict the changes in variables throughout a system. These diagrams are usually introduced in courses on feedback control systems, analysis of dynamic systems, or network analysis [3, 4,5]. An example of this kind of block diagram is shown in Figure 7.4. Mathematical expressions (called "transfer functions") relating the output of each block to its input are used in conjunction with these diagrams to determine whether a system will meet specifications. Transfer functions are generally based on linear differential equations, but nonlinearities can also be introduced,

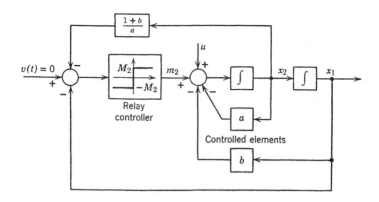

Figure 7.4 Block diagram of a feedback control system using a zero-hysteresis relay controller, and having desired inputs $v(t)$ and undesired disturbance u. (Reproduced with permission from Paul M. DeRusso et al., *State Variables for Engineers,* John Wiley & Sons, New York, 1965.)

as shown in the figure. For example, the relay controller is obviously a nonlinear element, but controlled elements "a" and "b" may be also.

7.3 NETWORK MODELS

One of the important tools by which engineers learn more about the devices they conceive and design is the use of a representation of the physical device by network elements. Electrical elements will be used in much of the discussion, but by the use of analogs [6] the method can be applied to design problems in other branches of engineering. This type of network is usually called an equivalent circuit in the literature, but as we will use it here the word "equivalent" means that the network acts as a model of the device being considered. Network models for many devices can be found in the publications of technical

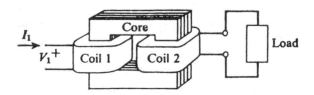

Figure 7.5 Two-winding core-type transformer with load.

MODELS

societies and in engineering textbooks. The techniques used in developing suitable network models are varied. Some begin with simple models and then add other elements to take care of effects that were ignored in the simple model. Others look at terminal conditions and how the variables at the terminals are related to each other.

As an example of the first technique, consider the transformer of Figure 7.5. In its simplest form, a transformer consists of two coils wound on a laminated steel core, as shown in that figure. The equivalent circuit is developed in Reference 7 in a series of steps that begins with the equivalent circuit of one coil (an inductor and a resistor in series), continues by adding the second coil assuming an ideal transformer, then modifies the equivalent circuit developed to that point by adding an inductive branch to take care of the magnetizing current needed for the core, adds a resistive branch to take care of the hysteresis and eddy current losses in the core, and finally takes into account the effects of different numbers of turns on the two coils. Although there are some approximations in this development, they are minor in nature. The resulting equivalent circuit is shown in Figure 7.6. Use of this circuit, given the input voltage and the circuit element values, results in values for the input current, the output voltage and current, the power lost in each winding and in the core, the efficiency of power transfer, and the voltage regulation.

As an example of the technique of considering the variables at the terminals of a block and taking into account the interrelationships among those variables, look at the diagrammatic representation of a transistor in Figure 7.7, in which the designator b refers to the base of the transistor, e refers to the emitter, and c refers to the collector. In the following, upper case subscripts B and C are used with voltages and currents having reference values of zero, while lower case subscripts b and c refer to variations in voltage or current around average values. The voltages v_B and v_C can be shown experimentally to be functions of only i_B and i_C. Hence

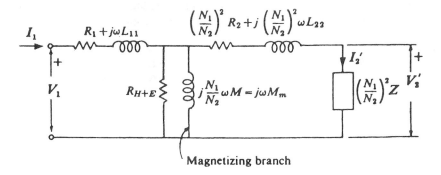

Figure 7.6 Equivalent network of a two-winding transformer.

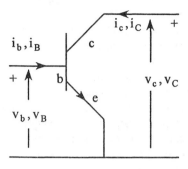

Figure 7.7 Diagrammatic representation of a transistor.

and
$$v_B = f_1(i_B, i_C)$$
$$v_C = f_2(i_B, i_C)$$

The total differentials expressed in terms of the independent variables i_B and i_C are

$$dv_B = \frac{\delta v_B}{\delta i_B} di_B + \frac{\delta v_B}{\delta i_C} di_C$$
$$dv_C = \frac{\delta v_C}{\delta i_B} di_B + \frac{\delta v_C}{\delta i_C} di_C \tag{1}$$

Equations such as these are frequently rewritten on the assumption that the changes in the variables, rather than being infinitesimal as the differential form assumes, will be finite. That is, differentials such as dv_C may be replaced by Δv_C. With the further stipulation that the finite changes to be considered are to be small and alternating in nature, and noting that the partial derivatives all have the dimensions of ohms, Equations (1) may be rewritten as

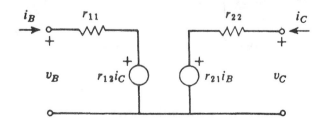

Figure 7.8 Small signal equivalent (model) circuit of a transistor.

MODELS

$$v_b = r_{11}i_b + r_{12}i_c$$
$$v_c = r_{21}i_b + r_{22}i_c \qquad (2)$$

In passive networks, $r_{12} = r_{21}$, and a simple T-network can be drawn in which the currents and voltages will satisfy these equations. A transistor, however, is an active device, and it can be shown experimentally that the necessary relationship cannot be satisfied for active devices. There is one rather obvious network for which Equations (2) will be satisfied, as shown in Figure 7.8. This is only one of several network models of the transistor.

When network models are used, it is always desirable to be able to associate elements of the model with physical aspects of the device itself, thus aiding the designer in his or her thinking. This is more easily done if the model has been built using the method shown in the example of the two-winding transformer. For example, if the efficiency of the transformer is too low as measured against the specification, knowledge of the losses in the windings individually and in the steel of the core will enable to designer to determine rather easily where the preponderance of the loss lies. If it is in the core, the solution will probably be to replace the core with a superior magnetic steel.

If the model has been built using a mathematical approach, such as was done with the transistor, there may be a dichotomy between the model and the physical device. For example, the resistors r_{11} and r_{22} may be associated with the ohmic resistances of the base and collector regions (although this is not strictly true), but what physical effect in the transistor accounts for resistors r_{12} and r_{21}? It is for this reason that a different model, one that uses what are called hybrid parameters, is favored by most engineers working with transistor circuitry.

However they are obtained and how closely or not they can be related to the actual device, network models are important because they aid in analysis. In Chapter 9 a technique called design by repeated analysis will be discussed. Because models facilitate analysis, they are important in design.

7.4 MATHEMATICAL MODELS

Of all the models discussed in this chapter, the mathematical model is probably the most important because it permits development of the most sophisticated design procedure, that called synthesis. It also permits one to use a computer to do laborious calculations (and do them correctly, if the programming has been done correctly) which can lead the designer to an optimum design. In the previous section, a mathematical model was developed for the transistor from consideration of the relationships of the variables at the terminals and the fact that experimentation had shown that the voltages at the base and the collector were dependent on the currents into those two terminals. A mathematical relationship was not developed for the two-winding transformer, but having the equivalent circuit the equations could now be written. That is, a mathematical model is established by determining how the phenomena involved in the operation of a device or system may be expressed by equations that relate the variables to each other. Much of a typical undergraduate engineering curriculum has a heavy emphasis on analysis and the development of the equations to carry out analysis.

There are several points that must be understood with respect to mathematical models, however. One of these is that the engineer is frequently forced to accept the use of approximations in developing the model. Recall that the transistor development included use of words such as "infinitesimal" and "finite," and that the caption of Figure 7.8 uses the words "small signal." There are approximations buried in the derivation that make the final model most useful and very accurate with small signals, but warn the designer or analyst to be cautious in using or accepting the results of analysis if the signals are large. The terms "small" and "large" cannot even be precisely defined. Other examples of approximations are the assumption that the resistance of connections between lumped elements in an electric circuit is zero; the assumption that heat flow is entirely longitudinal in an insulated bar of metal; and the assumption that a mass sliding along the ways of a large machine tool does so without friction. One does not ordinarily look on such impossible conditions as approximations, but in truth they are. Electrical connections always have some resistance (except for superconductors), heat flow through the insulation on the bar will occur, and a mass sliding on ways will have some friction. These are probably assumptions that will have negligible effect on the accuracy of your results, but there are others that may have a significant effect on accuracy.

Part of the job of a designer is to make the best judgment possible concerning which effects to neglect and which to include. If an attempt is made to include every effect that one can think of, the resulting mathematical model may become unwieldy even for a simple device. Worse, the inclusion of all the effects that can be thought of may so obscure the influence of the most important effects that no meaningful conclusions can be drawn from the final results of an analysis.

An example of the complications that may arise if one includes all the effects one can think of is the development of the equivalent circuit of a resistor of the type commonly used in electronic circuits. These resistors are commonly made of a cylinder of resistive material with metallic end caps which are connected to the externally available leads by soldering or welding, as shown in Figure 7.9(a). This assembly is encapsulated in an insulating cylinder, sealed at the ends so that only the leads are visible in the completed product. Because current flows in the resistor, a magnetic field is produced that encircles the resistor, thus "linking" the circuit. This is the phenomenon that describes inductance. (This same effect occurs in the interconnections between lumped elements.) Hence a resistor that is ordinarily thought of as having only resistance

MODELS

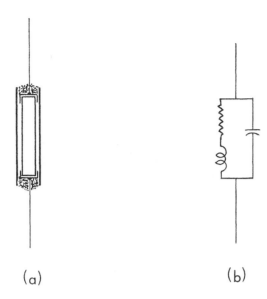

Figure 7.9 (a) Cross section of a type of resistor commonly used in electronic circuits. (b) Very high frequency model of the resistor.

should theoretically be thought of as a resistor and an inductor in series. In addition, the end caps, being metallic shapes, exhibit a capacitive effect. A "complete" equivalent circuit for a simple resistor therefore consists of two parallel branches, one having a resistor and an inductor in series, and the other having a capacitor. Should one use such an equivalent circuit and accept the complications that will arise in analysis? Or should one simply use a model consisting only of a resistor? It turns out that the inductor and the capacitor are so small that both can safely be ignored except at very high frequencies.

How can one proceed so as to be sure that only the important effects are included? One way to do this is to begin with assumptions that result in a simple model and an easy-to-solve set of equations. Then repeat the process several times, each time removing one or two of the simplifications. That is, for the second trial include what appear to be the most important of the secondary effects originally neglected, and determine the effect on the solution. If there is no noticeable difference, that secondary effect may safely be neglected. The complete solutions of the simple model give valuable insight into how to handle the more complicated cases and what may or may not be neglected.

Besides accepting the fact that certain effects need to be neglected to simplify the mathematical model, you must also realize that some of the important phenomena involved may be described by expressions that are themselves approximations. Some degree of approximation is always accepted in design and analysis. An example of the development of a mathematical model appears in Section 7.6.

7.5 PHYSICAL MODELS

Why are physical models made at all? There are a number of answers to this question. One reason for making a model is to obtain some information that can be obtained only from a physical device. Another reason is that it is always wise to see what the finished product will look like before going into full production. Moreover, a physical model can yield information on the population of devices to follow that cannot be obtained in any other way.

Often, as a project is begun, some pieces of information are needed that can be determined only from a physical model. Moreover, it is not always necessary that a complete model be built. Many times the information needed can be determined by the use of a model of part of the finished product. Automobile manufacturers routinely build full size mockups of proposed new models. These are made of wood and plaster, their only intended purpose being to fix the outward appearance of the car. The advantage of a mockup such as this is that some decisions can be made without involving a large expenditure of funds. As the design progresses, a complete model will finally be made, but this is not possible in the early stages. Since acceptable decisions cannot be made on all dimensions and materials at once, it is important to make as many decisions as possible in what appears to be a piecemeal fashion. In the automobile industry appearance of the car is an important design criterion that often drives the decisions on many interior dimensions and features.

Whenever incomplete models are to be used, the order in which new information is sought from these models is important. You need to obtain information from one partial model that will enable you to make the next partial model more nearly like the finished product. While doing this, it is important to avoid changing any part of the device or system that proved to be acceptable in a previous version. A technique for determining the order in which subsystems should be designed is presented in the next chapter.

The construction of a model represents a challenge to the designer, even though the actual work may be done by modelmakers. The modelmaker must have accurate information as to what is to be built, materials, dimensions, methods of assembly, and so forth. This is most often done by means of sketches or drawings with dimensions and materials specified, and with appropriate notes to cover other aspects. One of the best ways to assemble all of the relevant information is to go through the process in your mind's eye, writing down every step as you go along. The result will probably be much more voluminous than the modelmaker needs and you will probably edit it before passing the information along, but the chance of important information being omitted is reduced considerably. When deciding how much information to

MODELS

include and the form in which it is to be communicated to the technician/modelmaker, you must keep in mind that person's skills and knowledge, as well as their whims. A close working relationship with the person who will build the model can be very helpful. A good technician/modelmaker will make suggestions for improvement in the design or for better or easier ways of making parts, and thereby assist in arriving at a good design. Conversely, if you have a poor relationship with the same individual, you can suffer multiple frustrations and delays.

The modelmaker can also be helpful in learning about the population of devices to follow. The clues may be subtle in nature, but a good working relationship will lead to information as to whether the product will be easy to make or not. Was it necessary to hold close tolerances on some parts or to custom fit them? Mathematical models will never provide information as to ease of assembly, problems with sliding surfaces, the necessity of improving the finish on a given surface or the fact that a good finish on another surface is less important than the designer thought, which parts must be deburred, where chamfers would be desirable, and so forth. The modelmaker will have opinions (and probably sound reasons for them) on these and other points.

7.6 COMBINATIONS OF MODELS

While developing a product you may find that some decisions can be made more easily or with greater confidence using one type of model while others require another type. The story of Edison asking a young engineer to determine the volume of a rather oddly shaped light bulb is instructive. The young engineer is supposed to have sat down with adequate references so that he could write a mathematical expression for the surface of the bulb with the intention of integrating the differential volume in order to find the total volume. Edison, who was an extremely practical man and saw what the engineer was trying to do, simply picked up the bulb, submerged it in a partially filled beaker of water, and noted the rise in the liquid level on the scale on the beaker. The point of this story is that you want to select the type of model that gives the needed information with sufficient accuracy and in the most convenient way. Models exist for only one purpose: To provide information to aid you as a designer to make the best decision possible.

Another important idea in the use of models is that not all major components or parts of a device need be treated in the same way. Relatively few devices will allow a complete mathematical model. This being so, many designers do not develop mathematical models for many of the products they design. Instead of shunning mathematics entirely, however, you should use mathematical models where they can be used and resort to others where it is not possible. Moreover, it may not even be necessary to use mathematics in an overt fashion. One of our former students who went on to a successful career as a designer of feedback control systems for a major machine tool manufacturer once admitted that he used mathematics to design the system, but that when the first unit was built and did not completely meet specifications (a not uncommon occurrence) he resorted to his knowledge of the theory of feedback control systems to make modifications without doing a completely new analysis. He found that with one or two "cuts" at the problem he was almost always able to get the system within the specifications by drawing on his grasp of the theory.

In some cases, the product may need to be represented by a combination of models. An example of a familiar product that illustrates this necessity is the design of a door-closer of the type used on home screen and storm doors. Assume that the design concept, developed using the techniques of Chapter 6, is as shown in Figure 7.10. As the door is pulled open, the spring is compressed. The energy stored in the compressed spring is available to close the door. Movement of the door to the closed position after it is released is retarded by damper or dashpot action to avoid slamming. This action is provided by compressing the air in the chamber formed by the metal housing and the gasket attached to the movable spring seat and valving the compressed air to the atmosphere through a screw-adjustable escape valve to control the rate of closure. Moreover, on closure it is desirable to remove the damping action when the door is almost closed so that the final closure is rapid enough to insure engagement of the latch. In addition, some means must be provided to reduce or eliminate the damping when the door is being opened, otherwise the door will be difficult to open.

There are a number of questions to be answered as the process begins:

1. What is a suitable length and diameter for the closer?
2. What is the minimum difference in length of the spring with the door closed and with the door fully opened?
3. Can the door closer work acceptably for all possible locations of the hinge-side mounting bracket on the door frame?
4. May the friction of the hinges be neglected?
5. How does one solve the conflicting requirements for damping? Will an inexpensive gasket suffice?
6. What is the required stiffness constant for the spring (the spring constant)?
7. What is the required damping constant?

It should be evident at this point that there are really two design problems to be solved. The main problem will be solved if acceptable answers can be obtained to all of the questions except 5 and 7. Those questions relate to the secondary problem of the damper action that is desired. Leaving the damper problem for the moment, one may concentrate on the other questions.

As a first step, accurate and complete drawings were made of a typical door, door frame, and closer of various sizes and locations, the final combination being shown in Figure 7.11. Using this diagram, the first three questions can be

MODELS

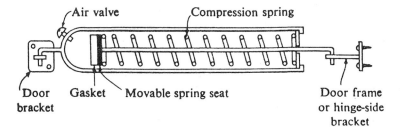

Figure 7.10 Door closer for a residential screen or storm door.

answered. Note that when the door is moved from the closed to the 90-degree-open position, the required extension of the closer does not change appreciably with the closer length but it is very sensitive to the location of the hinge-side mounting bracket. This is a consequence of the fact that displacement of the bracket on the door frame is nearly perpendicular to the axis of the closer with the door closed, but is very nearly in line with that axis when the door is open. The answer to question 3 is that the hinge-side bracket must be located with some precision. This may be done by use of a template, but a simpler solution is to make the bracket so that the edge of the bracket adjacent to the door should be lined up by the installer with the edge of the door frame. An instruction sheet included with the closer would specify this placement.

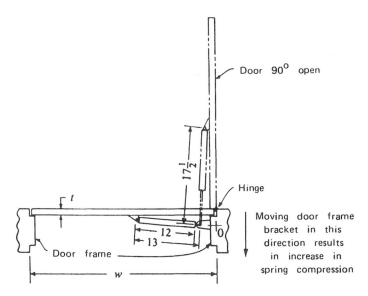

Figure 7.11 Top view of door closer with door and door frame. All dimensions are in inches.

The spring is the most important component in fixing the closer length. It is obvious that the compression of the spring is equal to the extension of the closer. For a spring to be reliable, its unstressed length should be several times the required compression. By simply assigning reasonable dimensions to the various parts, a tentative length of 12 inches is determined as being a suitable length for the closer cylinder. In a real project, this choice would be reconsidered later, possibly for the purpose of trying to reduce material costs. For the present, this is the best answer to part of question 1 and it also leads to the answer to question 2. From the layout, a 12-inch closer with the mounting bracket properly placed has a required spring compression of 4-1/2 inches. Next, consider question 4. Probably the most reliable answer to this question can be obtained from a physical model. Tests of typical hinges loaded with the weight of the door can be used to determine the frictional forces. As might be expected, these turn out to be negligible.

The second part of question 5 is also best answered by use of a physical model. Samples of typical gaskets should be obtained and mounted in cylinders with the fit to the side walls equal to that which could be maintained in production. Suitable tests can be devised to measure the damping constants for various choices of diameters. You will find that a closer of 1-inch diameter or greater will trap enough air to make the damping action insensitive to slight variations in the gaskets. Thus, as the answer to the second part of question 1, a closer diameter of 1 inch is chosen for further development. The answer to the first part of question 5 is still unanswered, but as noted above this is a secondary problem to be addressed after the main problem has been solved.

It remains now to attack questions 6 and 7. These could be addressed by use of a physical model, but to do so would require that many springs be available, running over a range of unstressed lengths and having a range of spring constants for each unstressed length. A better method is to make use of a mathematical model.

Figure 7.12 shows a simplified diagram of the door and the line of action of the closer. From this diagram, the spring compression is given by $\ell - \ell_c$,

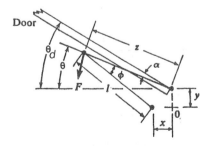

Figure 7.12 Force and displacement diagram of the door and closer.

MODELS

where

ℓ = length of closer with the door open
ℓ_c = length of closer with the door closed

The force exerted by the closer along its axis if the spring constant is K is then given by

$$K(\ell - \ell_c)$$

and the torque, T, applied by the closer to the door is found to be

$$T = K(\ell - \ell_c) z \sin\phi \qquad (3)$$

where

z = the length of the line drawn from the hinge pivot to the point of connection between the closer and the bracket mounted on the door

ϕ = the angle between the line z and the centerline of the closer

By the cosine law,

$$\phi = \cos^{-1}\left[\frac{\ell^2 + z^2 - (x^2 + y^2)}{2\ell z}\right] \qquad (4)$$

where

x = the distance in the plane of the door frame from the door hinge to the pin connecting the closer to the hinge-side bracket

y = the distance perpendicular to the plane of the door frame from the door hinge to the pin connecting the closer to the hinge-side bracket

The length ℓ may be found by constructing a right triangle with ℓ as the hypotenuse and having sides $z\sin\theta + y$ and $z\cos\theta - x$ leading to

$$\ell = \sqrt{x^2 + y^2 + z^2 + 2yz\sin\theta - 2xz\cos\theta} \qquad (5)$$

Also

$$\theta_d = \theta + \alpha \qquad (6)$$

where

θ_d = the angle by which the door stands ajar

θ = the angle between line z and the plane of the door frame through the hinge pin

α = the (constant) angle between line z and the outer surface of the door

Finally, the sum of the torques applied to the door must equal zero. Therefore

$$K(\ell - \ell_c)z \sin\phi = I\frac{d^2\theta}{dt^2} + D\frac{d\theta}{dt} \qquad (7)$$

where

D = the damping constant of the closer, and
I = the mass moment of inertia of a "typical" door, for

which

$$I = \frac{M}{3}(w^2 + \tau^2) \qquad (8)$$

where

M = the mass of the door
w = the width of the door
τ = the thickness of the door

Equations (4) through (8) can be combined into a single nonlinear second order differential equation in θ_d (the angle by which the door stands ajar) and time, t. Values of M, w, and τ can be determined for a typical door. Values of K and D can be determined by repeated solution of the differential equation, running through ranges of those parameters until solutions are found for which the door will close without slamming. These solutions would, of course, not be attempted by classical methods; program the equation into a computer.

Once values of K and D have been obtained, questions 6 and 7 have been answered, and the closer design is within grasp. The spring is specified by its free and compressed lengths and its stiffness constant. The damping mechanism already incorporated into the closer is specified by the damping constant. Reference to the data obtained while testing a physical model in order to answer the second part of question 5 will tell you whether the actual closer will be able to achieve the value of D obtained from the repetitive solutions of the differential equation. We now have enough information to determine materials and dimensions of the various parts, and the models have served their purposes.

This is a relatively simple example, but it illustrates the essence of the use of models in design. To summarize,

1. The example shows the need for drawings and for physical and mathematical models. A design project usually requires the use of a variety of models.

MODELS 231

2. The mathematical model was developed in a straightforward manner from the geometry and physics of the system.
3. Although this is a simple device and the design could be very well worked out if one goes through the simulation of the nonlinear differential equation, one should not have confidence in the final design until a complete physical model has been made and tested.

Having now arrived at a solution to the main problem, the first part of question 5 needs to be addressed. How does one obtain the damping needed on closure but reduce or eliminate it on opening? If one looks at the air flow patterns that are wanted, one sees that on closure air is to flow out of the chamber in which air is compressed for damping action through the adjustable air valve. On opening, one wants air to flow easily into that same chamber. This immediately suggests the use of a check valve. One could be placed in the cylinder wall near the adjustable valve, or one could be placed in the movable spring seat. It may even be possible to combine the check valve and the adjustable valve into one unit. There is a third solution, however. Suppose the gasket is made of a flexible material in the shape of a cup, the base of the cup being attached to the movable spring plate and the walls being oriented toward the chamber used for damping. On closure, the differential air pressure between the damping chamber and the spring side of the movable plate will cause the walls to flair out and form a good seal with the inside of the cylinder. On opening, however, the differential will be in the opposite sense and the gasket walls will be forced to move away from the inside of the cylinder. The question of removing damping just before the end of closure is still unsolved.

7.7 USE OF DIMENSIONAL ANALYSIS

The design of devices and systems sometimes leads an engineer to new areas of technology or to technology practiced as an art rather than as a science. In those situations, a technique known as dimensional analysis can be combined with experimental investigations to gain insight into the analysis and design of such products. There are at least three reasons for learning more about this method:

1. A reduction in the number of variables can be achieved by combining some of the system variables into dimensionless products. As a consequence, the amount of experimental investigation can be reduced, smaller databases are needed to store the essential data, and the results can be displayed using fewer and simpler graphical displays.
2. For simple products, it is possible to develop equations among the system variables. For more complex products, it is possible to develop and interpret the relationships that exist among the variables on a piecemeal basis.

3. It is possible to design products similar to those for which performance data are already available based on the fact that all systems with the same dimensionless products will perform alike if the dimensionless products are equal. This is an especially valuable facet of dimensionless analysis because most manufacturers make families of products that vary in size and performance specifications. Furthermore, it is sometimes uneconomical or even impossible to make full-size replicas of the final product. The designs of dams, bridges, or radio antennas for extremely long-distance transmission are cases in point. The solution is to make an off-scale model and interpret data from tests on it. (See, for example, the quotation from Osborne Reynolds at the beginning of this chapter.) Dimensional analysis leads to a set of dimensionless terms that are the basis for interpreting data from off-scale models in terms of equivalency to full-scale units.

7.7.1 The Buckingham π Terms

The development of the theory of dimensional analysis in its present form is credited to Dr. Edgar Buckingham, who published a paper on the subject in 1914 [8,9]. Although dimensional analysis had been used prior to the work of Buckingham by such well-known scientists as Rayleigh, Buckingham appears to have been the first to recognize that a search for dimensionless products was a worthwhile end in itself because it could lead in an orderly fashion to the form for equations relating variables in a system. The fundamental theorem of the method is:

> If an equation is dimensionally homogeneous, it can be reduced to a relationship among a complete set of dimensionless products.

The dimensionless products are called pi or π terms.

All equations that express the relationships among the variables of physical phenomena are dimensionally homogeneous. Hence the Buckingham π theorem applies to all engineering and scientific analysis. If an equation that relates physical phenomena is known, it can always be arranged so that only one term appears on one side of the equation. Division of both sides by that term will immediately result in dimensionless terms on both sides of the equation. This immediately makes obvious the correctness of a corollary of the Buckingham π theorem, which is stated as follows: If a complete set of dimensionless products can be found, the terms of the governing equations of the system will be known; however, dimensional analysis will not indicate the form of the equations. In some cases, the form can be established by experimentation. However, even if the governing equations cannot be developed, a great deal of

MODELS

valuable information can be obtained by studying the dimensionless products individually, as will be seen below.

A set of dimensionless products is complete when each product is independent and any other dimensionless product that can be formed from the variables is a product of powers (including negative powers) of the π terms in the set. This completeness is not obvious from casual observation. We need to know how many dimensionless products form a complete set and how to find dimensionless products.

To establish some insight into the questions of number of products needed to form a complete set and how to find those products, we will consider a specific example. Suppose that we wish to determine the drag force F on a smooth sphere in a stream of incompressible fluid having velocity v with respect to the body, body diameter D, mass density of fluid ρ, and viscosity of fluid μ. Assuming that this list includes all the variables involved in the phenomenon and none that are not involved, a typical dimensionless product will be

$$\pi = F^a v^b D^c \rho^d \mu^e \qquad (9)$$

where the exponents a through e are to be determined.

Using the SI system of units, dimensions for the variables can be obtained from Table 7.2 at the end of this chapter. Then, the units of π are given by

$$(kg \cdot m \cdot s^{-2})^a (m \cdot s^{-1})^b (m)^c (kg \cdot m^{-3})^d (kg \cdot m^{-1} \cdot s^{-1})^e \qquad (10)$$

Like terms may now be grouped so that the units of π may be shown in compact form as

$$kg^{(a+d+e)} \cdot m^{(a+b+c-3d-e)} \cdot s^{(-2a-b-e)}$$

But π is dimensionless, and therefore

$$\begin{aligned} a + d + e &= 0 \\ a + b + c - 3d - e &= 0 \\ -2a - b - e &= 0 \end{aligned} \qquad (11)$$

This set of simultaneous equations relates the exponents of the system variables. Any solution of these equations will result in a dimensionless π term. Furthermore, from matrix algebra we know that the number of independent solutions of a set of simultaneous equations equals the number of variables of the equation set minus the rank of the coefficients matrix [10].

The coefficient matrix can be written from Equations (11) as an array of the multipliers of the exponents a, b, c, d, and e. The matrix is

$$[C] = \begin{bmatrix} 1 & 0 & 0 & 1 & 1 \\ 1 & 1 & 1 & -3 & -1 \\ -2 & -1 & 0 & 0 & -1 \end{bmatrix} \qquad (12)$$

We notice that there is another way to arrive at this array. If one associates one of the system variables with each column and associates one of the dimensions with each row, one obtains

$$[C] = \begin{array}{c|ccccc} & F & v & D & \rho & \mu \\ \hline kg & 1 & 0 & 0 & 1 & 1 \\ m & 1 & 1 & 1 & -3 & -1 \\ s & -2 & -1 & 0 & 0 & -1 \end{array} \qquad (12)$$

the numbers in the array being picked out of Table 7.2 for each of the system variables. This is a very convenient way to arrive at the coefficient matrix and is less susceptible to error than the technique first shown.

The rank of a matrix is defined by Bell [11] as the order of the largest non-zero determinant that can be constructed from the rows and columns of the matrix. (See also pages 216 and 217 of [3].) For the matrix of Equation (12), one of the 10 possible 3 x 3 determinants is

$$\begin{vmatrix} 0 & 0 & 1 \\ 1 & 1 & -1 \\ -1 & 0 & -1 \end{vmatrix} = 1$$

Hence the rank of the matrix is 3. We now know that the number of independent solutions of Equations (12) (and hence the number of π terms) is given by the number of system variables less the rank of the coefficient matrix. That is, 5 – 3 = 2 π-terms.

How does one find π terms? Look back at Equations (12). There are only three equations and there are five unknowns. Three of the unknowns may be expressed in terms of the other two, which are called the excess variables. The excess variables may be chosen to be any of the unknowns. If d and e are chosen to be the excess variables, then

$$\begin{aligned} a &= -d - e \\ b &= -2a - e = 2d + 2e - e = 2d + e \\ c &= 3d + e - a - b = 3d + e + d + e - 2d - e = 2d + e \end{aligned} \qquad (13)$$

Any values can be chosen for d and e in order to obtain a, b, and c. A convenient set is to choose d = 1 and e = 0, yielding a = –1, b = 2, and c = 2. These are exponents for one of the π terms. If one then selects d = 0 and e = 1, the

MODELS

values for the other exponents are a = –1, b = 1, and c = 1. The two independent π terms are:

$$\pi_1 = F^{-1}v^2D^2\rho^1 = v^2D^2\rho/F$$

and (14)

$$\pi_2 = F^{-1}v^1D^1\mu^1 = vD\mu/F$$

Because both of the π terms are themselves dimensionless, it is possible to obtain other valid π terms by algebraic manipulation of these two. As an example,

$$\pi_3 = \frac{\pi_1}{\pi_2} = \frac{vD\rho}{\mu} \qquad (15)$$

which is the familiar Reynolds number. Any two of these three π terms is a complete set. π_1 is a useful form if one wishes to calculate the drag force as a function of the object's diameter, velocity of the stream, and the liquid density. On the other hand, if one wishes to determine the conditions at which nonlinear flow past the object would exist, π_3 is the term to select. However, it is necessary to determine the numerical values of the π terms before doing these calculations because their values may be other than unity.

One of the ways in which dimensionless terms may be utilized is when one is confronted with the necessity for gathering information about structures which may be difficult or impossible to test. For example, suppose that one needs to determine the drag force on a sphere having a diameter of 1 foot when moving through water at a speed of 25 miles/hour. Model basins are usually not equipped to tow at this speed. On the other hand, wind tunnels can easily provide a wide range of air speeds. The Reynolds number can be calculated for the 1-foot sphere in water at the speed desired, and one can then calculate pairs of velocities and diameters for a sphere to be placed in a wind tunnel in order to have the same Reynolds number. One pair of numbers that will satisfy the requirements is a diameter of 10 feet and an air velocity of 48 ft/sec, but there are an infinite number of combinations from which to choose.

There is a convenient way to write the π terms using the final form of Equations (13), which are:

$$\begin{aligned} a &= -d - e \\ b &= 2d + e \\ c &= 2d + e \end{aligned} \qquad (13)$$

A solution matrix can be written by associating a π term with each row and associating an exponent and the system variable on which the exponent operates with each column. For this example, the solution matrix is written by referring to the values previously obtained for the exponents. The matrix then appears in the following format:

	a	b	c	d	e
	F	v	D	ρ	μ
π_1	−1	2	2	1	0
π_2	−1	1	1	0	1

(16)

We note that columns 1, 2, and 3 are simply the coefficients of the three equations shown above as Equations (14), the first row being the coefficients of the d terms and the second row being the coefficients of the e terms. The solution matrix may be written by this simple technique whenever the values of the excess variables form a unit matrix, that is, one having values $a_{ij} = 1$ if $i = j$ and $a_{ij} = 0$ if $i \ne j$. We now know how to determine the number of π terms in the complete set for a given physical situation and how to determine what those terms are.

Although this example is instructive and leads to a result that is extremely useful, there are several questions that were not raised in the discussion that are of importance if one is to be able to apply the method with confidence. Among these questions are the following:

1. How are the system variables determined?
2. Must physical constants having dimensions be considered?
3. Can correct equations be derived by dimensional analysis without experimental or analytical verification?
4. What is to be done if the rank of the dimension matrix is less than the number of rows?

Unfortunately, there is no simple answer to the first question. Only a correct set of system variables assures a correct solution. Some preliminary experimentation may be of value at the beginning of an investigation to determine the variables that are involved. If too many variables are included, any that are extraneous may disappear (that is, have exponents of zero). Johnson [12] gives an example of determination of the π term for the velocity of sound in a gas in which he deliberately brings in the gas viscosity. After working through the solution, the viscosity has disappeared. However, he warns that one cannot always rely on the extraneous variable disappearing automatically.

There is another aspect to the question of system variables. When doing derivations of equations relating to a physical system, one frequently uses the same symbol with subscripts to distinguish between two variables having the same dimensions but which may play a different role. For example, if two lengths, ℓ_1 and ℓ_2, are important in a derivation, they will be so designated.

MODELS

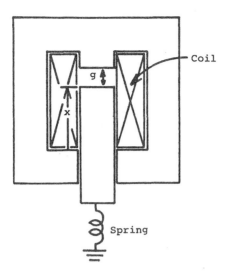

Figure 7.13 Iron-core solenoid.

Both having dimensions of length, however, some information is lost when one uses dimensional analysis. The same statement is true about areas, voltages, velocities, and so forth.

The second question above has to do with the necessity of including physical constants having dimensions. Any physical constant having units associated with it and that is associated with the phenomenon being considered must be included. Examples of such constants are the acceleration of gravity, permeability of free space, permittivity of a vacuum, Planck's constant, Boltzmann's constant, and so on. There are, however, many constants without dimensions, including Reynolds number, Poisson's ratio, specific gravity of a material, the 4π that appears in the rationalized system of units, and any multiplier required to establish equality between the two sides of an equation. These last appear frequently when a unit system other than the SI is used.

As a partial answer to the third question, let us look at a second example, chosen from a different field. The example we will consider is that of the force produced by the magnetic field of an electromagnet, such as the structure shown in Figure 7.13. The objective is to obtain a relationship for the force tending to reduce the gap length g in terms of the flux density in the air gap and other appropriate system variables. The variables which one must take into consideration are then the force F, the flux density in the air gap β, the cross section of the air gap A, and the permeability of the air μ. One may immediately write the dimension matrix as

	a	b	c	d
	F	β	A	μ
kg	1	1	0	1
m	1	0	2	1
s	-2	-2	0	-2
A	0	-1	0	-2

(17)

where dimension A is the ampere and the exponents are picked out of Table 7.2.

We note that the fourth-order determinant formed from the dimension matrix is equal to zero because rows 1 and 3 of the matrix are linearly dependent. Using numbers from the first three columns of rows 1, 2, and 4, a nonzero third order determinant is:

$$\begin{vmatrix} 1 & 1 & 0 \\ 1 & 0 & 2 \\ 0 & -1 & 0 \end{vmatrix} = 2$$

Therefore the rank of the matrix is 3 and there will be $4 - 3 = 1$ π-term. Using the rows (1, 2, and 4) of the matrix from which the nonzero determinant was obtained, the exponent equations can be written as:

$$a + b + d = 0$$
$$a + 2c + d = 0 \qquad (18)$$
$$-b - 2d = 0$$

The solutions of these equations in terms of the excess variable d are

$$a = d$$
$$b = -2d$$
$$c = -d$$

Selecting the value 1 for the excess variable d, we immediately have values for a, b, and c, so that the solution matrix may be written as

	a	b	c	d
	F	β	A	μ
π	1	-2	-1	1

resulting in

$$\pi = \frac{F\mu}{\beta^2 A}$$

or

MODELS

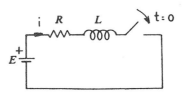

Figure 7.14 Circuit for which the influence of E, R, and L on i as a function of t is to be determined.

$$F = \pi \frac{\beta^2 A}{\mu} \tag{19}$$

This equation can also be derived from basic electromagnetic field concepts [13] leading to the same relationship among the variables and establishing the value of π as being 1/2, all quantities being in SI units. With proper instrumentation and with a structure adapted to be usable with the instrumentation, the same result for the value of π can be determined experimentally.

Finally, we will consider an example in which experimentation can play an important role. We begin by assuming that we know nothing about electrical network analysis, but we have been given the circuit of Figure 7.14 and assigned the task of determining how the current as a function of time is influenced by the voltage E, the resistance R, and the inductance L. One way to do this is by selecting a number of values for E, a number for R, and a number for L, and trying all possible combinations. Even if the number of values is limited to say 4 for each of these, one quickly sees that there will be a large number of trials to make (64 in this case). Rather than take that approach, let us instead try to obtain dimensionless terms using the Buckingham π theorem. We begin by using Table 7.2 to construct a coefficient matrix, as follows:

	a	b	c	d	e
	R	L	E	i	t
kg	1	1	1	0	0
m	2	2	2	0	0
s	−3	−2	−3	0	1
A	−2	−2	−1	1	0

We immediately note a relationship that often appears but is not generally so obvious as in this case: The first and second rows are linearly dependent. Hence a fourth-order nonzero determinant cannot be constructed. However, the elements in columns 3, 4, and 5 of rows 1, 3, and 4 form a nonzero determinant.

Therefore the rank of the matrix is three and the number of π terms will be two. Proceeding to form equations as before,

$$\begin{aligned} a + b + c &= 0 \\ -3a - 2b - 3c + e &= 0 \\ -2a - 2b - c + d &= 0 \end{aligned} \qquad (20)$$

Still assuming we know nothing about network analysis, we arbitrarily choose a and b as the excess variables. This choice will certainly force R and L to appear in separate π terms, which we may assume in our ignorance to be an advantage. Solving for c, d, and e, we obtain

$$\begin{aligned} c &= -a - b \\ d &= a + b \\ e &= -b \end{aligned} \qquad (21)$$

Choosing to let a = 1 and b = 0 for π_1 and a = 0 and b = 1 for π_2, the solution matrix may now be written as

	a	b	c	d	e
	R	L	E	i	t
π_1	1	0	-1	1	0
π_2	0	1	-1	1	-1

so that

$$\pi_1 = \frac{Ri}{E}$$

$$\pi_2 = \frac{Li}{Et}$$

Other values of the π terms may be obtained by using other sets of excess variables, or by manipulation of the π terms already in hand. In this case, we form

$$\pi_3 = \frac{\pi_1}{\pi_2} = \frac{Ri}{E} \cdot \frac{Et}{Li} = \frac{Rt}{L} = (L/R)t$$

which we will find to be more useful because we now have two π terms, one of which (π_1) is independent of t and the other (π_3) is independent of i.

We now set up the circuit of Figure 7.14 and obtain data to plot π_1 versus π_3. We hold E, R, and L constant and make a continuous time recording of current versus time. The results will be as shown in Figure 7.15. Recognizing that the curve along which the data lie is an exponential or determining the exponential nature of the curve by curve-fitting, the relationship is written as

MODELS

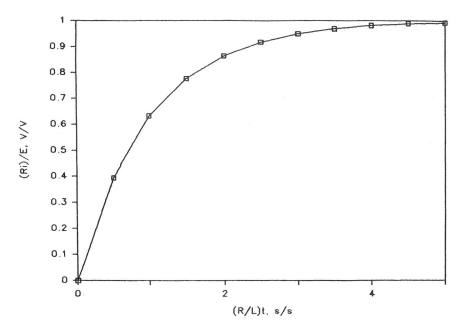

Figure 7.15 Variation of π_1 as a function of π_3.

$$\pi_1 = \left(1 - \varepsilon^{-\pi_3}\right) \tag{22}$$

That is

$$\frac{Ri}{E} = \left(1 - \varepsilon^{-(R/L)t}\right)$$

or

$$i = \frac{E}{R}\left(1 - \varepsilon^{-(R/L)t}\right)$$

This result is, of course, identical to that obtained by classical network analysis but, aside from the insight that helped to guide the solution toward the desired result, classical analysis was not used. Instead, the solution was reached by a combination of dimensional analysis and experimentation. This process is never as convenient as the derivation techniques typically used, but the theory behind some phenomena has not been so well developed that conventional derivation techniques can be used. In some situations, dimensional analysis may be all that is available to you.

Note that in setting up the experiment or handling data, the use of π terms allows a significant increase in convenience and efficiency. Instead of trying to understand the dynamics of the electric circuit by changing E, R, and L one at a time leading to 64 solutions if the earlier suggestion of using four different values for the driving voltage E and four values each for R and L had been adopted, this solution simply dealt with only two π terms for which only one graphical display was necessary. In general, allowing the π terms to act as the variables in the study of a phenomenon reduces the problem from k variables to k – r, where r is the rank of the dimension matrix. Other examples of the use of dimensionless analysis will be found in References 8 and 12.

7.7.2 Model Theory

The use of dimensional analysis in model work and design is even more important to the engineer than its use in developing equations. In the design area, it is very useful when extending a product line to larger or smaller devices or if it is necessary to predict the effect of minor changes on a product's performance. When dealing with models, one must keep in mind that models are not exact replicas of the final product.

The use of scaled and distorted models can be illustrated by referring again to the π terms for the circuit of Figure 7.14.

$$\pi_1 = \frac{Ri}{E}$$
$$\pi_3 = \frac{R}{L}t$$

Assume that the circuit is to be set up to verify the performance, but the only inductance available is 50% larger than the inductance we wish to use. The π terms can be kept unchanged by increasing R by 50% and by increasing E by the same percentage. (These changes can generally be accomplished much more easily than changing the inductance.) This is a "properly scaled" model and will have the same time response as the desired circuit because the π terms have been kept unchanged. The "sameness" does not, however, extend to all system variables. Because the applied voltage, resistance, and inductance all differ from the desired values, the voltages across the resistor and the inductor as functions of time will not be the same as in the desired circuit.

Another way to use dimensionless analysis in this case is to accept the inductance that is 50% too large, but to keep E and R unchanged. Examination of π_3 shows that it will be unchanged if t is also 50% larger. That is, if the desired circuit is known to reach a certain state (that is, i has reached a certain value) for a certain value of t, this "distorted" model will reach the same state for a new value of t that is 50% larger. In every case, the π terms can be used to interpret the data of a scaled or distorted model relative to the device being represented.

MODELS

As products resulting from a given design concept and materials are made larger or smaller, a point may be reached at which the π terms can no longer be kept equal to those of a working unit. The various sizes of paper airplanes described in Section 4.1 is a case in point. It was recognized in that discussion that a device having the same characteristics as the original model will not be physically realizable without a change more substantive than changing the size; it became obvious that the ten-times original size model would have to be made of a different material, and that it would have other shortcomings as well.

In mechanical devices or systems, complete similarity between the model and the product is impossible without geometric similarity. In some cases it may not be feasible to impose complete similarity on off-scale models; consequently some of the π terms that are thought to have only secondary influence on the test results or which will affect the phenomena in a known manner can be allowed to deviate from the correct value. This is more reasonable than it may at first seem. Models are often tested to destruction to determine only one or two characteristics. The fact that they do not operate in all respects like the final product may be fully acceptable. One caution with regard to neglecting dimensionless products for which model and product equality cannot be established must be mentioned. It may happen that variables that are negligible in the actual system do have significant effect on the off-scale model. A simple example of such a variable is surface roughness; this may be of little consequence in the full-scale device and of major importance in a miniature replica.

When it is necessary to determine similarity or dissimilarity between a product and its model, we need to consider the relationship among the π terms. For the original device $\pi_n = f(\pi_1, \pi_2, \ldots, \pi_{n-1})$ and for the model $\pi_{nm} = f_m(\pi_{1m}, \pi_{2m}, \ldots, \pi_{(n-1)m})$, where $f = f_m$ because the same phenomena are involved in both the product and the model. Now, if the model is made so that

$$\pi_1 = \pi_{1m}$$
$$\pi_2 = \pi_{2m}$$
$$\vdots$$
$$\pi_{(n-1)} = \pi_{(n-1)m}$$

(23)

it can easily be seen that

$$\pi_n = \pi_{nm}$$

because all of the π terms are equal and the functional relationship between the π terms for the device and the model is identical.

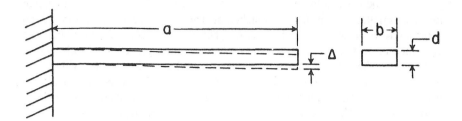

Figure 7.16 Off-scale model of a cantilever beam.

As an example of the problems that may arise, consider the cantilever beam of Figure 7.16. One set of π terms that can be developed can form the equation

$$\frac{\Delta}{a} = f\left(\frac{b}{a}, \frac{d}{a}, \frac{F}{Ea^2}\right) \qquad (24)$$

where F is the applied force causing the deflection Δ and E is the modulus of elasticity. Now if plans are made to change the physical size and the force to get data from an off-scale model, it makes most sense to change all of the dimensions in the same proportion. The designer therefore sets

$$k_1 = \frac{a_m}{a} = \frac{b_m}{b} = \frac{d_m}{d}$$

where k_1 is selected so as to bring the device dimensions down to (or up to) dimensions that one is comfortable working with. The force must also be scaled so that the force used on the model is sufficient to cause measurable deflections but not so large as to cause the elastic limit of the material to be exceeded. The designer therefore sets

$$k_f = \frac{F_m}{F} \qquad (25)$$

The first two π terms in the function of Equation (24) are unchanged. In order to force the third term to remain unchanged, one sets

$$\frac{F}{Ea^2} = \frac{F_m}{E_m a_m^2}$$

Use of the values of k above results in

MODELS

$$E_m = \frac{k_f}{k_l^2} E \qquad (26)$$

That is, if a material can be found to use in the model whose modulus of elasticity is given by the result of substitution into Equation (26), then the deflection of the model will be

$$\Delta_m = k_l \Delta$$

The likelihood that a material will be found to satisfy Equation (26) is rather remote. However, our thinking can now be directed along a reverse path. We know the modulus we want, we know that none of the materials worth consideration has this modulus, but we find others whose modulus is in the vicinity of this value. Beginning with one of those materials, one can select a reasonable value of k_f. From Equation (26), k_l is immediately known, and it is easy to determine whether a model of reasonable size results. There are other variations on this process. Rather than using the force relationship of Equation (25), one could use the relationship

$$k_e = \frac{E_m}{E}$$

between the actual device's modulus of elasticity and that of some other known material as a starting point.

Under some circumstances considerable deviation can exist between the model and the product without adversely affecting the desired data. For example, by choosing variables F, Δ, a, and the moment of inertia I for the cantilever beam, a single π term, Fa^2/EI, results. If the product EI is kept the same for the model as for the product, a type of similarity called quasisimilarity results. This can lead to simplifications of the model. For example, an I-beam can be represented by the simple cantilever beam with the rectangular cross section of Figure 7.16.

It is not at all unusual to encounter a situation in which it is not possible to satisfy all similarity requirements. This usually is the result of a need for values of physical constants in an off-scale model that are simply not available. Such constants include the gravitational constant, permeability of magnetic material or of free space, and permittivity of dielectric materials. One of the π terms of the model of Equations (23) will not equal the corresponding π term of the product, and as a result $\pi_{nm} \neq \pi_n$. The model is then said to be distorted. This problem can be handled by using algebraic manipulation to remove the unattainable system variable from all π terms except one, then making models that have

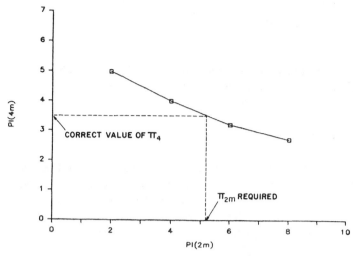

Figure 7.17 Determination of product characteristics using distorted models.

values of that π term that bracket the unattainable value, and finally finding an acceptable value of the π term by the use of interpolation. That is, if

$$\pi_4 = f(\pi_1, \pi_2, \pi_3)$$

and

$$\pi_{2m} \neq \pi_2$$

then make several models that satisfy the equality constraints for π_1 and π_3, but with various values of π_{2m} such that they bracket π_2. Figure 7.17 shows the graphical analysis leading to the value of π_4 that would result if complete similarity of model and product were possible.

7.7.3 An Engineering Problem

The examples so far have been chosen to illustrate various points concerning dimensionless analysis, but each has been solvable by classical analysis methods available in texts or in technical journals. Some engineering problems are, however, not susceptible of being analyzed by classical methods with any degree of confidence in the accuracy of the final results. An example of such a problem is that of the termination of electrical conductors by holding the conductor in place under a screw head or by clamping it by pressure from the tip of a screw. Typical connectors are shown in Figure 7.18.

Until the 1970s, this connection method was thought not to be a problem. However, at that time a number of fires occurred which were attributed to the

MODELS

Figure 7.18 A family of electrical connectors with a cutaway model.

failure of connectors such as those shown to maintain a tight grip on the wire when the wire and connector were subjected to cycles of heating and cooling over a long term. Such cycles are perfectly normal, because a plant may run on a one- or two-shift basis, and use of electric energy will then cycle in a daily pattern between a low value when the plant is not operating to a high value when it is, and similar patterns can be observed in residential use.

The incidence of these fires was directly related to the use of aluminum conductors; where copper was used, such fires did not occur. It was obvious that differences between copper and aluminum were key to the investigations that followed. Two differences in the material properties were known: The relatively high coefficient of thermal expansion of aluminum as compared to copper, and the fact that the aluminum alloys used for electrical conductors suffered plastic deformation more readily than copper. One additional factor is that, although the connectors were made of aluminum so that expansion and contraction would nearly match that of the conductors, the screw used for clamping was normally made of steel because of price advantage as well as to avoid the galling that frequently takes place between screw and connector threads if both are made of aluminum. In response to the concern for the integrity of connections of this type, Underwriters Laboratories, Inc., went through two major changes in their connector standards. The second of these mandated that by October 1982 all connectors for use at 100 A or less had to pass a 500-cycle heat/cool test and by October 1984 all connectors for use at currents greater than 100 A had to pass the test. This test is an accelerated life test that uses currents far in excess of the current for which the connector is designed.

Connectors have evolved in the past by repeated experimentation, and experimental verification is still necessary as mandated by the UL standards. However, the designer can improve the likelihood of early success by using dimensional analysis. The variables that are believed to be of importance are:

F = the force applied to the conductor by the holding screw
R = the contact resistance between the connector and the conductor
I = the current
ΔT = the temperature rise
ℓ = critical linear dimensions; for example, the diameter of the conductor, the length of the screw thread engaged in the connector, and so forth
E = Young's modulus of elasticity
β = the thermal bulk expansion coefficient
h = Newton's cooling coefficient

The coefficient matrix using the SI system (see Table 7.2) is:

	a	b	c	d	e	f	g	h
	F	R	I	ΔT	ℓ	E	β	h
kg	1	1	0	0	0	1	0	1
m	1	2	0	0	1	−1	0	0
s	−2	−3	0	0	0	−2	0	−3
K	0	0	0	1	0	0	−1	−1

(K is temperature measured in degrees Kelvin.)

A determinant may be formed from columns 1-5 that has a nonzero value. Hence the rank of the matrix is 5 and there will be three π terms. The equations represented by the matrix can be solved in terms of excess variables a, b, and d, giving:

$c = 2b$
$e = -2a - 2b$
$f = -a$
$g = b + d$
$h = -b$

MODELS

This immediately leads to the solution matrix given by:

	a	b	c	d	e	f	g	h
	F	R	I	ΔT	ℓ	E	β	h
π_1	1	0	0	0	−2	−1	0	0
π_2	0	1	2	0	−2	0	1	−1
π_3	0	0	0	1	0	0	1	0

That is,

$$\pi_1 = F/\ell^2 E$$
$$\pi_2 = RI^2\beta/\ell^2 h \tag{27}$$
$$\pi_3 = \Delta T \beta$$

After the π terms are developed, they are to be studied and their physical significance interpreted. In this case, the first π term indicates that the mechanical stress in all parts of the connector ($s = F/\ell^2$) divided by the modulus of elasticity must be the same in the model as it will be in the actual product, which may be either larger or smaller than the model, in order for the two to have the same operating characteristics. Finite-element analysis can be used to select the correct cross section areas to meet this requirement. The topic of finite element analysis is discussed in Chapter 11.

The second π term shows that the product of the electrical power dissipated in the connector (all of which is converted to heat energy) when multiplied by the coefficient of thermal expansion and divided by the power loss per unit temperature rise from the surface area (ℓ^2) that is available to dissipate the heat must be identical for the model and for the product. Some of the heat energy developed at the connector-conductor interface is transmitted into the conductor and some into the busbar to which the connector is attached. All of the heat energy is dissipated from the surfaces of the connector and of the conductor and busbar in the vicinity of the connector. Assuming that the connector dimensions increase or decrease in proportion to the wire size for which the connector is designed, the easy-to-measure surface of the connector may be used to represent all of the heat-dissipating surfaces. Since the connector dimensions do follow this "rule," this is a safe assumption. The only effect will be to increase the apparent cooling coefficient, h, above its actual value. Before looking at the last π term, it may be noted that both the first and the second π term contain ℓ^2, but in the first case it is a cross-section area and in the second case it is a surface area. This loss of information when using dimensional analysis was mentioned earlier in Section 7.7.1.

The third π term requires only that the temperature rise times the coefficient of expansion be unchanged.

Now suppose that a connector of a *given rating* is designed that performs satisfactorily under the latest UL test requirements. Or suppose a connector is available whose *appropriate rating* is determined by a series of tests. Either of these can then be used as an off-scale model for smaller or larger connectors needed to fill out the product line. This powerful technique enables the manufacturer to make a family of products having related ratings. If one finds when moving to larger or smaller units that it becomes impossible to keep the π terms unchanged, the designer has an indication that the limits of acceptable products are being reached, and that a more substantive change will be required in the design.

Because much of the work that was done on electrical connectors in response to the need to meet the revised UL standards was proprietary in nature, details of results using the technique above are not available in technical journals. However, the soundness of the method can be illustrated by studying the π_2 term for the connectors shown in Figure 7.19, which were used commercially and which met a previous UL standard. As a first step, the resistance between the connectors and the conductors was measured for various combinations using the torque specified by UL for the wire-holding screw. This resistance depends mainly on the diameter of the conductor and the torque. Figure 7.19 shows the results. Secondly, the area of the outer surface of the connector was determined excluding the area at the bottom of the connector where it is mounted to the busbar. The aluminum alloy is the same and the surface conditions are identical for all connectors. Hence β and h are the same for the three. This leads to the information displayed in Table 7.1. These con-

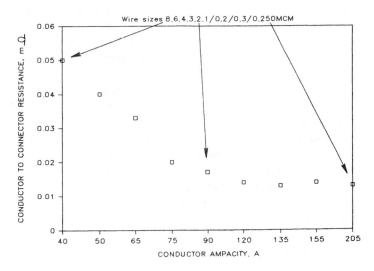

Figure 7.19 Aluminum connector to aluminum conductor resistance using UL standard values of tightening torque. (Ampacity = ampere capacity.)

MODELS

Table 7.1 Comparison of π Terms of Three Connectors

			Aluminum			
Nominal connector rating (A)	Conductor size (AWG)	Conductor diameter ($m \times 10^{-3}$)	Resistance from Fig. 7.19 ($\Omega \times 10^{-6}$)	Conductor ampacity at 75 C (A)	Connector outer surface ($m^2 \times 10^{-4}$)	$\pi_2(h,\beta)$ ($\Omega A^2/m^2$)
200	250 MCM	14.6	11.0	205	25.4	181.9
125	#0	9.4	14.0	120	11.4	176.8
70	#3	6.6	20.0	75	6.3	178.6

nectors, which were developed by experimentation and which met UL standards, have values of π_2 that are all equal to within experimental error. The π_1 and π_3 terms show similar correspondence. The engineers who developed these connectors did not use dimensionless analysis. Had they done so, the results could have been achieved with considerably less effort because success with the first connector would have led easily to the designs for the others.

To show how the development of these connectors could have proceeded, assume that samples of the midsize connector were made and subjected to a heat test wherein π_1 was held constant as temperature rise was determined for currents from 0 to 150 A. The test was then repeated using the same connector

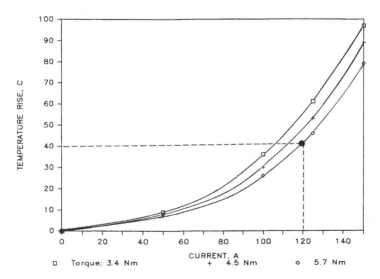

Figure 7.20 Connector temperature rise versus current for three values of torque applied to the wire-holding screw.

for two other values of π_1. The results are displayed in Figure 7.20. It is assumed that finite-element analysis will be used to verify the equality of the π term for other connectors.

To compute the numerical values of π_2 and π_3 a suitable condition of operation would be chosen on Figure 7.20, such as a 40 C rise at 120 A with the highest value of torque (indicated by the black dot). Other data needed are

$$\beta = 23 \times 10^{-6} \text{ m/m for the alloy chosen}$$
$$h = 9.3 \text{ W/m}^2\text{-C for a 40-50 C rise}$$
$$\ell^2 = 11.4 \times 10^{-4} \text{ m}^2 \text{ for the heat-dissipating area}$$
$$R = 14 \times 10^{-6} \, \Omega \text{ from Figure 7.18}$$

Substitution of values into Equations (26) yields:

$$\pi_2 = 4.374 \times 10^{-4}$$
$$\pi_3 = 9.2 \times 10^{-4}$$

On the basis that larger and smaller connectors will operate in the same satisfactory manner if designed to the same values as the π terms we now have, the heat-dissipating surface area of other ratings can be determined. For a connector rated at 205 A and having a connector-to-conductor resistance of $11.0 \times 10^{-6} \, \Omega$, the surface area computed from π_2 should be $26.14 \times 10^{-4} \text{ m}^2$. This will determine the outer dimensions of the connector, all other dimensions being determined by considerations of strength and/or economics. Similarly, the smaller connector rated at 75 A and having a connector-to-conductor resistance of $20.0 \times 10^{-6} \, \Omega$ is required to have an exposed area of $6.36 \times 10^{-4} \text{ m}^2$. Both of these calculated surface areas agree well with the measured surfaces listed in Table 7.1.

Before leaving this example of the use of π terms as an aid to design, it must be pointed out that the phenomena that lead to degradation of the connector-conductor interface include chemical processes, oxidation being the principal one. Present practice calls for the use of anti-oxidation compounds on assembly. However, note that by keeping temperature rise and thermal expansion constant in connectors of various sizes, these time-dependent chemical processes remain constant. That is, if a connector and conductor of a given size operate satisfactorily relative to these chemical processes, all units of the same materials for which this acts as a true model will also operate satisfactorily.

7.8 SUMMARY

The appropriate model to use when executing a design will depend on the situation in which the engineer works. Factors to take into account include the skills of associates and technicians, the ease of establishing a mathematical model, and the extent and quality of the test equipment. The best strategy is to use the least expensive model that will allow design decisions to be made with a high degree of confidence. As the design progresses, physical models must be made because, without realistic tests on a physical model, it is impossible to be certain that every eventuality has been anticipated. These models, more than any other design aid, can indicate just how trouble-free the final product will be. It is for this reason that as much emphasis has been put on physical models in this chapter as possible.

An important part of the use of physical models is the extension of the pool of information that is made possible by the use of dimensional analysis. This method has become more useful in recent years because of the growing acceptance of the SI system of units across all branches of engineering. The problem of the electrical connectors, simple as it appears at first glance, was seen to require some knowledge of electrical, mechanical, and thermal phenomena. Most engineering problems cut across the traditional boundary lines between disciplines.

One of the most important concepts developed in this chapter is that available products can be used as off-scale or distorted models of products to be designed. Many designers have used this concept in a quantitative way, even though they may not have formulated the thought. They have relied on experience and judgment to predict that a new design will probably be adequate, and so on. However, by expressing the relationship among variables as π terms, quantitative information can be developed for new products soon after the design concept to be used has been established.

REFERENCES

1. Blauth, R. E., and E. J. Preston. Computing the payback for CAD, *Machine Design,* 1981, 53(19): 91-95.
2. Magrab, Edweard B. *Computer Integrated Experimentation,* Springer-Verlag, Berlin, 1991.
3. DeRusso, Paul M., Rob J. Roy, and Charles M. Close. *State Variables for Engineers,* John Wiley & Sons, Inc., New York, 1965.
4. Cheng, D. K. *Analysis of Linear Systems,* Addison-Wesley, Reading MA, 1959, pp. 258-267.

5. Shearer, J. L., A. T. Murphy, and H. H. Richardson. *Introduction to System Dynamics,* Addison-Wesley, Reading MA, 1967, Chap. 7.
6. Cheng, D. K. *Analysis of Linear Systems,* Addison-Wesley, Reading MA, 1959, Chap. 4.
7. Engelmann, R. H. *Static and Rotating Electromagnetic Devices,* Marcel Dekker, New York, 1982, Chap. 4.
8. Langhaar, H. L. *Dimensional Analysis and Theory of Models,* John Wiley & Sons, New York, 1951.
9. Skoglund, V. J. *Similitude Theory and Applications,* International Textbook Company, Scranton PA, 1967, p. 46.
10. Bell, W. W. *Matrices for Scientists and Engineers,* Van Nostrand Reinhold Company, New York, 1975, p. 113.
11. Bell, W. W. *Matrices for Scientists and Engineers,* Van Nostrand Reinhold Company, New York, 1975, p. 109.
12. Johnson, Walter C. *Mathematical and Physical Principles of Engineering Analysis,* McGraw-Hill, New York, 1944.
13. Engelmann, R. H. *Static and Rotating Electromagnetic Devices,* Marcel Dekker, New York, 1982, Chap. 6.

REVIEW AND DISCUSSION

1. What is the definition of a model?
2. How many different kinds of models that fit the definition of question 1 can you name?
3. What is an orthographic projection? An isometric view?
4. How do computers help the engineer in modeling?
5. What kind of model would you use in designing an information-handling system?
6. What is meant by a mathematical model?
7. Can nonlinear elements be represented mathematically? Name a number of different methods of representation that might be useful.
8. Why do most projects involve more than one kind of model?
9. What is the basic theorem of dimensional analysis?
10. How do you determine the rank of a matrix?
11. What is the rule for determining the number of independent π terms?
12. Under what conditions will the data obtained by an off-scale model be identical to that obtained from the full-scale product?
13. What is meant by quasisimilarity?
14. When is a model said to be distorted?

MODELS

PRACTICE PROJECTS

1. Develop a block diagram of the money-handling system of a vending machine that accepts nickels, dimes, and quarters, rejects pennies and slugs, and delivers change. Assume that the product being sold has a price of 65 cents.
2. A transistor amplifier with resistive networks at both the input and the output and with an interconnection between those networks (that is, there is some feedback) can be represented by a three-terminal network as shown in Figure 7.21. Experimentation with the amplifier leads to the knowledge that

$$e_1 = f(e_2, i_2) \quad \text{and} \quad e_2 = f(e_1, i_1)$$

Develop the equivalent circuit for small signals. Note that you do not have to understand how a transistor works nor do you need to know anything about the connections of the resistive networks to find the equivalent circuit.

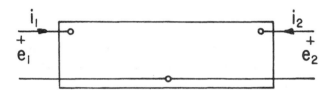

Figure 7.21 Block diagram of a transistor amplifier.

3. We left the door closer without solving the problem of allowing for quick closure as the door nears the end of its motion, the objective being to assure that the latch engages. The authors have two possible solutions to the quick closure question, but there are certainly others. See how many solutions you can find. Remember, you are looking for the elegant solution.
4. The door closer used as an illustration in this chapter is quite similar to many on the market today. All of them suffer from one major deficiency as far as the disabled person is concerned: There is no simple way for a disabled person to hold the door open while exiting or entering and yet permit normal operation of the door closer after the person is all the way in or all the way out. Design a door closer with these features. If at all possible, wipe out of your memory everything you already know about door closers, because a completely new approach may very well be necessary.
5. In Section 7.7.1, we investigated the movement of a body through an incompressible fluid. Using what was learned there, develop a viscosity measurement scheme that uses spheres having different weights that are

allowed to fall through the fluid whose viscosity is to be determined. Can you propose a method of obtaining a quick readout of viscosity using this device?

6. A mechanical system consists of a spring, a mass, and a dashpot. Determine the relationship among the variables using dimensional analysis. Use the following symbols for the variables:

　　　y = displacement　　　　　　K = spring constant
　　　M = mass　　　　　　　　　　t = time
　　　B = damping constant

7. Determine the π terms of the door closer.
8. Suppose the circuit of Figure 7.13 is to be constructed to operate at 480 V, but for the sake of safety you decide to use 12 V for a breadboard model. What steps must you take in order to get the *same* current versus time trace on your recording equipment from the model as the 480 V circuit should produce?
9. Determine the π terms of a series RLC circuit with applied voltage E and time-dependent current i.
10. The current density of holes (positive charge carriers in semiconductors) that arises from the drift of the holes in an electric field involves the variables listed below. Use the Buckingham π Theorem to derive the relationship among the variables. As with project 2 above, you do not need to understand the phenomena involved to find this relationship.

　　　J_p = hole current density, A/m^2
　　　E = electric field strength, V/m
　　　μ_p = hole mobility, $m^2/V\text{-sec}$
　　　q = charge of a hole, coulombs
　　　p = hole density, m^{-3}

11. Heat transfer to air from a pipe in which fluid is flowing depends on the following variables, for which the appropriate units are given. Again, note that you need not understand the phenomenon to find the relationship among the variables.

　　　D = pipe diameter, m
　　　ρ = fluid density, $kg\ m^{-3}$
　　　v = fluid velocity, $m\ s^{-1}$
　　　μ = fluid viscosity, $kg\ m^{-1}s^{-1}$
　　　C = specific heat of the fluid, $m^2 s^{-2} k^{-1}$
　　　K = thermal conductivity of the fluid, $kg\ m\ s^{-1} k^{-1}$
　　　h = Newton's cooling coefficient (pipe to air), $kg\ k^{-1} s^{-3}$

Determine the π terms.

MODELS

12. Determine the π terms of a yo-yo. You must understand the physics of a yo-yo in order to select the correct set of variables.
13. Prove that LC/t (inductance, capacitance, and time) cannot be a π term. Is it possible for M/Kt^2 (mass, spring constant, and time) to be a π term?
14. Use data from the example of Section 7.7.3 to design connectors having ratings of 90 A and 155 A if the midsize design chosen has a 50 C rise with 115 A when tightened to a torque of 3.4 Nm. Complete all details.

Table 7.2 Units of Variables Used in Engineering and Science

The SI system is a coherent system of units founded on seven base quantities, each having its own name and symbol, as shown in the following:

Quantity	Name	Symbol
length	meter	m
mass	kilogram	kg
time	second	s
electric current	ampere	A
temperature	kelvin	K
amount of substance	mole	mol
luminous intensity	candela	cd

There are also two supplementary units which may be regarded either as base units or as derived units. These are the radian and the steradian for the measure of plane and solid angles; they have the symbols rad and sr.

The remainder of this table is divided into four sections: (A) Statics, Dynamics, Strength of Materials; (B) Electricity, Magnetism, Light; (C) Thermodynamics, Heat Transfer; and (D) Physical Chemistry, Molecular Physics, Molar Transfer. Each section contains a list of important quantities, the symbol used for each quantity, the unit name, unit symbol (if any), and then the exponent(s) on the base quantities.

Table 7.2A Statics, Dynamics, Strength of Materials

Quantity name	Quantity symbol	Unit name	Unit symbol	kg	m	s	A	K	mol	cd	rad/sr
Acceleration, angular	α	radian/second squared				−2					1
Acceleration, gravitational	g	meter/second squared			1	−2					
Acceleration, linear	a	meter/second squared			1	−2					
Angle	θ	radian									1
Area	A	square meter			2						
Coefficient of friction	μ	newton/newton									
Energy	E	joule	J	1	2	−2					
Force	F	newton	N	1	1	−2					
Frequency	f	hertz	Hz			−1					
Impulse, linear	Ft	newton-second		1	1	−1					
Impulse, angular	Tt	newton-meter-second/rad		1	2	−1					−1
Inertia	M	kilogram		1							
Inertia, rotational	I	kilogram-meter squared/ radian squared		1	2						−2
Length	L,a,b	meter	m		1						
Mass	m	kilogram	kg	1							
Mass density	ρ	kilogram/cubic meter		1	−3						
Mass flow rate		kilogram/second		1		−1					
Mass moment of inertia	I	meter squared kilogram		1	2						
Modulus of elasticity	E	pascal	Pa	1	−1	−2					−2

MODELS

Modulus of rigidity	G	pascal	Pa	1	−1	−2
Momentum	mV	kilogram meter/second		1	1	−1
Momentum, angular	Iω	kilogram meter squared/ radian-second		1	2	−1
Polar second moment of area	J	meter to fourth power			4	
Power	P	watt	W	1	2	−3
Pressure	p	newton/meter squared	Pa	1	−1	−2
Section modulus	S	meter cubed			3	
Shear stress	σ_s	pascal	Pa	1	−1	−2
Spring modulus	k	newton/meter		1		−2
Surface tension	σ	newton/meter		1		−2
Tensile stress	σ_t	pascal	Pa	1	−1	−2
Time	t	second	s			1
Torque, moment of force	T	newton-meter/radian		1	2	−2
Velocity, angular	ω	radian/second				−1
Velocity, linear	v, V	meter/second			1	−1
Viscosity, dynamic	μ	pascal second		1	−1	−1
Viscosity, kinematic	nω	meter squared/second			2	−1
Volume	V	cubic meter			3	
Volumetric flow rate		cubic meter/second			3	−1
Weight	w	newton		1	1	−2
Work	W	joule	J	1	2	−2

Table 7.2B Electricity, Magnetism, Light

Quantity name	Quantity symbol	Unit name	Unit symbol	kg	m	s	A	K	mol	cd	rad/sr
Absorbed dose		gray			2	−2					
Activity of radionuclide		becquerel	Bq			−1					
Capacitance	C	farad	F	−1	−2	4	2				
Charge	Q,q	coulomb	C			1	1				
Conductance	G	siemens	S	−1	−2	3	2				
Conductivity	σ	siemens/meter		−1	−3	3	2				
Current	I,i	ampere	A				1				
Current density	J	ampere/meter squared			−2		1				
Electric field strength	$\mathcal{E}$	volt/meter		1	1	−3	−1				
Electric flux	ψ	coulomb				1	1				
Electric flux density	D	coulomb/meter squared			−2	1	1				
Electron volt	ev	coulomb volt		1	2	−2					
Energy	E	joule	J	1	2	−2					
Exposure (X & γ rays)		coulomb/kilogram		−1		1	1				
Frequency	f	hertz	Hz			−1					
Illuminance		lux	lx		−2					1	1

MODELS

Inductance	L	henry	H	1	2	−2	−2
Luminous flux		lumen	lm					1	1
Magnetic field strength	H	ampere/meter		−1			1
Magnetic flux	φ	weber	Wb				1
Magnetic flux density	β	tesla	T	1	−1	−2	−1
Magnetomotive force	$\mathcal{F}$	ampere (turns)					1
Permeability	μ	henry/meter		1	1	−2	−2
Permeance	$\mathcal{P}$	weber/ampere (turn)		1	2	−2	−2
Permittivity	ε	farad/meter		−1	−3	4	2
Power	P	watt	W	1	2	−3	
Reluctance	$\mathcal{R}$	ampere (turn)/weber		−1	−2	2	2
Resistance	R	ohm	Ω	1	2	−3	−2
Resistivity	ρ	ohm meter		1	3	−3	−2
Voltage, electromotive force, electric potential	V,v,E,e	volt	V	1	2	−3	−1
Work	W	Joule	J	1	2	−2	

Table 7.2C Thermodynamics, Heat Transfer

Quantity name	Quantity symbol	Unit name	Unit symbol	kg	m	s	A	K	mol	cd	rad/sr
Coefficient of heat transfer (convection & radiation)	h	watt/meter squared kelvin		1		−3		−1			
Coefficient of thermal expansion	β	meter/meter-kelvin						−1			
Conductance	c	watt/meter squared kelvin		1		−3		−1			
Conductivity, thermal	k	watt/meter kelvin		1	1	−3		−1			
Energy, quantity of heat	E	joule	J	1	2	−2					
Energy density		joule/cubic meter		1	−1	−2					
Energy per unit area		joule/meter squared		1		−2					
Enthalpy	H	joule	J	1	2	−2					
Entropy, heat capacity	S	joule/kelvin		1	2	−2		−1			
Heat transfer rate	q	watt	W	1	2	−3					
Heat transfer rate density	q"	watt/meter squared		1		−3					
Molar energy		joule/mole		1	2	−2			−1		
Molar entropy, heat capacity		joule/mole kelvin		1	2	−2		−1	−1		
Power	P	watt	W	1	2	−3					
Resistance	R	kelvin meter squared/watt		−1		3		1			
Specific energy	e	joule/kilogram			2	−2					
Specific enthalpy	h	joule/kilogram			2	−2					
Specific entropy, heat capacity	s	joule/kilogram kelvin			2	−2		−1			
Stefan-Boltzmann constant	σ = 5.67·10⁻⁸	watt/meter squared kelvin to the fourth power		1		−3		−4			
Temperature	T	degree Celsius of kelvin (C or K)						1			
Temperature gradient		kelvin/meter			−1			1			
Thermal diffusivity		meter/second			1	−1					
Time	t	second	s			1					

MODELS

Table 7.2D Physical Chemistry, Molecular Physics, Molar Transfer

Quantity name	Quantity symbol	Unit name	Unit symbol	kg	m	s	A	K	mol	cd	rad/sr
Coefficient of substance transfer		meter/second			1	−1					
Concentration gradient		mole/meter to 4th power			−4				1		
Density of substance flow rate		mole/meter squared second			−2	−1			1		
Diffusion coefficient		meter squared/second			2	−1					
Molar concentration		mole/meter cubed			−3				1		
Molar mass		kilogram/mole		1					−1		
Molar volume	V_m	meter cubed/mole			3				−1		
Mole		mole	mol						1		
Substance flow rate		mole/second				−1			1		

8

DECISIONS

Early in this work, we pointed out that engineering requires a great deal of decision making. It is, in fact, the foremost activity in design. How does one make the necessary decisions? The most important decision of all is probably that of deciding with which part of the device or system one should begin when making a design for a new product. As an example, consider an electric car for which specifications have been established as to the number of passengers, battery type, cruising speed, acceleration, and range. One can easily envision a number of subsystems of the vehicle, such as the drive motor system, suspension, drive train (if any), steering mechanism, and so forth. The designer must choose one of the subsystems as the one which will be designed first. Is there a logical way to make this choice?

Because the designer is faced with the necessity of making many decisions, knowledge of appropriate decision-making techniques can make the design process more efficient, thereby saving time and financial resources. This chapter is intended to give the reader a number of these techniques, with examples of their application. Some of these techniques may seem at first glance to be oriented more toward management decision making than toward engineering decision making, but many engineers are directly concerned with scheduling, meeting delivery dates, and lead times for purchasing components from others. Hence decision techniques that seem to be management tools are also of importance to the engineer.

The technique used to solve engineering problems is to abstract from the complex situation existing in real life those aspects of reality with which we wish to deal, being careful to include all aspects which are relevant. This leads to a somewhat idealized situation that is then described mathematically and which is manipulated to solve for the desired results. In other words, we model the engineering problems we wish to solve.

When dealing with decision problems, we resort to the use of other models. These models are often much less rigorous than those in engineering

DECISIONS

disciplines. Nevertheless, these models allow the decision maker to determine how a set of controllable variables should be manipulated in order to achieve some desirable result. The process is not perfect; no formal approach can guarantee that the best decision will be made. However, the use of the techniques that follow will make it more probable that the decision reached will be the best than will an intuitive selection among the several possibilities that may be listed. For example, we will see below how one of these techniques leads to the most logical selection of the first subsystem to be designed in the case of the electric vehicle.

The key step in most of the procedures is to divide the required decision into many subordinate decisions, each of which can be made more reliably than the ultimate decision because each of the subordinate decisions is simpler. The formal procedures merely manipulate these subordinate decisions to arrive in a logical way at the ultimate decision. If the subordinate decisions have been made correctly, then the final decision will be the best possible one given the uncertainty inherent in some of the evaluations required in the process.

8.1 ELEMENTS OF EVERY DECISION

Every situation requiring a decision has three characteristics:

1. There must be more than one alternative available to the decision maker.
2. There will be some condition or conditions over which the decision maker has no control but which will influence the benefits derived from the decision.
3. The benefits of the decision will vary depending on the alternative chosen and the conditions that prevail.

The conditions over which the decision maker has no control are called *states of nature* if they issue from environmental, political, economic, legal, or other considerations over which the decision maker has no control. They are called *opponent strategies* if they result from competition with an adversary.

The benefits to the decision maker are called the *outcomes* or the *payoffs,* and these must be expressed in quantitative terms. In some cases the outcomes may be financial gain the company can expect, or they may be physical values such as the stress in a beam or the horsepower of a motor. Outcomes such as these can be determined quantitatively using analytical techniques learned in engineering courses or from study of economics.

The elements of a decision can be displayed in several ways. A matrix form is especially useful in clarifying the relationship between the elements. Each row of the matrix represents one of the decision maker's alternatives; each column represents one of the states of nature that may influence the benefits to the decision maker. At the intersection of each row and column, an entry is

	Rain	No rain
Take umbrella	8	5
Don't take umbrella	2	9

Figure 8.1 Simple decision matrix.

made that represents the payoff to the decision maker. Obviously, all available alternatives and all pertinent states of nature must be included in the matrix. A simple example is that of deciding whether to take an umbrella with you on a cloudy day. This can be represented in matrix form as shown in Figure 8.1.

The payoffs in this figure are somewhat arbitrary, as is true in many cases. The numbers assigned to the payoffs may be either positive or negative, they will be different for each decision maker, and they will depend on the objective of the decision maker when making the decision. For example, if it does rain the payoff of 8 for taking an umbrella will be too low if the decision maker is keeping an appointment for an important interview; a more appropriate value would be 10. On the other hand, this payoff might be made lower than 8 if the decision maker is dressed in informal clothing and is making a quick run to the nearest mall. If it does not rain and the decision maker made the decision not to take an umbrella, this would engender a feeling of satisfaction, represented by the 9 in the matrix. A number of ways in which this kind of matrix can be used to arrive at a decision will be given later.

8.2 TYPES OF DECISION PROBLEMS

Decision problems can be characterized in terms of the information the decision maker has on the states of nature that will exist. In this sense, decisions may be categorized in one of four ways. That is, they are decisions made under *certainty, risk, uncertainty,* or *conflict*.

Certainty means that the state of nature that will exist at the time the decision becomes effective is known. In other words, one state of nature has 100% probability and all others have zero probability. This reduces every decision problem under certainty to a single column. There is, however, no limitation on the number of rows. If the local weather forecaster has proved to be very reliable and states on the morning news that "It will rain all day," Figure 8.1 reduces to the "Rain" column only.

The second type of decision, that is, those made under risk, are decisions in which the probabilities of the states of nature are known or can be determined. For example, if the weather forecast calls for a 70% chance of rain (equivalent,

DECISIONS 267

of course, to a 30% chance of no rain), this fact should be considered when making the umbrella-carrying decision displayed in matrix form in Figure 8.1.

The third type of decision, decisions made under uncertainty, are those in which the decision maker can identify the states of nature but has no information on the likelihood of occurrence. The fourth type of decision, those made under conflict, are closely related to decisions made under certainty. Decisions made under conflict are decisions made in those situations in which a rational opponent has an objective diametrically opposed to that of the decision maker, and therefore attempts to counter every strategy pursued by the decision maker.

One of the major tasks in solving decision problems is to determine the true value of the outcomes or payoffs. This is best done by assuming the existence of a certain state of nature for a given alternative and then using the analytical techniques of engineering, economics, or operations research to arrive at a value for the utility of that alternative. This must be repeated, of course, for each alternative for each state of nature. The wealth of problem-solving techniques included in typical undergraduate curricula is but a small part of the whole body of the techniques available. The designer should add to his or her set of tools convenient methods to be found in company memoranda, trade journals, single-product design books, handbooks, technical literature in the area, and so forth. Later in this chapter, examples of such techniques are shown that answer decision-related questions that are common to many design tasks.

8.3 DECISIONS UNDER CERTAINTY

As noted in the previous section, decision making under certainty occurs when the state of nature that will exist when the decision becomes effective is known. The solution is obtained by calculating payoffs for all alternatives available to the decision maker and choosing the most attractive. This can be a trivial problem, but if the number of alternatives is large it can also be difficult. The best examples of this type of decision problem are found in the area of operations research.

Operations research (or operations analysis) is defined as "the systematic and scientific analysis and evaluation of problems, as in government, military, or business operations." It had its genesis as a formal discipline during World War II to provide answers to such problems as determination of the best search pattern to locate an enemy submarine given that it was last sighted at a known latitude and longitude, that the nearest anti-submarine vessel would require a given number of hours to arrive at those coordinates and begin a search, and that the course and speed of the submarine were not known, other than that the speed of the submarine would be very low when compared to the speed of the anti-submarine vessel. This is obviously a very difficult problem because of the many uncertainties involved, such as the actual course or courses of the submarine and the fact that the speed could not be assumed constant.

Fortunately, problems facing engineers in which operations research can be utilized are usually more tractable because there are far fewer uncertainties.

Examples of problems in which operations research can be applied are the following. The first one, which is an intractable one if the numbers are evenly moderately large, is that of assigning people to do various jobs. Assuming that there are 10 people, all of whom can do any of 10 jobs that need to be done (but perhaps with different degrees of competence), a complete evaluation of all possible assignments would involve a list that is 10! long, or 3,628,800 rows long. A more tractable problem in this category is the transportation problem that considers the cost of shipping a commodity—say, coal—from any number of sources (mines) to any number of destinations (generating stations of an electrical utility). This problem will be further complicated if one is looking for the lowest cost of operating the entire system. One must then consider differences in the quality of the coal, the efficiency of each of the stations and sometimes of the units within each station, and the efficiency of transmission from the various stations to the points of utilization. Another problem is making the decision on production lot size that considers the optimum number of units to build for stock in view of set-up costs for a production run and the storage costs of completed units in stock. Analysis methods for these kinds of problems give the optimum solution without requiring the decision maker to look at every possible solution. In contrast to these problems, design problems usually involve enough explicit or implied specifications that the number of alternatives is limited. That is, engineering decision problems under certainty are usually made by simply comparing results from several applications of classic analytical procedures.

Decisions under certainty that the designer is more likely to find difficult are the following:

1. Which component of a product should be designed first?
2. How long will it take to do the job?
3. What tolerance can be assigned to the specification of a component?
4. How much is the production cost of this device likely to decrease in a year?
5. How can an evaluation function be devised which includes all of the important product characteristics?
6. How are the outcomes of the decision matrix found when selecting a component of a product from among several that are available?

The following subsections address questions such as these.

8.3.1 The Precedence Matrix

If you are assigned the task of designing a device or system, where do you start? There is *always* a best starting point. The reason that this is so is that the various

DECISIONS

parts of any device or system impose requirements on one another because they must work together. In an ideal situation, the part designed (or component chosen) first will be one that satisfies only the overall product specifications. The next part designed will satisfy the overall product specifications and the needs or requirements imposed by the first part designed. In such an ideal situation, there is a logical sequence in which the design of all of the parts proceeds without ever having to retrace one's steps to modify any of the earlier designs.

In nonideal (that is, practical) situations, parts are so dependent on each other that some must be designed based on assumptions of how other parts will interact. As the design proceeds, some of these assumptions will be found to be invalid and it will be necessary to backtrack and do some redesign on parts designed earlier. The final design will, then, be the result of a number of iterations. In order to minimize the redesign tasks and to reduce the number of iterations, we look for a method of deciding on an order in which to design the parts or subsystems that will tend to eliminate the necessity of redesign.

The model that determines the order in which information will become available to enable parts to be designed with a minimum of iterations is called a *precedence matrix*. An example is given in Figure 8.3. Development of the precedence matrix is preceded by the development of a precedence table, such as that in Table 8.1. All of the major parts of a device or system are listed in the table and assigned numbers for ease of identification. If a mathematical model is available for one of the parts, study of the equation or equations governing the model will allow the designer to name other parts that will affect its design. If a mathematical model is not available, consideration of the function of each part will still allow the designer to list the other parts affecting the design of that part. The list should reflect two or more levels of dependency, depending on whether the design of the part will depend in large measure on another part or parts or whether a variety of designs can work quite well with the part in question. It is unwise, however, to try to assign more than three levels of dependency, because the information available at the beginning of the process will probably not be accurate enough to warrant such an attempt at high precision.

At the beginning of the chapter, reference was made to the design of an electric car, and the question was raised as to whether there was a logical way of making a decision as to which subsystem to design first. Table 8.1 is the precedence table for the vehicle. The specifications are entries "a" through "e," and the possible subsystems are numbered from 1 through 9. Note that the general design concept has to be determined before creating this table. For example, it is possible to use a single motor for the drive, a motor driving each of the rear wheels, a motor driving each of the front wheels, or a motor driving each wheel. Only the first choice requires a drive train of some type to couple the motor to the driving wheels. Moreover, the choice of batteries in the front (or rear) may dictate the location of the drive motor in the rear (or front). For the purposes of our example, we assume that these choices have been made.

Table 8.1 Precedence Table for Electric Automobile

Specification	Major part #	Description	Predecessor level High	Medium	Low
a		No. of passengers			
b		Battery type			
c		Cruising speed			
d		Acceleration			
e		Range			
	1	Battery system	b,c,e,2,3	d,7	4
	2	Drive motor syst.	c,d,3,7		1
	3	Passenger comp.	a		
	4	Load-b'r'g struc.	1,2,3,6,7	8	
	5	Elect. dis. system	1,2		
	6	Suspension syst.	4,9		8
	7	Drive train	c,d,2		e
	8	Steering system	2,4		
	9	Wheels	1,2,3,6		4

With specifications, major subsystems, and the design concept determined, the designer must give careful consideration to each subsystem in order to identify those specifications and characteristics of interacting subsystems that will affect the subsystem under consideration. In Table 8.1, we note that the battery system (1) is shown to depend heavily on the battery type (b), the cruising speed (c), the range (e), the drive motor system (2), and the passenger compartment (3). Its design will also depend on the required acceleration (d), but to a lesser extent, and on the load-bearing structure (4) to a still smaller extent. When this type of analysis has been done for each subsystem, those with the least dependency on other subsystems are candidates for being placed first in the design sequence.

When the precedence table is constructed, it is helpful to make a list of all variables that are included in the specifications of the parts or subsystems. This enables you to look at each part as a black box and determine how it affects and is affected by other parts of the system or device. For example, the power output and speed of rotation of the drive motor are the variables that influence the design of the drive train, while the weight of the car (determined largely by the weight of the passenger compartment and of the batteries), aerodynamics of the passenger compartment, required acceleration, and cruising speed are the chief determinants of the drive motor.

DECISIONS

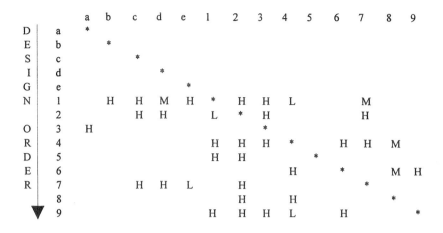

Figure 8.2 Matrix created directly from Table 8.1. Design information is deficient for numbered parts above the major diagonal.

Once the precedence table has been completed, the precedence matrix can be formed. The initial step is to form a matrix directly from the precedence table. This matrix shows the relative dependency of each of the subsystems on the specifications or on other subsystems, using symbols H for high dependency, M for medium, or L for low dependency. Figure 8.2 shows the initial matrix for the electric vehicle, the asterisks being shown merely to define the major diagonal. Keep in mind that the rows identify the subsystem or part to be designed; the entries in the columns identify specifications and other subsystems (yet to be designed if above the major diagonal) having an influence on the design of the subsystem being considered. For example, if the subsystems were to be designed in their numerical order, beginning with the battery system, one would have to make assumptions about subsystems 2, 3, 4, and 7, that is, about the drive motor system, the passenger compartment, the load-bearing structure, and the drive train. The designer will soon be hopelessly mired in a mass of assumptions on so many of the yet-to-be designed subsystems that closure of the design process will not be possible except in a very unsatisfactory way and with considerable time spent in redesign iterations. To avoid such an inefficient approach, the initial matrix is rearranged so that the number of H, M, and L designators above the major diagonal is minimized, resulting in the matrix of Figure 8.3. Note that during the rearrangement, moving any row up or down will simultaneously require a similar rearrangement of the columns because the sequence of the entries on the two axes must remain the same. Figure 8.3 was arrived at from the initial matrix of Figure 8.2 by several revisions, some of which resulted in less desirable situations than earlier attempts. Iterations will

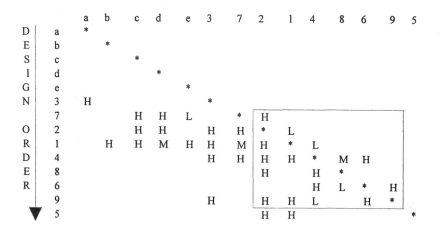

Figure 8.3 Precedence matrix for an electric automobile. Design information is deficient for numbered parts above the major diagonal.

have to be made in any case; take the time to iterate the matrix rather than the design.

Once the precedence matrix has been determined, it is immediately obvious in which order subsystems should be designed and which are likely to have to be iterated and the order of iteration. In this case, the passenger compartment is first. The subsystems that will require iteration begin with subsystem 7, the drive train, leading on to the drive motor system (2), the battery system (1), the load-bearing structure (4), the steering mechanism (8), the suspension system (6), and the wheels (9).

Whenever H, M, or L designators appear above the major diagonal of the matrix, it is necessary to assign certain needed values to the yet-to-be-designed part or subsystem named by the column as the subsystem designated by the row is being designed. For example, when designing the drive train (7), one must make assumptions about the drive motor system (2). The process continues: The design of the drive motor system requires some assumptions about the battery system (1); the design of the battery system requires some assumptions about the load-bearing structure (4); the design of the load-bearing structure requires assumptions about both the steering mechanism (8) and the suspension system (6); and the design of the suspension system requires assumptions about the wheels (9). Some of these decisions may be resolved by choices made by the designer or imposed on him or her; for example, the decision may be made to use a 240-volt battery system. It is important that, as the design proceeds, the actual value of any parameter is checked back against the value assumed earlier. If the values are not reasonably close, one must begin with new values and go through the process again. Some iteration is inevitable in this design.

DECISIONS

The possible extent of iteration in this example is indicated by the box in Figure 8.3 enclosing all subsystems except the passenger compartment (3) and the electrical distribution system (5). If one compares the possibilities and problems of the initial matrix of Figure 8.2 to those of the matrix in Figure 8.3, however, it becomes quite obvious that manipulation of the precedence matrix can result in a striking reduction in the number of subsystems for which assumptions are needed before starting the design of a subsystem or part. This, in turn, benefits the designer by reducing the number of iterations required. Note especially that for every one of the subsystem designs using the benefits of the precedence matrix of Figure 8.3, assumptions need be made about only one other subsystem except in one case, where two are required.

The example used above to illustrate the benefits of the precedence matrix is a system, and is therefore composed of a number of subsystems. The method is, however, general. It can be used equally as well when one is designing a device. For example, consider the design of a relay, such as that shown in Figure 8.4. In order to reduce the complexity of the example, certain parts of the relay will be ignored, including the wire terminals, the mounting bracket, insulation between the coil and the core, spacers, and so forth. The specifications and the main parts of the relay are listed in Table 8.2. The predecessor levels for each of the parts are determined from the following considerations:

1. The size and composition of the main contacts (1) are determined solely by the current and voltage ratings of the relay.
2. The armature (2) must be made of ferromagnetic material because it must carry the flux produced by the control coil. Its size will be determined by the force necessary to overcome the main spring (3).
3. The main spring (3) must provide rapid and positive action throughout a specified number of cycles of operation (e). The extension of the spring during each cycle is determined by the required break distance of the contacts, and this in turn depends on the voltage rating (b). The spring must provide sufficient force to seat the normally closed (NC) contact and to break minor contact welds on the normally open (NO) contact.
4. The movable contact support (4) is mounted on the armature and must carry the current from the contact to the pigtail (5). It is flexed slightly in both the NO and NC positions in order to assure positive contact. That is, it is a cantilever spring, and must provide sufficient contact pressure on the fixed contacts when the armature has been pulled down against the magnetic core and when the armature has been pulled back to the stop in the open position by the spring.
5. The pigtail (5) is a very flexible braided copper conductor capable of withstanding a large number of operations (e) without mechanical fatigue. It must also be large enough to carry rated contact current (a) without overheating.

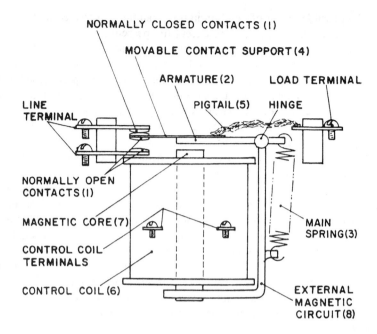

Figure 8.4 Basic elements of an electric relay.

6. The control coil (6) must be capable of establishing a magnetic field sufficient to attract the armature in opposition to the main spring. It must operate at the control voltage (d) and not exceed the specified current drain (c) on the power supply that operates the relay.
7. The magnetic core (7) is that part of the magnetic circuit inside the control coil. It is usually the part of the magnetic circuit (consisting of (2), (7), and (8)) that is smallest in cross section. Saturation must be avoided in order to maintain the necessary force for firm closure of the NO contacts. However, as the core is made larger, the control coil becomes larger due to the increase in the mean turn length of the coil turn and the larger diameter of wire necessary to maintain the correct value of coil resistance. The contact voltage rating (b) influences the break distance, and this in turn influences the operating force required from the magnet.
8. The complete magnetic circuit with the contacts in the NO position consists of the magnetic core (7), the external magnetic circuit (8), the armature (2), and the air gap between the core and the armature. The external magnetic circuit (8) must be made of ferromagnetic material and the cross section of all metal parts of the magnetic circuit must be sufficiently large to avoid magnetic saturation. The external magnetic circuit is that part of the relay on which the magnetic core is mounted and on which the armature is hinged.

DECISIONS

Table 8.2 Precedence Table for an Electric Relay

Specifi-cation	Major part#	Description	Predecessor level High	Medium	Low
a		Contact curr. rating			
b		Contact volt. rating			
c		Control coil current			
d		Control coil voltage			
e		Cycles of operation			
	1	Contacts	a,b		
	2	Armature	3,6		
	3	Main spring	1,b,c		
	4	Movable contact support	1,a	2	
	5	Pigtail	a,e		
	6	Control coil	c,d,1,7	b	2
	7	Magnetic core	b.1	3	
	8	External magnetic circuit	6	2	

In general, it is unlikely that an engineer can start a design project with a well-annotated drawing of the finished product and a description of the parts. The designer will probably begin with a design concept and make at least a crude sketch that identifies the major parts of the product. Although later refinement would be necessary, a fairly good description of the various parts similar to that above can be written based on the sketch and the design concept. In this case, relays being such common electrical devices, the engineer has the advantage that he or she can begin with an outline drawing such as that in Figure 8.4.

The precedence matrix of Figure 8.5 would then be constructed in a manner similar to that described for the electric car. The order of part design is very simple for this example. One begins with the contacts (1) based on the relay current and voltage ratings and the force necessary to insure good contact when the NO contacts are closed. This value of force would then be used to determine the flux density and the air gap cross section using Equation (18) from Chapter 7, which was derived from dimensional analysis. The cross section of the magnetic core (7) can then be determined so that the flux density is kept below saturation of the magnetic material. The control coil can be designed using an inner diameter slightly larger than that of the core. The length of the coil and the core will be fixed simultaneously based on the wire size and the number of turns necessary to produce the required magnetomotive force at the

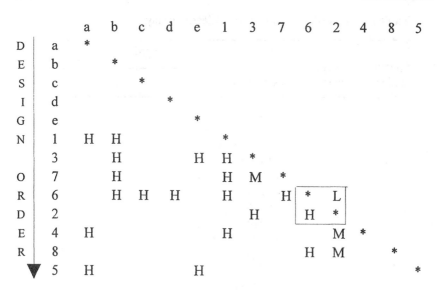

Figure 8.5 Precedence matrix for an electric relay.

rated coil current (c). The outer diameter of the coil will determine the width and the minimum length of the armature (2) and the movable contact support dimensions will be related to overall length of the armature and movable contact support. The material used for the movable contact support will typically be phosphor bronze, the thickness being determined by the desired contact force when the armature is pulled against the magnetic core. The external magnetic circuit requires sufficient cross section, but the longitudinal dimensions and shape are determined simply from the dimensions of the previously designed parts. The pigtail is selected for flexibility and current-carrying capacity. Finally, the spring is designed to provide a force sufficient to break minor contact welds, ensure mechanical movement to the upper contact, and of the correct length to fit between the hook formed on the hinge end of the armature and the hook formed on the vertical part of the external magnetic circuit. Note that minor iteration of coil size and armature dimensions may be needed.

8.3.2 Critical Path Method (CPM) and Program Evaluation and Review Technique (PERT)

"How long will it take to do this job? What delivery date can I promise the customer? Can't you make an earlier date?" These are questions that frequently arise in the development of new products. For example, if a book publisher is negotiating to buy a nipper and smasher from your firm and the machine is to be

DECISIONS

custom-made, the delivery date will be an important part of the purchase contract. If you are supplying components to another manufacturer, their buyer needs to know when the first units will arrive so that they can plan their production schedules. If your delivery time is too long, someone else may get the business. If you and a competitor are simultaneously developing products having identical functions, substantial advantages can accrue to the company that is earlier to market, as Figure 8.6 shows. Reduction in development time improves the profitability of the product because development costs are reduced, being first to market frequently puts the product in the commanding position of dominating the market and at a higher price, and the production run will be longer. Time estimation techniques are therefore important in order to be able to answer questions such as those above with confidence, as well as to be able to find a sequence of steps in the development process that will take as little time as possible.

The development of orderly procedures for estimating the time to complete a task has led to two popular methods. Both methods were developed in the 1950s, and both continue to be used today. Numerous variations

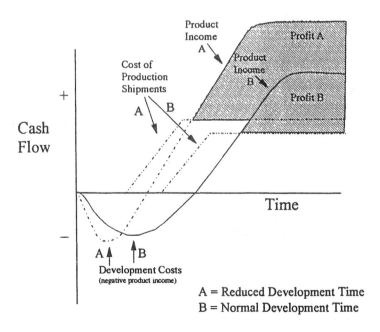

Figure 8.6 Product income versus time with reduced and normal development times.

of the original methods have been described; many of the variations are applicable only under special circumstances or are oriented toward a particular kind of business activity.

One of these methods is the critical path method, usually referred to as CPM. The other is PERT, an acronym that means either program evaluation and review technique (apparently the original meaning) or project estimating and resource tracking, depending on one's personal preference. These will be referred to henceforth as CPM and PERT. The methods are similar in many respects, the principal difference between them being the way in which estimates of activity time are decided. The basic tool is a work-flow chart made with two kinds of elements. They are:

1. An *event*, which must be recognizable in character and in time. A circle will be used to indicate an event on the chart.
2. An *activity*, which leads to an event and has a known duration. These are shown on the chart by lines joining two events with arrowheads pointing in the direction of progress toward completion of the project.

There are two simple (and rather obvious) rules that must be followed in construction of the chart:

1. An event is not reached until all activities leading to it are complete.
2. An activity cannot commence until its tail event is reached.

As an example of CPM, suppose that your design team is to submit a report to your supervisor. In addition to the engineers in your group, you have support personnel such as a word processing specialist and a person who does computer graphics. The computer graphics person is new and is still becoming accustomed to unfamiliar equipment. One of the engineers in your group has demonstrated great skill in checking reports for adherence to company format, spelling, and syntax. (We will call that person the wordsmith.) You are the only person who can write the report, and your style is to make the necessary sketches as you write the rough draft, completing the last of them after the rough draft is finished. One possible work-flow diagram is that shown in Figure 8.7, in which:

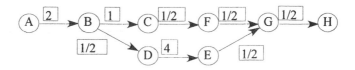

Figure 8.7 Work-flow diagram for report preparation.

DECISIONS

Event A = start of the process
Event B = rough draft is complete (and corrections begin)
Event C = corrections are complete (word processing begins)
Event D = rough sketches are complete (computer graphics begin)
Event E = computer graphics complete (proofing of graphics begins)
Event F = word processing is complete (proofreading text begins)
Event G = proofreading of complete report begins
Event H = final corrections of report made and report delivered

The time necessary to complete each activity is given by the following estimates and each estimated time is shown in a box alongside the arrow. The estimated times are:

A-B [writing the rough draft]: Two days
B-C [correcting the rough draft]: One day
C-F [doing word processing on the text]: One-half day
B-D [completing the rough sketches]: One-half day
D-E [doing the computer graphics]: Four days
F-G [proofreading the text]: One-half day
E-G [proofing the computer graphics]: One-half day
G-H [making final corrections and delivering the report]: One-half day

The diagram of Figure 8.7 immediately reveals some problems with the initial plan. The total time for preparation of the report is 7-1/2 days, 1-1/2 work weeks. This is the time along the path A-B-D-E-G-H. Moreover, we note that path B-C-F-G is only 2 days, whereas the path B-D-E-G is 5 days. The major part of that path is taken up by the production of the computer graphics. It is immediately obvious that production of the computer graphics is the bottleneck, and it is also obvious that there is no need to wait until the rough draft is completed and then to wait an additional half day for completion of the last rough sketches before beginning work on the computer graphics. Moreover, proofreading the text and proofing the computer graphics are each done independently, and only after those have been completed is the entire report gone over as a whole. It would be better if proofreading of text and computer graphics were done while reviewing the entire report.

If you begin to make rough sketches available to the person who does the computer graphics while you are still writing the report, the diagram must be modified in some way to indicate that this process is going on before event B has been reached. That is, activities that in Figure 8.7 are planned to occur sequentially will now be planned to occur at least partially concurrently. This is shown by means of a "ladder" diagram. On the left in Figure 8.8, if an extra event E' is interposed between events D' and E and at the same level as E in the diagram, it can be seen that AB, D'D, and E'E would constitute three rungs of a (skewed) ladder. Event E' has been omitted from Figure 8.8, however, because it is the total time from event D' (the point at which the first rough sketches are delivered to the computer graphics person) to event E (completion of the computer graphics) that is important. (Activity D'-E takes 4 days, and times for activities D'-E' and E'-E would be arbitrary, requiring only that their sum is 4 days.) It is assumed that you will deliver the rough sketches to the computer graphics person at a rate that will keep that person busy.

It was noted above that it would be desirable to proof the text and computer graphics as part of your final review of the report. Moreover, the wordsmith should have copies of the rough sketches with the text material. To insure that these activities occur in the desired manner, it is necessary to introduce *dummy* activities. A dummy activity is defined as an activity that requires no time to complete and no resources. Since event D is the completion of the rough sketches, they must be delivered to the wordsmith before that person's review begins. The dummy activity D-B (shown by a dashed line) insures that this will occur, and the dummy activity E-G insures that the computer graphics and the text are assembled before you review and proof the final report. (Note that event F has disappeared.) The time from start to completion is now 5 days, 1 work week. The advantage was gained largely by the simple step of beginning to feed rough sketches to the computer graphics person concurrently with the writing of the rough draft.

With a work-flow chart as simple as those of Figures 8.7 and 8.8, the *critical path* is easily determined. The critical path is that path between the beginning event and the event that terminates the diagram that requires the longest time to complete. As we have seen, the time for Figure 8.7 is 7-1/2 days, the critical path being A-B-D-E-G-H; the time for Figure 8.8 is 5 days, the critical path being A-D'-E-G-H. If the work-flow chart is more complicated, a more formal procedure is needed to determine the critical path. Figure 8.8 will be used illustrate how this can be done.

The first step is to determine the *earliest event time* by assigning zero time to the starting event and proceeding through the diagram to the completion event. If several activities lead *to* an event, the earliest time is fixed by the long-

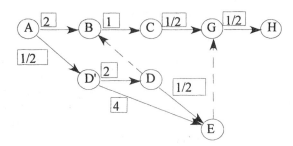

Figure 8.8 Final work-flow diagram of report preparation.

DECISIONS

est chain of activities leading to that point. This concept merely reinforces the concept that an event is not reached until *all* activities leading to it have been completed. For Figure 8.8, the earliest times may be tabulated as follows:

Event	Earliest time
A	0
B	2-1/2 (by way of D' and D)
C	3-1/2
D'	1/2
D	2-1/2
E	4-1/2
G	4-1/2 (by way of D' and E)
H	5

The next step in finding the critical path is to compute the *latest event times* by assigning the earliest event time from the previous tabulation to the last event, and then proceeding backward through the diagram to the first event. This will give information on the latest times the events can be reached if the project is to be done as quickly as possible. Keeping in mind that an activity cannot begin until its tail event has been reached, we see that if several activities follow *from* an event, the latest acceptable time of the event is determined by the longest time chain extending from it. For Figure 8.8, these times are:

Event	Latest time
H	5
G	4-1/2
E	4-1/2
D	3 (by way of G, C, and B)
D'	1/2 (by way of G and E)
C	4
B	3
A	0

Finally, the two listings are combined, having reversed the order of the second tabulation so that earliest and latest times can be easily compared. This gives us:

Event	Earliest time	Latest time
A	0	0
B	2-1/2	3
C	3-1/2	4
D'	1/2	1/2
D	2-1/2	3
E	4-1/2	4-1/2
G	4-1/2	4-1/2
H	5	5

Those events having the same values of earliest and latest event times are on the critical path. It is convenient to show the earliest and latest times on the work-flow diagram. Figure 8.8 has been redone as in Figure 8.9 to include this information. In this figure, attention is drawn to the critical path A-D'-E-G-H by the use of double arrows. Along the critical path the times associated with head or tail events differ by exactly the activity duration. This is not the case off the critical path, as will be seen by considering events A and B, for example. That is, the duration of activities off the critical path can be increased by various amounts without increasing the project time.

Since off-critical path activity times may be increased, it is wise to transfer effort from activities off the critical path to those that are on the critical path if it is possible to do so, thus reducing the duration time for some activities on the critical path and speeding up the entire project. If this concept is viable and is pursued, one needs to know how the maximum permissible duration times of the off-critical path activities are calculated.

The length of time by which an activity duration can be increased without disturbing the progress of the project is known as the *float*. For a given activity, float is the difference between the latest event time of the head event and the

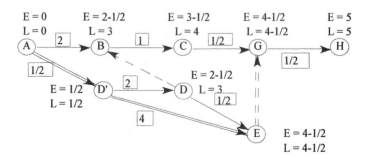

Figure 8.9 Report preparation schedule.

DECISIONS

earliest event time of the tail event less the duration time of the activity. Hence it is always zero for activities on the critical path. Off the critical path, we can find values for the float in Figure 8.9. For example, the float for activity A-B is 3 – 0 – 2 = 1 day. However, it soon becomes evident that extension of the duration of various activities by the float time will delay the project. For example, the float for activity D'-D = 3 – 1/2 – 2 = 1/2 day, and that for D-E = 4-1/2 – 2-1/2 – 1/2 = 1-1/2 days. If these activities are increased in duration by their respective floats, path A-D'-D-E requires 5 days, and the overall duration of the project has increased by 1/2 day.

The concept of float may be improved in usefulness if the concept of *free float* is introduced. Free float is defined as the difference between the earliest time of the head event and the earliest time of the tail event less the activity duration. For activity A-B in Figure 8.9, the free float is 2-1/2 – 0 – 2 = 1/2 day. Extending the duration of an activity by the free float will not affect succeeding activities but will in general place the constraint on previous activities of meeting the earliest event time as a completion date. In this figure, the free float for D'-D is 2-1/2 – 1/2 – 2 = 0. For D-E it is 4-1/2 – 2-1/2 – 1/2 = 1-1/2 days. If D-E's duration were to be increased by the free float, a second critical path, A-D'-D-E-G-H, would appear in parallel with the first one, A-D'-E-G-H.

The time by which the duration of an activity can always be increased is called *independent float*. It is the difference between the earliest time of a head event and the latest time of the tail event less the duration time of the activity. In Figure 8.9 the independent float for A-B is 1/2 day, and that for D-E is 1 day. For other activities, such as C-G, the independent float is zero.

Design is inherently an iterative process; any diagram showing the activities of a design project will be laced with feedback loops. On the other hand, work-flow diagrams are based on the concept that a project proceeds in a continuous flow from start to completion. Thus the two seem to be antithetical in nature. However, the work-flow diagram can still be used as a model for a design project if suitable activities to review, verify, test, and improve the design are included. Figure 8.10 shows a work-flow chart for a typical design project.

One of the main advantages of using CPM is that it forces the design leader to examine each job that must be done to complete a project. Without this type of formal approach, there is a tendency to feel that mere involvement with a project is equivalent to progress. There *may* and probably will be progress, but very likely far less in any given time period than if a clear plan has been devised setting forth a timetable and frequent and clearly identifiable goals. The comparisons drawn earlier of the results of the plans of Figures 8.7 and 8.8 (or 8.9) make this quite clear.

CPM lends itself very well to planning, scheduling, and controlling a project. An example of the advantages to be gained from the planning idea have already been seen with respect to the report project. As soon as Figure 8.7 was drawn, it was immediately apparent that there was a bottleneck in the computer graphics area, and the decision to send rough sketches to the computer graphics

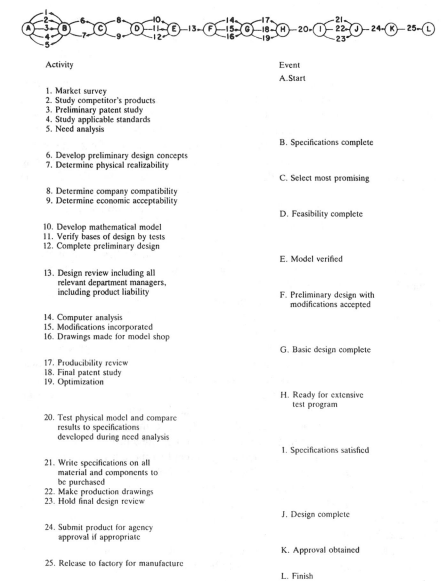

Activity	Event
	A. Start
1. Market survey	
2. Study competitor's products	
3. Preliminary patent study	
4. Study applicable standards	
5. Need analysis	
	B. Specifications complete
6. Develop preliminary design concepts	
7. Determine physical realizability	
	C. Select most promising
8. Determine company compatibility	
9. Determine economic acceptability	
	D. Feasibility complete
10. Develop mathematical model	
11. Verify bases of design by tests	
12. Complete preliminary design	
	E. Model verified
13. Design review including all relevant department managers, including product liability	
	F. Preliminary design with modifications accepted
14. Computer analysis	
15. Modifications incorporated	
16. Drawings made for model shop	
	G. Basic design complete
17. Producibility review	
18. Final patent study	
19. Optimization	
	H. Ready for extensive test program
20. Test physical model and compare results to specifications developed during need analysis	
	I. Specifications satisfied
21. Write specifications on all material and components to be purchased	
22. Make production drawings	
23. Hold final design review	
	J. Design complete
24. Submit product for agency approval if appropriate	
	K. Approval obtained
25. Release to factory for manufacture	
	L. Finish

Figure 8.10 Work-flow diagram of a typical design project.

DECISIONS

person while the rough draft was still in the process of being written was an obvious one.

Scheduling is facilitated by the use of work-flow charts. The supervisor of a group of design teams, each of which is working on a different project, can ensure the availability of technical assistants, draftsmen, modelmakers, and other kinds of support personnel when they are needed if he or she has a work-flow chart for each project. Without coordination of the several plans, there is a high probability that one category of support personnel will be needed to supply their services virtually simultaneously, while at the same time another category of personnel has no work to do.

Finally, the problem of controlling a project is substantially reduced by the availability of the work-flow chart. Anticipated earliest and latest dates of reaching various events can be added to the diagram as soon as the starting event is reached, and actual event dates can be then be inserted as the events occur. As long as the actual date is between the earliest and latest dates on the work-plan, the project can be said to be "on schedule." The difference between the earliest and latest dates to reach an event is referred to as *slack*. If event times fall outside their slack, decisions must be made as to how to get the project back on schedule if the actual date is after the latest date on the plan, or what advantage can be taken of an actual date that is before the earliest date for reaching the event on the plan. Since events are specific in nature, there is no ambiguity as to the date to report. In contrast, reporting that a certain activity is "75% complete" involves judgment, and cannot be as precise as reporting that a particular event has occurred.

One of the most difficult tasks of the designer is making estimates of the duration of any activity. If there are tests to be run, there may be uncertainty about the number of tests that will be required or about technician availability, and test equipment failure is always a possibility. If drawings are to be made, the designer may be able to enumerate exactly how many and have a feel for how long it should take to do each drawing, but there is always uncertainty about these numbers. In order to be able to give the designer the ability to state with some confidence an estimation of the time for an activity, the technique referred to earlier as PERT was devised for the U.S. Navy Special Projects Office. Using PERT, the designer makes one estimate of how quickly each activity can be completed if everything goes exactly as it should, another estimate of how long it should take if things go wrong, and a third of the most likely time. With the assumption that the actual activity times follow a statistical distribution called the beta-distribution, the *expected time* is

$$T_e = \frac{\text{optimistic time} + 4 \times (\text{most likely time}) + \text{pessimistic time}}{6}$$

Moreover, the tolerance that should be assigned to the expected time under the assumption of a beta-distribution is given by the standard deviation σ_β where

$$\sigma_\beta = \frac{\text{pessimistic time - optimistic time}}{6}$$

In PERT, the expected time is used for each activity in computing the earliest and latest event times, just as in CPM. The accumulated standard deviation along any path is simply the square root of the sum of the deviations squared (the variances) along the path. That is,

$$\sigma_{\beta p} = \sqrt{\sum_{x=1}^{n} \sigma_{\beta x}^2}$$

where

$\sigma_{\beta p}$ = accumulated standard deviation along a path,

and

$\sigma_{\beta x}$ = the standard deviation for activity x

The standard deviation is used in a very interesting way. Suppose you wish to find the probability of completing a job at a certain scheduled time T_s. First, express the difference between the scheduled time and the critical path time, T_c, in terms of the number of standard deviations of the critical path. That is, let

$$z = \frac{T_s - T_c}{\sigma_{\beta p}}$$

Table 8.3 Cumulative Probability versus Number of Standard Deviations from Mean

z	p
3.0	0.9987
2.5	0.9938
2.0	0.9772
1.5	0.9932
1.0	0.8413
0.5	0.6915
0.0	0.5000
−0.5	0.3085
−1.0	0.1587
−1.5	0.0668
−2.0	0.0228
−2.5	0.0037
−3.0	0.0013

DECISIONS

Table 8.3 gives the cumulative probabilities of the normal probability distribution corresponding to the z-factors. A more extensive table can be found in most books on probability and statistics. For example, suppose the critical path time to complete a project is computed to be 50 weeks with a standard deviation of 4 weeks. The probability of meeting a scheduled completion of 48 weeks (without increasing personnel, changing the work plan, and so forth) is found from

$$z = \frac{48-50}{4} = -0.5$$

Entering Table 8.3 with this value of z we find that the probability is 30.85%.

For an expanded discussion of CPM and PERT, the reader may wish to consult Reference 1.

8.3.3 Tolerances

The foundation of mass production is the ability to produce parts for a device that are interchangeable with each other. Yet we know that even with great care and complete disregard for cost it is impossible to make two parts that are completely identical. Hence the designer must think in terms of tolerances on parts. How great a difference is allowable between parts without coming to the unpleasant realization that the device cannot always be successfully assembled from parts selected at random? Conversely, if the parts of a device have tolerances that the designer knows or can set, what can be learned about the tolerance of the device as a whole?

The following technique is useful whenever an analytical expression can be devised that relates the performance to component values and their tolerances. The end result of application of the technique is the tolerance of the device performance. The technique is based upon the assumptions that (1) the part tolerances are small compared to the nominal values, (2) there is no correlation between the deviations from nominal of any two variables, and (3) the parts are randomly assembled. Then, if $y = f(x_1, x_2, \ldots, x_n)$ the standard deviation is given by

$$\sigma_y^2 = \left(\frac{\partial y}{\partial x_1}\right)^2 \sigma_{x_1}^2 + \left(\frac{\partial y}{\partial x_2}\right)^2 \sigma_{x_2}^2 + \ldots + \left(\frac{\partial y}{\partial x_n}\right)^2 \sigma_{x_n}^2 \qquad (1)$$

If the deviations of the parts have a normal distribution, three standard deviations will include 99.74% of the population, as can be seen from Table 8.3. If we therefore let $T = \pm 3\sigma$, multiplication of both sides of Equation (1) by 9 will allow us to rewrite the equation as

$$T_y^2 = \left(\frac{\partial y}{\partial x_1}\right)^2 T_{x_1}^2 + \left(\frac{\partial y}{\partial x_2}\right)^2 T_{x_2}^2 + \ldots + \left(\frac{\partial y}{\partial x_n}\right)^2 T_{x_n}^2 \qquad (2)$$

where the tolerances are each expressed in units of the respective variables.

To see the value of this approach, we will look at some design problems. Suppose that an electronic circuit design requires a resistance of 30,000 ohms (Ω) and that it must dissipate 1 watt (W) of power. The standard (off-the-shelf) series of resistor values does not include a 30,000-Ω value, but 10,000-Ω resistors capable of dissipating 1/2 W each are stock items. The designer proposes to use three 10,000-Ω resistors in series. A standard tolerance on resistors used in electronic circuits is ±10%, and the designer wants to know what the tolerance will be on the series combination. This situation fits a model described by the equation $y = x_1 + x_2 + x_3$ because the total resistance is simply the sum of the three resistances. The partial derivatives of y needed in Equation (2) are all equal to unity. Each of the T's on the right side of the equation is equal to ±1,000 Ω. Substitution into the equation yields a tolerance for the combination of three resistors in series of ±1,730 Ω, or ±5.8%. That is, we can expect that 99.74% of all three-resistor combinations will fall between 28,270 and 31,730 Ω.

Although we will not investigate the problem, you should be aware of the fact that the population of resistors having a tolerance of 10% may in fact have no members that fall into the ±5% band because the manufacturer has used automatic testing equipment to select out the ±5% group for sale at a higher price per unit. The same statement is true about resistors having a stated tolerance of 5%; there may be no 1% resistors in their population.

Suppose that the 30,000-Ω resistor combination is to be used as the load resistor in a transistor amplifier. The current amplification A_i for the amplifier can be expressed as

$$A_i = \frac{h_{21}}{1 + h_{22} R_L} \qquad (3)$$

where

h_{21} = the short-circuit forward-transfer current ratio (dimensionless)
h_{22} = the open-circuit output driving-point admittance (siemens, or reciprocal ohms)
R_L = the load resistance (ohms)

The manufacturer of the transistor gives the following values and tolerances for the h-parameters:

$h_{21} = 49 \pm 20\%$
$h_{22} = 5 \times 10^{-5} S \pm 10\%$

DECISIONS

We already know that $R_L = 30{,}000 \, \Omega \, \pm 5.8\%$.

The nominal value of amplification is

$$A_i = 49/(1 + 5 \times 10^{-5} \times 3 \times 10^4) = 19.6$$

From Equation (2),

$$T_{A_i}^2 = \left(\frac{\partial A_i}{\partial h_{21}}\right)^2 T_{h_{21}}^2 + \left(\frac{\partial A_i}{\partial h_{22}}\right)^2 T_{h_{22}}^2 + \left(\frac{\partial A_i}{\partial R_L}\right)^2 T_{R_L}^2$$

$$= \left(\frac{1}{1 + h_{22}R_L}\right)^2 T_{h_{21}}^2 + \left(\frac{h_{21}R_L}{(1 + h_{22}R_L)^2}\right)^2 T_{h_{22}}^2 + \left(\frac{h_{21}h_{22}}{(1 + h_{22}R_L)^2}\right)^2 T_{R_L}^2$$

(Some of the partial derivatives are negative, but those signs are ignored because of the squaring.) Substitution of numerical values yields

$$T_{A_i}^2 = \left(\frac{1}{2.5}\right)^2 (9.8)^2 + \left(\frac{49 \times 30{,}000}{(2.5)^2}\right)^2 (5 \times 10^{-6})^2 + \left(\frac{49 \times 5 \times 10^{-5}}{(2.5)^2}\right)^2 (1730)^2$$

$$= 15.36 + 1.38 + 0.46 = 17.20, \text{ leading to} \qquad T_{A_i} = 4.15 = 4.2.$$

That is, the amplification for 98.74% of the amplifiers assembled from these components will be 19.6 ± 4.2, or $19.6 \pm 21.4\%$. Without going into a detailed analysis, it is easily seen that the major part of the variations to be expected in the current gain may be ascribed to the tolerance in h_{21}. Some manufacturers, faced with this problem, have put all incoming transistors through a test process in order to sort them into narrower tolerance bands.

As another example, consider a mechanical system consisting of a mass $m = 10$ kg $\pm 5\%$ hanging on a coil spring having a modulus $k = 600$ N/m $\pm 7\%$. If the mass is displaced vertically and then released, the system will oscillate. We want to know the natural frequency of oscillation of the nominal system, and the range of frequency that we might expect if we assemble a number of these systems.

The angular frequency is given by $\omega = \sqrt{k/m} = 7.746$ rad/s, or 1.232 Hz. Application of Equation (2) leads to

$$T_\omega^2 = \left(\frac{\partial \omega}{\partial k}\right)^2 T_k^2 + \left(\frac{\partial \omega}{\partial m}\right)^2 T_m^2$$

$$= \left(\frac{1}{2\sqrt{km}}\right)^2 T_k^2 + \left(\frac{-1}{2m}\sqrt{k/m}\right)^2 T_m^2$$

$$= \left(\frac{1}{2\sqrt{6000}}\right)^2 (42)^2 + \left(\frac{-1}{20}\sqrt{\frac{600}{10}}\right)^2 (0.5)^2$$

$$= 0.0735 + 0.0375 = 0.111$$

Then $T_\omega = 0.333$ rad/s, or 0.053 Hz. The variation in frequency to be expected of randomly assembled systems is no more than 4.3% of the nominal frequency.

When picking parts at random for assembly, one would expect to pick parts having parameter values that are above nominal as often as one would pick parts having values below nominal. Should there not, therefore, be a tendency to cancel out variations? That is, doesn't intuition tell us that assembly of systems from components having known tolerances should result in systems having tolerances smaller than the largest of the tolerances of the individual parts? We see in these examples that intuition can easily lead us astray; the hypothesis above is true in two of the three cases we have examined, but in the example of the transistor amplifier the tolerance of the amplifier current gain is larger than the largest tolerance of any of the parts.

We are accustomed in engineering analysis to working through a problem and arriving at a single result—*the answer*. We see from the discussion of this section that there will always be some uncertainty about the precision of that result. If the tolerances are known for the parameters appearing in any equation you have studied in engineering, you have a tool for determining how much variation can actually be expected.

8.3.4 The Learning Curve

When production begins on a new device or system, the time required for the production of the first few units almost invariably turns out to be so long that the expectation of profitability seems to evaporate because of the labor costs. However, as production continues, the time to produce each unit falls as the assemblers learn the task. There is generally a steep decline in production time per unit early in the production, but the slope of the curve will be reduced as time goes on, the end result being an asymptotic approach to some fixed value. This may be followed by an increase in production time as tools wear, the assemblers become a bit careless, and so forth.

Unless the production process was designed very astutely, a second reduction in production time will be possible because experience has indicated favorable ways to reorganize the work flow, a difficult-to-assemble subassembly

DECISIONS

has been redesigned, some operations can be combined, and so forth. This pattern of productivity is evident from data accumulated on a wide variety of commodities, from steel to airplanes to personal computers, although reversals of the trends occur from time to time.

It should be understood that other factors may influence production rates. The famous Hawthorne experiment of Western Electric in the 1920s made it clear that attempts to measure the effect on productivity of changes in working conditions could easily lead to misleading results. In this experiment, productivity rose whenever a change in working conditions was made, even when the change was expected to result in a decrease in productivity. The conclusion of the study was that the effects observed were due to the fact that the workers involved knew that an experiment was going on, and were stimulated in their activities by the mere fact that they knew that what they were doing was being evaluated. That is, the psychological effects of the way in which the experiment was conducted overrode the changes in the work situation.

The trend of decreasing production time may be modeled mathematically if it is assumed to follow a smooth curve in which each doubling of the number of units produced results in a constant percentage decrease in the cumulative average time necessary to produce the product. For example, if production of the first unit requires 10 hours, the cumulative time and the cumulative average time for that unit are both 10 hours. If the second unit is produced in 6 hours, the cumulative time to produce two units is 16 hours, and the cumulative average time is 8 hours for two units. The ratio of the cumulative average times is 0.8; this is an 80% learning rate. Extension of these values to later (and larger) groups of production units results in the third column of Table 8.4. Assuming this ideal situation, the cumulative average time of production C(X) as production continues can be calculated from

$$C(X) = C(1) \cdot X^B \qquad (4)$$

where

X = the total number of units produced
$C(1)$ = the production time of the first unit
B = the slope of the learning curve on log-log paper

The value of B depends on the learning rate. Consider two values of cumulative production related by an integer multiple. If that multiple is 2 and the learning rate is 80%, then the ratio of any value of C to its predecessor is 0.8, and the ratio of the corresponding X's is 2. Hence

$$2^B = 0.8, \qquad \text{so that} \qquad B \log 2 = \log(0.8)$$

and B is then –0.322.

Table 8.4 Learning Curve Data (All times in hours)

Group number	Unit number(s) produced	Cum. avg. unit time	Total time	Total group time	Avg. time of units in group	Est. time for last unit of the group
1	1	10.00	10.00	10.00	10.00	10.00
2	2	8.00	16.00	6.00	6.00	5.42
3	3,4	6.40	25.60	9.60	4.80	4.33
4	5 to 8	5.12	40.96	15.36	3.84	3.47
5	9 to 16	4.10	65.60	24.64	3.08	2.78
6	17 to 32	3.28	104.96	39.36	2.46	2.22
7	33 to 64	2.64	168.96	64.00	2.00	1.78
8	65 to 128	2.09	267.52	98.56	1.54	1.42

The total time to produce all units from the beginning of production to the end of any group is obtained by multiplying the cumulative average unit time by the total number of units produced, giving the data in column 4. The total time to produce the units in any group is obtained from the data of column 4 by subtracting from the total time to the end of that group the total time to the end of the previous group, resulting in the data of column 5. The average production time for units in each group, shown in column 6, is the ratio of the total time for the group to the number of units in the group.

Since production time is assumed to fall on a continuous basis, the production time per unit shown in column 5 is accurate only for a unit produced about midway in the group produced in that production run. The production time U(X) for any unit can be calculated from Equation (5):

$$U(X) = (1+B) \cdot C(1) \cdot X^B \qquad (5)$$

The data in column 7 of Table 8.4 were calculated using this equation. Assuming that the production run is expected to be a very long one and that no major changes are to be made in the production line or in the product itself, the equation may be used to predict production times to be expected in the future. If 10,000 units are to be produced, the time to produce the last unit can be expected to be 0.349 hour.

The cumulative average unit production time and the average production time of units in the various groups are plotted in Figure 8.11.

The discussion above and the data in Table 8.4 are predictive in nature. *If* production of the first unit requires 10 hours and *if* the learning rate is actually 80%, then the data of the table and the curves of the figure may be expected.

DECISIONS

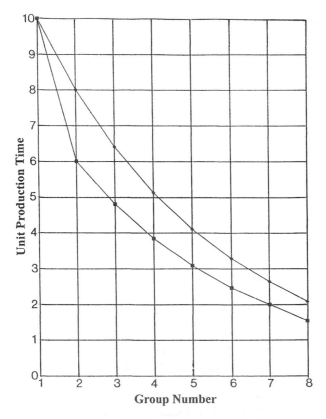

Figure 8.11 Learning curve (80%) for a hypothetical product. ♦ - Cumulative average unit time. ■ - Unit time.

The techniques are most useful in the early stages of production if they are used for both a retrospective analysis and for predictive purposes. If careful records are kept of the production time for each unit or group of units early in the production run, a retrospective analysis of these early data will yield a value for the learning rate which will in turn allow reasonable predictions to be made of the production times later. As production proceeds and more data become available, refinement of the value of the learning rate will allow the engineer to predict production times with increasing confidence. Learning rates that can be used even before production begins can be obtained from the data from previous production runs on similar products. These data may also allow an estimate of the production time of early units.

The reduction of 20% in average labor time (80% learning rate) for each doubling of units produced is not unusual for a labor-intensive manufacturing process. Industries that are capital-equipment-intensive, such as steel rolling mills, are more likely to see reductions of only 5 to 10%, corresponding to learning rates of 95% to 90%. In those industries, production time improvements may be difficult to achieve.

8.3.5 Criterion Functions

As a design progresses, there will be situations in which a choice must be made among several materials, components, or alternative designs. This usually involves comparing quantities expressed in different units, thus making direct comparison impossible. One way of making the necessary comparison is to form a criterion function [2,3,4] which includes a set of constants that normalize the units for the range involved in the comparison. It is also possible to include a set of weighting factors. Criterion functions are also known as objective functions, utility functions, optimization functions, payoff functions, and response functions, depending on the author or the industry terminology.

Whatever the name used, a criterion function may be written as

$$CF = \sum_{i=1}^{n} a_i K_i x_i \qquad (6)$$

where

a_i = the weighting factor of variable i
K_i = a constant used to normalize the dimensions of variable i
x_i = a criterion variable, such as a parameter, cost, and so forth

and the function CF must either be dimensionless or be in homogeneous units.

An example will be used to illustrate the method. Consider the design of a transformer using scrapless E-I laminations as shown in Figure 8.12. The important quantities for the designer, the sales department, and the company administration to consider are the cost, the weight, and the power loss. These quantities are functions of the center leg width D of the lamination, as shown by the three curves of Figure 8.12. They are also functions of the cost of magnet wire in various sizes, the weight of the steel in the core, the kind of material chosen for the core and its cost, and the allowable change in output voltage of the transformer secondary from no load to full load. One factor that is kept constant throughout the design process is the flux density in the steel. That constraint causes the cross-section area of the steel to be a constant, so that as the center leg width D varies from a small value to a large value, the depth of the stack of laminations will vary from a large value to a small value, and the length of a turn of wire on the coil will vary from large to small and back to large. The data for the curves were obtained by repeating the transformer design process using different assumed values of D.

The criterion function for the transformer is

$$CF = a_C K_C C + a_W K_W W + a_P K_P P$$

DECISIONS

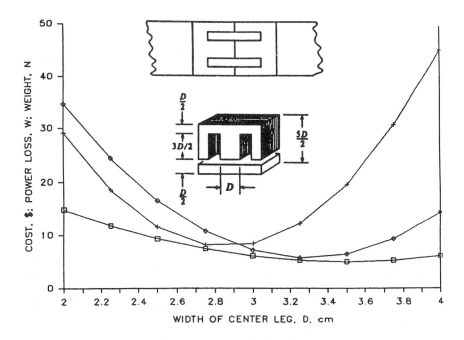

Figure 8.12 Cost (□), power loss (+), and weight (◊) versus core size for a transformer using scrapless laminations.

where C, W, and P are the cost, weight, and power loss functions in dollars, newtons, and watts, respectively.

Curve-fitting techniques may be used to obtain equations for the three curves in terms of center leg width D:

$$C = 4.41(D^2 - 6.99D + 13.34)$$
$$W = 17.37(D^2 - 6.58D + 11.15) \qquad (7)$$
$$P = 28.85(D^2 - 5.72D + 8.45)$$

These equations may be substituted into the criterion function, but before doing so values must be selected for the weighting factors. The sales department agrees that weight of the transformer is the least important factor, but that cost should weigh heavily. However, they also agree that the company needs a line of high-efficiency transformers. After some discussion, a consensus is reached that the weighting factors to choose are $a_C = a_P = 1$ and $a_W = 0.25$. With these values,

$$CF = K_C C + 0.25 K_W W + K_P P$$

The criterion function will be dimensionless if the K's have values that are reciprocal dollars, reciprocal newtons, and reciprocal watts. One way to do this is to select reasonable values for the three dependent variables, C, W, and P, and then use their reciprocals for the K's. We choose values in the vicinity of the minima of the three curves, which is where we expect the criterion function to lead us. The K's become 1/$6, 1/8N and 1/10W respectively.

Using these values of the K's and Equations (7) the criterion function becomes

$$CF = \frac{4.41}{6}\left(D^2 - 6.99D + 13.34\right) + \frac{0.25 \times 17.37}{8}\left(D^2 - 6.58D + 11.15\right)$$
$$+ \frac{28.85}{10}\left(D^2 - 5.72D + 8.45\right)$$
$$= 4.16 D^2 - 25.21 D + 40.24 \qquad (8)$$

Upon differentiation with respect to D and setting the derivative to zero, we find that D = 3.03 cm. This value can be substituted back into Equation (8) to find the lowest value of the composite criterion function possible with this design. This value (2.06) may then be compared against the criterion function minima resulting from other designs. With multiple designs, identical values must be used for the weighting factors and the K multipliers.

The optimum value of D was obtained in this case by differentiation of the criterion function on the assumption that D can be varied continuously. In actuality, laminations are readily available only in discrete sizes, one of which would be 3 cm. In this case, the value obtained for D is so close to an available size that the designer would not be concerned about being just off a minimum, and it is common practice to select the nearest available size. If the variables had not been represented by parabolic functions but left as discrete data, the criterion function could have been developed in the same way, but a value for the criterion function would have been found for each value of D. The resulting table of values could then have been easily examined to locate the smallest value of the function.

8.3.6 Quantifying the Outcomes

The development of a criterion function as discussed in the preceding section allows the decision maker to express combinations of different product characteristics by a single number which can be easily compared to a like number for various alternatives. We found that we could do this even though the units of the characteristics may be different, as they were in the example. However, this method does not take into account the possibility of nonlinearity

DECISIONS

of the function that expresses the value or utility of a characteristic. As an example, a component that fails to meet a specification is of no value and may even be assigned a negative value. One that just meets the specification is of doubtful value. One that meets the specification with a reasonable factor of safety is all that is required; one whose factor of safety exceeds that reasonable value has no added utility because of that excess capacity.

If a functional relationship can be established between utility and each characteristic of importance, it is possible to more nearly express the value of each alternative by a criterion function. For example, a study was made [5] to compare aluminum and copper transmission lines for installations in Tennessee. The principal criteria were cost and avoiding downed lines during the ice storms that occur occasionally. Calculations showed that the expected ice load, 6.5 mm thick, and an 18 N wind load would result in a tension of 20.36 kN in #4/0 copper cable and 16.53 kN in 336 MCM aluminum cable, which is allowed to have a greater sag. These conductors have equal resistance values, and the pole span (92 m) was the same for both analyses. The breaking strength of the copper cable is 40.72 kN, which is twice the anticipated loading, and 26.44 kN for aluminum, which is 1.6 times the anticipated loading.

The horizontal scale of the utility curve of Figure 8.13(a) is the nondimensional ratio of the cable strength to the tension. The vertical axis of the utility curve was normalized so that it runs from zero to unity. Because a strength to tension ratio of 1.0 or below means that the cable will break, the utility curve intersects the horizontal axis at unity. It then rises in a nearly linear fashion, gradually curving over so as to become asymptotic to the upper limit of utility, unity. That is, the utility function is nonlinear, values of the strength to tension ratio far in excess of a reasonable value being of no additional value to the utility. Entering the curve with the breaking strength to tension ratios of the two cables results in a utility factor of 0.70 for the copper and 0.48 for the aluminum.

The utility curve of Figure 8.13(b) was constructed using the information that an annualized cost of $1,500 per km for construction was the highest value at which the line could be constructed economically. That dollar value therefore is the point of zero utility. As costs per km decrease, the line has increasing utility, but as the cost per km approaches zero, one finds that it makes little difference whether the cost is $100 per km or $200; thus the curve becomes asymptotic to the normalized value of unity as the cost per km nears zero. If the equivalent annual cost for copper is $1,100 per km and for aluminum is $785 per km, the copper has a utility value of 0.46 for cost and the aluminum has a utility value of 0.73. The total utility for the alternatives is 1.16 for the copper and 1.21 for the aluminum. From the analysis to this point, one can conclude that there is little reason for choosing one material over the other. The copper is more likely to resist mechanical failure, but this fact is offset by the lower cost of the aluminum.

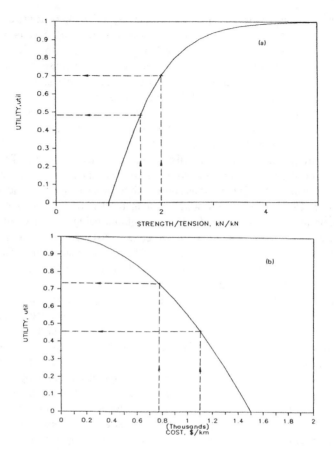

Figure 8.13 Utility curves for a transmission line. (a) Strength/tension utility factor. (b) Installed cost per km utility factor.

The expected ice load was based on 6.5 mm of ice, but long-term historical data show that severe ice storms in the region have coated the lines with from 5.5 to 7.5 mm of ice. The wind load also depends on total cable-plus-ice diameter. Calculations show that the total tension in the copper will be between 16.28 kN and 24.63 kN at the extremes, the mean value still being 20.36 kN. It is also found from tests that the breaking strength varies, probably due to unavoidable installation damage. The limits of breaking strength were found to be 32.57 kN at the lower end and 48.86 kN at the upper end, still with the mean of 40.72 kN.

Assuming that these limits are the three-standard-deviation limits of a normal distribution, new strength utility curves can be drawn for the alternatives. In Figure 8.14(a), which is drawn for copper, the utility curve intersects the horizontal axis at 26.05 kN, which corresponds to 4σ above the mean loading

DECISIONS

due to ice and wind. The loading distribution curve, shown to the left, is centered at 20.36 kN. The breaking strength distribution curve appears to the right, being centered at 40.72 kN. Using that mean, the utility factor for the copper cable is 0.7.

Figure 8.14(b) shows the new utility curve for aluminum. The high end of the ice and wind load distribution curve (at the left) is at 19.94 kN and the low end of the breaking strength distribution curve (at the right) is at 21.25 kN. These distribution curves meet at a point about three and one-half standard deviations from both means. It is clear that the use of aluminum cable is much less attractive now that a more in-depth analysis has been made.

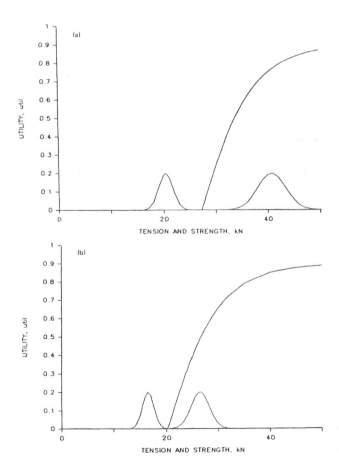

Figure 8.14 Utility curves for (a) the copper transmission line and (b) the aluminum transmission line taking into account the statistical variations in loading and in breaking strength.

CHAPTER 8

Utility curves can take on many different shapes, depending on the characteristic being considered. More (or less) is not always better. For example, in recent years considerable effort has gone into reduction of the noise level from domestic gasoline-engine driven lawn mowers, and the striking improvements that have been made are very welcome. However, a completely silent engine (if one could be achieved) would not be as desirable as one that produces audible sound at some level. After all, the sound emanating from the engine gives some indication of whether the engine is running properly or not, and it serves as a warning to children and pets that the mower engine is running. One of the unexpected effects encountered with electric vehicles is that pedestrians tend to be unaware of their presence unless they are directly in the field of vision, even when they are very close. Investigation has shown that the absence of a certain minimum level of sound from these vehicles is the critical factor for this situation. The curve of Figure 8.15 represents a reasonable utility curve for the mower situation, although the placement of the peak of the curve on the intensity-level scale is debatable. There is probably a similar curve for electric vehicles. In fact, sound generators for the benefit of pedestrians have been proposed for electric cars with sound levels proportional to speed (although not linearly so).

The examples of the preceding paragraph are intended to make the point that, by gathering data on the particular characteristic involved and with sufficient thought about the situation, it is possible to develop a reasonable utility

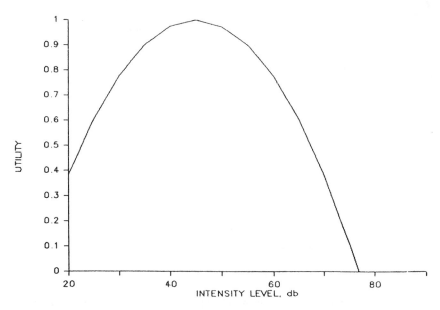

Figure 8.15 Utility curve for the sound level of a domestic gasoline-engine driven lawn mower.

DECISIONS

curve for any given characteristic of any product. Even if the scales on the axes cannot be quantified with precision, these curves will serve as valuable guides in decision making.

8.4 DECISIONS UNDER RISK

There are decision-making techniques that apply to much broader classes of problems than those given in the preceding section. The techniques presented in this section are useful chiefly when the probabilities of the states of nature are known or can be determined with sufficient confidence to warrant their use.

8.4.1 The Decision Tree [6]

A decision tree is used to depict a scenario of multiple decisions. It is made up of three principal elements: Decision events, chance events, and actions.

Decision events are points at which selections are made among alternatives under the control of the decision maker. They will be represented in the decision tree by squares. *Chance events* are possible occurrences over which the decision maker has no control. They will be represented in the decision tree by circles. Chance events may be the result of a competitor's activity, a change in technology, the enactment of a law that imposes new constraints on the product or its use, a change in a product standard, or any other event that may influence the ultimate decision. An example of a chance event to be included in decision trees being constructed by producers of "cold remedies" is that a medical research team will find a complete cure for the common cold. *Actions* are the results of decisions or of chance events. These are indicated in the decision tree by branches radiating from the boxes of the decision events or the circles of the chance events. Time is conventionally shown increasing to the right in these diagrams, but there is no time scale implied.

The fact that time and the sequence of decisions are displayed is of great advantage in constructing and using the decision tree. Another advantage is that it allows the states of nature to have different probabilities depending on the alternative chosen. This is often the case in real-life situations.

The concept of a decision tree and the way in which it is used is best illustrated by means of an example. Suppose your company (XYZ) is considering the production of a device that is to be sold in competition with another company, ABC. The device is intended to be incorporated by other manufacturers into a system. ABC has been in the business for some time, and its product is well accepted. XYZ must decide, before going any farther, whether to design the device so as to be interchangeable with ABC's device, or not. This is the single decision for which the decision tree of Figure 8.16 has been constructed.

There are many considerations to be taken into account in making this decision. For example, XYZ's merchandising methods have been superior to those of ABC. Hence, making an interchangeable device would probably allow XYZ to make inroads on ABC's market. However, the specification of interchangeability restricts XYZ's design choices and can easily result in a more complicated internal mechanism than would be necessary if interchangeability were not made a requirement. Furthermore, ABC holds patents on the internal mechanism of their product, and there is some doubt in your engineering and legal departments that an interchangeable device can be made that avoids infringement. Another possibility to be considered is that of negotiating an arrangement with ABC to buy the device with XYZ's name and logo on the nameplate, and then resell to the user. The penalty, of course, is reduced profit, but there is considerable avoidance of cost and risk if this avenue is chosen.

This recitation of various considerations, although admittedly incomplete, gives some indication of the conflicting arguments to be considered in many business decisions having to do with new product development. In spite of all of the uncertainty, the managers of a successful business must somehow reach a correct decision more often than not. The mere act of drawing a decision tree clarifies the problem, and probably exposes more areas of uncertainty.

Figure 8.16 illustrates several features of decision trees. Note that a decision usually sets the stage for a chance event. Although it is unusual for one decision to lead immediately to another, such a situation is possible. Conversely, the occurrence of two chance events in succession is not at all uncommon. The figure shows this possibility. Another feature of decision trees is that decision or chance events have only a single action leading to the event, but can have any number of actions leading away from it. The present example is restricted to two alternative actions only.

There is a considerable amount of detailed information shown on the tree. The decisions and chance events are described briefly. Estimates of the investments required by each decision are indicated. It should be noted that investment outlays result only from decisions, never from chance events, even though chance events are the original motivators of the decisions. Another bit of information is the estimate of the probability of a particular action ensuing from a chance event. Each probability is shown inside the chance event circles adjacent to the relevant outgoing branch. The sum of the probabilities in each chance event circle must, of course, be 1.00. Another important set of numbers is that associated with the outgoing branches of each decision event. These are shown inside the decision event squares. Except for the first decision (the one for which the entire tree has been constructed), a probability equal to the reciprocal of the number of alternatives is assigned to each outgoing branch. These numbers indicate that if a true decision is to be made at some later time each alternative is equally likely. In this case, since only two alternatives are

DECISIONS

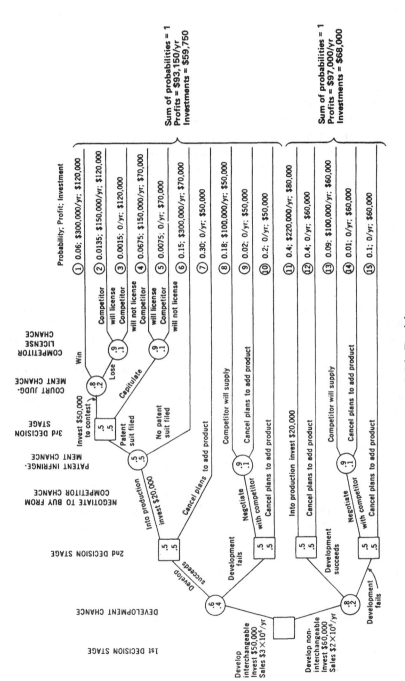

Figure 8.16 Decision tree.

shown, the probability is 0.5 for each alternative. The probabilities in the decision event squares and the chance event circles are used to compute the probability of each outcome.

Probabilities associated with chance events must be determined with care. A feasibility study can provide the basis for judging the probability of successful completion of product development. The chance of ABC suing XYZ is very difficult to assess. This may depend as much on the temperament of ABC's officers as it does on the legal facts of the case. The probabilities of 0.5 for the two actions originating at this chance event indicate the difficulty of this assessment. If ABC files a suit, no one can be certain in whose favor the court will rule. A probability of 0.2 for an adverse decision is used because XYZ's officers believe that it can be shown that ABC's patents were anticipated by others or that the mechanism that XYZ will use will be substantially different, and that ABC's patents will not be infringed. The "Win" probability must therefore be 0.8. Note that, in the lower part of the tree, in which the noninterchangeable option is considered, the chance event of a patent infringement suit being filed is not even shown, reflecting the confidence of XYZ's personnel that the final device will not infringe any patent, either one of ABC's or a patent held by anyone else.

This tree leads to fifteen different possible outcomes. Three numbers are associated with each outcome. The first expresses the cumulative probability of that outcome being the end point of the project. It is obtained by multiplying the probabilities along the path leading to that outcome. The sum of the probabilities of the outcomes for each of the alternatives stemming from the primary decision equals 1.00. The second number associated with each outcome is the outcome itself, expressed in this tree as yearly profit. The third is the sum of the investments required to reach this outcome.

The profit figures were obtained by assuming 10% of sales if XYZ makes the interchangeable product without paying royalties, 5% if royalties are paid, and 3-1/2% if the devices are purchased from ABC for resale. Because the noninterchangeable device will be somewhat less complicated than that designed for interchangeability, an 11% profit on sales is expected if that alternative is chosen. Estimates of total sales, required investment, and anticipated profits are usually available from data accumulated during the feasibility study.

It is important that a high degree of confidence be established that all probabilities and money values are correct and that the tree is complete. There is no point in using a decision tree with known omissions or data that are not believable. However, if the tree of Figure 8.16 is complete and the data are reasonable, the company is in a position to make a decision. Keep in mind that the tree was developed with only one aim in mind: Should XYZ make an interchangeable device or a noninterchangeable device? It is inappropriate for second- or third-stage decisions, such as seeking a license or defending a patent suit, to be made on the basis of this tree. If a project proceeds to one of these later decision events, a new tree should be developed.

DECISIONS

Once the tree of Figure 8.16 is complete, XYZ can make the primary decision. How is the final comparison made? For each major section of the tree, multiply each outcome (the yearly profit) by the probability of arriving at that outcome, and sum these values. For the interchangeable device approach, this sum is $93,150 per year. This is the expected return from this choice. For the noninterchangeable choice, the sum is $97,000 per year. The expected investment is calculated in the same way, giving values of $59,750 and $68,000 respectively.

The final decision is made by comparing expected revenues and expected investments for the alternatives. In this example, making the noninterchangeable device will require a larger investment and will yield a larger annual profit than would the interchangeable device, even though the sales of the interchangeable device would be expected to result in a larger sales volume.

Although most engineers will not be involved with a decision tree such as that of the example, designers are confronted with decisions and chance events occurring sequentially in time as a consequence of the fact that designers are managers of projects. In that position, the designer may have to recommend one of two alternative designs with chance events of success or failure of preliminary experiments, the ability or nonability to find certain vital components, acceptance or rejection by a group having authority to approve or reject a product, such as UL, unacceptably high manufacturing cost, and so forth. Suitable decisions would, of course, follow each chance event.

A large percentage of practicing designers do not draw out a decision tree to make such decisions, but if they are asked for their reasons for a particular choice it frequently becomes apparent that they use a similar method. Actually drawing out the decision tree helps the designer to anticipate the decisions that will be needed during a design project and to advance reasons for making one decision as against another. The decision tree is therefore a useful technique for improving the efficiency of the design process and to avoid high cost and high risk activities.

The reader may also wish to consult Section 2.7.1 of Reference 4 for further discussion and another example. Reference 6 is also of interest, although the notation is somewhat different.

8.4.2 The Decision Matrix

A simple decision matrix was shown in Figure 8.1. The question in that case was whether to take an umbrella when going out on a cloudy day. Even though the elements of the matrix enter into the decision, no one will create a matrix for such a situation. More complicated situations can, however, be dealt with very successfully using a decision matrix, such as that in Figure 8.17.

In the decision matrix, the rows are associated with the strategies or alternatives, S. That is, the strategies are those elements of the decision problem

States of nature		N(1)	N(2)	.	.	N(m)
Probability p(n)		p(1)	p(2)	.	.	p(m)
Strategies	S(1)	O(11)	O(12)	.	.	O(1m)
	S(2)	O(21)	O(22)	.	.	O(2m)
	S(3)	O(31)	O(32)	.	.	O(3m)
	.	.	.	.	.	.
	.	.	.	.	.	.
	S(n)	O(n1)	O(n2)	.	.	O(nm)

Figure 8.17 General form of the decision matrix.

that are under the control of the decision maker, such as alternative designs, the materials that may be considered for use in a particular device, or possible decisions relating to financial matters. The columns are associated with the set of chance events or states of nature, N. These are not under control of the decision maker. They may be properties of the materials listed under the strategies, they may be wins or losses in court cases, or they could be environments in which a prospective user might wish to use the device. In many cases N may be an extremely large number, and the states of nature may not even be discrete. The designer discards all but the most likely ones and, if they vary in a continuous manner, he or she selects a typical set for use in the matrix. In any case, the distinguishing feature between the strategies (S) and the states of nature (N) is that the designer has control in the first of these and no control in the second.

Associated with each of the states of nature is a weighting factor, p. If the states of nature are mutually exclusive, the weighting factors will be the probabilities of occurrence of each state of nature, and their sum will be 1.00. If the states of nature are properties of materials, the strategies being materials being considered for use in a design, the factor p may a weighting factor. For example, if two of the states of nature are normalized values of the strength to density ratio and the modulus of elasticity, but the strength to density ratio is more important than the modulus, the value of p selected could be 4 for the strength to density ratio and 3 for the modulus.

DECISIONS

The outcomes, O, may be a variety of functions. For example, if the decision matrix is being used to make a financial decision, the outcomes may be estimated profits per year. If the matrix is being used to select a material, the outcome may be a strength to density ratio, modulus of elasticity, and so forth.

How does one use the decision matrix once it has been formed? A tempting approach is to use the column with the maximum value for p and select from that column the maximum value O of the outcomes in the column. The strategy (or design) is then known. However, if the expected state of nature does not occur, we may find that the outcome for the state of nature that actually occurred is very low; we may have chosen the worst of the strategies or designs. The best approach is to use the maximum expected utility defined by the equation

$$E(i) = \sum_{j=1}^{m} p(i) \times O(ij) \qquad (9)$$

In the decision tree of Figure 8.16, assume that the decision was made to proceed with an interchangeable device, that it is now in production, and the first units have reached the market. At this point the attorneys for ABC send XYZ a letter with the following major points:

1. ABC believes that XYZ is infringing one of their patents.
2. ABC intends to sue XYZ unless a licensing agreement paying 5% royalties to ABC is signed immediately.
3. If ABC sues and wins, ABC will not permit a licensing agreement to be made.

This is a different situation from that used when Figure 8.16 was developed. When that decision tree was drawn, before the project had ever started, the best judgment was that the odds were equal on the possibility of litigation or nonlitigation on the part of ABC. You now know that a suit will be filed unless a licensing agreement is signed at once. Moreover, when the tree was drawn it was assumed that if a suit was lost, there was a probability of 0.9 that ABC would still license XYZ to produce the product. ABC is now telling you that the probability is zero. A new decision must be made based on the situation as it exists now. In making this decision, a new estimate of the profit for each alternative must be made. This is done using Equation (9).

The decision matrix is shown in Figure 8.18. This is also called a payoff matrix because the outcomes express the utility (the payoff) to XYZ of the alternative strategies if the states of nature indicated by the column headings exist. The expected revenues have been added at the right, using Equation (9). One of the first things to note is that the third alternative, terminating the project, has payoff values equal to or less than the payoff values of each of the other alternatives. If an alternative has payoff values meeting this condition, it is said

States of nature		Win	Lose	
Probability		0.8	0.2	Expected revenue
	Accept license	$150,000 per year	$150,000 per year	(0.8 + 0.2) x $150,000 = $150,000 per year
	Refuse license (1)	$300,000 per year	0	0.8 x $300,000 + 0.2 x 0 = $240,000 per year
	Terminate project	0	0	0.8 x 0 + 0.2 x 0 = 0

(1) Requires an estimated $50,000 for litigation.

Figure 8.18 Decision matrix for patent infringement suit.

to be dominated by another alternative and should be removed from consideration. In this case, it is dominated by each of the others.

The first alternative, that of accepting a license, will cause the suit not to be filed, and the win-lose question remains moot. The second alternative is better by $90,000 per year *expected revenue*. We note that, if the suit is won, it is better by $150,000 per year, but thoughts of that sort are not permitted when using a decision matrix, being the result of a subconscious shifting of the probabilities to 1.00 for the win situation and zero for the loss. However, this is the time to make a reassessment of the probability estimates. Are the 4:1 odds on winning still valid? If the odds are even, then the expected revenues under the second alternative equal those under the first. If ABC's case now seems stronger than when the decision tree was first drawn, licensing should be given serious consideration.

A more extended discussion of decision matrices will be found in Section 3.7 of Reference 3.

8.5 DECISIONS UNDER UNCERTAINTY

Many decisions must be made without any information or at least any *reliable* information as to the probability of the state of nature that will occur to influence the payoff. Such decisions are said to be made under uncertainty. There are several criteria that have been used to make decisions in this undesirable situation. We will examine a few of these. Some of them have the names of the persons who introduced them attached to the criterion; others have highly descriptive names, as we will see. Regardless of the criterion used, one important lesson to be learned is that decision makers may make different decisions because, either consciously or unconsciously, they are using different criteria.

DECISIONS

8.5.1 The Laplace Criterion

The Laplace criterion is the oldest and easiest to understand. The premise on which it is based is simply that, if there is no way by use of data or by subjective evaluation of arriving at probabilities of occurrence for the several states of nature, one chooses equal probabilities for each. In the matrix of Figure 8.17, each probability is simply the reciprocal of the total number of states. The expectations can then be calculated using Equation (9) and the best of the resulting values leads to the strategy to be selected. Because equal values of the probabilities are selected, this criterion is also known as the "equal likelihood criterion." It is also referred to as the Bayes-Laplace criterion. Applying this criterion to the matrix of Figure 8.18, the third alternative having been discarded, one arrives at the result previously noted for equal odds: The expected revenues are the same for either alternative 1 or 2.

8.5.2 The Maximin or Pessimist Criterion

The decision maker may decide that the assumption of equal probabilities is too risky because the distribution of probabilities might be much less favorable than even that criterion assumes. There may actually be some reason to believe that this is true. In any case, being a pessimist, he or she is cautious enough to make this assumption. The reasoning now would be as follows: Whichever alternative is selected, one should expect the poorest outcome to result. This is the only safe assumption. Hence the best solution is to select the best of the worst outcomes. Application of this criterion to the matrix of Figure 8.18 (alternative 3 having been eliminated) immediately leads to the decision to accept a license.

8.5.3 The Maximax or Optimist Criterion

This criterion is the opposite of the previous one. The decision maker, being an optimist, selects the alternative corresponding to the best outcome. This leads immediately to selection of refusal of a license and proceeding to litigation. If the decision maker is not a wholehearted optimist, he or she might choose the strategy of selecting the best utility in each row of the matrix, but then choosing the worst of these. With alternative 3 again eliminated, this leads in this simple example to the decision to accept a license.

8.5.4 The Hurwicz Criterion

A generalization scheme of the two previous criteria has been proposed by Hurwicz. (See Reference 3.) To apply this criterion, one begins by defining a number, α, which is an inverse measure of the optimism of the decision maker. The value of α will vary between 0 and 1. The next step is to form a decision index, defined by

$$\text{Decision index} = \alpha \cdot o(i) + (1 - \alpha) \cdot O(i)$$

for each of the alternatives, where $o(i)$ is the minimum outcome (payoff) in the *i*th row of the decision matrix and $O(i)$ is the maximum outcome in that row. One then selects the alternative having the most favorable decision index. We note that if $\alpha = 1$, the maximin criterion (the pessimist criterion) results; if $\alpha = 0$, the result is the maximax criterion (the optimist criterion). The value of α is therefore a measure of pessimism, with $\alpha = 1$ being the complete pessimist and $\alpha = 0$ being the complete optimist. Siddall [3] suggests a theoretical experiment for the decision maker for the determination of the designer's personal value of α.

8.5.5 The Savage Criterion; the Minimax Criterion

The Savage criterion, also known as the minimax risk criterion, is thought by some to be the best criterion to use under uncertainty. It takes into consideration outcomes that are lost by selecting the wrong alternative. Its objective is to minimize the *regret* that will occur at some future date when the payoff to the decision maker becomes evident. This does not mean that the decision maker or the company suffers; it may mean only that the positive result of making a certain decision is less than it would have been had a different decision been made.

The criterion is applied by setting up a decision matrix as in Figure 8.17 and then subtracting the payoffs (outcomes) of each column from the maximum payoff of *that column*; the resulting values express the regret of each combination of alternative and state of nature. These values are placed in a new matrix, called a regret matrix. As an example, suppose the decision matrix is as given in Figure 8.19(a). The resulting regret matrix appears in Figure 8.19(b).

Savage's contribution was to point out that future regret is a valid basis for decision making and that it can be identified using the simple manipulation shown in this example. There is no definite rule for selecting the "best" alternative from the regret matrix. The decision maker may wish to avoid large regret. In that case, the alternative having the smallest maximum regret would be selected. In Figure 8.19(b), this is S(3), having a regret value of 7. On the other hand, the decision maker may wish to follow the Laplace approach and use equal probabilities of each state of nature. In that case application of Equation (9) shows that the least expected regret value $(3 + 8 + 0)/3 = 3\text{-}2/3$ results from the choice of S(2). (Note that using a weighting factor of 1.00 for each state of nature will lead to the same conclusion as to which alternative to select.) This

DECISIONS

	N(1)	N(2)	N(3)
S(1)	1	9	6
S(2)	8	4	10
S(3)	11	7	3
S(4)	2	12	5

(a)

	N(1)	N(2)	N(3)
S(1)	10	3	4
S(2)	3	8	0
S(3)	0	5	7
S(4)	9	0	5

(b)

Figure 8.19 Development of a regret matrix.

technique always results in at least one zero in each column, although there may be more than one zero in a column.

Application of this technique to the matrix of Figure 8.18 (with the third alternative neglected) is of no value to the decision maker because each of the remaining alternatives exhibits a regret value of $150,000 per year. If the probabilities of that matrix are used, the result will clearly be a decision to accept a license from the ABC company.

8.5.6 Choice of Criterion

Which of the various criteria described above should one choose? There is no clear answer to that question. Decision making under uncertainty is inherently subjective in nature. Moreover, it is doubtful whether complete uncertainty really exists in real life. For example, if you are taking a true-false test and do not know the correct response to a particular statement, you may tell someone else after the test that you "guessed" the answer. In all probability, what you really mean is that you didn't have complete confidence in your answer. Having learned the general background of the material on which you are being tested (and maybe having some insight into the characteristics of the instructor in the course), a true guess is an impossibility unless you made a selection of true or false without even reading the statement.

Most decisions are made in a state of partial uncertainty. Given that condition, if it is at all possible it is probably best to make estimates of the probabilities of the various states of nature, and to use the decision-under-risk techniques.

8.6 COMBINATION OF MODELS

Models for decision making can be viewed as techniques that can be applied piecemeal to a problem situation, just as was done in Chapter 7 with product models. To illustrate the utility of this point of view, consider the situation in which an engineer must decide among various grades of electrical insulation for a particular product, one to be subjected to the following adverse conditions:

1. High humidity, which lowers the volume resistivity and increases the dielectric constant and the dissipation factor.
2. High-voltage gradient due to switching transients.
3. Mechanical stress in a flexural mode.

Table 8.5 gives the data available in NEMA Standard L1-1-1988 for four grades of insulation. These are industry-wide minimum performance data. Design calculations for the product itself indicate the following limits on various parameters of the insulation.

$$\text{Flexural strength} > 68.9 \times 10^6 \, N/m^2$$
$$\text{Dissipation factor} < 0.060$$
$$\text{Dielectric constant} < 6.0$$
$$\text{Voltage breakdown strength} > 7.5 \, kV/mm$$
$$\text{Volume resistivity} > 10^3 \, M\Omega \, mm$$

Table 8.5 Manufacturer's Data for Candidate Electrical Insulations

Grade	Flexural strength (N/mE2)	Dissipation factor (a)	Dielectric constant (a)	Voltage breakdown (kV/mm)	Volume resistivity (b) (MΩ mm)
CEM1	240E6	0.040	4.6	10.0	10E6
FR3	137E6	0.040	4.8	8.9	10E5
XXP	97E6	0.050	5.2	13.3	10E5
XXXPC	83E6	0.040	4.8	17.3	10E4

a) Measured after immersion in distilled water for 24 hours at 23 C.
b) Measured after 96 hours at 35 C, 90% RH.

DECISIONS

The first step in the process is to select suitable insulations for entry into Table 8.5. The limiting values of various parameters as listed in the inequalities above constitute specifications for the insulation. It is then easy to run through published insulation characteristics and select those having parameters that satisfy the inequalities and thus meet the specifications. Any insulation that fails to meet any one of the specifications, even though it may be far superior in some other specification, is not considered because the designer cannot compensate for a material's inability to meet all of the design specifications.

The next step is to obtain a criterion function, CF_H, for the effects of humidity on the dissipation factor, the dielectric constant, and the volume resistivity. Looking ahead, this criterion function will then be used with a utility curve to obtain utility values for each of the candidate insulations. We will then obtain utility values for each insulation for the voltage and flexural strength conditions. Having this set of twelve numbers, we will be able to construct a decision matrix.

The criterion function for humidity is given by

$$CF_H = -a_1 K_1 (DF) - a_2 K_2 (DC) + a_3 K_3 R \qquad (10)$$

in which the dissipation factor term DF and the dielectric constant term DC are preceded by minus signs because low values are desirable, but the volume resistivity term R is positive because a large value is desirable. The better the material, the higher will be the resulting value of CF_H.

Average values of the material characteristics in Equation (10) are found from the data of Table 8.5 so that we have values for each of the K's which can be used in normalizing the equation's terms. These values are:

DF = 0.0425
DC = 4.85
R = 10^5

Assuming that DF, DC, and R are equally important for good insulation performance under humid conditions, the weighting functions (a) are chosen to be equal, and a value of 1 is assigned.

The criterion function may now be written as it was earlier, using the reciprocals of the average values above as the values of the respective K's. However, in order to simplify the calculations, we will set K_1 to unity, and multiply the other two K's by the average value of DF. (This multiplies CF by the same factor, but the relative values of the CF's for the four insulations and for the specifications will be unchanged.) The multiplier on DC will then be $0.0425/4.85 = 0.00876$ and on R it will be $0.0425/10^5 = 4.25 \times 10^{-7}$.

The value of the criterion function corresponding to the design specifications is:

$$CF_{spec} = -0.060 - 8.76 \times 10^{-3} + 4.25 \times 10^{-7} \times 10^3 = -0.1121$$

For the candidate insulations,

$$CF_{CEM1} = -0.040 - 8.76 \times 10^{-3} \times 4.6 + 4.25 \times 10^{-7} \times 10^6 = 0.3447$$
$$CF_{FR3} = -0.040 - 8.76 \times 10^{-3} \times 4.8 + 4.25 \times 10^{-7} \times 10^5 = -0.0395$$
$$CF_{XXP} = -0.050 - 8.76 \times 10^{-3} \times 5.2 + 4.25 \times 10^{-7} \times 10^5 = -0.0531$$
$$CF_{XXXPC} = -0.040 - 8.76 \times 10^{-3} \times 4.8 + 4.25 \times 10^{-7} \times 10^4 = -0.0778$$

The criterion function for the specifications is smaller than that for any of the candidate insulations, as might be expected since all of them were selected to exceed the specifications originally determined. Moreover, their relative values will not be changed if the same constant is added to each of them. If we add 0.1121, then the criterion function for the specifications has a value of zero and all others are larger than zero. We then have

$$CF_{spec} = 0$$
$$CF_{CEM1} = 0.4568$$
$$CF_{FR3} = 0.0726$$
$$CF_{XXP} = 0.0590$$
$$CF_{XXXPC} = 0.0343$$

The general shape of the utility curve will be a curve passing through the origin and rising with increasing criterion function, but approaching an asymptote (which we arbitrarily set at unity) for large values of the criterion function. The reason for the asymptote is, as before, that materials having criterion functions far in excess of what is needed should not have undue weight placed on that fact. If you have a beam which has a factor of safety of 6 in a particular building, it is not sound engineering practice to specify one which would have a factor of safety of 12 in its place, and it would actually be a very bad decision from the viewpoint of economics.

In order to establish the curve more firmly, we calculate a value for the criterion function assuming a material twice as good as the specifications (DF and DC divided by 2 and R multiplied by 2), remembering to add in the shift of 0.1121. This yields a value of 0.0567, which is in the vicinity of the criterion function value for the last three candidate insulations. We assign a value of the utility function of 0.7 to this value of the criterion function, and the utility curve can now be drawn as in Figure 8.20.

Before proceeding with the example, a word should be said about the apparently high-handed way in which numbers have been thrown around and the

DECISIONS

intermediate point on the curve established. Remember that we are looking for utility values that will allow us to distinguish between alternatives and that will not place undue emphasis on any alternative that has no really substantial advantage over others. A utility curve that rises more steeply from the origin will tend to crowd all the utility values toward unity, whereas one that rises very slowly from the origin will cause the three insulations with low values of the criterion function to have utility values that are all low and nearly indistinguishable. The curve of the figure separates the utility values for three of the insulations and does not give an inordinately large utility value to the fourth.

From the curve and the values of the criterion functions, we pick off the utility values for each insulation.

Material	Utility
CEM1	1.0
FR3	0.76
XXP	0.70
XXXPC	0.54

The two remaining states of nature are the flexural strength and the voltage stress. Each of these is dependent on only one parameter of the insulation characteristics. Hence utility curves can be drawn by assigning a utility value of zero to the design specifications and 0.7 to a material whose flexural

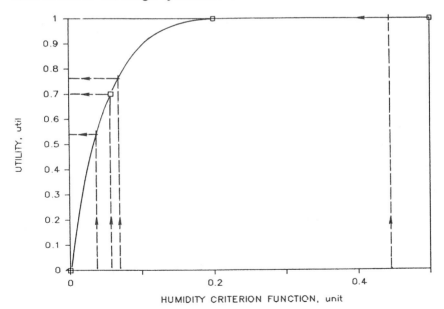

Figure 8.20 Utility of four candidate electrical insulations under humid conditions.

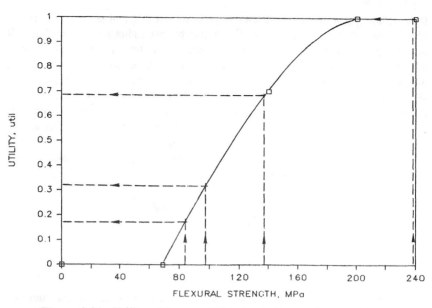

Figure 8.21 Utility of four candidate insulations under conditions of flexural stress.

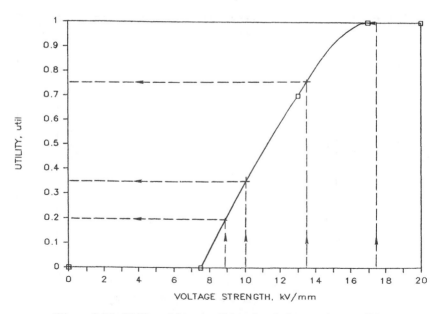

Figure 8.22 Utility of four candidate insulations under condition of voltage stress.

DECISIONS

	States of nature		
	High voltage	Humidity	Mechanical stress
CEM1	0.35	1.00	1.00
FR3	0.20	0.76	0.68
XXP	0.75	0.70	0.32
XXXPC	1.00	0.54	0.16

Figure 8.23 Decision matrix for candidate insulations considering humidity, flexural stress, and voltage breakdown.

strength and voltage breakdown strength is sufficiently high to assure generally reliable performance. Figures 8.21 and 8.22 show these two utility curves. The utility values are obtained by entering the horizontal axes with the parameters of the materials as given in Table 8.5. All of the utility values are now entered into the decision matrix of Figure 8.23.

With the decision matrix completed, any of the techniques described earlier may be used to select the insulation to be used in the design. If the Laplace criterion is used, a probability of 1/3 can be assigned to each of the states of nature, yielding expected values as follows:

Material	Expected value
CEM1	0.783
FR3	0.547
XXP	0.590
XXXPC	0.567

Using this criterion, CEM1 would be the insulation of choice. The maximin (pessimist's) criterion leads to the same choice. The maximax (optimist's) criterion allows either CEM1 or XXXPC to be chosen, although the presence of

	States of nature		
	High voltage	Humidity	Mechanical stress
CEM1	0.65	0.00	0.00
FR3	0.80	0.24	0.32
XXP	0.25	0.30	0.68
XXXPC	0.00	0.46	0.84

Figure 8.24 Regret matrix for four candidate insulations considering humidity, flexural stress, and voltage breakdown.

two 1.00's in the CEM1 row would prejudice the decision maker in that direction.

The Savage (or minimax) criterion requires the construction of a regret matrix. This matrix for this problem is shown in Figure 8.24. Once again, CEM1 is the clear choice. It not only has the lowest total regret, but two of the entries for that alternative are zero.

8.7 SUMMARY

The key to making good decisions is careful research. That research must precede the decision. The research should include background information so that one has confidence that all alternatives and states of nature or opponent's strategies are identified. This is perhaps the part that demands the highest level of creativity. Repeated searches for available alternatives should not be neglected.

Once the alternatives and states of nature are determined, calculation of relevant payoffs or outcomes can begin. This usually involves extensive problem solving, and the person having the greatest knowledge of analytical techniques should carry this phase of the work.

In spite of the development of decision theory, most individuals still make decisions in an unstructured way. One of the objectives of this chapter is to encourage you to begin the practice of making important decisions using the models described above. If you do so, you will be forced to think about the decision more before you make it, you will understand the ramifications more completely, and you will discover which criterion you have been using when making decisions.

REFERENCES

1. Ruskin, Arnold M., and W. Eugene Estes. *What Every Engineer Should Know About Project Management*, Marcel Dekker, New York, Chapter 3, 1995.
2. Woodson, Thomas T. *Introduction to Engineering Design*, McGraw-Hill, New York, Chapter 13, 1966.
3. Siddall, James N. *Analytical Decision-Making in Engineering Design*, Prentice-Hall, Englewood Cliffs, New Jersey, 1972.
4. Ertas, Atila, and Jesse C. Jones. *The Engineering Design Process*, John Wiley & Sons, New York, 1993.
5. Tankersley, A. B. Copper vs. Aluminum as Electrical Conductors, paper presented to the Eastern Division Power Distributor's Association, Gatlinburg, TN, September 9, 1966.
6. Tribus, Myron. *Rational Descriptions, Decisions and Designs*, Pergamon Press, New York, 1969, pp. 332-335.

DECISIONS

REVIEW AND DISCUSSION

1. What three elements are necessary for a decision opportunity to arise?
2. How does the number of states of nature or information about them influence the solution of a decision problem?
3. What is the objective of the decision maker when creating a precedence matrix?
4. What is the planner's objective when using the critical path method (CPM)?
5. How does the program evaluation and review technique (PERT) differ from CPM? What additional information does it yield?
6. What are the restrictions on starting an activity? When can you say that an event has occurred?
7. What are dummy activities? Why are they used? How are they shown?
8. What is a critical path?
9. Why are float, free float, independent float, and slack important to the decision maker?
10. What three assumptions were made in the determination of tolerance using the technique of Section 8.3.3? For the examples given, are you satisfied that the conditions stated in those examples met all of the assumptions? If not, which one is suspect, and why?
11. Where could you apply the technique for finding tolerances? Why would it be useful in that situation?
12. What information does the designer gain from the learning curve?
13. The criterion function, such as Equation (6), yields numerical results. Do those numbers have any physical significance? Is this true in every case?
14. In what ways or by what methods can values of the outcomes be determined?
15. Under what circumstances do decisions under certainty present formidable problems?
16. What techniques are used when confronted with decision problems under risk? Which of the two techniques contains an element of reality that is lacking in the other?
17. What is the key element in the Laplace criterion for solution of a decision problem under uncertainty?
18. To which criterion does the phrase "the best of the worst" apply? Which can be defined as "the best of the best?"
19. Can you suggest a phrase similar to those in the previous question that applies to the criterion of least regret?

PRACTICE PROJECTS

1. A child's wagon consists of a formed steel bed, fifth wheel (the movable structure that supports the front axle and rotates on the bottom of the bed),

steering tongue, rear axle supports, front and rear axles, and four wheels. You are to design such a wagon so that it weighs no more than 10 lb, rolls freely down a smooth one percent grade when carrying a 50-lb child, and is strong enough to support a 200-lb adult. Using a precedence matrix, determine an order of designing the major parts listed with the minimum amount of iteration.

2. Design variables are related as shown in the following table. Determine the order in which they should be found to minimize iteration. Which variables will be unavoidably involved in the iteration steps?

Variable number	Predecessors Most significant	Less significant
1	2,4	3
2	3,9,1	None
3	None	None
4	3,9	2
5	4	2,1
6	4,9,5	None
7	5,4,6	2,9
8	9,7,2	4
9	2,3,1	None
10	5,3,2	None

3. The mathematical model of a compression spring is given by the following set of equations:

Hooke's Law: $$\frac{F_1}{x_1} = \frac{F_2}{x_2} = \frac{F}{x}$$

Stiffness constant: $$\frac{F}{x} = \frac{Gd^4}{8D^3N}$$

Final stress: $$S_2 = \frac{8DF_2}{\pi d^3}$$

Stroke: $$s = x_2 - x_1$$

Volume of spring material: $$V = \frac{\pi^2 d^2 DN}{4}$$

where
    F = force at any compression x, N
    x = compression, m

DECISIONS

D = mean coil diameter, m
G = modulus of rigidity, N/m^2
V = volume of spring material, m^3
N = number of turns on the coil
d = diameter of the spring wire, m
S_2 = stress in the compressed condition, N/m^2

Figure 8.25 shows an unstressed spring and springs in two stages of compression. Suppose that the specifications are the initial force F_1, the mean coil diameter D, and the stroke s. The modulus of rigidity, G, is of course a constant for a given material, and the final stress S_2 is not to exceed a certain value, say 0.70×10^9. Use the set of equations above to develop a precedence matrix so that you can determine the order of calculating the final stress S_2, the wire diameter d, the number of turns N, the volume V, and the unstressed length l_f with minimum iteration.

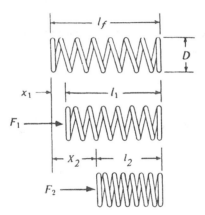

Figure 8.25 Compression spring.

4. What is the probability of getting the job related to the PERT chart of Figure 8.26 to completion in 26 days? The first number on each activity is the expected completion time and the second is the standard deviation. Identify and show the critical path.
5. Reduce the expected time to complete the job of Figure 8.26 by reducing activity times along the critical path. Keep account of the days removed from each activity. What is the final result of this procedure?

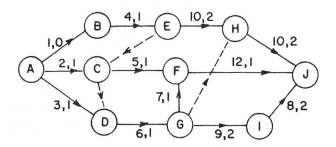

Figure 8.26 Work-flow diagram for problems 4 and 5.

6. Determine the critical path and the probability of completing the job of Figure 8.27 in 28 days. As in the preceding problems, the first number adjacent to each activity is the time in days and the second number is the standard deviation. Show the critical path.
7. Determine the float, free float, independent float, and slack for activities and events in Figure 8.27.

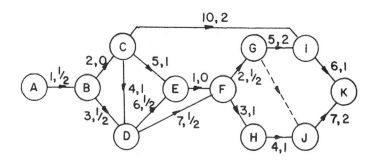

Figure 8.27 Work-flow diagram for problems 6 and 7.

8. In Chapter 13, the reliability R is shown to be $R = e^{-\lambda t}$, where λ is the per unit failure rate. Unfortunately, this number is never known precisely. Suppose that the number is estimated to have a $\pm 10\%$ tolerance. What is the corresponding tolerance of R if $\lambda = 10^{-6}$ failures per hour and the time, t, is 1000 hours? Repeat the problem with $\lambda = 10^{-6} \pm 20\%$. Does the tolerance of R double with a doubling of the tolerance on λ?

DECISIONS

9. The tolerance (equivalent to ± 3σ limits) of each resistor shown in Figure 8.28 is ±10% and of the applied voltage is ±5%. Determine the tolerance of the current, I.

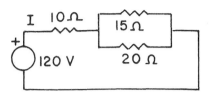

Figure 8.28 Simple resistor network.

10. A mass, spring, and damper are connected as shown in Figure 8.29. The spring constant is 10 N/m, the mass is 2 kg, and the viscous friction is 0.5 Ns/m, all values being known to an accuracy of 5%. What is the nominal resonant frequency and the expected tolerance?

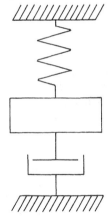

Figure 8.29 Simple mechanical system.

11. Your company has put a new device into production, and now has a year's worth of data on production, as shown in Table 8.6 on the next page. The production time (average labor per unit) required for a single unit was determined by dividing total production time each month by the number of units produced during that month. Draw the monthly average production time and the cumulative average time versus total quantity produced. What is the number in percent that would be used in the literature to describe the learning curve? If 60,000 units are to be produced in the following year at the rate of 5,000 per month, what is the expected average unit labor time during December of that year?

Table 8.6 Productivity Data

Month	Quantity	Avg. labor per unit
********	1	10.00
January	500	7.00
February	1,000	5.00
March	700	4.50
April	800	4.20
May	1,500	3.90
June	2,000	3.40
July	2,500	3.20
August	3,000	3.10
September	3,000	3.00
October	5,000	2.95
November	10,000	2.82
December	10,000	2.66

12. You are designing a product and wish to compare it to a competitor's product. According to your evaluation, the following statements are true:
    a) The industry standard for 10-year life requires reliability R = 0.8. Your product has R = 0.95; the competitor's is 0.85.
    b) Your product has the advantage of smaller size, having a footprint of 3× 6 in., whereas the competitor's is 5 × 7 in. The average industry product is 4 × 6 in.
    c) Your product costs more. It sells for $279, whereas the competitor's product sells for $225 and the industry average is $190.

    Use this information to compare the overall utility of your product to that of the competitive product. Size is considered to be the least important of the three characteristics. Cost is twice as important as size, and reliability is three times as important as size.

    Make the decision using the criterion function method given in Section 8.3.5 and using the utility curves of Section 8.3.6. Are the results of the two methods consistent?

13. In developing the criterion function in Section 8.3.5 for optimization of the transformer, "average values" of cost, weight, and power loss were selected in the vicinity of the minima of the three curves of Figure 8.12. These values were C = $6, W = 8 N, and P = 10 W. But suppose that other values had been chosen, such as C = $10, W = 20 N, and P = 12 W. Using those values, obtain a new criterion function and determine the theoretical "best"

value of D. Can you come to any conclusion as to the sensitivity of the procedure to arbitrary choices of the "average values" of the variables?

14. One of the authors (WHM) lives in northern Kentucky. There are a number of routes from his home to the University of Cincinnati, but only two (one of which itself has two branches) are reasonably direct. One route is by way of I-75, crossing the Ohio River on the Brent Spence Bridge and exiting at Hopple Street. The other route is by way of US 25, crossing the river on the Bailey Bridge or on the Suspension Bridge, depending on traffic conditions when the fork leading to those two bridges was reached, and then on through downtown Cincinnati. Some years ago WHM followed these routes for a month, selecting which route to follow on a random basis. The data accumulated during this experiment are shown in Table 8.7. As with most commuters, the least time from home to office is wanted. Use a decision tree to reach a decision as to which route to make the "permanent" one. How frequently should the experiment be run? (This question, of course, has no precise answer.)

Table 8.7   Traffic Travel Data

Route	Condition	Probability	Avg. travel time (min)
I-75	Light traffic	0.5	18
	Heavy traffic or other adverse conditions	0.4	29
	Accident	0.1	75
US 25	Light traffic	0.7	22
	Heavy traffic or other adverse conditions	0.25	27
	Accident	0.05	40

15. Use the cost/benefit analysis of profits and investments developed by the decision tree of Figure 8.16 to decide which option, interchangeability or noninterchangeability, to recommend to management.

16. You have developed a new product and, since you know the equipment very well, you are assigned the project of selecting test equipment to determine its reliability. You must choose between a relay control and a solid-state control for the test equipment. The relay equipment will cost $300, but there is some doubt as to whether it will work acceptably. The supplier agrees that if it does not work you can return it and get $150 credit toward a better unit, one that sells for $500. The solid-state equipment has an original cost of $450. Moreover, the supplier agrees that if the control does

not do the job it will be replaced by an advanced model at no additional cost. Because you are paying money out, the prices for the equipment should be negative in your payoff matrix. Which choice would you make? Which criterion or criteria did you use? Did they lead to different results?

17. A designer located three package control systems that will function in a product under development. Maintainability, availability, cost, and expected life are important to get the project completed on time, to be competitive in price, and to ensure customer satisfaction. From the data available, utility numbers are assigned to the three systems as given in Figure 8.30. Which system would you choose? Why?

	Maintainability	Availability	Cost	Expected Life
System A	6	8	5	6
System B	10	2	6	10
System C	1	9	10	3

9

THE DESIGN OF SYSTEMS

In Section 1.1.2, it was pointed out that four levels of design complexity can be identified. In order of increasing complexity, they are the component level, the product level, the system level, and the community level. It was also pointed out that system-level and community-level design projects have essentially identical design procedures and engineering aspects, and that component-level and product-level designs are quite similar to each other. For brevity, we have generally referred to the two higher design levels as systems; we have referred to the two lower design levels randomly as devices, components, or products. In all likelihood, you have probably thought about devices or products while studying the previous chapters. This is a natural thing to do; it is easier to comprehend simple cases than the more complex ones.

You are now familiar with a number of design procedures. One of the major points of this chapter is that planning must precede those design steps when dealing with systems. Another point that you have probably already grasped is that systems require the use of many subsystems or components. Those subsystems or components that are available may not meet your specifications, but come as close to your needs as any that can be found in the market. Systems usually require substantially more compromise than is necessary when designing a component.

In this chapter, we will treat system design, leaving component and device design to the following chapter. Pedagogically, this appears to be the reverse of the technique that should be followed—going from the simple to the complex. However, we will see that we already have a grasp on much of the material and that techniques presented in previous chapters are used in this one as part of the system design process. The system design technique that follows includes orderly methods of breaking down the whole into component parts and then designing at the component level. It is at the component level that precise and reasonably simple mathematical models may be available for use; at the system level mathematical models may be available for each component, but a mathematical description of the entire system will be so unwieldy as to be useless. Typically, components are designed, but systems are evaluated.

9.1 WHAT IS A SYSTEM?

To illustrate what is meant by a system, consider the field of transportation. The first means of transportation that can be considered to be the result of design was a simple two-wheel cart, drawn by a horse or one of the other beasts of burden. The cart evolved into other types of wheeled vehicles, including war chariots, buggies, wagons, and coaches. As technology advanced, other means of transportation emerged, such as the steam-driven train, requiring extensive rail trackage to be useful, and the automobile, which relies for its usefulness on roads and highways. The automobile was followed closely by the early and primitive airplane.

In this list, it is clear that the earlier vehicles fit the definition of a device; the later vehicles fit the definition of a system. For the early vehicles, a single individual, the designer, was able to select materials, sizes, and configurations to fit whatever specifications were determined to be necessary. The early steam locomotive and the first automobiles can also be classified as devices, many of them having been designed by one man. But it is also clear that the later vehicles must be classified as systems, and they are useful only because community-level systems, such as streets and highways for the automobile, rail systems for trains, and air traffic control systems for the aircraft industry, have been developed and are in place.

The automobile, for example, now has such complexity that the design team consists of engineers having skills in the mechanical, electrical, air-conditioning, chemical, hydraulic, and aerodynamic fields, and to this part of the team must be added experts in medical, physiological, and ergonomic areas, as well as other areas of expertise. Many of the components used in automobiles are made by outside suppliers to the automobile manufacturer's specifications, and may have been designed by the supplier's engineers, who have only imperfect communication with the design team of the automobile manufacturer. The necessary interchange of information between the automobile manufacturer and the supplier can take place only with an orderly procedure of planning, such as that to be described in this chapter.

This example illustrates the following necessary characteristics of a system:

1. A system has distinct parts.
2. The parts involve a broad range of physical phenomena.
3. The parts have observable attributes.
4. The parts are designed by many different individuals.
5. The parts operate interactively.

THE DESIGN OF SYSTEMS

9.2 THE SYSTEM DESIGNER

As the examples in the preceding section have shown, with the evolution of larger and more complex systems it became necessary for designers to work as teams, with each person assuming responsibility for components or subsystems within his or her area of expertise and working closely with others to assure compatibility with other parts. Constraints often limit the ability to effect complete compatibility, and a compromise must be struck that will work best from the overall system point of view. That is, the engineer in charge of the system design team needs to be a scientific and engineering generalist, and as such cannot be an expert in every area of technology necessary to complete the complex systems facing designers today. The supervising engineer (or system designer) should have mastery of his or her own branch of engineering, but should in addition comprehend the principles of such broad subjects as feedback control and probability and statistics. The focus of attention must be on the effectiveness of the system as a whole rather than on the details of any particular part of it. Hence that individual's characteristics must include the ability to be objective in making judgments, the ability to take constructive criticism, and the capability of looking at all facets of the problem. Creativity is a desirable asset, but the willingness to recognize and nurture the creative abilities of those in the design team is probably more important.

In addition to these characteristics, many of which relate to engineering, the supervising engineer must have the skill to deal with many subordinates, some of whom will be at odds with others from time to time because of differences of opinion, inability to provide the part or component that another engineer insists is absolutely required, or because one engineer has failed to deliver a solution to a problem when expected. The supervising engineer must, therefore, have some knowledge of psychology, and must be able to lead the engineers in the team to a compromise solution. Being a respected engineer is not enough—the supervising engineer must also have considerable skill as a politician.

The system designer must have broad knowledge relative to the availability of components likely to be used as well as of their input and output characteristics, and also of the advantages and disadvantages of alternatives. An extensive, up-to-date design file is an important source of information, not only for the system designer but for the members of the design team. The design file will include a collection of catalogs, standards relevant to the kind of system likely to be the subject of design, instruction manuals, technical papers, various charts and graphs, and engineering literature having thorough descriptions of component parts, assemblies, and materials.

A design file will be useful only if it is organized in a manner that makes it easy to find the required information. Some files are arranged in alphabetical order by component, others use various product classification systems (perhaps based on the U.S. Patent Classification System), and some are arranged by type of engineering project. Whatever the classification system used, it is important that everyone in the design team understand how to access the files in an efficient manner. Moreover, some method must be in place so that, when material is removed from the file for use, it will be easy for others to determine that the material is not present and who has it. Gradual erosion of the file resulting from engineers removing material for their own use and forgetting to replace it must be avoided.

An inference that can be drawn from the preceding discussion is that there should be a single design file accessible to every member of the team. Although this is generally true, most design engineers will build their own files which relate to the parts, components, or subsystems most likely to be the focus of their own work. These files may duplicate, at least in part, materials in the central file.

9.3 BRINGING THE SYSTEM INTO FOCUS

In Chapter 1, five steps of a design procedure were given. They are: Problem definition, problem evaluation, synthesis, analysis, and communication for manufacture. For a device (and perhaps also for a simple system), these steps give sufficient opportunity for the information gathering and the decision making necessary to bring a product from recognition of the need at the beginning of the procedure to manufacture at the end. However, faced with the complexity of system designs, the list of problem-solving steps must be given more general interpretation and applied to every phase of the program, that is, from the inception of the activity that exposes the need for the system through the life cycle of the system until it is retired after serving its purpose. These ideas have been presented in the literature [1] and with modification are shown in Figure 9.1. Although the logic steps applied to each phase of the program may have somewhat different meanings depending on which phase is being considered, the advantages of using the same designation for all phases is sufficiently great that the less-than-perfect designations are tolerated. These advantages include recognition of the similarity that does exist in the procedure regardless of the phase of the program and the ability to use a simple display showing the various steps in the procedure.

To understand the design of a system, refer to Figure 9.1. During the program planning phase of a community-level project, the overall problem is defined as the first logical step, cell 11 of the matrix. This is followed by the development of a scenario (cell 12) that sets down in writing as much history and data as are needed to describe the program and to act as a basis for subsequent evaluation of all activities and products that result from it. The synthesis step, cell 13, is used to develop alternative programs that might accomplish the desired results. In a program involving engineering, this will usually identify

THE DESIGN OF SYSTEMS 331

LOGIC STEPS APPLIED TO TEMPORAL PHASE OF PROGRAM / EACH PHASE	1 PROBLEM EVALUATION	2 DEVELOPMENT OF EVALUATION	3 SYSTEMS SYNTHESIS & SELECTION	4 SYSTEMS ANALYSIS & OPTIMIZATION	5 PREPARATION FOR NEXT PHASE
1 PROGRAM PLANNING					
2 PROJECT PLANNING					
3 SYSTEM DESIGN					
4 PRODUCTION					
5 DISTRIBUTION					
6 CONSUMPTION					
7 RETIREMENT					

Figure 9.1 Task matrix of a community-level project.

buildings, roads, traffic control systems, and so on that are describable by name or by function. During the fourth step, cell 14, the alternative programs are analyzed for their effects on people and the environment, any foreseeable legal ramifications, and so forth. The best program, as determined by the scenario developed in cell 12, is then selected. The last step, preparation for the next phase, cell 15, will include listing the projects that must be carried out to accomplish the selected program.

Following this development procedure, the same steps are applied to each project identified by the program development (cells 21 through 25). The result is identification of the best alternative of the various ways to accomplish the project's goals, and within that alternative, the needed engineering systems.

The program planning and project planning phases generally refer to community-level systems. These often involve political, legal, geographic, and demographic considerations much more than engineering principles. Yet those considerations have important influences on the engineering activities that follow.

The third phase of the program as exhibited in Figure 9.1 is system design. This is the point in the program where the design activities begin that are similar to, but more complex than, those which have been considered in the previous chapters.

Before discussing system development in more detail, note that phases 4-7 of Figure 9.1 are the phases of the life cycle of a product as discussed in Chapter 1. Here, however, the concepts of production, distribution, consumption, and retirement are much more complex than when applied to the simple product or device we have probably had in mind previously. Whereas earlier it may have been sufficient to consider product retirement only from the point of view of not endangering people and not contributing to environmental degradation, the large program development now being considered might involve retiring an interurban transportation system when it is no longer economically feasible to operate it. This deserves more planning than the simple-minded solutions of letting the system deteriorate in place or of tearing it down and putting everything into a landfill. It is for this reason that the same steps, that is, problem definition, problem evaluation, synthesis, analysis, and preparation for the next phase, must be followed in planning for system retirement. These same considerations apply, of course, to all other phases.

It was noted above that there are many sociological areas that may have to be considered when carrying out the steps shown in Figure 9.1. These include, in addition to engineering considerations, medical, legal, political, and environmental aspects. To emphasize this dependence, Figure 9.1 may be extended as in Figure 9.2. The obvious impossibility of getting inside this cube does not obviate the fact that the considerations along the third axis must be kept in mind when using Figure 9.1.

9.4 SYSTEM DESIGN

The activity central to the designer in a community-level program is that of system development or design. The first steps in the procedure call for problem definition. This must be carried out keeping a focus on what is to be accomplished, but one must also look at the scenarios of the two previous phases (program planning and project planning) for constraints that may be imposed on the system by the community and also at the constraints that may be imposed by production, distribution, consumption (use), and retirement. In each step of the system development phase, the information gained by planning the entire program must be factored into the decisions affecting the system design.

The second step in system development is to make a need analysis and to determine the specifications. Feasibility and existing patents should be explored, at least in a preliminary way, and any problems that seem insurmountable should be brought to the attention of the project planners.

The third step, system synthesis, is usually begun by developing a block diagram model naming the functions and the lines of processing information, energy, or material that must be provided to arrive at the specified output for a given input. (Figure 9.3 is a simple example of this type of diagram.) The block diagram is the result of a trial-and-error exercise that is based on the designer's practical experience and theoretical knowledge. It has the economic advantage of being strictly a pencil and paper activity. The end product is a functional description of the subunits that will make up the system. A second model is now developed, based on the first, with the subunits expanded into functional blocks (hardware) which together provide the outputs expected from the blocks of the

THE DESIGN OF SYSTEMS

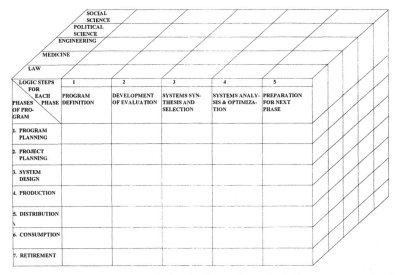

Figure 9.2 Morphological cube for an interdisciplinary community-level project. (Adapted from Hall, Reference 1.)

previous diagram. (See Figure 9.6.) Note the similarity of these two steps to steps 4 and 5 of the functional synthesis method of stimulating invention, as described in Section 6.7.1. The mental activity of developing a new system is not unlike inventing. It is true that it does not demand invention, although inventions often result.

The two diagrams are now used to develop a list of subunits and the specifications for each. The overall system specifications determined by the need analysis allow the designer to work from the output of the system toward the input to determine the output/input relationship of each subunit and from that information to determine the subunit specifications. These specifications must be so well written and complete that different individuals or teams can carry out the subunit development with confidence that assembly into the whole will result in a working product.

As examples of the concepts illustrated by the two block diagrams discussed above, consider the development of a speed controller for a three-phase induction motor having a rated horsepower output at a specified voltage and frequency. The controller and motor are to be used to drive an electric vehicle. The motor horsepower, rated voltage, full load current, and the inrush current if the motor is started across the line are all known values. Since the speed of induction motors is closely related to the frequency of the supply voltage, one approach is to convert the supply voltage to direct voltage and then to construct an alternating voltage at any desired frequency from the direct

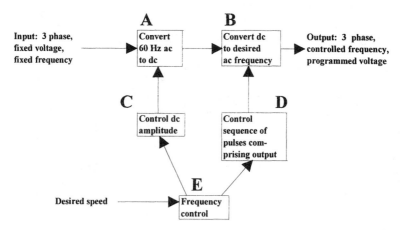

Figure 9.3 Preliminary block diagram of a motor speed controller.

voltage. If one looks into the theory of induction motors, one also finds that the (constructed) voltage at the motor terminals must have a frequency-dependent amplitude. This information leads to the block diagram of Figure 9.3, which names each function that must be provided.

As silicon-controlled rectifiers (SCR's) were developed, it became evident that the functions of the preceding paragraph and of Figure 9.3 could be provided by a circuit shown in elementary fashion in Figure 9.4. (There are other circuit configurations that will provide the same functions, and that avoid some of the problems, such as harmonic content, inherent in the output of this circuit.) The output unit, called an inverter, uses six SCR's which are fired

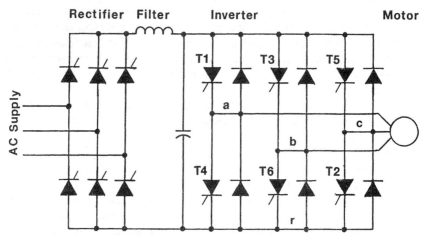

Figure 9.4 Six-step inverter circuit.

THE DESIGN OF SYSTEMS

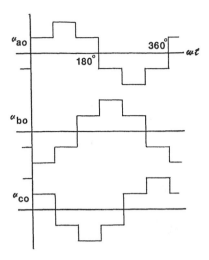

Figure 9.5 Phase voltages for the motor of Figure 9.4.

(turned on) in a sequence that generates a waveform such as that in Figure 9.5 at a desired frequency. The amplitude of the output wave is determined by the direct voltage at the input to the inverter, and that voltage is controlled by the firing point of the six SCR's in the rectifier section of the circuit.

From the block diagram of Figure 9.3 and an understanding of the requirements and limitations of the motor and of the elementary circuit of Figure 9.4, a working block diagram of the controller can be devised, as shown in Figure 9.6. In this diagram, the units identified by letters A through E in Figure 9.3 are further subdivided into subunits, each of which is an identifiable entity needed in the final controller, and each of which can be specified by its output/input characteristics. The subunits of the system are:

1. A rectifier unit which changes ac to variable voltage dc.
2. A main inverter circuit with six SCR's which are fired in the proper sequence to generate the waveforms of Figure 9.5.
3. An auxiliary inverter which generates negative voltage pulses to turn off the SCR's in the main inverter.
4. A variable frequency oscillator having a stable output frequency at any setting of the frequency control.
5. A speed control to set the frequency of the oscillator.
6. A logic block to control the firing of the SCR's in the main and auxiliary inverters.
7. A voltage control which responds to the difference between the feedback signal and the reference voltage to initiate programmed firing by the bridge firing control.

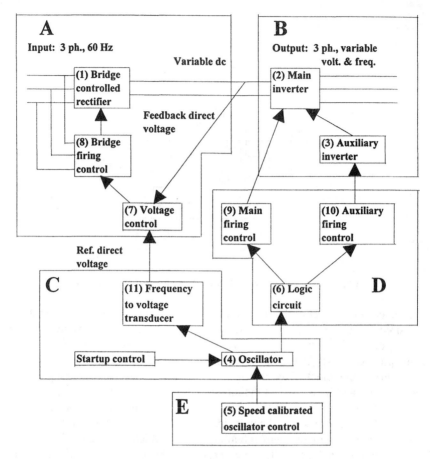

Figure 9.6 Working block diagram of motor speed controller.

8. Bridge firing control.
9. Main inverter firing control.
10. Auxiliary inverter firing control.
11. Frequency-to-voltage transducer.

The block diagram of Figure 9.6 now leads on to the construction of a precedence matrix, as discussed in Section 8.3.1. The precedence matrix follows from the precedence table, Table 9.1, for the system.

The precedence table lists the subunits and the dependence of each of these on the specifications and/or the characteristics of other subunits. The entries in the predecessor level columns depend on an unstated assumption: That every subunit can be selected or designed to meet the requirements of the contiguous subunits. This may or may not be true, but assuming that it is true, the

THE DESIGN OF SYSTEMS

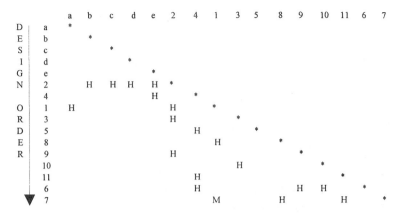

Figure 9.7 Precedence matrix for the speed controller.

subunit precedence matrix of Figure 9.7 can be devised. It is clear that all parts can be designed in the indicated sequence without any further assumptions or iterations.

As pointed out in Chapter 8, the precedence matrix simply indicates that sufficient information about other devices and subsystems will be available to fully define the specifications of the unit being considered if the indicated sequence of design is followed. However, the procedure cannot guarantee that any unit will be physically realizable simply because the best order of design is being used. A simple example of nonrealizability would be arriving at an inverter design that requires SCR's having capabilities that exceed those of any unit available on the market.

If any of the subunits are to be purchased, added constraints are placed on the design team. For example, suppose that Unit 11, the frequency/voltage transducer, is to be a purchased device, that there is only one manufacturer in the market, and that manufacturer has only one model available. This situation would not normally be encountered, but it is a usual condition in system design that the number of off-the-shelf items from which the designer must select the required components is limited. Hence there will be compromises in order to accommodate the specifications of connected subunits on the basis of the best possible match, rather than on meeting all specifications in full. If one or more subunits are to be purchased, the precedence table and the precedence matrix must be modified. The characteristics of the purchased subunits become in essence specifications; as a result, the precedence matrix will no longer be the same.

Although there may be compromises if subunits are purchased, there are also advantages. The "purchase" may be from within the company, the unit being already made for another purpose. If the subunit is bought from outside, design and development time and costs are avoided, and there are no production line equipment costs to consider. Decisions as to whether to purchase or to develop will be strongly influenced by predictions as to the number of identical or nearly identical systems to be produced. Large numbers may swing the decision toward doing the necessary development, but if only a few are to be built, it will almost invariably be less expensive to purchase. In many cases, however, some of the units will need to be designed specifically for the system.

Once the synthesis activity has been completed (cell 33 of Figure 9.1), there is an analysis of the system performance using appropriate models (cell 34). This may be strictly mathematical in nature, or it may involve physical modeling, or a combination of the two. In the case of the speed control, a prototype of the entire controller would be the most sensible final "analysis" of the system. If specifications are not met, suitable changes will need to be made to bring the system performance within specified tolerances. Once this has been done, the system engineer is ready to formulate plans for cell 35, the production phase.

9.5 SUMMARY

System design is, by its very nature, a complex task. The system design team will include individuals with many different forms of expertise. Engineers are accustomed to thinking in terms of such teams as being composed of engineers from various branches of the discipline, but it is important to keep in mind that large systems will have sociological, environmental, political, and legal implications and/or effects. There is a considerable advantage in planning every design project using the procedure presented or a similar procedure. The simple tasks done in this way give practice in use of the procedure, and application to more complex tasks is then facilitated.

A fairly simple system has been used as an example in this chapter, but it presents many of the principles needed for design of much larger systems. For most engineers, involvement with design of a system begins below the community level. It often begins at the system design phase, as did our example of the motor speed control.

The reader confronted by the assignment of designing a complex system, especially a first-of-a-kind system, will find a great deal of thought-provoking material in Reference 2.

REFERENCES

1. Hall, A. D. Three-dimensional morphology of systems engineering, *IEEE Transactions on Systems Science and Cybernetics,* 1969, SSC-5(2): 156-60.
2. Rechtin, Eberhardt. The synthesis of complex systems, *IEEE Spectrum,* 1997, 34(7): 50-55.

THE DESIGN OF SYSTEMS

REVIEW AND DISCUSSION

1. What are the two principal ways in which system design differs from the design of simpler products?
2. List the characteristics of a system.
3. An automated factory is to be designed. Enumerate the characteristics you believe should be sought in the person to be designated as the chief system designer.
4. A number of different kinds of materials are listed in the chapter that should be included in a design file. Add to this list any additional materials that may have been omitted.
5. What are the phases of system development? How do these relate to the life cycle of a product?
6. List the logic steps for each phase of system development.
7. One of the problems faced by many American cities in the past half-century was the obsolescence of their street railway systems, and the retirement phase of those systems was probably not handled very well. Poles were left standing, rails were covered by blacktop, stations for converting alternating current to direct current were simply abandoned, and so forth. Follow the logic steps of Figure 9.1 for the retirement phase, listing under each step what you would propose doing in this situation.
8. What is the main reason for constructing block diagrams of proposed systems?
9. How do the design of a system and the invention of a new device resemble each other?
10. Iteration has not been stressed in this chapter on system design. Is it important to do this for systems? Give reasons for and against.

PRACTICE PROJECTS

1. If possible, devise a system in which the product you are developing would be a component.
2. Apply the matrix of Figure 9.1 to the system development of a small factory to produce the product you are developing.
3. Set up an indexing system for a design file for the product you are developing.
4. The frequency-to-voltage transducer of the example is not an easy component to design, especially if the relationship is not to be a completely linear one. Hence the decision is made to purchase this component. For this new situation, formulate the precedence table and the precedence matrix.

10

DETAILED DESIGN OF DEVICES AND SYSTEMS

The components, products, and small systems that are the objects of design projects for manufacture as salable items or that comprise portions of a larger system have now been identified and their specifications determined, as explained in the previous chapter. For example, a gear or a resistor is a component that may be needed; a variable-pitch pulley speed reducer or an amplifier is a product or possibly part of a larger system; and a transmission system or a public address system can be recognized as a small system. Design methods for each of these items are known or may be devised from the knowledge already in the engineer's tool kit. The task of the designer now is to decide every detail of these devices so that the device may be made in such a way that the specifications are met. These details include dimensions, power ratings, materials, alloy numbers, tolerances, specifications of components used in a system, and so forth. We shall use the term "design parameter" to designate the value or other identifying features for each of these details.

10.1 EVOLUTION, REPEATED ANALYSIS, AND SYNTHESIS

The way in which a designer proceeds to determine design parameters depends on the designer's success in establishing various types of models. For some devices it is impossible to establish mathematical models, or at least models that are simple enough that their manipulation is tractable. An example of a component for which no mathematical model can be easily established is a lamination that constitutes part of a motor. The magnetic flux pattern depends on such factors as the material, the dimensions of various parts of the magnetic circuit, the shape of slots, position and length of air gaps, and, especially for high flux densities, the hysteresis loops of the steel. A global mathematical

DETAILED DESIGN OF DEVICES AND SYSTEMS

formulation (that is, one that attempts to use a single equation or a series of equations to account for all of the physical phenomena involved) would be highly nonlinear. For other devices a mathematical model may be possible, but physical models must wait for some technological developments. The principle of facsimile transmission (FAX) was known and demonstrated by 1920 and a mathematical model was available for the important elements, but practical use had to wait until development of efficient mechanical paper-handling and dry printing techniques had occurred. The first step in deciding how to proceed is therefore a determination of what kinds of models are available or can be devised.

If mathematical models are not possible, the designer must resort to drawings and physical models. A drawing is made, a physical model is built and tested, changes are made based on the intuition of the designer as to what to change, in which direction, and how far, and the process is repeated. The repetition continues until the specifications are met. An appropriate name for this technique is *device evolution*.

The evolution method has many disadvantages. Unless the model can be made by simply combining components that are readily available, device evolution is usually very expensive. It involves a major amount of work by skilled craftsmen to build the first model, and the designer may inadvertently decide to change one of the design parameters in the wrong direction for the second model. Even when one has made a series of models and arrived at one that meets the specifications, a decision must be made as to whether to accept what will probably be a marginally successful design, or to proceed with additional models in order to try to optimize the design in some sense, whether cost, performance, size, or some other attribute. For these reasons, device evolution should be used only as a last resort.

A more desirable method of determining the design parameters is possible if one of the analytical models can be developed. Such models include the functional block diagram, the network model, and the mathematical model. Other kinds of models have been developed for particular kinds of products, and the designer should seek out suitable models when he or she is confronted by an unfamiliar problem. No matter what the product, if some kind of analytical model is available, the design can be accomplished by assigning values to the design parameters and determining by analysis whether the specifications are met. If the specifications are not met, it is necessary to make a new selection of the design parameters and repeat the process. Even if they are met, repetition of the process should still occur in an attempt to improve the design. This method is, for obvious reasons, referred to as *design by repeated analysis,* or *design by iteration.* The discussion of the design of the door-closer in Chapter 7 pointed toward this technique.

Design by repeated analysis, or design by iteration, is much more desirable than device evolution for a number of reasons. First of all, any device that can be analyzed can be designed by this technique. This broad applicability makes design by repeated analysis the most useful of the techniques presented here, but there are other reasons that this method deserves attention. A second reason is that much of the cost, boredom, and potential for errors in computation has been removed by the availability of computers. Many products previously designed by the evolution technique can now be designed by repeated analysis. A third reason is that this method often allows the designer to determine the "best" design by the simple expedient of repeating the cycle of parameter selection and analysis for a wide range of parameters. This ability to optimize is a desirable feature of any design method.

A third method, one that is far superior to either of the other two, is called *synthesis* [1,2]. Synthesis is defined as a technique wherein knowledge of the specifications, perhaps together with some easy-to-make (and obvious) choices on the part of the designer, leads directly to the final design without iteration. This method is not applicable to all products or components. A necessary condition for the application of this method is that a complete mathematical model can be developed. Although this is a necessary condition, it is not a sufficient condition. The mathematical relationships must also be simple enough to permit explicit solutions for the design parameters.

There are therefore three methods—design evolution, design by repeated analysis, and synthesis—that the designer must consider when approaching the task of determining components, materials, dimensions, and other parameters of the device to be designed. The methods are not mutually exclusive; many times the most efficient way to proceed is to use more than one method in designing a product, especially when it is a complicated product. If the product is divided into components or subunits, evolution may be the only approach possible on one part, repeated analysis can be applied to another, and synthesis may be possible on a third. Select the best mix of procedures to make the necessary decisions with as high a degree of confidence as possible without unduly increasing development time or cost.

In order to help you develop deeper insight into these design procedures (except for evolution), several examples will be used. Because analysis is well understood by engineers, design by repeated analysis will not receive strong emphasis; synthesis is a less familiar procedure, and will receive more stress. However, it should be obvious that the objective of the examples that follow is *how* to develop the necessary procedure; the specific procedures are intended only to provide examples to clarify the lines of thought to be pursued by the designer.

10.2 DESIGN BY REPEATED ANALYSIS

The first step in design by repeated analysis is to gather data that are available relative to material characteristics, limitations imposed by manufacturing processes, costs of materials and labor, and so forth. A mathematical model is

DETAILED DESIGN OF DEVICES AND SYSTEMS

then constructed, consisting of equations that relate the important variables to the specifications, and consideration must be given to the order in which the design will be tested against the specifications. The order in which evaluation for compliance with specifications will be made is especially important for complex devices.

For example, in one program developed to design induction motors in standard NEMA frame sizes, it was found that the deviation of full-load speed from synchronous speed must be determined first, then the starting current, and then the starting torque. The deviation of full-load speed from synchronous speed is usually expressed as the dimensionless ratio (called the "slip") of the deviation to the synchronous speed. The slip is almost always small (0.035 to 0.05 is a typical range). Specifications usually give an upper limit for slip, an upper limit for starting current, and a lower limit for starting torque. The design should, of course, approach these limits as closely as possible, allowing for normal manufacturing variations. For most sets of specifications, these three are the critical specifications, and are sometimes known as *absolute specifications*. Other specifications, frequently established as important goals, need not be met completely. The second example below, illustrating the design of a three-phase induction motor, will elaborate on some of these terms.

The second step is to develop a logic-flow diagram in which choices of dimensions and materials are made in some orderly fashion to arrive at a design. Typically, the procedure begins by purposely choosing design parameters known to result in an inadequate device and then increasing values selectively until the specifications are met with a comfortable margin of safety. The extensive calculations required for design by repeated analysis make the use of a computer mandatory.

As a first example, consider the design of a flywheel that is to be used with a 5-hp, 1750-rpm electric motor to reduce speed variations due to momentary

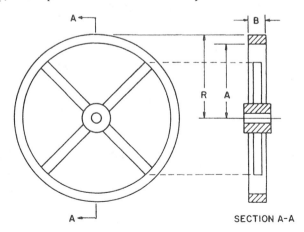

Figure 10.1 Flywheel sketch.

CHAPTER 10

overloads, such as those encountered with a punch press. A typical flywheel for this application is shown in the sketch of Figure 10.1 The flywheel is to have 20,000 J of stored kinetic energy at rated speed and an outer radius R of 0.5 m or less. This requires a rotational inertia of 1.2 kg-m² / rad². The flywheel is to have a weight less than 100 N and the width B of the rim is to be four times its thickness, (R − A). Four possible metals, with their important properties, are listed in Table 10.1. The least costly design is to be selected for manufacture. Assuming a rim that is thin compared to the radius and neglecting the spokes and hub, the mathematical model is given by the following set of equations:

$$\sigma_t = \rho R^2 \omega^2 \qquad (1)$$

$$I = K^2 \frac{WT}{g} \qquad (2)$$

$$WT = g\pi B\rho\left(R^2 - A^2\right) \qquad (3)$$

$$K = \sqrt{\left(R^2 + A^2\right)/2} \qquad (4)$$

$$B = 4(R - A) \qquad (5)$$

where
σ_t = tensile stress, N/m^2
R = outer radius, m
A = inner radius, m
B = width of rim, m
WT = weight of rim, N
I = rotational inertia, kg m² / rad²
K = radius of gyration, m/rad
ρ = mass density, kg/m^3
ω = angular velocity, rad/sec
g = gravitational constant, m/s^2 = 9.81 m/s^2

Table 10.1 Data for Flywheel Design

Material	Maximum tensile strength (Mpa)	Density (kg/m^3)	Cost ($/N)
Cast iron	20	7200	0.23
Cast steel	68	7850	0.33
Brass	60	8300	0.45
Monel™	80	8520	0.90

DETAILED DESIGN OF DEVICES AND SYSTEMS

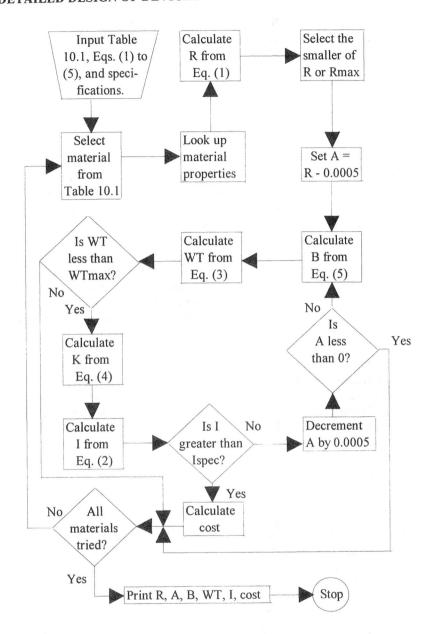

Figure 10.2 Logic diagram for flywheel design by repeated analysis.

Material	σ	R	A	B	WT	I	Cost
Cast steel	0.66E+08	0.500	0.493	0.30E-01	54.0	1.356	$17.82
Brass	0.60E+08	0.464	0.456	0.32E-01	60.2	1.298	$27.09
Monel	0.72E+08	0.500	0.493	0.28E-01	51.1	1.284	$45.99

Figure 10.3 Parameters for possible flywheel designs. (Dimensions of the parameters as given in the text.)

The logic diagram is shown in Figure 10.2. The program written by the engineer to carry out the calculations will depend on the programming language to be used. A program in Microsoft Basic is shown in Reference 3, and the resultant design parameters are shown in Figure 10.3. Because cast iron cannot meet the specifications, it is not even listed in the results. Of those that can meet specifications, cast steel is obviously the most economic choice.

The second example shows some of the data generated in the design of a 7.5-hp, 220-volt, 3400-rpm, 3-phase induction motor [4]. Before discussing the data, some additional points should be mentioned. The term *absolute specifications* was introduced earlier. These are specifications that *must* be met if the design is to be accepted. There are *other specifications* that it is desirable to meet, although it is not absolutely necessary. There will be, in general, more than one design that will meet the absolute specifications. How closely the other specifications are met by a particular design may be evaluated by use of a function that measures the absolute deviation from the standard set by the other specifications. In Reference 4, this function is referred to as the *error*. It is obtained by normalizing the difference of each of the actual parameters from the [other] specification, squaring these normalized differences, summing them, and then taking the square root. This process prevents cancellation of a positive error by a negative one, and places more weight on a single large error than on a number of small ones. If one wishes, the various errors may be weighted, so that the more important of the other specifications contribute more to the error value than those having less importance.

For this design, the absolute specifications chosen were the breakdown torque (BDT) and the allowable slot fullness (FUL), which is a measure of how much of the slot cross section area is used by the conductors of the winding. The other specifications chosen were the full load power (FLW) in watts, the full load current (FLA) in amperes, the full load speed (FLS) in rpm, the allowable full load stator tooth flux density (STD) in teslas, the locked rotor torque (LKTQ) in lb-ft, and the locked rotor current (LA) in amperes. The design parameters are based on a standard lamination, on the number of conductors (N1) in a stator slot, stator wire size using the American Wire Gage (AWG), and the height of the stack of laminations (STAK) in the design.

The design procedure is as follows: Select a reference motor (one that the designer knows is reasonably close to what is wanted), and calculate its

DETAILED DESIGN OF DEVICES AND SYSTEMS

Table 10.2 Example of Repeated Analysis for a 7.5-hp, 220-V, 3400 rpm, 3-phase Motor

Parameter	Unit	Specifications	Ref motor	Trial 1	Trial 2	Trial 3	Trial 4	Trial 5	Trial 6
N1			43	38	40	40	39	41	42
Wire AWG			18.5	18.5	18.5	18.0	18.0	18.0	18.0
STAK	cm		12.7	12.7	12.7	12.7	12.7	11.4	11.4
BDT	lb-ft	31.0–34.0	26.6	34.0	31.0	31.8	33.4	33.1	31.6
FUL		0.7	0.645	0.570	0.600	0.666	0.650	0.683	0.700
FLW	W	6325	6448	6332	6370	6332	6314	6331	6348
FLA	A	18.0	*18.2*	18.0	18.0	17.9	17.9	18.0	18.0
FLS	rpm	3400	*3353*	3417	*3393*	*3395*	3407	3403	*3391*
STD	T	2.02	1.39	1.57	*1.49*	*1.49*	1.53	1.62	1.58
LKTQ	lb-ft	24.5	20.0	25.9	23.3	*24.1*	25.5	25.2	23.9
LA	A	110	88	*113*	102	103	109	108	102
ERROR				0.0104	0.0202	0.0067	0.0000	0.0004	0.0100
COST	$			32.30	32.93	34.58	34.19	32.34	32.72

Notes:
1. Specifications shown for FUL, FLW, FLA, STD, and LA are maxima; values shown for FLS and LKTQ are minima.
2. Values shown in italics are unacceptable, being either too high or too low.

347

performance. Adjust the number of conductors N1 to meet the absolute specifications, calculate its performance, and tabulate the data. If the specifications are met in the initial trial, successively smaller wire sizes (that is, larger AWG numbers) are tried until specifications are met. If specifications are not met in the initial trial, the largest wire size that will fill the slot is selected and if specifications are then met successively smaller sizes are tried. If specifications are not met when using the largest wire size, the next longer stack height (STAK) is investigated beginning with a full slot. Table 10.2 (adapted from Table 4.7 of the reference) shows the results of this process. Note the sequence that was followed. From the reference motor, the number of conductors N1 was adjusted to meet the absolute specifications, but two of the other specifications (FLW and LA) were outside allowable limits. To reduce the locked rotor current LA, the number of conductors was increased in trial 2, bringing LA within specifications, but FLW remains outside limits and LKTQ is now too small. For trial 3 the wire size was increased, keeping N1 unchanged, resulting in FLW, FLS, and LKTQ being outside limits. Reduction of N1 in trial 4 results in a design that meets all specifications. Can a shorter stack height (STAK) be used? Trials 5 and 6 demonstrate that this is not a possibility.

As noted earlier, these examples are intended to give an insight into how to design by repeated analysis or by iteration. The lengthy details of calculation have been omitted because they add nothing to understanding of the process.

10.3 DESIGN BY SYNTHESIS

The key to design by synthesis is the development of a complete mathematical model, which we shall define as a model having the number of independent equations equal to the number of design parameters. This is equivalent to the classical problem in algebra in which one has a set of N equations in N unknowns. There can be several sets of values of the unknowns that will satisfy all of the equations, although generally there are only one or two, and there may be none. Our objective here is to demonstrate how a designer can develop a complete mathematical model.

As a first step, develop a set of equations that involve every relevant physical relationship that can be expressed mathematically. Once this has been done, place the literal terms of the equations into four categories: (1) Specifications, (2) values to be chosen by the designer, such as a dimension or a ratio of dimensions, (3) physical constants, such as the resistivity of copper or Young's modulus for steel, and (4) design parameters. The design parameters being the unknown quantities, the equations are solved for those values.

Three conditions can exist. One is that there are fewer independent equations than unknown design parameters. In this case, it may be possible to develop additional equations by considering physical limitations that the product will encounter during the production, distribution, or consumption periods. At

DETAILED DESIGN OF DEVICES AND SYSTEMS 349

least one additional equation can always be generated by optimizing the design with respect to some criterion. For example, in the case of the flywheel of the previous section, another equation can be written for the cost of the material, such as $C = WT \cdot \$$, where C is the cost and $\$$ is the cost in dollars per newton. Elimination of unknowns enables one to express C in terms of either A or B, and the most economical design can theoretically be obtained by differentiating the resulting equations with respect to A and B respectively and setting the derivatives to zero. Having A and B, R can be calculated, as can I, K, and C. Referring back to Figure 10.3, however, in which the results of the design are displayed, one notes that in two cases (cast steel and Monel™) the value of R used in the design is the maximum allowable value, and the process just described will no doubt result in larger values for R. One can then use the maximum value of R as a specification, and begin the process anew.

If the excess of unknowns to equations persists, it will be necessary to choose enough design parameters to reduce the number of unknowns to the number of equations. When doing so, one should select if possible those design parameters that appear to have the least influence on the overall design, assigning the most reasonable values possible to each of them. Alternatively, as in the flywheel case, a limiting value of a design parameter may have to be chosen. Once this has been done, the equations can be solved for the remaining unknowns.

A second condition is an excess of equations. This condition does not often occur, but if it does, the design task is overspecified. In this case, select those specifications of lesser importance and eliminate them to bring the number of unknowns and the number of equations into balance.

The third condition exists when the designer can obtain a mathematical model having the same number of independent equations as the number of unknowns. Since the unknowns are the design parameters, these equations can be solved to form the basis for a noniterative design procedure.

This is the ideal situation. Since each of the design parameters is expressed in terms of specifications, physical constants, and values chosen by the designer, the solutions for the design parameters constitute a set of design equations. These remain the same for all future designs of the product having the same set of named specifications, even though the values of the specifications may be changed. Since the products of a given company tend to be closely related, the advantage of developing a synthesis procedure for components and devices within the product line should be obvious.

Unfortunately, the development of a complete mathematical model becomes less probable as the complexity of the product increases. For example, the stator tooth magnetic flux density in the induction motor design cannot be expressed easily in terms of other parameters because of the inherent nonlinearity of the saturation curve of the steel. In general, the limitation on design by synthesis is determined by the designer's ability to determine all relevant physical phenomena and to express them mathematically. This task is a natural one for the engineer because a large part of every engineering course has been directed toward the objective of developing these mathematical relationships. You already have a stock of useful equations, or you know where to look for them. Additional sources will appear in the following examples.

A final comment before proceeding to look at examples. Even if the number of equations equals the number of design parameters sought, one must still deal with the fact that certain decisions must be made which are somewhat arbitrary. For example, wire comes in discrete sizes, and the theoretical value developed in the design process will probably not be precisely a size appearing in the American Wire Gage (AWG) series. One must usually pick a standard size, and make other adjustments as the design proceeds. Bearings come with certain internal diameters, which may or may not coincide with the diameter of a shaft that the design process has determined from a set of equations. The horsepower required by the design will probably fall between two standard horsepower ratings. The designer must be prepared for these events, and learn to adjust the order in which design parameters are determined to accommodate.

As a first example of design by synthesis, the parameters of a wire-wound resistor will be determined. These resistors are used in electrical networks where certain resistance values are needed and the resulting power dissipation will be quite large. Such resistors are off-the-shelf items, and are offered by a number of manufacturers. The construction is shown in Figure 10.4, and consists of a tube with a winding of bare wire overlaid with porcelain. Because the power dissipation is high, an operating temperature of 325 C will be used in the design. This mandates that the tube be of ceramic.

Because of the wide temperature range experienced by the resistor and because of the desirability of maintaining the resistance value in narrow limits throughout the anticipated temperature variations, the wire chosen must have a very low temperature coefficient of resistivity. Moreover, in order to maintain

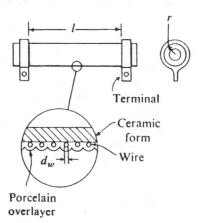

Figure 10.4 Configuration of a wire-wound resistor.

DETAILED DESIGN OF DEVICES AND SYSTEMS

the structural integrity of the completed unit over the wide swings of temperature to which it is subjected, the temperature coefficient of expansion must also be small. Suitable conductors are Nichrome™, Chromel™, Advance™, Constantan™, and others.

The equations needed to express the physical relationships are basically quite simple:

$$\Delta T = \frac{P}{hA} \tag{6}$$

$$R = \frac{\rho \ell_w}{(\pi/4)d_w^2} \tag{7}$$

$$A = 2\pi r \ell \tag{8}$$

where the literal terms may be categorized as follows:

Specifications:
  R = resistance, Ω
  P = power dissipation in the resistor, W
  ΔT = temperature rise, °C

Parameter chosen by the designer:
  ρ = resistivity of the resistance wire, Ωm

Physical constant available in the literature:
  h = cooling coefficient, W/m² °C

Design parameters:
  A = heat-dissipating surface of the resistor, m²
  r = outer radius of resistor, m
  ℓ = length of resistor between terminals, m
  ℓ_w = length of resistance wire, m
  d_w = diameter of resistance wire, m

We note that we have five design parameters, the unknowns, but only three equations. The designer's first reaction is frequently to select two of the five unknowns, and solve Equations (6) through (8) for the others. However, as was pointed out above, it is always possible to develop another equation that can be used to optimize the design in some agreed-upon sense, whether cost, size, or some other feature.

Another way of developing additional equations is to look at manufacturing or environmental constraints, or at constraints imposed by the physics of the situation, and express those constraints in mathematical form. In this case, there are two such constraints that have been ignored so far. One of these is that no thought has been given to whether the adjacent turns of the bare wire on the tube will touch each other, or possibly even "pile up" to form a multilayer winding. If the turns touch each other, they will short out and the resistance of the completed unit will be below the design value. If they are too far apart, the uneven heating of the porcelain overlay will result in crazing of the surface and eventual loss of an insulating cover over the entire resistor. Experience has shown that coverage by the wire of 60 to 80% of the total surface area will provide adequate spacing of the wires without their being so far apart that porcelain crazing occurs. In equation form,

$$c = (d_w \ell_w)/A \tag{9}$$

where c = the coefficient of coverage in decimal form.

A second constraint may not be so obvious. These resistors are rarely used in an absolutely clean environment. As time goes on, dirt settles on the surface, moisture or other air-borne gaseous contaminants may be absorbed by the dirt, and leakage current paths will develop between the terminals over the surface. Experience has shown that there will be little difficulty with leakage currents if the voltage gradient over the length of the resistor is kept below 20 kV/m. In mathematical form,

$$\frac{V}{\ell} \leq 20{,}000 \tag{10}$$

The voltage between the terminals can be related to two of the specified values, P and R, by the equation

$$P = V^2/R, \text{ or } V = \sqrt{PR} \tag{11}$$

If we add the limitation on voltage gradient to the set of specifications and add the coverage coefficient, c, to the list of the items to be chosen by the designer, we retain the set of five design parameters listed above, but we have added two equations. A noniterative design is now theoretically possible.

Algebraic manipulation of the several equations leads to a set of design equations, as follows:

$$d_w = \sqrt[3]{\frac{4Pc\rho}{\pi hR\Delta T}} \tag{12}$$

$$\ell_w = \frac{\pi R d_w^2}{4\rho} \tag{13}$$

$$A = d_w \ell_w / c \tag{14}$$

DETAILED DESIGN OF DEVICES AND SYSTEMS

$$\ell \geq \frac{\sqrt{PR}}{20{,}000} \tag{15}$$

$$r = \frac{A}{2\pi\ell} \tag{16}$$

Which design parameter does one solve for first? Examination of the equations shows that only Equations (12) and (15) will have known values for all terms on the right side of the equation, and Equation (15) only sets a lower limit for ℓ, a limit that may be met very easily. That is, the voltage gradient specification may be only a very weak limitation. In this case, it seems obvious that one should solve for d_w first, using Equation 12, and then use the remaining equations in sequence.

There is another, and more important, reason for solving for d_w first. As pointed out above, wire comes only in discrete sizes, generally those shown in wire tables available in many handbooks. While it is true that, for large quantities, wire may be had in half-sizes (as in the second example in the preceding section), nevertheless the diameters are discrete. It is better to solve for such parameters first, adjusting from the theoretical value to the nearest practical value, and then using that value in succeeding equations. Another design parameter in the present example that is available only in discrete sizes is the ceramic tube on which the resistance wire is wound.

As an example of the procedure, assume specifications of 100 W, 10 Ω, and a temperature rise of 300 C above a 25 C ambient (thus conforming to NEMA standards). The designer chooses a coverage coefficient of 0.7, the midpoint of the usual range. Assume that Chromel A is selected as the wire to be used. This is an 80% nickel-20% chromium alloy, having a room temperature resistivity of 1.08 μΩm, and a temperature coefficient of resistivity of 0.0000937 per °C in the range from 25 C to 427 C. Hence the resistivity at the working temperature is

$$\rho = 1.08\left[1 + 0.0000937(300)\right] = 1.11 \; \mu\Omega m$$

We now have all of the values for the right side of Equation (12) except for the cooling coefficient, h. This value is not precisely known. If you are assigned the task of designing a line of wire-wound resistors, it would be wise to do an experimental investigation using partial models in order to establish this value or a range of values for various length to diameter ratios. A cooling coefficient of $20 W/m^2 \; °C$ is frequently accepted as suitable, and will be used in this design.

Using these values,

$$d_w = \sqrt[3]{\frac{4 \times 100 \times 0.70 \times 1.11 \times 10^{-6}}{\pi \times 20 \times 300 \times 10}} = 0.00118 \text{ m, or } 46.51 \text{ mils}$$

This is less than 3% larger than AWG #17, which has a diameter of 45.26 mils. The next larger size, #16, which has a diameter of 50.82 mils, would be a little over 9% larger. If #17 is chosen, the length of the wire will have to be reduced to meet the resistance specification, and the surface area from Equation (14) will be reduced from the theoretical value needed to meet the allowable temperature rise. On the other hand, if #16 is selected, the length of the wire will be increased, as will the surface area, leading to a resistor that will operate at a lower temperature than allowed.

This is an example of the kind of dilemma with which the designer is frequently faced. One solution is to proceed with two designs, one using #17 and the other using #16, and to compare the results when the designs are completed. Instead of doing that, let us proceed using #17, and see if adjustments elsewhere can take care of the possibility of overheating.

Equation (13) is now used to obtain the length of resistance wire needed. Using the value of d_w for #17 wire, which is 0.00115 m, we obtain

$$\ell_w = \frac{\pi \times 10 \times 0.00115^2}{4 \times 1.11 \times 10^{-6}} = 9.36 \text{ m}$$

From the discussion earlier as to whether to use #16 or #17 wire, we already know that use of Equation (14) will result in a surface area that is too small to meet the temperature rise specification. We can satisfy that requirement, however, if we return to Equation (6), and solve it for A.

$$A = \frac{P}{h\Delta T} = \frac{100}{20 \times 300} = 0.0167 \text{ m}^2$$

Before accepting this value, we need to check the coverage of the ceramic tube by the wire to insure that we have not strayed outside the limits that experience has shown to be necessary. Equation (14), rearranged to solve for c, yields

$$c = \frac{d_w \ell_w}{A} = \frac{0.00115 \times 9.36}{0.0167} = 0.64$$

This value is still well within the limits.

We can now check the minimum length between terminals using Equation (15).

DETAILED DESIGN OF DEVICES AND SYSTEMS

$$\ell = \frac{\sqrt{PR}}{20,000} = \frac{\sqrt{100 \times 10}}{20,000} = 0.0016 \text{ m}$$

This value (1.6 mm) is so small that it is obvious that the voltage gradient specification will be satisfied by any practical design.

We have now reached the point at which we know the surface area of the finished resistor and the minimum distance between terminals, although the latter datum is of no real significance. We proceed to look up standard diameters of ceramic tubes and make rough calculations of lengths of tube required. This leads us to a tube having an outside diameter of 2.5 cm as a likely candidate. Using this tube diameter and adding twice the wire diameter, we arrive at an overall resistor diameter of 0.0273 m, and thence to a tube length between terminals of 0.195 m. The resistor design is complete, except for a decision on the type of terminal to be used.

Has a reasonable design resulted from this process? Examination of commercially available power resistors shows that those having ratings that match the specifications set down at the beginning of the process bear a strong resemblance to the resistor design just completed.

As a second example, and one that requires fewer decisions than the previous example, consider the design of a helical compression spring. These springs, using any specified alloy, can be made using any of a large number of wire sizes, coil diameters, and coil pitches, all of which will produce the same deflection when subjected to a given force. Figure 10.5 shows some possibilities. We will see below that a simple design procedure will allow the engineer to design acceptable springs having any of the configurations shown.

A common procedure in spring design is:

1. Assume an allowable stress that is based on the material, severity of loading, and estimated wire size.
2. Assume a reasonable spring index (ratio of mean diameter of coil to wire diameter) and determine all dimensions of the spring.

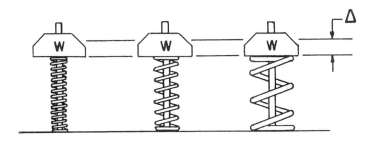

Figure 10.5 Helical compression springs having equal spring constants.

3. If wire size does not fall in the assumed range, assume a new spring index and repeat the cycle.

This is clearly design by repeated analysis or iteration.

Several excellent synthesis procedures for springs appear in the literature [5,6]. To begin this development, the mathematical model for a close-coiled helical compression spring, such as that in Figure 10.6, is given as follows:

Hooke's Law $$\frac{F_1}{x_1} = \frac{F_2}{x_2} \tag{17}$$

Stiffness constant $$\frac{F_1}{x_1} = \frac{Gd^4}{8D^3 N} \tag{18}$$

Final stress $$\sigma_2 = \frac{8DF_2}{\pi d^3} \tag{19}$$

Stroke $$s = x_2 - x_1 \tag{20}$$

where

F_1 = applied initial force, N
F_2 = applied final force, N
x_1 = initial deflection, m
x_2 = final deflection, m
D = mean coil diameter, m
G = modulus of rigidity, N/m^2
σ_2 = stress in compressed state, N/m^2
V = volume of spring material, m^3
N = number of active turns of wire
d = diameter of spring wire, m
s = stroke, m

The volume of spring material is given to a close approximation by:

$$V = \frac{\pi^2 d^2 DN}{4} \tag{21}$$

Typical specifications are the initial force F_1, the mean coil diameter D, and the stroke s. These would be appropriate specifications for the spring used in a door-closer such as that shown in Section 7.6, where some force is required to keep the door in the closed position and a fixed stroke is required to permit opening the door. When the designer selects the spring material, the modulus of rigidity G is established, and the designer can also decide on a reasonable maximum value of stress, σ_2.

DETAILED DESIGN OF DEVICES AND SYSTEMS

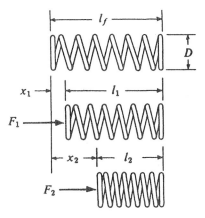

Figure 10.6 Helical spring showing relevant dimensions and forces.

There are eleven quantities involved in the five equations of the model. As described in the preceding paragraph, five of these quantities are known, either as specifications or as properties of the selected material. If no other equations are possible, the design can proceed using repeated analysis, resulting in a variety of springs, such as those shown in Figure 10.5. However, as was pointed out earlier, an additional equation can always be obtained by optimizing some property. Since we already have an equation for the volume of the material in the spring (Equation (21)), and since it would be desirable to minimize this volume, one approach is to differentiate the volume with respect to the wire diameter, d, and set the derivative to zero. However, Equation (21) involves some of the other design parameters. Using Equations (17) to (21) to eliminate all of the design parameters except V and d, we obtain

$$V = \frac{\pi^2 s d^6 G}{4(\pi d^3 \sigma_2 D - 8F_1 D^2)} \qquad (22)$$

To find the wire diameter d that results in minimum volume, Equation (22) is differentiated with respect to d, yielding

$$d = \sqrt[3]{\frac{16 F_1 D}{\pi \sigma_2}} \qquad (23)$$

Substitution into Equation (22) results in

$$V_{min} = \frac{8sF_1G}{\sigma_2^2} \qquad (24)$$

We proceed to design a spring that satisfies the following specifications:

$F_1 = 20$ N
$D = 25$ mm
$s = 75$ mm

If music spring wire is chosen, appropriate values for maximum stress and modulus of rigidity are:

$\sigma_2 = 0.689$ GN/m^2
$G = 79.3$ GN/m^2

From Equation (23), the wire diameter is:

$$d = \sqrt[3]{\frac{16 \times 20 \times 0.025}{\pi \times 0.689 \times 10^9}} = 0.00155 \text{ m, or } 60.9 \text{ mils}$$

This wire size corresponds to music wire gauge 28. Substituting this value in Equation (24) yields

$$V_{min} = \frac{8 \times 0.075 \times 20 \times 79.3 \times 10^9}{(0.689 \times 10^9)^2} = 2.00 \times 10^{-6} \text{ m}^3$$

If Equation (21) is now rearranged to solve for N, we obtain

$$N = \frac{4V}{\pi^2 d^2 D} = \frac{4 \times 2.00 \times 10^{-6}}{\pi^2 \times 0.00155^2 \times 0.025} = 13.6 \text{ turns}$$

Partial turns are permissible.

At this point, we have selected the material to be used, and we have determined the wire diameter and the number of turns. The mean diameter of the spring was specified initially. The only remaining value to be determined is the free length of the spring, that is, the length as manufactured.

If the spring is fully compressed, the closed length will be Nd. However, it is desirable not to fully compress the spring when in use because minor variations may cause that compressed length to be slightly too long. If we allow a 10% clearance factor, the compressed length will be given by

DETAILED DESIGN OF DEVICES AND SYSTEMS

$$L_2 = 1.1Nd$$

and the initial length by

$$L_1 = 1.1Nd + s = 1.1 \times 13.6 \times 0.00155 + 0.075 = 0.098 \text{ m}$$

If we have the stiffness constant, we can obtain the value of the initial compression x_1 from that constant and the specified initial force F_1. From Equation (18),

$$\frac{F_1}{x_1} = \frac{79.3 \times 10^9 \times 0.00155^4}{8 \times 0.025^3 \times 13.6} = 269 \text{ N/m}$$

and $\quad x_1 = 20/269 = 0.074 \text{ m}$

so that the as-manufactured length is 0.172 m.
The design is now complete.

10.4 SYSTEM DESIGN; REDUCTION TO COMPONENTS

As a case study of system design, consider the task of designing equipment to meet the test requirements of Underwriters Laboratories (UL) for circuit breakers, especially those in the 15 A to 20 A range. UL requires circuit breakers to undergo an endurance test as part of the extensive evaluation made before the product is permitted to carry the UL label. If the product passes the first tests, a smaller number of tests are repeated four times each year in order to continue the listing. These tests are not run just on a single unit; a number of units are tested in order to ascertain that a representative sample has been tested and has passed.

The rationale behind the requirement for repeated tests is easy to understand: The manufacturer may have made a minor design modification that is thought not to have any effect on performance; a change may have been made in the way the breaker is produced, again with the belief that there will be no effect on the quality of the product; or the manufacturer may have changed suppliers for one of the materials or components, and the effect of the change has been to incorporate substandard parts or materials. These comments are not intended to denigrate the manufacturer, because the manufacturer is very interested in maintaining quality and continuing to meet standards so that it can continue to market the product and as a protection in the case of litigation.

The endurance test depends on the breaker rating. For the 15 A and 20 A sizes, which are very common in residential wiring, the test requires 6,000 operations at six times per minute with rated current and voltage as well as 4,000 operations without current. At higher current ratings, the number and frequency of operations decrease. The load for the rated current test must be inductive, with a power factor between 0.75 and 0.80. The inductor is constructed with an air core, and a resistor is connected in parallel with the inductor.

The design concept for the test equipment is shown in elementary block diagram form in Figure 10.7, and Figure 10.8 shows an expanded block diagram. The system breaks naturally into two major parts, the mechanical system and the electrical load. The two parts can be designed separately, the circuit breaker(s) under test being the only component common to the two subsystems. This becomes clearer if the precedence table (Table 10.3) and the precedence matrix (Figure 10.9) are examined. The precedence table lists the specifications and the major parts to be designed. The precedence matrix, as before, shows the order of major part design running vertically down the left-hand column. Examination of the matrix shows that the circuit breaker (CB) operator (3) is the first unit to be designed, and that selection of the speed reducer (2) and motor (1) follow, with some iteration being necessary during the selection process. Moreover, the precedence matrix shows that the design or selection of the series resistor (4), inductor (5), and shunt resistor (6) have no dependence whatever on the design and selection of the mechanical elements, as might be expected.

The complete cycle of testing to be performed by this test equipment extends for 10,000 operations. At six operations per minute, this is 1,667 minutes, or 27.8 hours. Because a number of breakers are to be tested, it is evident that design of the circuit breaker operator so that several breakers can be tested virtually simultaneously will reduce the overall time required for testing. The mechanical engineer charged with the design of the mechanical subsystem determined that it was possible to test six breakers at once, arranging them in a symmetrical pattern around a cam operator, driven through the speed reducer from the motor, as shown in Figure 10.10.

The mechanical engineer had to determine the forces required to open and close the breakers, translate those forces into the torques on the cam and its shaft, determine the torque losses in the speed reducer, and work backward to the horsepower required of the motor. The supply voltage of 120 volts was selected for convenience. Most of the elements of the mechanical subsystem were off-the-shelf items, but the mechanical engineer had to design the cam mechanism, which was built by an in-house machinist.

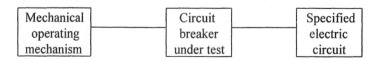

Figure 10.7 Elementary block diagram for circuit breaker test equipment.

DETAILED DESIGN OF DEVICES AND SYSTEMS

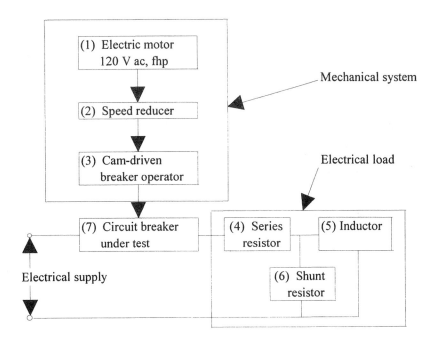

Figure 10.8 Block diagram identifying test equipment components.

Table 10.3 Precedence Table for Circuit Breaker Test Equipment

Specifi-cation	Major part #	Description	Predecessor level High	Medium	Low
a		Rated voltage			
b		Rated current			
c		Power factor			
d		Frequency of operation			
	1	Electric motor	a,2,3	7	
	2	Speed reducer	d		1
	3	CB operator	7		
	4	Series resistor	a,b,c,5		
	5	Inductor	a,b,c		
	6	Shunt resistor	a,b		
	7	Test breaker	Specified by UL		

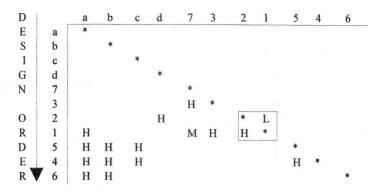

Figure 10.9 Precedence matrix for circuit breaker test equipment.

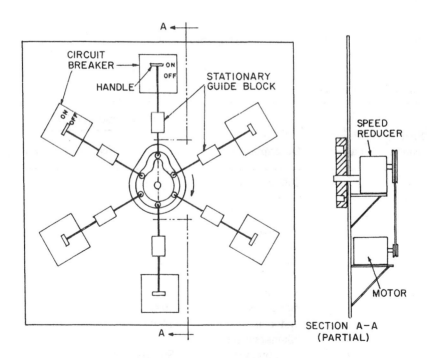

Figure 10.10 Mechanical operating system for circuit breaker test.

DETAILED DESIGN OF DEVICES AND SYSTEMS

The main current path is through the circuit breaker (7), the series resistor (4), and the inductor (5). The shunt resistor (6) is intended to reduce arcing at the circuit breaker contacts; its value is determined from an equation given in Underwriters Laboratories Standard UL489. In this case, the resistance is to be 52 times the rated voltage divided by the rated current. For a 15-A breaker designed for use in 120 volt circuits, the shunt resistance is to be 416 Ω. The circuit breaker (7) may not be changed in any way to accommodate the testing machine. The key element in the electrical load is the inductor, and it provided the main challenge in design of the load.

Conventional inductors are designed so as to have internal resistance that is as small as possible. Here, however, since resistance is needed in order to achieve the specified power factor, given in general by $R/\sqrt{R^2+(\omega L)^2}$, a much more economical design was possible by building as much resistance as possible into the inductors. This was done by reducing the cross section of the copper conductors used, which also reduced the length of the mean turn of the inductor. This in turn made a substantial reduction in the amount of wire needed. The limit of the resistance built into each inductor was that value that would increase the power factor above 0.78 (just above the midpoint of the allowable power factor range), or which would cause a heat rise that exceeded 50 C. The limitation on temperature rise allowed the use of cotton-covered wire, one of the simpler and least expensive forms of insulation for wire. In all cases, the heat criterion set the limit on the amount of resistance that could be built into the inductor.

A noniterative procedure for design of the inductor depends on the ability to formulate a complete mathematical model, as indicated earlier. The first step is to obtain an equation that relates the inductance to the dimensions of the coil, which is shown diagrammatically in Figure 10.11. Reference 7 gives an equation for inductance L as

$$L = \frac{31.5a^2n^2}{6a+9b+10c} \qquad (25)$$

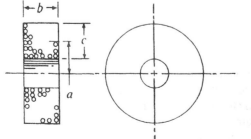

Figure 10.11 Air-core inductor configuration.

Other equations of the mathematical model are:

Newton's Law of Cooling	$P = hA\Delta T$	(26)
Surface area	$A = 2\pi ab + \pi bc + 4\pi ac$	(27)
Dissipated power	$P = F_t RI^2$	(28)
Coil resistance	$R = \rho \ell_w / A_w$	(29)
Wire length	$n\ell_m = \ell_w$	(30)
Mean turn length	$\ell_m = 2\pi a$	(31)
Winding space factor	$F_s = nA_w / bc$	(32)
Wire cross section	$A_w = \pi d_w^2 / 4$	(33)

These equations include specifications, values chosen by the designer, constants available in the literature, and the variables to be determined in the design process as follows:

Specifications:
  L = inductance, μH
  I = current, A

Chosen by the designer:
  F_t = fraction of time the circuit is closed (measured using the cam mechanism and found to be 0.25)
  ΔT = temperature rise of the inductor surface, °C

Constants available in the literature:
  h = cooling coefficient, 9.3 W/m² °C for a 40 C rise
  ρ = resistivity of copper wire at 40 C, 1.727×10^{-8} Ωm
  F_s = winding space factor, that is, the fraction of the coil cross section used by the conductors

Variables to be determined (design parameters):
  n = number of turns
  a = radius of the mean turn, m
  b = width of the coil, m
  c = depth of the coil, m
  A = the radiating surface, assumed to be the entire outside surface of the coil, but with the inside surface of the hole not included, m²
  P = dissipated power, W
  R = resistance of the inductor, Ω
  ℓ_w = length of the inductor wire, m
  d_w = diameter of the inductor wire, m
  ℓ_m = length of the mean turn (the one halfway out in the coil), m
  A_w = cross-sectional area of the wire, m²

DETAILED DESIGN OF DEVICES AND SYSTEMS 365

The list of variables to be determined includes eleven terms, but reference back to the list of equations in the mathematical model reveals that there are only nine equations. As will be seen in Chapter 12, it is possible to optimize the a, b, and c dimensions of the coil to give the largest value of inductance for a given length of wire. As shown there, the coil is optimized if

$$a = 1.5b \quad \text{and} \quad c = 0.9b$$

With these relations among a, b, and c, the mathematical model may now be written using the following nine equations, a and c having been eliminated using the relationships above. As will be noted, many of these equations are identical to those of the original set.

$$L = 2.625bn^2 \tag{34}$$
$$P = hA\Delta T \tag{35}$$
$$A = 9.30\pi b^2 \tag{36}$$
$$P = F_t RI^2 \tag{37}$$
$$R = \rho \ell_w / A_w \tag{38}$$
$$n\ell_m = \ell_w \tag{39}$$
$$\ell_m = 3\pi b \tag{40}$$
$$F_s = nA_w / 0.9b^2 \tag{41}$$
$$A_w = \pi d_w^2 / 4 \tag{42}$$

The model is now complete, and a noniterative design is possible. Algebraic manipulation leads to a set of design equations that yield all of the information needed to wind the coils. These equations are:

$$R = \sqrt{\frac{11.81\pi^2 \rho h L \Delta T}{F_s F_t I^2}} \tag{43}$$

$$b = \sqrt{\frac{1.27\pi \rho L}{F_s R}} \tag{44}$$

$$n = \sqrt{\frac{0.3bF_s R}{\pi \rho}} \tag{45}$$

$$A_w = \frac{0.9b^2 F_s}{n} \tag{46}$$

$$\ell_w = \frac{RA_w}{\rho} \tag{47}$$

Equation (43) yields a value of R that is determined by the allowable temperature rise. It is possible that this value of R will be higher than that allowed by the necessary phase relationship between the voltage and the current (the power factor limitation). Hence in the design process R was calculated first from Equation (43) and then calculated again from the phase relationships. The smaller of the two values was used in the remaining equations. The equations are used in the order given.

As an example of the use of the design procedure, assume that the test circuit for a 120-V circuit breaker rated at 15 A is to be assembled. Including the effect of the 416-Ω shunt resistor, the series combination of R and L that will produce a power factor of 0.78 inductive at 60 Hz is 6.27 Ω of resistance and an inductance with a value of 13,685 µH. All of the other values needed in Equations (43) through (47) have been listed earlier except for the winding space factor, F_s. This value, which must be less than $\pi/4$, is taken as 0.70. From Equation (43),

$$R = \sqrt{\frac{11.81 \times \pi^2 \times 1.727 \times 10^{-8} \times 9.3 \times 13,685 \times 50}{0.70 \times 0.25 \times 15^2}} = 0.570\,\Omega$$

This is the maximum amount of resistance that may be built into the coil without exceeding the 50 C rise. It is smaller than the resistance (in series with the inductance) required to limit the breaker current to the rated value, and is therefore the value used in the remaining equations.

The dimensions can now be found, starting with b from Equation (44).

$$b = \sqrt{\frac{1.27 \times \pi \times 1.727 \times 10^{-8} \times 13,685}{0.70 \times 0.570}} = 0.0309\text{ m}$$

The optimization procedure referred to earlier results in a = 1.5b = 0.0464 m, and c = 0.9b = 0.0278 m.

The number of turns is calculated from Equation (45).

$$n = \sqrt{\frac{0.3 \times 0.0309 \times 0.7 \times 0.570}{\pi \times 1.727 \times 10^{-8}}} = 261\text{ turns}$$

Calculation of the cross-sectional area of the wire from Equation (46) yields

$$A_w = \frac{0.9 \times 0.0309^2 \times 0.7}{261} = 2.305 \times 10^{-6}\text{ m}^2$$

This area, when converted into inch units, gives a diameter in mils of 67.5. This diameter is between #13 AWG and #14 AWG. At this point the designer will

DETAILED DESIGN OF DEVICES AND SYSTEMS

have to select one or the other, and make other adjustments as necessary. If the temperature rise limitation is to be met, the larger wire size must be selected.

Finally, the length of wire needed for the coil (using the theoretical wire size) is, from Equation (47),

$$\ell_w = \frac{0.570 \times 2.305 \times 10^{-6}}{1.727 \times 10^{-8}} = 76 \text{ m}$$

For the project for which this design procedure was developed, inductors were made covering the entire range of current ratings from 15 A to 100 A, and they met the specifications without adjustment.

10.5 SUMMARY

We have seen that product design can proceed by trial and error, by repeated analysis, or by synthesis. A modern engineering office will avoid trial and error design, except for final "fine tuning" that may be needed to adjust for approximations, availability of materials in discrete sizes, and secondary effects that may have been neglected in the models. We now have better models of hardware with which to work and far better computational equipment to analyze those models than even a few years ago. However, much more is now expected from the designer than previously. It was once a tedious task to compute the stresses in a beam with several loads; it can now be done almost immediately. The civil engineer can do "cut and fill" studies in a mere fraction of the time that was once required, and the end result is far superior to what was once accomplished. It is important, however, to document the calculations on a product design or redesign for product liability reasons, if for no other.

The emphasis in this chapter has been on the great advantages the designer derives from the use of mathematical models. Even if it is not possible to achieve a synthesis procedure, design by repeated analysis (iteration) is so easily done using computers that it yields results that are accurate and that may be relied on. If a complete set of equations can be obtained and solved for the design parameters, noniterative procedures can be established. Consistent pursuit by the designer of the strategy of using mathematical models will lead over time to the realization that the important components and subsystems of products made by the company are repeatedly redesigned, and that each time effort is reduced and confidence in the design rises. Every company, or at least every major division of a company, tends to develop a product line of closely related items. Hence in practice each product development is not a totally new task.

The last sentence in the preceding paragraph needs to be accompanied by a warning. Petroski [8], using bridges as his principal examples, points out that as civil engineering developments have occurred and materials have been improved, there has been a tendency to keep extending the scope or size of a design, expanding on earlier successful designs, and that catastrophic failures have occurred because some limitations or assumptions that had no importance in the early stages of the development became very important in later designs— and were forgotten. Although Petroski uses bridges as his main examples, his point is that the reasonable assumptions made in early designs may become completely invalid later, and that the design and the assumptions on which it is based must be thoroughly understood.

The last example in this chapter has been included to demonstrate again the applicability of the precedence matrix to the design process, and how the process may, in some cases, be partitioned into easily managed units. It is also intended to reinforce the concept of synthesis, a noniterative procedure.

REFERENCES

1. Middendorf, W. H. An approach to induction motor synthesis, *Transactions AIEE*, 1962, III(59):64-69.
2. Middendorf, W. H. Methods for improving design procedures, *IEEE Transactions on Education*, 1976, E-19(4):148-153.
3. Wehmeyer, K. R. *What Every Engineer Should Know About Microcomputer Program Design*, Marcel Dekker, New York, 1984.
4. Engelmann, Richard H., and William H. Middendorf, Eds. *Handbook of Electric Motors*, Section 4.9, page 307 ff., Marcel Dekker, New York, 1995.
5. Swieskowski, H. Optimum helical spring, *Product Engineering*, 1966, 37(3):67-69.
6. Hinkle, R. T., and Ivan E. Morse, Jr. Design of helical springs for minimum weight, volume, and length, *Journal of Engineering for Industry*, 1959. 81(1):37-39.
7. Pender, H., and K. McIlwain. *Electrical Engineer's Handbook*, 4th ed., Wiley, New York, 1950, pp. 3-39.
8. Petroski, Henry. *Design Paradigms: Case Histories of Error and Judgment in Engineering*, Cambridge University Press, Cambridge, U.K., 1994.

REVIEW AND DISCUSSION

1. What is the meaning of "design parameter"?
2. List the advantages and disadvantages of each of the three design methods discussed in this chapter.
3. What necessary condition must be met in order to establish a synthesis procedure?
4. When creating a mathematical model, the designer may be able to express some relationships by means of curves (such as magnetization curves) or sets of data (such as wire tables), but cannot obtain a simple mathematical

DETAILED DESIGN OF DEVICES AND SYSTEMS

relationship between the variables. Is a "table lookup" approach admissible as a mathematical relationship when deciding whether one can do a design by synthesis?

5. List as many approaches as you can for use if the number of design parameters exceeds the number of equations of the mathematical model by one.
6. Suppose, instead of a difference of one in the preceding question, there were a difference of two or more. Can you suggest approaches to use in this case? Can combinations be made of approaches listed under (5)?
7. What do you do if the number of unknown design parameters is one less than the number of independent equations? Suppose that one or more of the design parameters is a variable that is available only in discrete values, such as copper wire or the internal diameter of a bearing. Could you use this fact to your advantage in this situation of overspecification?
8. What practical factors were used to provide the necessary balance of equations and unknowns in the resistor design?
9. How would the synthesis procedure for the spring be changed if one more specification were given? As an example, the free length of the spring might be specified.
10. Suppose you have as many design parameters (unknowns) as independent equations, and there is no solution possible for the unknowns. What is the physical significance of this condition?

PRACTICE PROJECTS

1. Select a component whose mathematical model can be accurately developed. This may be a component required in your design project. Establish a synthesis procedure for a reasonable set of specifications for this component.
2. Using the metals listed in Table 10.1, design a slower, heavier flywheel having 2500 J of kinetic energy at an operation speed of 150 rpm. The maximum radius is still limited to 0.5 m, and the weight is to be 1000 N or less.
3. Develop a design procedure for a self-resonant coil given that the specifications are the inductance, the self-resonant frequency, and the impedance between coil terminals at resonance. A self-resonant coil is one for which the distributed capacity of the coil has a value that makes the coil appear completely resistive at the resonant frequency. Formulas may be found in electrical engineering handbooks.
4. Data for four different spring materials are given in Table 10.4. Determine which material gives the least costly compression spring as developed in Section 10.3, using as specifications an initial force of 20 N, mean spring

diameter of 25 mm, and a 75 mm stroke. Use the theoretical wire sizes determined in the design process.

Table 10.4 Data for Spring Design

Material	Modulus of rigidity (Gpa)	Elastic limit (Gpa)	Specific gravity	Cost ($/N)
Beryllium copper	48	0.635	8.25	0.50
Inconel	76	0.550	8.30	0.70
Nickel silver	38	0.450	8.80	0.40
Permanickel	73	0.620	8.10	0.75

5. Using materials listed in Table 10.4, design the least costly spring if the initial force is to be 20 N, the mean spring diameter is to be 25 mm, and the final force is to be 80 N. Note that these are not the specifications used in Section 10.3, thereby invalidating the design procedure there.
6. Develop a synthesis procedure for the attenuator shown in Figure 10.12. The load resistance R_0 is known, and the resistance "looking into the terminals on the left" is to have the same value. The ratio of the input voltage to the voltage across the load, V_1 / V_2, is also specified.

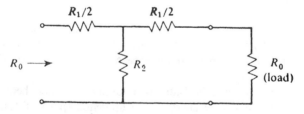

Figure 10.12 Attenuator.

7. The synthesis procedure given earlier for wire-wound resistors is applicable to resistors carrying current continuously. Many resistors, however, are used in circuits in which they carry current only for a short time, and then carry no current for a long time. A typical application having this kind of duty is in motor starters, where the resistor is used to limit the current during starting, but is then shorted out for running. Develop a design procedure for this kind of application. Assume that during the on-period all of the heat developed goes only into heating the wire, and none is transferred to the ceramic tube, porcelain covering, or to the air. The specifications are the

DETAILED DESIGN OF DEVICES AND SYSTEMS

resistance, the current, the on-time, and the allowable temperature rise. Note that this problem has little to do with electrical engineering.

8. The L-network of Figure 10.13 is used to make a given load resistor appear to have a higher apparent value so that there can be an optimum transfer of power. The specifications are R_1, R_2, and f, the fixed frequency of the voltage source. Find the values of L and C.

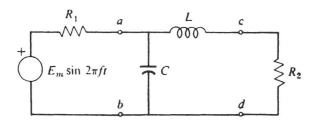

Figure 10.13 L matching network.

9. Change the synthesis procedure for springs given earlier to give minimum weight instead of minimum volume.
10. Develop a synthesis procedure for springs if the specifications are final applied force, mean coil diameter, and total deflection.
11. The energy stored in a compressed spring is $E = s(F_1 + F_2)/2$, where s is the stroke, F_1 is the initial force, and F_2 is the final force. Develop a synthesis procedure if the specifications are energy, stroke, and mean coil diameter.
12. The relay described in Section 8.3.1 and shown in Figure 8.4 requires a tension spring. Although there are special problems in the design of the end turns, which are set at 90° to the axis of the spring, those problems are to be ignored in the following. The mathematical model is similar to that of the compression spring. The largest extended length x_o and force F_o exist when the relay is held in the closed position by the electromagnet. This force must be sufficient to break minor welds at the normally open contacts. A smaller force and smaller extended length, F_c and x_c respectively, exist when the spring holds the relay so that the normally closed contacts are closed. In this case, the force must be at least equal to the force required by the contacts for good electrical conduction. The specifications are the two forces and the two values of x. Develop a synthesis procedure for extension springs of this type.
13. Develop a yo-yo made of aluminum rather than wood or plastic. Yo-yos typically have diameters of about 3 inches and axial dimensions of 1-1/4 inches, and your design should not deviate too far from those dimensions.
14. A column 4 m long, such as that shown in Figure 10.14, is restrained from sideways movement at the top and at the bottom, but is not restrained

otherwise. Columns subjected to excessive compressional loads will buckle, as shown in the figure. The column is made of mild structural steel, and is to support a load of 200 kN. Design a simple, four-sided beam to have minimum volume with a safety factor of 2-1/2. The mathematical model is:

$$F = \frac{\pi^2 EI}{L^2}$$
$$I = bd^3/12$$
$$V = bdL$$

where

F = critical buckling load, N
E = modulus of elasticity, 200×10^9 N/m^2
I = second moment of cross-section area, m^4
L = unconstrained length, m

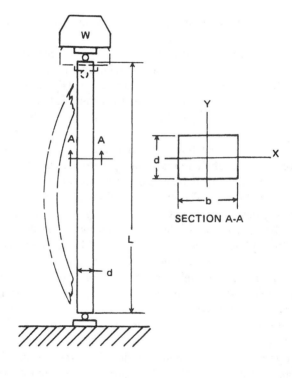

Figure 10.14 Steel column subjected to a buckling load.

DETAILED DESIGN OF DEVICES AND SYSTEMS 373

15. Examine a circuit breaker used in your home or in your engineering college laboratory. Design a mechanical system to operate this circuit breaker in the circuit breaker endurance test described earlier. Important parameters are the force required to open the breaker, the force required to close the breaker, and the stroke of the handle. If six of these breakers are each to be operated six times per minute, what horsepower is required?

16. Select a rather simple piece of measurement or test equipment used in your engineering college laboratory. Examples of possible equipment are a function generator used to provide signals having various waveshapes or a digital voltmeter. Using the instruction booklets, determine how the equipment operates, and develop a block diagram. From the block diagram, create the precedence table and the precedence matrix you would use if you were assigned the design task.

17. In Section 6.8.4, a water-powered chair for easy bathtub entry and exit is shown as Figure 6.9. Design the mechanical/hydraulic system if the chair is to lift and lower a 100-kg person if the water pressure is 0.1 Mpa. Use measurements from a typical bathtub to establish the vertical movement required.

18. An unspoken assumption in the design of the column in Problem 14 is that the column is to have the same cross section from top to bottom. Since the column will buckle about mid-length if subjected to too great a load, it seems that the cross section is larger than necessary as one considers the column farther away from the middle, and that the cross section could safely be reduced as one approaches either the top or bottom. Is this sketchy thinking correct? If it is, there must be some other limiting factor that will dictate minimum cross section at top and bottom, because extension of the line of thinking we have been following would lead to an infinitesimal cross section at each end. What is the limiting factor?

19. The water valves in clothes washers, dishwashers, and ice makers in freezer compartments of refrigerators are operated by solenoids, excited by alternating current. The solenoid plunger and the valve mechanism are totally enclosed, the plunger being housed in a cylindrical extension from the valve body. The solenoid surrounds this extension. The plunger is spring loaded to close the valve when the solenoid is not energized. On an extension of the plunger there is a flat plate with a flexible sealing disk that covers an orifice in the valve body to shut off flow of water when the valve closes. List as many variables, limitations, and assumptions as you can that the designer must take into account when designing valves of this type.

11

PRODUCT DESIGN USING COMPUTERS

In almost all of the preceding chapters, there have been passing references to the use of computers for the various tasks discussed. These have included:

The use of computers for simulation in Chapter 1 and again in Chapter 4.
Need analysis in Chapter 3.
Economic analysis in Chapter 4.
Patent searches in Chapter 5.
Determination of dimensionless constants and their use, Chapter 7.
Analysis of PERT and CPM diagrams and evaluation of criterion functions in Chapter 8.
Analysis of performance in Chapter 9.
Design by repeated analysis or by synthesis in Chapter 10.

Other uses for computers will appear in later chapters. Among the topics discussed in those chapters are the use of search techniques for optimization, reliability calculations, and analysis of data from accelerated life tests or attribute comparison tests. The chapter immediately preceding this one is on design by repeated analysis or by synthesis; in that chapter the use of computers was identified as being of major value for the necessary repeated calculations because it reduces the time required of the engineer or technician, and the possibility of human error (assuming the programming has been done correctly) in the calculations is eliminated. In this chapter we will take a more detailed look at the various uses of computers in engineering design offices.

11.1 HARDWARE

No other field of technology has ever experienced the rate of technical development found in the computer field. For example, in a period of six years

(1988 to 1994) [for which we happen to have easily accessible data], internal memory for a personal computer (PC) went from a range of 640 kilobytes to 1 megabyte (MB) to a range of 8 to 16 MB, expandable to 128 MB. CPU speeds went from 2 MHz to 100 MHz. Hard disk storage went from a few tens of megabytes to 1 gigabyte. CD-ROM drives, introduced during the time frame mentioned above, went from the original standard drive speed to quad-speed. At the beginning of the six-year period, many printers sold with PC's were dot-matrix printers; those sold at the end were largely ink jet printers, many with color capabilities, and many laser printers were sold as part of a personal computer package. Every one of the numerical specifications mentioned above has since been exceeded by factors of two or three at least. While the products are substantially better, costs have in general remained the same or have actually decreased. Central to the improvements in hardware have been the advances in very large scale integrated circuit technology, but every component of computer systems has been the subject of intense research and development that has resulted in ever-increasing performance and simultaneous cost reduction.

The usefulness of computers has also been enhanced by the ability to connect to various networks. There are a number of advantages to networking, even if the network is restricted to interconnection of computers within one organization. For example, the designer needs to be certain that the product being designed will conform to certain standards specified by outside organizations, such as NEMA or UL. He or she needs to be certain that components such as machine screws to be incorporated into a product meet the company's internal standards. When starting a new design, it is common practice to begin with an existing design and adapt as necessary to meet the new specifications. That is, there is a tremendous amount of information to which the engineer needs to have access, but it is not efficient to place all of that information in the memory of each engineer's computer. If it is kept in one central computer, accessible through a local area network, each engineer can access the information when needed. Moreover, by having the databases located in one memory bank, it is possible to allow all engineers to access the needed information, but simultaneously to limit to those having the appropriate password the ability to modify the database. File management is made much easier.

Networks outside an organization can also be useful. For example, at the University of Cincinnati, PC's in the offices of faculty members in the College of Engineering are all linked to a local area network, this in turn is linked to a campus-wide network, and that in turn is linked to a supercomputer in Columbus. The faculty member requiring the capabilities of a supercomputer for research thus has access to that machine from his or her office via the networks. The networking just described is not available to the general public, but there are a number of networks that can be readily accessed that have nationwide or worldwide coverage. Although one major use of these networks is for the transmission of e-mail, they can also be used for a variety of searches for wanted information. The patent searches mentioned in an earlier chapter are but one example.

Networks, it must be pointed out, also have disadvantages. The news reports of computer viruses introduced by individuals with destructive tendencies are too common. These vandals have also been able to access both governmental and industrial files and cause great damage. Web pages have been altered in various ways by inclusion of views or information at odds with those of the sponsor. Even the stand-alone PC is not completely immune to problems of this kind; hostile programs have been inadvertently loaded from floppy disks.

11.2 SOFTWARE

There is a wide variety of software available to the engineer for solving technical problems. Some of this software is useful in solving numerical problems for which only a single numerical result is wanted; other software is useful if time-dependent solutions are the goal of the engineer. Some software is written by the engineer using one of the many computer languages; other software is available on floppy disk or on CD-ROM. Limitations on choice of software or programming language may be dictated by company policy or by cost of commercially available software. If an engineer learned one computer language in college, he or she may feel very comfortable programming in that language, and may not wish to learn another if there seems to be no particular advantage to a second language. The following sections are intended to be a general guide to the designer as to what is available, but the reader should remember that, as with hardware, change is rapid.

11.2.1 Languages

Table 11.1 lists some of the well-known computer languages, their primary applications, and remarks on suitability for certain kinds of applications. The table is divided into two parts, one on high-level languages and the other on object-oriented languages, which are becoming important parts of the programming language sphere. Languages that seem to be of historical significance only, such as APL and SNOBOL, have been omitted. A third category of languages, which fall under the rubric of visual programming, is not listed here. The two most important of the visual programming languages are Visual BASIC and Visual C++, which itself has a number of variations. Visual BASIC has two different implementations, one for MS-DOS and the other for Windows. There are two implementations of Visual C++ for Windows 3.1 or higher, one implementation being a standard edition and the other being a professional edition. Visual C++ requires far more disk space and somewhat more RAM than

Visual BASIC. These languages, or others having the same general characteristics, will no doubt eventually displace many of those in Table 11.1.

Any programming language can do what any other language can do [1]. One important question is "How efficiently can it be done?" But even that question should be broken down into two questions: "How convenient is it to do the programming?" and "How much time will be required?" The second ques-

Table 11.1 Software Languages

Language	Main Applications	Remarks
High-Level Languages (HLL)		
FORTRAN	Scientific problems	Strong supercomputing, vector and parallel processing support.
BASIC	General purpose	Widely used on PC's. Easy to learn and experiment with.
COBOL	Business-oriented applications	Widely used in business data processing.
LISP	Artificial intelligence and expert system research	Useful for list processing and symbolic expressions.
PROLOG	Artificial intelligence applications	Logic programming for theorem proving.
Pascal	Educational tool for teaching language concepts	Useful as a general purpose language.
C	Scientific problems, word processing, system programming	Widely used in expert system development.
Ada	Systems programming, embedded applications	Required by the Department of of Defense for embedded programs. Military standard since 1983.
Object-Oriented Languages (OOL)		
Smalltalk	Mainly a research tool	
Eiffel	Production software	Not widely known.
C++	Object-oriented extension of C	Potentially only future OOL.

tion itself can be answered only if one looks at the more fundamental question of whether quick source code development is of major importance or whether fast execution of the program is the important parameter. A general evaluation of various languages can be found in Appendix B of Reference 1, in which the author has assigned *tentative* letter grades to nineteen features (plus "other") that are desirable in programming languages. These features include simplicity, naturalness, rigor, data typing, modularity, portability, efficiency, compactness, and generality.

In any case, every designer needs to have a good foundation in at least one language and must know the major syntactic and semantic concepts common to most programming languages. Having that background, he or she will find it relatively easy to program in any of a wide variety of languages and will therefore be better equipped to choose the language most suitable for a given problem.

Developing a program using a high-level language is similar to developing a hardware product. If the program is to meet the expectations of multiple users, the programmer should follow a structured procedure, such as that outlined in Reference 2:

1. *Need Analysis and Specification:* The function of the program is to control the hardware in an agreed-upon manner. The requirements, goals, and preferences of all parties involved should be recorded for review during the program development.
2. *Designing:* The program logic is developed and a flow chart constructed similar to that shown in Figure 10.2. The flow chart can use verbal statements or equations or a mixture of the two. In any case, it shows what the program is to do and specifies the sequence of operations.
3. *Programming:* The flow diagram is converted to program code using the computer language chosen. While doing this, a record should be made of the decisions made in writing the program. In many cases, this is done by inserting comments into the program. Once the program has been written, the programmer should "walk through" it in order to discover obvious errors. When writing the program, consideration should also be given to the ease with which the program can be modified without having to make major revisions.
4. *Verifying and Testing:* The program must be checked against the specifications from (1) above to make sure that the requirements and goals have been met. If errors or omissions are found, program code must be modified as necessary. The program must be verified for all input data the user is expected to have. Special situations, such as a sequence of calculations leading to division by zero or the accidental introduction of indeterminate values, must be recognized and avoided.
5. *Performance Appraisal:* Besides being functionally acceptable, a program should be evaluated for economy of memory usage and speed of execution.

PRODUCT DESIGN USING COMPUTERS 379

It must allow for changes of data, be readily adaptable for use with other situations similar to those for which the program was written, and be portable so that it can be used easily on other machines.

6. *Operation and Maintenance:* At this point the operations and instruction manuals are prepared. In the process, it will become evident that minor revisions will improve the program or make it more user-friendly. Any revisions made must be tested to insure that the program continues to yield correct results. The manuals themselves must be completely self-explanatory; if one must ask the programmer to clarify anything in a manual, the manual must be revised to guard against confusion should the programmer leave the company.

7. *Configuration Management:* Revisions requested by users should be documented by change notices, just as is done with hardware. Such notices should record in detail what the change was, the reason for the change, the name of the person who requested the change, and the date of implementation.

The software-development procedure just described need not be carried out for every program. However, if the program is expected to be used by all members of a design or engineering team, the procedure should be followed relatively closely. If only the programmer is expected to use the program, many of the steps will be quite short or may even be omitted. However, as one of the authors can testify, inadequate documentation may lead to a substantial amount of effort even on the part of the programmer if the program is not used for some time, and is then to be reactivated. *Memory is no substitute for records!*

11.2.2 Spreadsheet Analysis

If one has a relatively small design problem, such as the flywheel design of the preceding chapter, calculations can often be carried out quickly using spreadsheet analysis. The technique used is to enter equations into the various blocks of the spreadsheet. These equations are formulated so that they utilize original data and data generated by preceding equations. Table 11.2, which is discussed more fully below, is an example of the output obtained by this method.

Spreadsheet analysis has both advantages and disadvantages. Among the advantages is the fact that it is simpler than using one of the programming languages discussed in the preceding section. There are few rules to master. Once the hurdle of how to write the equations has been passed, the way in which the spreadsheet is used is straightforward. Numerical values of all the quantities of interest are generated and displayed. If the engineer makes an estimate of the numerical values to be expected and of their expected trends in the spreadsheet, errors in equation formulation are relatively easy to detect.

The disadvantages include the fact that there are no IF, GOTO, and similar commands available to terminate one set of calculations if a particular variable goes beyond a predetermined value. Reference to Figure 10.2 shows a number of blocks with statements of the type "Is X less than Xmax?", with alternate paths to follow depending on whether the answer to the question is "Yes" or "No." With spreadsheets, the user asks these questions by observing the results as they appear in the spreadsheet, and then takes appropriate action depending on the response. For example, for each of the four materials considered in the flywheel design, the values of WT and I were observed and the spreadsheet was limited so that when I exceeded 1.2 the array was terminated by intervention. There is one advantage to this process. One can also intervene to "tweak" the process so as to find limiting values. In the continuation of Table 11.2, the final value of A was adjusted for each material so that I was only slightly larger than the desired value, and the final results in every case were underscored.

As noted above, spreadsheets are most useful for small problems, about the size and complexity of the one shown here. If the problem is long and complex, if it is expected to be used frequently, and if it has many branches selected by IF statements, one of the programming languages above should be used.

In the spreadsheet analysis of Table 11.2, data from Table 10.1 is shown at the top, as is Rmax calculated from Equation (1) of Chapter 10, $\sigma_t = \rho R^2 \omega^2$. Also spread across the top of the tabulation are the limiting values for R, WT, and I. The column headings are outer radius R, inner radius A, width of rim B, all as shown in Figure 10.1, and radius of gyration K, weight WT, polar moment of inertia of the rim I, and cost of material CM. In the first column (R), the value used for cast iron is that shown above and calculated from Equation (1). This can be entered by an array equation which simply copies the value from the upper part of the table into the first column. The first entry for A was calculated from the first value of R, subtracting 0.001, and the remaining values of A were then calculated by an array equation written to subtract 0.001 from each of the preceding values in the column. Equation (5) from Chapter 10 is B = 4(R − A), and a simple array equation will then calculate all the values in the B column from the two previous columns. K is then calculated from Equation (4), WT is calculated from Equation (3), and I is calculated from Equation (2), all by the use of array equations. CM is the product of the cost of cast iron in the upper part of the table and WT. In this example, after data were generated using a longer table, the user intervened and removed all the data after the eighteenth line because the upper limit of WT had been exceeded before reaching the lower limit of I.

Values were calculated in a similar manner for cast steel, brass, and Monel™. The upper limits of R in the case of cast steel and Monel™ are set by the specification on R rather than the maximum value given by Equation (1). Once again, the user intervened to limit the length of the table, and also inter-

PRODUCT DESIGN USING COMPUTERS

Table 11.2 Spreadsheet for Flywheel Design

	Material	Maximum tensile strength (Pa)	Density Rho (kg/m^3)	Cost ($/N)	Rmax	
	Cast iron	20000000	7200	0.23	0.287595	
	Cast steel	68000000	7850	0.33	0.507870	
	Brass	60000000	8300	0.45	0.463949	
	Monel	80000000	8520	0.90	0.528760	
R<0.5				WT <100	I>1.2	
R	A	B	K	WT	I	CM
Cast Iron						
0.287595	0.286595	0.004	0.287095	0.509644	0.004282	0.117218
0.287595	0.285595	0.008	0.286597	2.035026	0.017039	0.468056
0.287595	0.284595	0.012	0.286099	4.570820	0.038138	1.051289
0.287595	0.283595	0.016	0.285602	8.111701	0.067447	1.865691
0.287595	0.282595	0.020	0.285106	12.65234	0.104837	2.910039
0.287595	0.281595	0.024	0.284611	18.18742	0.150178	4.183107
0.287595	0.280595	0.028	0.284117	24.71161	0.203341	5.683670
0.287595	0.279595	0.032	0.283623	32.21958	0.264201	7.410504
0.287595	0.278595	0.036	0.283131	40.70601	0.332632	9.362383
0.287595	0.277595	0.040	0.282639	50.16558	0.408509	11.53808
0.287595	0.276595	0.044	0.282149	60.59295	0.491710	13.93638
0.287595	0.275595	0.048	0.281659	71.98281	0.582112	16.55605
0.287595	0.274595	0.052	0.281170	84.32982	0.679596	19.39586
0.287595	0.273595	0.056	0.280682	97.62867	0.784040	22.45459
0.287595	0.272595	0.060	0.280195	111.8740	0.895328	25.73102
0.287595	0.271595	0.064	0.279709	127.0605	1.013342	29.22393
0.287595	0.270595	0.068	0.279224	143.1829	1.137965	32.93208
0.287595	0.269595	0.072	0.278740	160.2359	1.269084	36.85425

vened in the last line for each material by adjusting the value of A by trial-and-error in order to make I exceed the specified value by a negligible amount. It will be noted that the final values in this table do not agree with those given in Figure 10.3, one reason being that the final values of A have been adjusted to yield a value of I closer to the specification.

The Microsoft BASIC program written to obtain the solution shown in Figure 10.3 was 54 lines in length. It is certainly easier to do the spreadsheet

Table 11.2 (Continued) Spreadsheet for Flywheel Design

R	A	B	K	WT	I	CM
Cast Steel						
0.5	0.499	0.004	0.499500	0.96675	0.024588	0.319027
0.5	0.498	0.008	0.499001	3.863128	0.098056	1.274832
0.5	0.497	0.012	0.498502	8.683328	0.219964	2.865498
0.5	0.496	0.016	0.498004	15.42154	0.389874	5.089109
0.5	0.495	0.020	0.497506	24.07197	0.607351	7.943750
0.5	0.494	0.024	0.497009	34.62880	0.871961	11.42750
0.5	0.493	0.028	0.496512	47.08622	1.183273	15.53845
0.5	0.49295	0.0282	0.496488	47.75888	1.200057	15.76043
Brass						
0.463949	0.462949	0.004	0.463449	0.948394	0.020765	0.426777
0.463949	0.461949	0.008	0.462950	3.789484	0.08279	1.705268
0.463949	0.460949	0.012	0.462451	8.517131	0.185676	3.832709
0.463949	0.459949	0.016	0.461953	15.12519	0.329024	6.806338
0.463949	0.458949	0.020	0.461456	23.60754	0.512439	10.62339
0.463949	0.457949	0.024	0.460959	33.95802	0.735525	15.28111
0.463949	0.456949	0.028	0.460462	46.17050	0.997892	20.77672
0.463949	0.456264	0.03074	0.460123	55.60746	1.200082	25.02336
Monel						
0.5	0.499	0.004	0.499500	1.049262	0.026686	0.944336
0.5	0.498	0.008	0.499001	4.192847	0.106425	3.773562
0.5	0.497	0.012	0.498502	9.424453	0.238738	8.482007
0.5	0.496	0.016	0.498004	16.73778	0.423150	15.06400
0.5	0.495	0.020	0.497506	26.12652	0.659189	23.51387
0.5	0.494	0.024	0.497009	37.58438	0.946383	33.82594
0.5	0.49323	0.02708	0.496627	47.81296	1.202089	43.03166

analysis, but the BASIC program can be saved and run at any time a new design is to be made.

Finally, it should be pointed out that the flywheel design is not complete at this point. We have arrived at a rim design that will meet the specifications originally set forth, but the hub and spokes will add to the weight, inertia, and cost or, alternately, the addition of their weight and inertia can be used to reduce the cross-section of the rim. (These may and probably will be second-order effects.) However, it must be kept in mind that the purpose of the flywheel is to reduce speed variations in the driven unit by transferring some of the rotational kinetic energy stored in the flywheel rim to the shaft when additional load comes

PRODUCT DESIGN USING COMPUTERS

on the shaft. This energy is transferred through the spokes and hub. The designer must be certain that the torques to be anticipated will not cause movement of the hub on the shaft (possible use of keys and keyways?) and must be cognizant of the fact that the spokes act as cantilever beams, constrained against rotation at both ends and subject to internal tension because the relative motion of the rim with respect to the hub will cause a slight increase in the spoke length.

11.2.3 Commercially Available General-Purpose Programs

There is an ever-growing availability of general-purpose software available for doing mathematical analysis and engineering design in various fields. A list of such programs is shown in Table 11.3, but this is by no means a complete list, and the characteristics shown in the table are only summaries of the capabilities of the various programs. The designer having an interest in one or more of these programs must do a thorough investigation of the capabilities before committing to any one of them, taking into account the needs of the company for which he or she is doing design tasks, and how any one of these programs will complement those tasks. Attention to current technical literature will reveal many more programs than those tabulated here.

As will be noted, some of the programs have narrow orientations, such as toward mechanical engineering, printed circuit board design, and simulation of electronic circuits, while others are much broader, being very powerful mathematical tools and some having capabilities outside the strictly engineering sphere. Many have additional support packages that can extend the capabilities of the base program.

11.2.4 Finite-Element Analysis

Much of what an engineer does is related to the macroscopic nature of the elements or components with which he or she deals. We measure resistances from one terminal to another, we calculate moments of inertia and their influence on acceleration, we make gross assumptions about the flow of water in streams, and about the vibrational characteristics of airplane wings. By contrast, we have paid relatively little attention until recent years to the microscopic characteristics of the elements of our designs. In most cases, we arrive at reasonable results and suitable designs in spite of this inattention, but there are cases in which the microscopic view cannot be ignored, but must be investigated carefully. Failure of a mechanical part is often the result of localized stress that is not predicted by a macroscopic view of that same part. A magnetic circuit, considered macroscopically, may never approach saturation, but when looked at microscopically, it may be found that some parts of the magnetic circuit are indeed in a saturated

Table 11.3 Commercially Available General-Purpose Programs

Program & Source	Characteristics
TK Solver Universal Technical Systems, Inc.	Mechanics and mechanical engineering applications, including machine design, dynamics and statics, thermodynamics, mechanics of materials, heat transfer, Roark's formulas, vibration analysis, gear design.
Mathcad MathSoft, Inc.	Numerical and symbolic calculations of derivatives and integrals, wavelet and Fast Fourier Transforms, polynomial root finder, simultaneous equation solver, matrix calculations, various static plots and animation, statistics and data analysis, programming operations.
MicroStation Modeler™ Bentley Systems, Inc.	Mechanical modeling, including wireframes and solids, assembly, and parametric, explicit, and feature-based design.
WinBoard™ and WinDraft™ NTE Electronics, Inc.	Printed circuit boards, including routing, library of several hundred module footprints, wiring, drawing, printing, bill of materials generation.
Electronics Workbench® Interactive Image Technologies Ltd.	Analog, digital, and mixed-mode simulations of electronic circuits, simulation of faults and leakage paths, signal waveforms.
Maple V® Waterloo Maple Inc.	Solves equations analytically, matrix manipulation, investigation of design parameters, tensor manipulation, piecewise functions, matrix decomposition, three-dimensional system response plots.
Mathematica® Wolfram Research, Inc.	Useful in the physical sciences, engineering, and mathematics, as well as financial modeling and general planning and analysis, computer science, and software development. Large number of supporting packages and publications. Runs on over 20 platforms.

PRODUCT DESIGN USING COMPUTERS   385

condition, and we tend to make only rough approximations in our calculations to take care of magnetic anisotropy. Investigation of these and similar phenomena is done today using finite-element analysis (FEA). Except for very simple problems, some of which can be handled by mathematical transformations, as is shown in Section 1.4.2 of Reference 3, this can be done only by use of computers, and there are a number of programs that have been developed for this purpose.

The method of curvilinear squares, a "hand and eyeball" process, will serve as an introduction to the thinking process behind FEA. An example from the days when only mainframe computers, run in a batch mode, were available, and programming was done in machine language, will demonstrate the kind of problem to which the method can be applied. One of the authors (RHE) was approached by a company making grid-type resistors having very high power-dissipating capabilities. These resistors are used in a variety of applications, such as in series with the armature of a large dc motor to limit current during the starting mode or during dynamic braking. For many years, such resistors were made of cast iron grids, but the company was switching over to the use of stainless steel plate, which could be punched out in grid form using a nearly-scrapless pattern such as that shown in Figure 11.1, which can be repeated horizontally in order to get a long grid, and which repeats in the vertical direction to produce identical grids. The current path was along each grid to one end, through a short transverse connection to the next grid, and so on in a zigzag pattern from one terminal to the other. Very early in the production of the new resistors, it was discovered that the resistance of a new-design resistor was

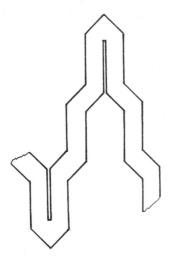

Figure 11.1 Nearly-scrapless punching using stainless steel for a line of high power resistors.

higher than had been predicted using a method that had worked very well with the cast iron grids. That method was to use the centerline length of the grid, together with the cross section area and the resistivity, to predict resistance. Because of mechanical properties of cast iron, the grids had small cross-section areas and there were no sharp curves, certainly none such as those at the top and bottom in Figure 11.1.

In order to resolve the question as to the reason for the resistance difference, the method of curvilinear squares was used. The method is illustrated for a simpler problem in Figure 11.2, which is a hand-drawn curvilinear square plot for a thermal field. In this figure, heat flow q enters the panhandle at the left, and all of the heat is assumed to exit at the right side and bottom of the "pan," the other faces being insulated. The left edge is constrained to be at temperature θ_1 and the temperature on the exit surfaces is constrained to be at temperature θ_2. In such a plot, when done correctly, the heat flow densities between all pairs of adjacent stream lines are equal and the potential differences between all adjacent equipotential lines are equal. From this plot, engineers could determine that the temperature gradient is greater in the panhandle region (which has the smaller squares), and that more heat exits through the bottom of the "pan" than through the right edge. Although the labeling on this figure is for a thermal field, it is just as applicable if the nomenclature is changed from heat flow to current flow, temperature to voltage, and heat flow densities to current densities. Guidance for making these plots appears in Reference 4.

Returning to the problem of the grid resistor, the first step was to take an actual grid, enlarge its image by use of an overhead projector, trace out the resulting image, and then apply the method of curvilinear squares to the resulting plot. When one does this, it is necessary to begin the squares from a transverse

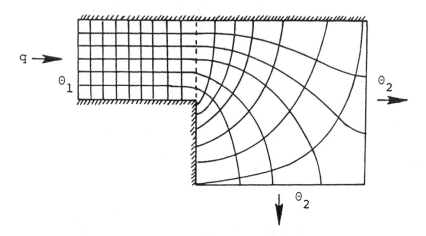

Figure 11.2 Hand-drawn curvilinear square plot of a thermal field.

PRODUCT DESIGN USING COMPUTERS

section for which the shape of an equipotential curve is known. This is almost invariably a straight line, although occasionally it may be part of a circle. If the diagram has symmetry, these transverse sections are easily found. In Figure 11.1, symmetry occurs along the vertical lines joining the upper and lower tips of the grid to the adjacent slits which create the narrow U at the bottom and inverted U at the top. It also occurs along the horizontal center line of the grid. Using these equipotentials as points of departure, curvilinear squares were plotted for the grid. As one would expect, considerable crowding of the streamlines appeared at the root of the slits at top and bottom, showing that the resistance especially in those areas would be increased over that to be expected from the "center line of the grid" approach. The curvilinear plot approach predicted an increase in resistance of 10% over that computed by the previous method; the manufacturer had measured increases ranging from 9% to 11%. If carefully done in such a simple case as this, very good accuracy can be obtained. Because the manufacturer was using the same proportions in all of the grids, merely scaling up or down dimensionally as necessary, it was now certain that new designs would meet specifications without trial and error.

In all probability, no one would use curvilinear squares today because FEA programs are readily available. There are additional reasons for use of FEA programs; these are the inability in the curvilinear squares approach to take

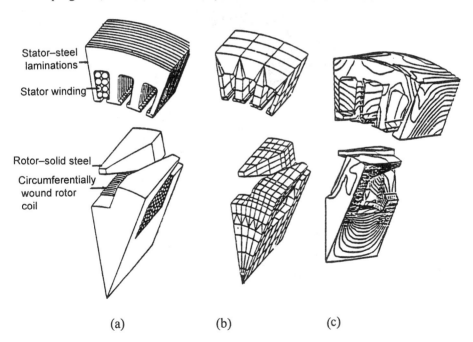

Figure 11.3 Finite-element analysis of a Lundell machine. a) Configuration of the materials; b) finite elements; c) flux plot.

into account the nonlinear characteristics of the material, such as saturation in magnetic circuits and the anisotropy of many materials, and the inability to construct meaningful plots for three-dimensional systems. For these reasons, there has been considerable activity in the development of FEA programs [5]. Section 1.5 of Reference 3 discusses the theory behind the programs, and several later sections in Reference 3 give examples of FEA plots for various types of electric motors, including the three-dimensional plots shown in Figure 11.3 for the Lundell machine, the configuration used in automobile alternators. Figure 11.4 shows the wire-frame model of a steel rim automobile wheel, and the stress contours of one of the elements along a flange. As these figures illustrate, it is now possible to do finite-element analysis for three-dimensional irregularly shaped field problems. Areas of exceptionally high temperatures, mechanical

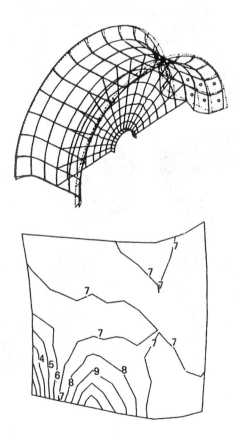

Figure 11.4 Wire-frame model of a steel automobile wheel, together with stress contours of an element along the flange. (From J. C. Lange, *Solving Mechanical Design Problems with Computer Graphics,* Marcel Dekker, New York, 1986.)

PRODUCT DESIGN USING COMPUTERS

stresses, intense electric fields, saturated magnetic fields, and nonlinear fluid flows can be located by this technique while the product is still in the design stage, and correction can be made in the early stages of development.

Finite-element analysis obtains temperatures, stresses, flux densities, and so forth by minimizing the energy function which consists of all the energies associated with the finite element model. That is, the solution satisfies the law of conservation of energy. In developing the finite element model, the geometric shape is subdivided into small elements, the "finite" elements of the method's name. In two-dimensional cases, these are generally triangles or rectangles. In three-dimensional cases, these may be parallelepipeds, tetrahedrons, pentahedrons, and hexahedrons. The elements are fitted into the figure in such a way that each corner (or grid point) of one finite element coincides with a corner (grid point) of an adjacent element. Accuracy is enhanced by decreasing the size of the elements, but at the expense of increased computation time. Typically, large elements are used initially in order to locate areas in which large field gradients or densities occur, and the model is then modified to use smaller elements in those areas to improve the accuracy.

The number of unknowns in the matrix equation for one element equals the number of grid points of the element times the number of degrees of freedom per grid point. Scalar fields have one degree of freedom; vector fields on the other hand have up to six, the three orthogonal axes and rotation in either direction about those axes. A vector field being investigated using parallelepipeds will then have 48 unknowns in the matrix equation for each finite element. It is easily seen that the number of unknowns of the matrix equations written for all of the finite elements will range into the several thousands. Software capable of handling fewer than a thousand unknowns will solve only the simplest of problems.

Commercial finite element software often includes pre- and postprocessor accessories. The preprocessor lays out the mesh of finite elements within the model, and checks for missing or duplicated elements, warped elements, and so forth. The post processor presents the results in easy-to-understand ways, such as graphically with color coding.

11.3 COMPUTER INTEGRATED MANUFACTURING (CIM) [6]

Computer integrated manufacturing (CIM) may be defined as the integration of all computer-based techniques applied to all manufacturing functions throughout a company. Figure 11.5 shows one possible interrelationship among these functions, the various blocks being defined for our purposes as follows:

1. Computer-aided engineering (CAE) includes all computer-based techniques used in design engineering and in manufacturing planning areas.

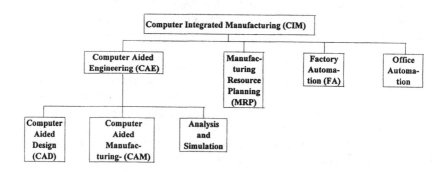

Figure 11.5 Block diagram showing one possible interrelationship among CIM functions.

2. Computer-aided design (CAD) is computer-aided design engineering. Computer-aided drafting may or may not be a part of CAD. If no part model exists in the database and the intention is to use the drawing only to manually input geometry data into a manufacturing program for a specific application, this process is simply computer-aided design drafting, and would be better identified by an acronym such as CADD. However, if the new part design is to become part of the database (or if a part's geometry has already been modeled and stored in a database and can be used again in manufacturing engineering), then drafting of the part is an application area of CAD. Since CAD defines what the product is and our purpose is to be able to design products and to communicate the necessary information to manufacturing, this topic and the next are explored in greater detail in the next section.
3. Computer-aided manufacturing engineering (CAM) defines how to make the product. In the diagram of Figure 11.5, there is an implication that this function can be isolated from the CAD function. However, if one is pursuing the philosophy of design for manufacturing (DFM) or design for assembly (DFA), one cannot separate CAD and CAM in as neat a manner as the figure would suggest.
4. The block labeled "Analysis and Simulation" includes a variety of tools used by the engineer, including the languages discussed in Section 11.2.1 as well as others, spreadsheet analysis discussed in Section 11.2.2, commercially available general purpose programs such as those in Section 11.2.3, the finite-element analysis of Section 11.2.4, electric circuit analysis, both in steady-state and transient conditions, analysis of mechanical system response, "cuts and fills" in civil engineering projects, and many others.

PRODUCT DESIGN USING COMPUTERS

5. Manufacturing resource planning (MRP) is used to decide when to make the product, and depends on availability of manufacturing equipment, schedules for other products, and so forth.
6. Factory automation (FA) relates to the kinds of tools to be used in manufacturing, such as turret lathes, robots, machines that automatically insert parts in circuit boards, and numerically controlled (NC) machines. NC programming may or may not be part of the general area of CAD/CAM. If the part is one that exists in the database and the part programmer uses the part geometry information to specify the operations of the NC machine, the programming may be considered to be part of CAD/CAM and would be added to the database. If, however, the programmer receives a manufacturing drawing on a physical medium and develops the control program from that information, the program would not be considered to be part of CAD/CAM, and this is true even if the program is saved for future use on the NC machine.
7. "Office Automation" is a separate function, although it can support all of the activities above for purposes of assessing costs, purchasing, technical manuals, and scheduling.

Although Figure 11.5 shows a set of separate functions identifiable by various blocks, in more progressive companies there has been for some years a trend toward forming a continuum throughout the entire computer integrated manufacturing area. With the present emphasis on concurrent engineering (CE), the boundaries between the blocks of Figure 11.5 are rapidly being blurred and will eventually be eliminated completely. Nevertheless, the blocks of the figure

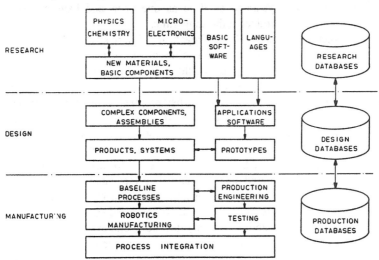

Figure 11.6 An expanded view of computer aided manufacturing (CIM). (Reproduced from Reference 7, by permission.)

are useful in identifying the kinds of functions that are inherent in computer integrated manufacturing.

As the use of computers in company operations have expanded, more global views of the future of large companies have developed, as shown in *Engineering Productivity Through CAD/CAM* [7]. Still more global views can easily be drawn.

11.4 COMPUTER-AIDED DESIGN/COMPUTER-AIDED MANUFACTURING

The overall aim of CAD/CAM is to improve productivity in design engineering and manufacturing, and it is on this basis that the idea has often been "sold" to upper management. Improvements in productivity can result in reductions in cost, better quality, elimination of late engineering changes, and reduced time from the beginning of design to first production. Unfortunately, improvements in productivity have proved to be difficult to quantify, probably because with the introduction of better techniques into design and manufacturing there has simultaneously been a recognition that many more highly desirable and even necessary functions can be added without much difficulty. Those possibilities were not contemplated when companies started down the CAD route, and their addition has tended to confuse the metrics of productivity measurement. Although quantification has been difficult, that improvement in productivity has occurred is clear from the fact that large-scale manufacturing entities, such as the automobile companies, have reduced the traditional design-to-production cycle time from four or five years to as little as one year, and have substantially upgraded quality of the product at the same time.

Computer-aided design (CAD) began principally as a method by which a designer could quickly run through alternative shapes and positions of components so that the designer could select the best arrangement and dimensions of parts, add tolerances, and produce hard-copy engineering drawings. Representations were two-dimensional, with orthographic projections to aid in visualization. This use of CAD is still a valid one if a single product is to be built, such as a roll lathe that is to meet the requirements of a single steel mill, but even in this case accessibility in a suitable database to previous designs may show the designer that parts that were designed for previous lathes may be incorporated, either unchanged or with relatively minor modifications.

CAD software is still two-dimensional in some cases, and is perfectly adequate for such tasks as electric circuit design. However, in most cases a three-dimensional representation is highly desirable, and even mandatory. Some software will produce images that appear three-dimensional, but which do not store all three dimensions for all pertinent points. Such software is referred to as 2-1/2D. True three-dimensional software stores the x, y, and z coordinates for each point.

PRODUCT DESIGN USING COMPUTERS

In addition to dimensionality characteristics of a CAD software program, there are a number of ways in which the screen image may be represented. The lowest level of these is the wireframe model, sometimes called a stick figure. Figure 7.1 shows such an example, both with all lines shown and with the hidden lines shown dashed. Figure 11.4 shows a wireframe model for a wheel. The next higher level of representation is the surface model, in which the corners and edges of the object are specified in the data base, but the surfaces as well. The lowest diagram in Figure 7.1 is a surface model. The highest level representation is the solid model. There are different methods used to reach a solid model, one stemming from the construction of a solid by combinations of solid primitives and the other coming from the surface modeling technique. In either solid modeling case, the program recognizes that the part being designed has volume, and is not simply a collection of points and surfaces. Obviously, solid modeling is the best representation of an object.

Regardless of the dimensionality of the program or how the modeling is done, the software that is now available permits many more functions to be performed than originally contemplated. Among these are the ability to access existing designs as mentioned above, but presently available software has broadened the range of options to include the following, among others:

- Mass properties, such as weight and moments of inertia about rotational axes.
- Static and dynamic analysis in a macroscopic sense.
- Kinematic analysis, so that the motions of parts relative to each other can be analyzed to insure that parts do not interfere with each other.
- Preparation of the mesh for finite-element analysis, and analysis for mechanical or electrical stresses, heat flows, flux patterns, and deformation patterns. (The second aspect of this option may be considered to be in the "Analysis and Simulation" block.)
- Analysis of part designs which have been modified because low-stress regions are found in which the part might be cored, thus reducing the amount of material needed, weight, and inertias.
- Effect of possible substitution of materials.
- Rapid investigation of alternative designs.
- The ability to design for reduced numbers of parts.
- Design of printed circuit boards.
- Very large scale integrated (VLSI) circuit design, a task that would be impossible without CAD.
- Vibration analysis, including the modes of vibration, frequencies of objectionable vibrations, and amplitudes. (Again, this could be considered as "Analysis and Simulation.")
- Group technology, in which the underlying philosophy is that both time and cost can be saved if parts can be placed in groups based on common characteristics, such as size, shape, ratio of width to length, material,

manufacturing technology used, and so forth, and then searched for possible parts for a new product, or for a part which might be easily modified.
- Production of parts lists.
- Assembly and exploded views of a machine or product for use in maintenance manuals or in catalogs. (The catalogs, of course, belong in the sales department, which does not even appear in Figure 11.5.)
- Flow sheets.
- General layout and arrangement drawings, such as piping diagrams.

The options listed above tend to be in the CAD block with possible exceptions as noted, but other options exist in available systems that tend to merge into other blocks, such as the CAM, MRP, and FA blocks of Figure 11.5. These additional options include:

- Programming of NC machines.
- Generation of bills of materials.
- Rapid prototyping of components by various means, such as stereolithography, so that the designer has a physical model of a proposed part or product.
- Tool and fixture design.
- Determination of cutter paths.
- Costing.
- Programming inspection equipment.
- Definition of the robot path using a robot programming application package.

It should be evident from all of the above that there is no single universal system for CAD/CAM work. The engineer designing circuit boards needs far different capabilities than does the engineer designing an automobile suspension, the kind of information in the data bank for the circuit board designer is obviously of no use to the suspension designer, and the form into which the output information is placed for communication to others will probably be different also. If a company is only now entering the CAD/CAM field, its personnel must do careful research into the various systems presently available and into the vendors offering them for sale. If a company has been using CAD/CAM for a number of years, its personnel will probably have sufficient insight that they can easily evaluate the shortcomings of their present system and the advantages or disadvantages of adding various packages to the system or of replacing it.

One of the major components of a CAD/CAM system is the database. As pointed out several times earlier, most companies tend to develop and manufacture closely related families of products. They therefore have data available from previous designs which can be helpful in future design projects. These data may range from all kinds of technical data, such as efficiencies, ratio of hot-spot temperature to surface temperature, and reliability of a particular component, to operational data, such as the time to complete the drawings for a design, the fraction of the total manufacturing cost represented by the cost of the materials for a product, and the time to produce a given part. If one can express the data in the form of the π-terms of Chapter 7, every similar product can act as an offscale or distorted model of the product to be designed.

In addition to design data available from previous products, test data from one's own products and from those of competitors can be added to the data bank. Information is also available in technical papers published by various professional and technical societies. Experience of the individual engineer provides an additional data source. Considering all these sources together with data from previous designs, we have a historical data bank. Engineers have always used such data, but generally in a rather personal way, probably because there was no simple way to share some of the information with others. This personal data bank could be called on when the engineer had, for example, to make a "judgment call" as to the probability of a part performing well or not. If the experienced engineer can structure the procedure he or she uses in such cases so that the basis of the judgment is available for storage in the database, then the organization will be able to benefit from that experience past the time when the engineer is moved into a different position or leaves the organization.

Having a good database, an analysis of a bare-bones model of a proposed design of a new member of the company's family of products can be made on the basis of a coarse approximation founded on such scant information as approximate size, an estimate of the required output power, and a preferred voltage. An initial evaluation of such functional characteristics as vibration, noise, voltage gradients, temperature rises, probable hot spot temperatures, estimates of product life, and cost can be made. With this information in hand, the next design steps can be taken with far more assurance. In the design of an induction motor discussed in Section 10.2, a design which resulted in the data of Table 10.2, the coarse approximation was the reference motor—one that was reasonably close, but the designer knew going in that it almost certainly would not be the final design.

Finally, the concept of testing during manufacturing should be added to the roster of activities under CAD/CAM. Testing has traditionally been thought of as an activity during development and possibly in early manufacturing runs. However, if testing continues during manufacturing, the output of the sensors measuring the variables of interest can be fed back to the computer to identify deviations from specified performance. This information is certainly needed for GO/NOGO decisions, but a much more important point is that one has the ability to detect subtle and slow-moving changes in the manufacturing process so that corrective action can be taken before the characteristics of the product drift outside an acceptable range. For example, when manufacturing large numbers of supposedly identical parts, measurement of key dimensions may indicate that all of the parts are within tolerance. If, however, those measurements also show that the dimensions are moving slowly toward one tolerance limit, that information needs to be acted on. Testing done for such purposes is often referred to as statistical process control. An example of production line testing possibilities for electric motors appears in Section 7.7 of Reference 3, especially in Section 7.7.5.3 on computerized data acquisition.

For some products, such as those for which health, safety, or substantial property damage may be involved, testing now continues on into the consumption phase of the life cycle. A common example is the on-board computer in automobiles, which will warn of a need for engine tune-up and will store computer codes to indicate problem areas.

Finally, one should be cognizant of the fact that, even with CAD/CAM, errors will occur. One of the authors (RHE), when discussing computerized data transfer from design to manufacturing in a jet engine manufacturing plant, was told by a manufacturing engineer about one case in which the design engineers had transferred data for a turbine disk to manufacturing, but the transfer bus "went down" after the data had been transferred. Design then found that some changes were necessary, but manufacturing did not receive the revised data before machining began. "Well," he sighed, "we produced $25,000 worth of scrap that afternoon."

11.5 EXPERT SYSTEMS

The December 16, 1994, revision of UL943 on ground fault circuit interrupters contains a section that is quoted in part as follows:

> 8.8 A rainproof enclosure made of sheet steel having a thickness of less than 0.120 inch (3.05 mm) shall be protected against corrosion by one of the following coatings:
>
> a) Hot dipped mill galvanized sheet steel conforming with the coating Designation G90 in Table 1 of ASTM A525-87, with not less than 40 percent of the zinc on any side, based on the minimum single spot test requirement in this ASTM Specification. The weight of zinc coating may be determined by any suitable method; however, in case of question the weight of coating shall be established in accordance with the test method of ASTM A90-81 (1991).

The article then goes on in b), c), d), and e) to other allowable coatings, some in zinc, others in cadmium, and one which is a combination of zinc and paint. (The preceding article, 8.7, pertains to enclosures made of heavier gages of steel.)

There is a large amount of information in the four column-inches of the complete article in UL943, and many other articles as replete with information could have been selected for this example. This article lists some limitations on

PRODUCT DESIGN USING COMPUTERS 397

allowable choices, imposed by a body external to the company, and there will be other limitations imposed internally as well. The point being made here is that one needs an *expert* if a rainproof enclosure for a ground fault circuit interrupter is to be designed. The *expert* has traditionally been an engineer who has been assigned the task of keeping all relevant information of the kind that appears in Article 8.8 of UL943 in an accessible form. The portion of the article quoted requires that the engineer know and understand the contents of two ASTM standards; the remainder of the article, as well as Article 8.7, refers to other coatings, other test procedures, and other articles within UL943 itself. The knowledge one obtains from this brief quote falls into two categories, one of which is quite *factual* in nature ("sheet steel having a thickness less than 0.120 inch") and susceptible of being entered into memory in the form "< 0.120in," the other being the *information* that, if Article 8.8a is followed, the coating must conform to Designation G90 and that ASTM A525-87 applies and A90-81 may apply.

The logical process the engineer follows in thinking through the design in order to make rational choices, given constraints and different kinds of knowledge such as those illustrated in this brief excerpt, is not susceptible of reduction to a simple algorithm or of programming in FORTRAN. What is needed is a logic structure that can represent and process what is, to the untrained eye, an amorphous mass of material, and can propose (and even defend) one or more design choices. We refer to such structures and systems as *expert systems*.

The preceding discussion should be convincing that we are not simply talking about an expanded database nor the processing of numerical information by use of prescribed algorithms. What we are talking about is the ability to take large amounts of *knowledge*, some of which is hard numeric data, others of which consists of pieces of information which specify in one way or another limitations of various kinds or which list possible choices, and to process the knowledge in symbolic and often heuristic ways. It is obvious that the acquisition of the necessary knowledge is one key to a successful system and that this acquisition can be a major undertaking. It should also be evident that whatever software is developed for an expert system should be user-oriented; the software should not optimize the way in which machines work, but rather the way in which people think. That is, the artificial intelligence (AI) viewpoint must be uppermost in the mind of the expert system designer.

The kinds of knowledge mentioned so far have fallen into two categories, factual knowledge, such as material properties, allowable material thickness, and so on, and information as to the relationship between objects and attributes, such as the allowable coatings for rainproof enclosures. Sheet steel being considered for an enclosure may be selected or not depending upon whether its coating meets the ASTM specification(s) in UL943. The accumulation of these kinds of knowledge is not as onerous as one might assume at first glance because of company's limitations on product lines. Factual knowledge should be in existing databases. The information concerning relationships will come from those standards that are applicable to the products being made, and in many cases only parts of those standards will be relevant. (If one is making ground fault circuit interrupters for domestic use, then the article quoted above is irrelevant.)

There are two other kinds of knowledge of great value. Experience and rules of thumb are frequently as important in decision making as other forms of knowledge. Some rules of thumb are well established. For example, if the current density in a single TW-insulated #10 AWG copper conductor in air is less than $487 A/cm^2$, the temperature rise will be less than 30 C. This piece of information was found (although not in this form) in another standard, the National Electric Code. Other rules are not so definite: If a molded-case circuit breaker trips on short-circuit current of 10,000 A in 20 ms or less, it is *not likely* to be damaged. This kind of rule comes from the experience of engineers who have run many tests under controlled conditions. If a degree of confidence can be attached to the statement on likelihood of damage under short-circuit trip conditions, the expert system will be able to respond to questions with a stated degree of confidence.

The last type of knowledge needed is evaluation knowledge. How is reasoning applied, which facts, procedures, and rules shall be used, and how shall the decision or list of alternatives be interpreted? This is the kind of knowledge that a designer builds up slowly from one project to the next and which is frequently buried in the subconscious mind. Yet the reduction to a written form is extremely important, not only because it represents one way of retaining the expertise of the designer if the engineer leaves the position for any reason, but because expert systems are rule-based, and the set of "rules" followed by a successful designer can be used as a model for the rules needed by the expert system. Acquisition of these general rules from the subconscious has proved to be one of the most difficult tasks faced by the designer of the expert system.

Another way of looking at expert systems is to divide the system into two parts [8]:

- The first part consists of a set of rules representing the area of expertise. This is the *knowledge base* of the system. The knowledge base must have access to factual information, that is, the database or -bases relating to the product.
- The second part consists of a mechanism that governs the way in which the rules are used. This is referred to as the *inference engine*.

The function of the inference engine is to reach new conclusions, using inputs from the user, factual data and rules in the knowledge base, and conclusions that may have been reached previously and stored. Figure 11.7 illustrates this process diagrammatically. It also shows some of the ways in which IF . . . , IF

PRODUCT DESIGN USING COMPUTERS

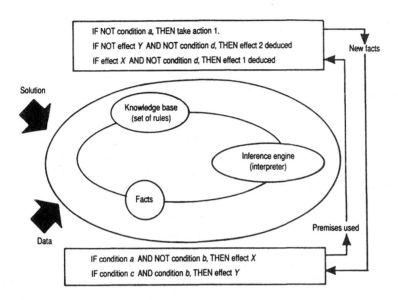

Figure 11.7 Diagrammatic representation of the elements of an expert system and the interrelationships among them, with examples of situation-action rules. (Reproduced from Reference 8, Jean-Baptiste Waldner, copyright holder.)

NOT..., AND..., AND NOT..., and THEN... statements appear. These rules are situation-action pairs. If a situation is recognized, then the specified action is carried out.

The first-generation languages that have been used in expert systems include LISP and PROLOG. There are a number of second-generation languages, and there is considerable progress being made in the development of "expert system shells," for which knowledge of the languages is not normally necessary. The reader interested in the area of expert systems may wish to consult References 9 and 10.

11.6 COMPREHENSIVE CAE SYSTEMS

The technology of computer-aided engineering (CAE) systems has advanced at a rapid pace. Keeping in mind that the acronym CAE is not restricted to design activities only, one expects that a true CAE system will have capabilities ranging from design all the way to production, with the built-in advantage for the user of being able to take advantage of the techniques of concurrent engineering. Hence these systems require computers with far more capability than the PC computers on which some of the programs listed in Table 11.3 can be run.

For our purposes, we will consider only one of the CAE systems that are available, that marketed by Structural Dynamics Research Corporation (SDRC™) under the designation I-DEAS™, an acronym standing for Integrated Design Engineering Analysis Systems. SDRC, which grew out of research on mechanical vibrations at the University of Cincinnati in the mid-1960s, has over 15,000 customer sites and 234,000 application licenses worldwide in a broad spectrum of industries, including the aerospace, defense, automotive, consumer electronics, optical microscopes, cameras, communications, electrical switchgear, power hand tools, wafer processing for semiconductors, and farm machinery industries. I-DEAS is available on a wide range of industry workstations and PC's running Windows to meet the needs of SDRC customers. It is an integrated package of mechanical engineering software tools.

Some understanding of the scope of the I-DEAS system can be obtained by perusal of the SDRC Product Catalog [11]. The list of products in that catalog is so extensive that it is feasible to give only a sampling of the coverage of various engineering disciplines. The foundation for SDRC's system is based on modeling, drafting, and test programs under the heading Core Functionality. This foundation, although it can be used alone, is almost invariably augmented by more specialized programs. Under their Application Sets, for example, one finds surfacing, assembly, mechanism design, tolerance analysis, finite-element modeling, simulation, and model frequency response, as well as the capability for capturing the manufacturing planning, tooling, and NC programming tasks in a single unified system. The Application Products group is divided into design, simulation, and test sections. The design section includes sheet metal design, electronic harness design, and assembly visualization. The simulation packages include nonlinear simulation capabilities, as well as programs that will handle laminates, electronic system cooling, and mold filling and cooling. The test section includes standard and transient measurement programs, acoustic intensity

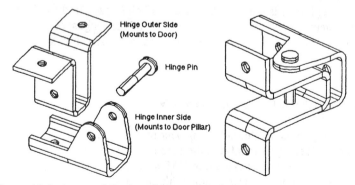

Figure 11.8 Automobile door hinge assembly. (Courtesy of Structural Dynamics Research Corporation.)

PRODUCT DESIGN USING COMPUTERS

measurements, fatigue, and collection of synchronously sampled data from rotating machinery. Other sections include programs on Data Products (for example, drafting symbols and standard parts), Data Exchange Products (for example, importation and exportation of product data in ISO 10303 format), and Product Data Management. Standard programs are available for importation of data already available in other systems, such as NASTRAN and ABAQUS, and for exportation to those systems. The breadth of capabilities of I-DEAS is so wide that only a few examples can be described. We shall look at one very simple example of how the system can be used in the design of a small mechanical part, and then look at a few applications that demonstrate capability in other technical areas.

Consider the hinge for the left front door of an automobile, as shown in Figure 11.8, in which an assembly view is shown on the right and an exploded view on the left, the assembly having been rotated clockwise through 90° for the exploded view. The U-shaped piece is formed from a rectangular piece of steel, and is the part of the hinge that is attached to the door. The other part of the hinge is the fixed part, and attaches to the door pillar. The design task to be considered is that of designing this second portion of the hinge. Only the major steps are shown in the following. For example, the holes used to mount the part to the door pillar appear in some of the following figures, but their generation is omitted from the discussion.

The design begins by drawing a closed figure using straight lines and then modifying it by creating the upper rounded curve and inserting a hole for the hinge pin, so that a surface has been created as shown in Figure 11.9(a). This surface is then "extruded" (that is, moved perpendicular to itself), thus generating the solid of Figure 11.9(b). By generating a rectangle parallel to the face at the right end and extruding it through the solid, the central portion of the solid is removed, leaving the shape shown in Figure 11.9(c).

This is a shape that would be satisfactory for the fixed portion of the hinge, but the question immediately arises as to how to make it. Because of the sharp internal and external corners, it cannot be made by cutting the correct shape out of a heavy sheet steel, and then bending it as necessary to form the required final shape. The cutting and bending process is the preferred approach, however, because it is the most economical method. To produce the shape shown in Figure 11.9(c), the part would have to be cast or machined out of a large block of steel. The machining process would be especially wasteful because it would result in a larger volume of scrap in the form of chips than the volume of the final product, to say nothing about the production time per part.

Because the part is to be made by cutting the correct shape out of flat stock and then bending it, it becomes necessary to allow for fillets at the corners where the bending takes place. The solid of Figure 11.9(c) is therefore modified by addition of the necessary fillets, resulting in the final design of Figure 11.10.

The part is now unfolded, resulting in Figure 11.11. The unfolding process takes into account the fact that, when the stock is bent to the desired

(a)

(b)

(c)

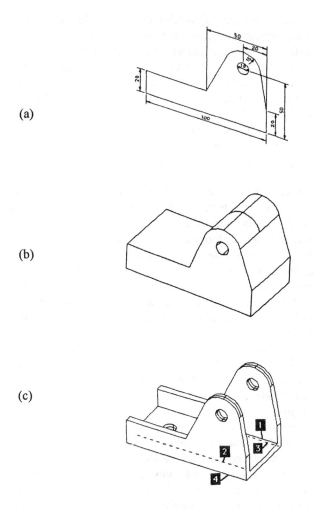

Figure 11.9 Development of the basic shape for part of the automobile door hinge. (Courtesy of Structural Dynamics Research Corporation.)

shape, the steel will stretch along the outside of the bends. The designer now has a layout for the fixed portion of the hinge, and this can be used to specify the machining operations. Drilling of the holes would be done first, and then the profile around the outside would be cut, using either a milling machine or possibly by flame cutting. Finally, the stress and deflection of the completed piece due to a load parallel to the hinge pin can be simulated using finite-element analysis, as shown in Figure 11.12. This simple example is intended only to show the power of the system. Use of similar techniques allows the designer to

PRODUCT DESIGN USING COMPUTERS 403

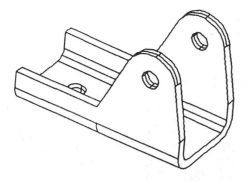

Figure 11.10 Final design of the fixed portion of the automobile door hinge. (Courtesy of Structural Dynamics Research Corporation.)

generate any other shape needed, such as the housing and the impeller rotor shown in Figure 11.13.

In addition to the static deflection information that can be obtained, dynamic response of parts can be predicted. Figure 11.14 shows an example of the dynamic response of a part subjected to an impulse of force. The I-DEAS

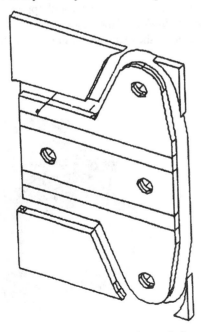

Figure 11.11 Unfolded hinge part, showing the path for cutting the part out of flat stock. (Courtesy of Structural Dynamics Research Corporation.)

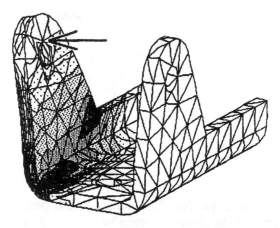

Figure 11.12 Stress and deflection of the hinge subjected to a transverse load. (Courtesy of Structural Dynamics Research Corporation.)

Electronic System Cooling package can be used to simulate 3D air flow, convection, conduction, and radiation in order to predict electronic component temperatures, as shown in Figure 11.15. Finally, the I-DEAS Order Tracking software can be used with rotating machinery and appropriate sensors to obtain the time response and amplitudes of the fundamental and various orders of the harmonics at desired points in the machine, as shown in Figure 11.16.

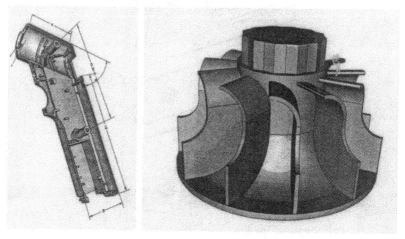

Figure 11.13 Examples of solids modeled using I-DEAS. (Courtesy of Structural Dynamics Research Corporation.)

PRODUCT DESIGN USING COMPUTERS 405

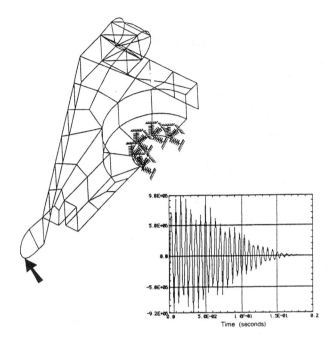

Figure 11.14 Dynamic response of a part subjected to an impulse of force. (Courtesy of Structural Dynamics Research Corporation.)

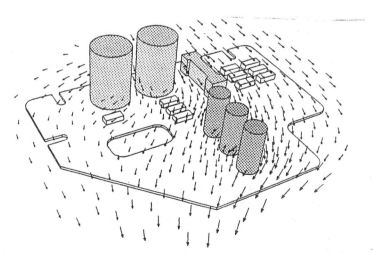

Figure 11.15 Simulated air flow patterns used in the prediction of electronic component temperatures. (Courtesy of Structural Dynamics Research Corporation.)

406   CHAPTER 11

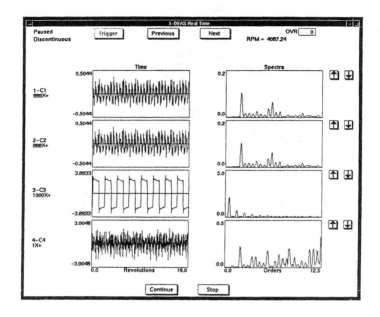

Figure 11.16 Time and frequency responses at various points in a rotating machine. (Courtesy of Structural Dynamics Research Corporation.)

This cursory and highly selective review of the I-DEAS capabilities is intended only to indicate the wide range of possibilities. Further information is available directly from SDRC at "http://www.sdrc.com/pub/catalog/ideas" or at the address given with Reference 11.

11.7 SUMMARY

The engineering student will be expected to arrive at his or her first job with sufficient skill in the use of computers to be comfortable in learning to use the hardware and software systems the employer happens to have. The broader the student's experience with various kinds of equipment and types of problems, the more likely he or she is to have the requisite skill. For this reason you should take every opportunity you have to work problems by computer, even when the problem could be solved easily in other ways. The practice projects at the end of this chapter suggest selected problems in preceding chapters to be revisited for computer solutions. In addition, problems in the chapters to follow should be solved using computer techniques when possible and reasonably feasible.

REFERENCES

1. Cezzar, Ruknet. *A Guide to Programming Languages: Overview and Comparison,* Artech House, Boston, 1995.
2. Wehmeyer, K. R. *What Every Engineer Should Know About Microcomputer Program Design,* Marcel Dekker, New York, 1984, pp. 5-10.
3. Engelmann, Richard H., and William H. Middendorf. *Handbook of Electric Motors,* Marcel Dekker, New York, 1995.
4. Seely, S., and A. D. Poularikas. *Electromagnetics,* Marcel Dekker, New York, 1979, pp. 64-71.
5. Brauer, J. R., Ed. *What Every Engineer Should Know About Finite Element Analysis,* Marcel Dekker, New York, 1988.
6. Stark, John. *What Every Engineer Should Know About Practical CAD/CAM Applications,* Marcel Dekker, New York, 1986.
7. Chorafas, Dimitris N. *Engineering Productivity Through CAD/CAM,* Butterworths, London, 1987.
8. Waldner, Jean-Baptiste. *CIM—Principles of Computer-integrated Manufacturing,* John Wiley & Sons, Chichester, England, 1992.
9. Chorafas, Dimitris N. *Expert Systems in Manufacturing,* Van Nostrand Reinhold, New York, 1992.
10. Hopwood, Adrian A. *Knowledge-Based Systems for Engineers and Scientists,* CRC Press, Boca Raton, FL 1993.
11. SDRC Product Catalog, May 1996. Accessible on the World Wide Web at "http://www.sdrc.com/pub/catalog/ideas/" or from SDRC, 2000 Eastman Drive, Milford OH 45150.

REVIEW AND DISCUSSION

1. Networks are generally conceded to be desirable and indeed necessary. If you have had personal experience with the use of networks, list as many advantages as you can that you discovered. What disadvantages were encountered, and what might have been done to obviate their effects?
2. Table 11.3 lists commercially available general-purpose software programs. If you have used any of these, explain to the class the advantages and difficulties you found in their use. If you used any other programs of a similar nature, which did you use and did you find them worthwhile? Specifically, why or why not?
3. Figures 11.5 and 11.6 shows an elementary diagram of computer-integrated manufacturing in 11.5 and a more complex diagram in 11.6. An interesting in-class project is to construct a CIM diagram that is still more global in nature than that of Figure 11.6.
4. Name as many devices or processes as you can in which internal testing is being used or in which it could easily be used. These may be products with

domestic or commercial use, and the test may be initiated either internally by the device itself, or may be initiated by the user. Can you think of reasons why automatic initiation would be more or less desirable than user initiation?

5. Formulate a problem in which finite-element analysis would be required for a good understanding of important phenomena. To trigger your thinking, consider the cycle time for a domestic ice maker, from inflow of water to ejection of the ice. What are the heat flow patterns from the water to the heat sink initially, and what happens as ice begins to form? Can you find in physics or engineering handbooks the bulk properties of water and ice that are needed to reach a solution? If a short cycle time is important, what might you look for in the solution to indicate ways to speed up the process?

PRACTICE PROJECTS

1. Develop a program to calculate the probability of occurrence of the top event of a generalized model of fault tree analysis.
2. In Figure 4.4, it is evident that a shift from double-declining balance depreciation to straight-line is advantageous toward the end of the depreciation cycle. Write a computer program that will change from double-declining depreciation to straight-line when the straight-line method gives a more rapid depreciation of the remaining book value.
3. A number of the problems at the end of Chapter 4 can be done using spreadsheet analysis. Identify those problems that appear to be suitable for this approach, and work two or three of them. Are there problems for which you would not use spreadsheets? Why?
4. If you have facilities available for patent searching, if you are doing a project for which patents *may* be important, and you have not done a search as yet, carry out a patent search using your chosen product class.
5. Table 8.1 is the precedence table for an electric automobile design. The matrix of Figure 8.2 was created directly from that table, and the precedence table of Figure 8.3 was created by hand manipulation of the matrix of Figure 8.2. Develop a program to get from a precedence table directly to the precedence matrix. If the program is used for the electric automobile, does a better precedence matrix result?
6. Develop a program to determine the critical path of a generalized PERT diagram.
7. Expand the program of (6) to include the determination of probability of a specified completion time.
8. Design a family of air-core inductors for continuous operation using copper wire, and having inductances of 2, 4, 8, 16, and 32 mH with a current rating of 10 A. Allow for a temperature rise of 75 C above a 25 C ambient.

9. You are to design a family of wire-wound resistors using the equations derived in Chapter 10, with the objective of achieving minimum cost. The resistance values are to be 1, 10, 100, and 1000 ohms, and two sets are to be designed, one for an operating temperature of 100 C and the other for an operating temperature of 200 C, using a 20 C ambient. Within each of these two sets, subsets are to be designed with power ratings of 5, 25, 50, and 500 watts. The cost of resistance wire depends mainly on the wire diameter, not upon the alloy. Data for various resistance wires are shown below, as are data relating AWG gage number to wire diameter. For this problem assume that AWG #40 costs $75.00/N and that costs decrease by $2.00 for each lower gage number. That is, AWG #10 costs $15.00/N. All alloys weigh 0.07564 N/cm^3.

Ceramic cores are available in length increments of 5 mm and diameter increments of 2 mm. The cores cost $0.05/$cm^2$ of outer surface. The two terminals to which the resistance wire is connected cost $0.005/A each, rounded to the next higher decade. That is, if the current required is, say, 3 A, the cost will be that of a 10-A terminal.

Material	Resistivity at 20°C ($\mu\Omega$m)	Temp. coeff. at 20°C ($\Omega/\Omega°C$)
NiCr (75-20)	1.33	0.00002
NiCr (60-15)	1.12	0.00015
NiCrFe (30-2-66)	0.830	0.00070
CuNi (55-45)	0.489	±0.00002
18% NiAg	0.316	0.00019
NiCu (12-88)	0.158	0.00038

AWG	Wire diam. (cm)	AWG	Wire diam. (cm)	AWG	Wire diam. (cm)
10	0.259	21	0.0711	31	0.0226
11	0.231	22	0.0635	32	0.0201
12	0.206	23	0.0584	33	0.0178
13	0.183	24	0.0508	34	0.0160
14	0.163	25	0.0457	35	0.0142
15	0.145	26	0.0406	36	0.0127
16	0.130	27	0.0358	37	0.0113
17	0.114	28	0.0320	38	0.0101
18	0.102	29	0.0284	39	0.0090
19	0.0914	30	0.0254	40	0.0080
20	0.0813				

Since the project obviously requires a large amount of repetitive calculation, it is wise to investigate several approaches before beginning those calculations.

Some ratings may not be feasible; list those ratings with a statement for each of the reason(s) that the rating is not feasible.

12

OPTIMUM DESIGN

One can state the following general principle. If one is looking for the maximum or minimum of some function of many variables subject to the condition that these variables are related by a constraint given by one or more equations, then one should add to the function whose extremum is sought the functions that yield the constraint equations each multiplied by undetermined multipliers and seek the maximum or minimum of the resulting sum as if the variables were independent. The resulting equations, combined with the constraint equations, will serve to determine all unknowns.

J. Lagrange

We have touched on the notion of optimum design in various places earlier in this work. In this chapter, we will look at a few procedures for choosing the design parameters to produce an optimum design. The word "optimum," coined by Leibniz, may be defined as the point at which the condition, degree, or amount of something is the most favorable. The "amount of something [that] is the most favorable" in this definition may be cost of production, net profit (thus taking into account such factors as distribution costs), weight, volume, or some weighted combination of these factors. Because the difficulty of the procedures varies, the presentation will in some cases include mathematical proofs, but others are presented without such proofs. The interested reader will wish to consult the references at the end of the chapter, including those in the concluding section that are not referenced in the discussion but which are of interest and value if the subject is to be explored in depth.

In any case, the distinguishing mark of a skillful designer is the ability to develop designs that are close to optimum—that is, the best products from some stated point of view.

12.1 PRELIMINARY CONSIDERATIONS

The study of optimization problems began at least 2,500 years ago. An early example appears in Virgil's *Aeneid,* and is known as Dido's problem. Dido was fleeing from her brother, and negotiated with a local landowner on the shore of what is now the Bay of Tunis to purchase as much land as could be encircled by a bull's hide. Her problem can then be simply stated as follows: How does one encircle the largest amount of land possible with a bull's hide? Dido solved the problem by cutting the hide into very narrow strips, tying them together, and enclosing a large tract of land in a semicircle, with the ends of the resulting rope reaching the sea. The *Aeneid* tells us that she then built a fortress and the city of Carthage on the land. Other problems were solved in ancient times by geometers, such as Zenodorus, who proved (using the standards of his time) such theorems as that, if there exists a plane n-gon having the largest area among all n-gons of a given perimeter, then it must have equal sides and equal angles. Over the intervening centuries, others have solved specific optimization problems. One example of such solutions is Snell's law of refraction. Beginning in the 1940's, there was a considerable amount of research into the subject using a variety of mathematical techniques.

Formal optimization procedures can discover the optimum design only for a given design alternative. For example, if one is to produce at lowest cost a four-passenger automobile with specified acceleration and fuel efficiency using an internal combustion engine, there is an optimum design that specifies all dimensions and materials. This optimum design obviously has no direct relationship to the design of a four-passenger automobile if the motive power is to be changed from an internal combustion engine to electric drive.

If substantial gains in utility are the goal, one does not look to optimization techniques. Rather, one looks at unused or neglected design alternatives. If one can be found that results in a substantial gain in utility, we refer to this as a design breakthrough. This is the thrust of the discussion in Chapter 6. Optimization techniques will not result in substantial gains, but will rather result in smaller gains in desirable characteristics that are obtained by choosing "just right" design parameters, but only after the general design concept or design alternative has been chosen.

The likelihood of establishing an optimum design is enhanced if a mathematical model can be developed. If a mathematical model is not developed, designers with considerable design experience sometimes arrive at a near-optimum design by drawing on their experience. Without a mathematical model the designer is restricted to the use of physical models. Making physical models is usually very expensive, and the process is usually terminated long before a truly optimum model has been reached. Making models using CAD is a more modern approach, and has the advantage that dimensions, for example, can be varied in a systematic manner in order to achieve sufficiently low stress, for example, or some other desirable or necessary feature. One of the major bene-

OPTIMUM DESIGN

fits of development of a mathematical model is that, with it, formal procedures are available to determine the optimum.

The optimization problem is complicated by the fact that there are usually several design characteristics to be considered. For example, Figure 12.1 (which is a repeat of Figure 8.12) shows how the cost, weight, and internal power loss of a control transformer vary with the important dimension D of scrapless E-I laminations. If the design is to be optimized on the basis of cost, D = 3.5 cm would be selected. Is this the best design? Probably not, because the cost difference between the D = 3.5 cm design is only slightly less than if one were to select D = 3.25 cm or D = 3.75 cm. Intuition tells us that selection of the 3.25 cm value is the best design because it results in slightly lower weight and substantially reduced internal losses.

What is shown by this example is generally true. Repeated modelmaking, repeated analysis as was used to develop the data and curves of Figure 12.1, or the formal optimization procedures to be studied—all have in common the fact that the optimum value of one characteristic is obtained for each application of the procedure. The procedure can be repeated to find the optima of other characteristics, and finally a decision can be made based on the relative merits of

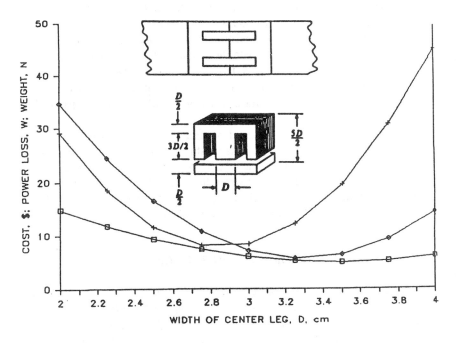

Figure 12.1 Cost (□), power loss (+), and weight (◊) versus core size for a control transformer using scrapless laminations.

the optimized characteristics with respect to what the designer, production engineers, and sales department believe will result in a successful product.

A formal way of reaching a decision in a case such as this was presented in the discussion of criterion functions in Section 8.3.5. Criterion functions depend on the relative desirability of the characteristics to be optimized. For example, in the transformer design attaining maximum efficiency may be twice as desirable as attaining minimum cost or weight. The weighting factors a_i of that section were introduced in order to establish the relative importance of the various terms. As was noted in that discussion, the relative weight of the terms is not the only complication that arises in choosing the optimum combination. Typically, every term in the criterion function inherently has different units, and the utility of one unit as compared to another is difficult to decide. How many watts of power loss does one dollar of cost reduction equal? This problem was taken care of by the introduction of the K terms.

There is still another factor in thinking about optimum design that must be emphasized. It was noted earlier that the designer is often faced with a situation in which only part of a system or device can be expressed as a mathematical model, with resort to other kinds of models for other parts. Each part may then be optimized in one way or another, but the optimum design of each part of a device or system is not sufficient to ensure the optimum design of the whole. To take a trivial example, suppose that one were to optimize the transformer design considering only the losses in the steel (the hysteresis and eddy current losses). The resulting curve would approximate the shape of the cost versus D curve of Figure 12.1, leading one to select a large value of D. However, when the "copper" losses of the windings (the I^2R losses) are added, the U-shaped curve of the figure results and one would select a value of D that is considerably smaller.

Despite this difficulty, piecemeal optimization is often accepted simply because nothing better can be done. It is most acceptable if the components for which a mathematical model are not available are those having little influence on the components being optimized. However, the fact that in general the device or system as a whole may gain by changing the design parameters from those indicated by the piecemeal procedure should be kept firmly in mind.

12.2 THE GENERAL OPTIMIZATION PROBLEM

The general optimization problem involves three types of functional relationships among the specifications and the design parameters. The first type of relationship is called the *criterion function* or *objective function*. It is the mathematical expression of the quantity whose maximum or minimum is to be found as a function of the design parameters. As pointed out in the preceding section, this may be a single characteristic or a weighted combination of several characteristics. There can be only one criterion function.

OPTIMUM DESIGN

The second type of function is called a *functional constraint*. This consists of equations that are expressions of the physical laws involved in the proposed product, for example, $F = ma$ or $E = RI$. The equations forming the functional constraint constitute the mathematical model. The number of equations in this set must be less than the number of design parameters. If these equations are independent and their number equals the number of design parameters to be determined, the design is completely fixed, and the parameters are known. On the other hand, if the number of equations is substantially smaller than the number of design parameters to be determined, some of the optimization techniques that follow are not easily applied. Multidimensional sequential methods [1] can be applied, however.

The third type of function is called a *regional constraint*. These are always expressed as inequalities, and there are no limits to the number of such constraints. These functions simply define the limits within which the design parameters must lie. In the design of the control transformer whose characteristics are shown in Figure 12.1, the designer set a lower limit of D as 2 cm and an upper limit as 4 cm. That is, $2 \leq D \leq 4$, D being in cm.

These functional relationships may be placed in a more formal mathematical form as follows. The criterion function

$$C = c(x_1, x_2, \ldots, x_n) \tag{1}$$

is to be maximized or minimized by choice of values for the variables in the function. Functional constraints limiting the choice of variable values may be written in the form

$$\begin{bmatrix} F_1 = f_1(x_1, \ldots, x_n) = 0 \\ \ldots\ldots\ldots\ldots\ldots\ldots\ldots\ldots\ldots \\ F_m = f_m(x_1, \ldots, x_n) = 0 \end{bmatrix} \tag{2}$$

where m < n. Regional constraints defining the regions of acceptability in the n-dimensional variable space are

$$\begin{bmatrix} R_1 < r_1(x_1, \ldots, x_n) < R_1' \\ \ldots\ldots\ldots\ldots\ldots\ldots\ldots\ldots\ldots \\ R_p < r_p(x_1, \ldots, x_n) < R_p' \end{bmatrix} \tag{3}$$

where p is any number.

It should be self-evident that the criterion function is always present in an optimization problem. The choice of procedure to solve a given problem depends on whether the functional and regional constraints are present or not.

12.3 OPTIMIZATION WITH ONLY A CRITERION FUNCTION

In some optimization problems, there are no functional or regional constraints. In that case, a very simple method of obtaining the largest or smallest value of a criterion function can be used. This is the *derivative method*. To use this method, simply set the first derivative of the criterion function with respect to an independent variable equal to zero. This method was used in Chapter 10 for the design of a spring. We shall review this method briefly here and present some additional pieces of information that are important.

Elementary calculus demonstrates that if the first derivative is zero, the function is at a maximum or a minimum, or at least at a stationary point. That is, $dC/dx = 0$ at each of the five points designated in Figure 12.2. The word "extremum" is used to refer to maximum and minimum points, and the optimum design is usually at one of the extrema. A stationary point is usually of no significance. To determine which of the extrema has been located or whether a stationary point has been found, one method is to evaluate the second derivative of the function. If the second derivative is positive, the extremum is at the least a relative minimum. If the second derivative is negative, the extremum is a maximum, but again it may be only a relative maximum, as shown in the figure. If the second derivative is zero, a stationary point has been located.

Another way to determine whether a maximum or minimum has been found is to evaluate the function at the values of x for which dC/dx is zero. The designer may wish to know these values in any case, and this is usually an easy calculation. Stationary points of the function will always have values between the values of the function at the extrema, and hence are easily identified.

What about the value of C in Figure 12.2 when $x = 0$? This value is larger than the value of C at any of the extrema, but the derivative method will not locate this point because the first derivative is not zero. That is, if $x = 0$ is

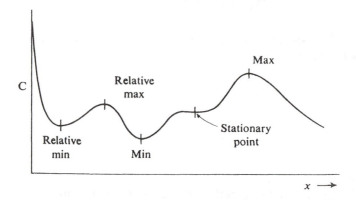

Figure 12.2 Extremum points of an arbitrary criterion function.

OPTIMUM DESIGN

an allowable value of the independent variable and this point is missed because of fixation on the derivative approach, the optimum value of x has been missed. It is always necessary that C be calculated at the endpoints of the acceptable range of values as well as at the extrema.

12.3.1 The Derivative Method

The use of the derivative method was shown earlier in the spring design problem. We consider now a different example. The armature (the rotating member) of medium and large size direct current generators requires that the coils be wound off the machine, using machines known as coil-winders, and placed into the slots after being wound. Hence the slots must have parallel sides, as shown in Figure 12.3, which shows three armature punchings having slots of equal cross-sectional area but widely differing depth-to-width ratios. (Various means are used to insure that the coil sides are retained in the slots when subjected to centrifugal force during rotation, but the means used is irrelevant to this example.) The center holes for the motor shaft are not shown in this figure.

The number of slots, the number of conductors in each slot, and the cross section of the conductors are determined by the voltage and current specifications. The voltage rating also determines the thickness of the insulation placed between the coil itself and the steel of the armature. Hence the area of each slot and the number of slots are fixed before the designer chooses the slot dimensions. There are other considerations the designer must take into account relative to the magnetic path and mechanical construction which are discussed briefly below.

The optimization problem before us becomes evident when Figure 12.3 is studied with some care. All of the slots have the same cross-section area, but the slots in Figure 12.3(a) are shallow, forcing the width of the tooth at its root,

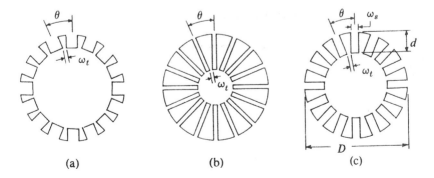

Figure 12.3 Typical armature lamination punchings.

designated as w_t, to be small. In Figure 12.3(b), the slots are very deep, again forcing w_t to be small. The punching of Figure 12.3(c), which has an intermediate depth, has a width of the tooth root that is somewhat larger. Small widths at the root of the teeth cause magnetic saturation of the steel in those areas, resulting in reduction of flux in the magnetic circuit and inefficient use of the steel (in a magnetic sense) elsewhere. There are also mechanical problems associated with narrow root widths. None of these structures can be completely rigid, and the magnetic forces acting on the teeth during part of a revolution have tangential components that give rise to cyclical stress concentrations at the base of the tooth. There are documented cases in which fatigue in these regions has resulted in teeth breaking off and being thrown out of the armature, either singly or in groups, resulting in major damage to the stator. Moreover, deep slots increase the inductance of the coils and hence have adverse effects on the life of the brushes and commutator. Deep slots also reduce the allowable diameter of the shaft on which the armature steel is placed.

In this case and in many other designs, very large or very small choices for the independent variable (d in this case) result in undesirable situations. We expect, therefore, that the optimum design often exists at some intermediate point, and intuition would indicate that the punching of Figure 12.3(c) should be selected. The optimization procedure using the derivative method will, however, be pursued in order to demonstrate the derivative technique.

In Figure 12.3, the known values are

D = outside diameter of the armature
A = cross-sectional area of the slot
S = the number of slots

The unknown values are

d = slot depth
w_s = slot width
w_t = tooth width at the root

From the earlier discussion, we want to maximize w_t. We write the following equations:

$$w_s = \frac{A}{d} \tag{4}$$

$$w_t \cong \frac{\pi(D-2d)}{S} - w_s \tag{5}$$

OPTIMUM DESIGN

Equation (5) is an approximation, based on the fact that a circle may be approximated by an n-gon if the number of sides is large. Substituting from Equation (4) into (5), we obtain the criterion function

$$C = w_t = \frac{\pi D}{S} - \frac{2\pi d}{S} - \frac{A}{d} \tag{6}$$

which, on differentiation with respect to d, yields

$$\frac{dC}{dd} = \frac{dw_t}{dd} = -\frac{2\pi}{S} + \frac{A}{d^2} \tag{7}$$

Setting the derivative equal to zero, we obtain

$$d = \sqrt{\frac{SA}{2\pi}} \tag{8}$$

Since the second derivative, $d^2 w_t / dd^2$, is negative, the value of $C = w_t$ is a maximum.

In general, because of the conflicting requirements discussed above, the slots of a direct current generator are not made as deep as the value calculated from Equation (8), resulting in a reduction of tooth width and higher flux densities at the root of the tooth from the maximum. The reason is that the designer is driven toward reduced inductance, thus improving brush life. That is, maintenance considerations are allowed to skew the design away from the theoretical optimum point obtained by this analysis. A wise compromise must be made. If an expression relating brush life to inductance, and therefore to slot depth, could be found, that expression could be made part of a composite criterion function. Unfortunately, as discussed in Section 13.3.2 of Reference 2, "Satisfactory brush life is a comparative index usually established by comparing past and present performance on a particular machine or by comparing similar machines," and there are many other factors relating to the environment in which the machine is operated.

Criterion functions may be functions of more than one variable. In these cases, the extremum is found by taking the partial derivative of the criterion function with respect to each variable in turn, setting those derivatives to zero, and solving the resulting set of equations simultaneously. For example, how does one make a rectangular lidless box of a given volume using the smallest amount of material? Let the volume of the box be V, the base dimensions be a and b, and the height be h. Then

$$V = abh \tag{9}$$

The surface area is

$$S = ab + 2bh + 2ah \qquad (10)$$

Solving Equation (9) for h and substituting into Equation (10) yields

$$S = ab + \frac{2V}{a} + \frac{2V}{b} \qquad (11)$$

which is the criterion function in the variables a and b. Taking the partial derivatives and setting each of them to zero gives

$$\frac{\partial S}{\partial a} = b - \frac{2V}{a^2} = 0 \qquad (12)$$

$$\frac{\partial S}{\partial b} = a - \frac{2V}{b^2} = 0 \qquad (13)$$

Equations (12) and (13) may be solved simultaneously for V, resulting in

$$a = b$$

and h may be found to be a/2. That is, the base should be square and the height is one-half the length of a side on the base.

Note that, had we drawn on results given us thousands of years ago by Greek geometers, we would have known *a priori* that the base should be square (a 4-gon), and the criterion function could have been made a function of h only.

The process shown here can be extended to find the optimum value of the criterion function for any number of independent variables.

12.4 OPTIMIZATION WITH FUNCTIONAL CONSTRAINTS

Suppose that in addition to a criterion function as in Equation (1), we have functional constraints as shown in Equation (2). One approach to optimization is to solve the functional constraints for each of the variables, substituting the resulting relationships into the criterion function. If done correctly, m variables (that is, the number of functional constraints) will be eliminated from the criterion function. The extremum is then found by applying the techniques of the preceding section. As a matter of fact, this technique was the one used in the optimum armature slot problem. Equation (5) can be considered to be the criterion function in two variables, Equation (4) is the functional constraint, and Equation (6) is then the criterion function in one variable. A similar procedure was used to find the optimum dimensions of the box.

OPTIMUM DESIGN

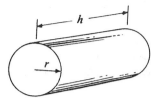

Figure 12.4 Right cylindrical tank.

As another example, suppose that you wish to find the values of r and h that will result in use of the smallest amount of material for the right cylindrical tank shown in Figure 12.4, where it is assumed that the thickness of the material is the same everywhere and is independent of the r/h ratio. The tank is to have a capacity of 100 m^3. Since the least amount of material is to be used and the thickness is a constant, the variable of interest is the surface area, the minimum surface being required. The criterion function is therefore

$$C(r, h) = S = 2\pi rh + 2\pi r^2 \qquad (14)$$

and the functional constraint is the equation for the volume of a cylinder. That is,

$$f(r, h) = \pi r^2 h - 100 = 0 \qquad (15)$$

or

$$h = \frac{100}{\pi r^2} \qquad (16)$$

Substituting Equation (16) into Equation (14) yields

$$C(r) = S = \frac{200}{r} + 2\pi r^2 \qquad (17)$$

We have thus reduced the criterion function to a function of one variable, r, that now contains the information expressed by the functional constraint. We may now differentiate to obtain

$$\frac{dC(r)}{dr} = -\frac{200}{r^2} + 4\pi r = 0 \qquad (18)$$

Hence $\quad r^3 = \dfrac{200}{4\pi} = 15.9$

yielding $r = 2.5 \text{ m}$ and $h = \dfrac{100}{\pi(2.5^2)} = 5 \text{ m}$

Another approach to the optimization problem with functional constraints was devised by Lagrange. Before illustrating Lagrange's method, consider the function $C = c(x,y)$. By the chain rule, the total differential is

$$dC = \frac{\partial C}{\partial x} dx + \frac{\partial C}{\partial y} dy \qquad (19)$$

From geometric considerations, it can be shown that, for an extremum, $dC = 0$ whether x and y are independent or not.

Now if x and y are independent, dx and dy can be chosen as arbitrary changes in those variables. A necessary and sufficient condition for $dC = 0$ is therefore that

$$\frac{\partial C}{\partial x} = 0 \quad \text{and} \quad \frac{\partial C}{\partial y} = 0$$

Suppose, however, that x and y are not independent. That is, suppose that $C = c(x,y)$ and $F = f(x,y) = 0$, the second relationship being a functional constraint. From these relationships,

$$dC = \frac{\partial C}{\partial x} dx + \frac{\partial C}{\partial y} dy = 0 \qquad (19)$$

$$dF = \frac{\partial F}{\partial x} dx + \frac{\partial F}{\partial y} dy = 0 \qquad (20)$$

Because dx and dy cannot be arbitrary changes in the variables as a consequence of their dependence, the partial derivatives cannot be equated to zero as a necessary condition of these two equations.

The approach of Lagrange, as stated in his words in the quotation at the beginning of the chapter, was to multiply Equation (20) above by λ (called the *Lagrange multiplier*) and add it to Equation (19). Then

$$\left(\frac{\partial C}{\partial x} + \lambda \frac{\partial F}{\partial x}\right) dx + \left(\frac{\partial C}{\partial y} + \lambda \frac{\partial F}{\partial y}\right) dy = 0 \qquad (21)$$

where λ is that value that makes the coefficients of dx and dy equal to zero. That is,

OPTIMUM DESIGN

$$\frac{\partial C}{\partial x} + \lambda \frac{\partial F}{\partial x} = 0 \tag{22}$$

$$\frac{\partial C}{\partial y} + \lambda \frac{\partial F}{\partial y} = 0 \tag{23}$$

and the functional constraint,

$$f(x,y) = 0 \tag{24}$$

The description of Lagrange's method can now be illustrated using an example already at hand, that of the 100-m³ cylindrical tank. The criterion function is given as Equation (14) and the functional constraint is given by Equation (15). To apply Lagrange's method, we need

$$\frac{\partial C}{\partial r} = 2\pi h + 4\pi r \qquad \frac{\partial C}{\partial h} = 2\pi r$$

$$\frac{\partial F}{\partial r} = 2\pi r h \qquad \frac{\partial F}{\partial h} = \pi r^2$$

The simultaneous equations to be solved are:

From Equation (22), $\quad 2\pi h + 4\pi r + \lambda \times 2\pi r h = 0 \tag{25}$

From Equation (23), $\quad 2\pi r + \lambda \pi r^2 = 0 \tag{26}$

From Equation (24), $\quad \pi r^2 h - 100 = 0 \tag{27}$

Equation (26) immediately yields

$$\lambda = \frac{-2}{r} \tag{28}$$

Substitution into Equation (25) gives

$$h + 2r - 2h = 0$$

or

$$h = 2r \tag{29}$$

Hence, from Equation (27),

$$r^3 = \frac{100}{2\pi} = 15.9$$

from which

$$r = 2.5 \text{ m} \quad \text{and} \quad h = 5 \text{ m}$$

as before.

One of the advantages of this method is that the simultaneous equations can usually be solved to give the general relationship among the design parameters, as was done here for r and h (Equation (29)). We therefore have a general solution for all cylindrical tanks that have a minimum surface area for a given volume, that is, that $h = 2r$.

We will now apply the method of Lagrangian multipliers to the problem of finding the dimensions of a randomly wound coil so that it has maximum inductance for a wire of a given length and cross-sectional area. The results were already used in Chapter 10 in the synthesis procedure for a resistive inductor. The dimensions to be used are shown in Figure 12.5. Establishment of the correct set of functional relationships is important in all problems such as these.

Because the length of wire and the cross-sectional area of the wire have been specified, the volume of the conductor material is fixed. However, when winding a coil in a random manner, there are always voids in the coil due to one turn crossing over another and the fact that the wire and its insulation is round, leaving spaces between adjacent wires. The ratio of the net cross-sectional area of the conductors to the gross cross-sectional area of the coil is known as the *space factor,* and experiment has shown that this ratio is very nearly constant independent of the dimensions of a coil. Moreover, since the total length of wire has been specified, the length of the mean turn (that turn lying half-way between the inner and the outer radii of the coil) times the number of turns, designated as n, is fixed.

OPTIMUM DESIGN

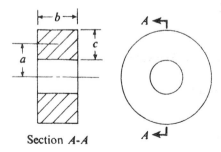

Figure 12.5 Air-core inductance coil.

OPTIMUM DESIGN

From this discussion, we can now write two functional constraints using the dimensions a, b, and c from the figure, and using V as the volume of the coil itself and K as the length of wire. These constraints are:

$$F_1 = 2\pi abc - V = 0 \tag{30}$$

$$F_2 = 2\pi an - K = 0 \tag{31}$$

The variable to be optimized is the inductance. That is, the criterion function is given by Equation (25) of Chapter 10, which is

$$L = \frac{31.5a^2 n^2}{6a + 9b + 10c} \tag{32}$$

as shown in Reference 3. For simplicity in writing the equations below, we will use the symbol D to represent the denominator in this equation. Equations (22) and (23), generalized to make use of two Lagrange multipliers, one for each of the functional constraints, and for four variables, leads to the following set of equations.

$$\frac{\partial L}{\partial a} + \lambda_1 \frac{\partial F_1}{\partial a} + \lambda_2 \frac{\partial F_2}{\partial a} = \frac{(6a + 9b + 10c) \times 2 \times 31.5an^2 - 6(31.5an^2)}{D^2}$$

$$+ 2\pi\lambda_1 + 2\pi\lambda_2 = 0 \tag{33}$$

$$\frac{\partial L}{\partial b} + \lambda_1 \frac{\partial F_1}{\partial b} + \lambda_2 \frac{\partial F_2}{\partial b} = -\frac{31.5a^2 n^2 \times 9}{D^2} + 2\pi\lambda_1 ac = 0 \tag{34}$$

$$\frac{\partial L}{\partial c} + \lambda_1 \frac{\partial F_1}{\partial c} + \lambda_2 \frac{\partial F_2}{\partial c} = -\frac{31.5a^2 n^2 \times 10}{D^2} + 2\pi\lambda_1 ab = 0 \tag{35}$$

$$\frac{\partial L}{\partial n} + \lambda_1 \frac{\partial F_1}{\partial n} + \lambda_2 \frac{\partial F_2}{\partial n} = \frac{2 \times 31.5a^2 n}{D} + 2\pi\lambda_2 a = 0 \tag{36}$$

From Equations (34) and (35),

$$\lambda_1 D^2 = \frac{9 \times 31.5a^2 n^2}{2\pi ac} = \frac{10 \times 31.5a^2 n^2}{2\pi ab}$$

which leads immediately to the fact that

$$c = 0.9b \tag{37}$$

Solution of Equation (36) for λ_2 followed by substitution into Equation (33), along with the relationship between b and c of Equation (37), leads after simplification to

$$a = 1.5b \tag{38}$$

Equations (37) and (38) specify the geometric relationships that yield the maximum inductance for the given conditions.

Note that these results were obtained in two pages of text, including the "on-paper" thinking about the problem and the relevant equations. By contrast, if one looks into the technical literature on the subject, one finds a paper by Brooks [4], written in 1931, in which he reaches the conclusion, based on experimental evidence, that b = c and a = 1.48b for the optimum design. Coils of these dimensions are sometimes referred to as *Brooks coils*. Why the differences between results, small though they may be? Certainly experimental error enters the mind as one possibility. Another reason for the difference may be that the criterion function (Equation (32) for inductance) may not be completely accurate.

The important point is that the optimization procedure applied to the mathematical model reached the optimum configuration very quickly and at little expense, whereas the experimental method required that a large number of carefully made coils be made and evaluated in order to arrive at an essentially identical conclusion.

In this example, there were two Lagrange multipliers. If more functional constraints exist, the number of multipliers is increased accordingly. Reference 5 shows an example for four multipliers.

12.5 OPTIMIZATION WITH REGIONAL CONSTRAINTS

Suppose that there are no functional constraints. The criterion function must, of course, be available, and there will be regional constraints. The criterion function was given in general terms in Equation (1) as

$$C = c(x_1, x_2, \ldots, x_n) \tag{1}$$

and the regional constraints were given in general terms in Equation (3) in the form of inequalities, such as

OPTIMUM DESIGN

$$R_1 < r_1(x_1, x_2, \ldots, x_n) < R_1'$$
$$\ldots\ldots\ldots\ldots\ldots\ldots\ldots\ldots\ldots\ldots\ldots\ldots \quad (3)$$
$$R_p < r_p(x_1, x_2, \ldots, x_n) < R_p'$$

In the simplest form, the criterion function and the constraints are linear equations or inequalities in the variables $x_1, x_2, \ldots, x_n$. The regional constraints are expressed in terms of only one type of inequality, that is either "[equal to or] greater than" or "[equal to or] less than." All of the variables must be nonnegative. If these conditions are met, the problem is referred to as a *linear programming problem*.

The term "programming" does not refer to computer programming but to scheduling and planning. It is most likely to be of interest to industrial, chemical, and civil engineers. It was originally developed in the late 1940's, long before commercial computers were in widespread use. As one might anticipate, with the advent of commercial computers, programs were developed to solve linear programming problems, and it is now possible to solve problems having several thousand independent variables and constraint equations.

As an example of the technique, suppose that

$$x_1 \geq 0 \quad \text{and} \quad x_2 \geq 0$$

Regional constraints are given by

$$0 \leq 3x_1 + 5x_2 \leq 15$$

$$0 \leq 5x_1 + 2x_2 \leq 10$$

The maximum value of the criterion function C is to be found, where

$$C = 4x_1 + 3x_2$$

The problem can be solved graphically, as shown in Figure 12.6, if there are only two variables. The regional constraints may be written as functions of x_1 and x_2. Being linear in these two variables, the plots are straight lines which intersect the axes at the limits shown by the regional constraint inequalities, thus insuring the nonnegative character of the variables. Any set of values of the variables within the shaded area shown in the figure will satisfy the regional constraints.

The criterion function is also shown plotted in the figure for C = 0. It is obvious that for C > 0, the line representing the criterion function will move upward and to the right, parallel to the line shown, falling first into the shaded

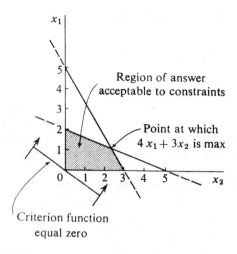

Figure 12.6 Linear programming problem with a single solution.

area and, if C is increased too far, outside the shaded area. Hence it is clear that the point at which the criterion function line leaves the shaded area is the upper right-hand point of the shaded trapezium. This is the extremum sought.

If one of the constraint lines has the same slope as the criterion function, then for a given value of C the two lines will coincide. This simply means that an arbitrary choice can be made for the values of the variables along the line. In general, however, the solution of a linear programming problem is always found at a corner of the region described by the constraints. That is, a unique answer is usually obtained, and only a relatively few possible points need to be investigated to find the extremum.

As an example, consider the problem faced by a manufacturer of small solid-fuel rocket motors. The company makes two types; motor A produces a profit of $3.00 per motor, and motor B's profit is $4.00 per motor. (These are not realistic numbers for the profit, but the objective here is to demonstrate the principle.) A total processing time of 80 hours per week is available to produce both motors. Motor A takes an average of 4 hours per motor, but motor B takes only 2 hours. Motor A contains a less hazardous material than B, so that a preparation time of 2 hours per motor is required for A, but B has a preparation time of 5 hours per motor. The preparation time available each week is 120 hours, and may be divided by the production manager between the two motors. How many motors of each type should be produced each week in order to maximize the profit?

The criterion function for this problem is

$$P = 3A + 4B \tag{39}$$

OPTIMUM DESIGN

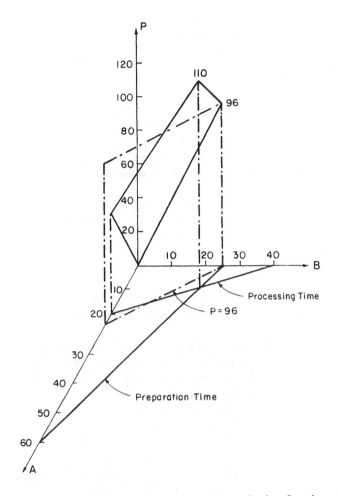

Figure 12.7 Geometric representation of the criterion function and the regional constraints for maximizing rocket motor profit. (With permission from Pike [6].)

The regional constraint for the processing time is

$$4A + 2B \leq 80 \tag{40}$$

and for the preparation time is

$$2A + 5B \leq 120 \tag{41}$$

In addition, both A and B are equal to or greater than zero.

This problem may be solved in the same manner as has already been shown, leading to the decision to produce 10 A motors and 20 B motors, creating a profit of $110.00, rather than the $96.00 if only B motors were produced (24 units) or $60.00 if only A motors were made (20 units). A geometric representation of the solution is shown in Figure 12.7.

This figure shows, in addition to the boundaries set by the regional constraints, an inclined plane (which is seen from below and to the right) for the criterion function. This plane leads to the results given above, and also illustrates that smaller profits will be generated if one is forced away from the optimum by, for example, a reduction in the volume of sales of one motor or the other.

If more than two variables are involved, the graphical method is of little use. A method known as the *simplex method* is available to solve problems in any number of variables. A comprehensive discussion of this method appears in Sections 6.4 to 6.6 of Cooper [1] and in Sections 6.3 to 6.5 of Siddall [5].

12.6 OPTIMIZATION WITH FUNCTIONAL AND REGIONAL CONSTRAINTS—LAGRANGE MULTIPLIERS EXTENDED

Optimization problems with functional constraints were discussed in Section 12.4, and the Lagrange method was introduced as one method of optimization under those conditions. The method can be extended to include problems having regional constraints by the introduction of *slack variables* in order to change the regional constraints into pseudofunctional constraints. Slack variables are additional unknowns that take on any values necessary to satisfy the equations to which they are assigned. They are always introduced as squared quantities because partial derivatives of the slack variables are taken with respect to each slack variable, as will be seen below. Using the square insures that the slack variable will not disappear in the differentiation process. Moreover, using a squared quantity ensures that the slack effect is always positive even though the value of the slack variable itself may be negative.

In the development of the Lagrange method, only two variables, x and y, were used. In the following, only Equation (22) is used, but the variable x in that equation is replaced by x_i so that all variables from 1 to n enter into the solution. The equation is then expanded to include the regional constraints with the criterion function and the functional constraints.

The procedure can be described as follows. We begin with the basic equations of Section 12.1, with the criterion function

$$C = c(x_1, x_2, \ldots, x_i, \ldots, x_n) \tag{1}$$

OPTIMUM DESIGN

the functional constraints of Equation (2), which may be written in general terms as

$$F_j = f_j(x_1, x_2, \ldots, x_i, \ldots, x_n) \tag{2}$$

where $\quad j = 1, 2, \ldots, m \quad$ and $\quad m < n$

and the regional constraints, again written in general terms as

$$R_p = r_r(x_1, x_2, \ldots, x_i, \ldots, x_n) \leq R_p' \tag{3}$$

where $\quad p = 1, 2, 3, \ldots, k, \ldots, q \quad$ and q can be any number.

By introduction of slack variables φ_p, the regional constraints are converted into equalities. That is,

$$R_p - R_p' + \varphi_p^2 = r_p(x_1, x_2, \ldots, x_i, \ldots, x_n) - R_p' + \phi_p^2 = 0 \tag{42}$$

Equation (22) is now expanded to yield

$$\frac{\partial C}{\partial x_i} + \sum_{j=1}^{m} \lambda_j \frac{\partial F_j}{\partial x_i} + \sum_{p=1}^{q} \Lambda_p \frac{\partial}{\partial x_i}(R_p - R_p' + \varphi_p^2) = 0 \tag{43}$$

and Equation (23) is expanded to yield

$$\frac{\partial C}{\partial \varphi_k} + \sum_{j=1}^{m} \lambda_j \frac{\partial F_j}{\partial \varphi_k} + \sum_{p=1}^{q} \Lambda_p \frac{\partial}{\partial \varphi_k}(R_p - R_p' + \varphi_p^2) = 0 \tag{44}$$

the slack variables φ_k replacing y in the earlier formulation.

Examination of Equation (44) reveals that it is far simpler than the imposing form appears to be initially. The criterion function and all of the functional constraints are independent of the slack variables; hence the first two terms in (44) are zero. In the third term, all of the R_p's are independent of the slack variables as well, and all of the R_p' values are constants. Hence the third term is the only one that will appear at all, and it will appear only when k = p, yielding q equations of the form

$$\Lambda_k \varphi_k = 0 \tag{45}$$

The multiplier 2 that appears on differentiation of the squared term was suppressed because it is simply a scalar multiplier.

It is important to keep in mind that restrictions on the values of the variables, such as $x_i \geq 0$, are also constraints. Because the optimization procedure rarely leads to an extremum at a boundary and yet the criterion function may yield a superior value at a boundary, it is always important to test the criterion function at the boundaries. Problem 15 in the Practice Projects is an example of such a situation.

To illustrate the use of the method, consider an optimization problem for which

$$C = x_1 + 1.5x_2 \tag{46}$$

There is only one functional constraint, which is given by

$$F = (x_1 - 1)^2 + (x_2 - 1)^2 = x_1^2 - 2x_1 + x_2^2 - 2x_2 + 2 = 1 \tag{47}$$

which may be rewritten as

$$x_1^2 - 2x_1 + x_2^2 - 2x_2 + 1 = 0 \tag{48}$$

There are two regional constraints,

$$R_1 = x_1 + 0.8x_2 \leq 3.5 \tag{49}$$

and

$$R_2 = x_1 x_2 \leq 2.25 \tag{50}$$

Both variables, x_1 and x_2, are greater than zero, and the maximum value of C is to be found.

We note, before proceeding farther, that we are no longer dealing with linear programming. The functional constraint is quadratic in both variables, and the second of the regional constraints depends on the product of the two variables. There are general remarks on nonlinear programming at the end of this section.

The regional constraints are changed to equalities by introducing slack variables, giving

$$x_1 + 0.8x_2 - 3.5 + \varphi_1^2 = 0 \tag{51}$$

$$x_1 x_2 - 2.25 + \varphi_2^2 = 0 \tag{52}$$

OPTIMUM DESIGN

Substituting Equations (46), (48), (51), and (52) into Equation (43) yields a pair of equations,

$$1 + \lambda(2x_1 - 2) + \Lambda_1 + \Lambda_2 x_2 = 0 \tag{53}$$

and

$$1.5 + \lambda(2x_2 - 2) + 0.8\Lambda_1 + \Lambda_2 x_1 = 0 \tag{54}$$

As shown earlier, Equation (44) reduces to Equation (45), so that

$$\Lambda_1 \varphi_1 = 0 \tag{55}$$

$$\Lambda_2 \varphi_2 = 0 \tag{56}$$

Equations (48) and (51) through (56) form a set of seven equations in the seven unknowns $x_1, x_2, \lambda, \Lambda_1, \Lambda_2, \varphi_1$, and φ_2. The solution to these equations is found by examining all possible combinations of values of the Λ's and the φ's of Equations (55) and (56). There are four possible combinations.

Solution 1:
Assume that $\Lambda_1 = \Lambda_2 = 0$. From Equations (53) and (54),

$$\lambda = \frac{-1}{2(x_1 - 1)} = \frac{-1.5}{2(x_2 - 1)}$$

leading to

$$x_1 = \frac{x_2 + 0.5}{1.5}$$

On substitution into Equation (48), we obtain the quadratic form

$$3.25x_2^2 - 6.5x_2 + 1 = 0$$

which yields two values for x_2, 1.8321 and 0.1679. The corresponding values for x_1 are 1.5547 and 0.4453. [Important results for all four solutions appear below in tabular form.]

From Equations (51) and (52), we obtain values for the slack variables. For the first set of values of x_1 and x_2, that is, 1.5547 and 1.8321, we find $\varphi_1^2 = 0.4796$ and $\varphi_2^2 = -0.5984$. The negative value is not permissible, and this avenue of investigation therefore terminates at this point.

However, for the other set of values of the x's, we find

$$\varphi_1^2 = 2.2920 \quad \text{and} \quad \varphi_2^2 = 2.1752$$

These values are acceptable, and we may now evaluate the criterion function.

$$C(x_1 = 0.4453, x_2 = 0.1679) = 0.697$$

Solution 2:

Assume Λ_1 and φ_2^2 are both zero. Then, from Equation (52),

$$x_1 x_2 = 2.25 \quad \text{or} \quad x_1 = 2.25/x_2$$

Substitution into Equation (48) leads to the fourth-order polynomial

$$x_2^4 - 2x_2^3 + x_2^2 - 4.5x_2 + 5.0625 = 0$$

for which the four roots are:

$$x_2 = 1.1298, \text{ for which } x_1 = 1.9915$$
$$x_2 = 1.9915, \text{ for which } x_1 = 1.1298$$
$$x_2 = -0.5607 \pm j1.3913$$

Since the x's must be real, the complex roots are neglected.

Equations (53) and (54) are now used to solve for λ and for Λ_2. For the first pair of x's,

$$\Lambda_2 = -0.7426 \quad \text{and} \quad \lambda = -0.0812$$

For the second pair,

$$\Lambda_2 = -0.4359 \quad \text{and} \quad \lambda = -0.5077$$

The values of λ's and Λ's may be either positive or negative. Hence these are admissible values, and the criterion function may be evaluated, resulting in:

$$C(x_1 = 1.9915, x_2 = 1.1298) = 3.6862$$

and

$$C(x_1 = 1.1298, x_2 = 1.9915) = 4.1171$$

We suspect that the latter value of C is the maximum value of the criterion function, but we pursue the remaining two solutions before attempting to draw any final conclusions.

OPTIMUM DESIGN

Solution 3:
Let $\Lambda_2 = 0$ and $\varphi_1^2 = 0$. Equation (51) may be solved for x_1 and substituted into Equation (48), leading to a quadratic in x_2. The resulting roots are complex, and the solution is therefore not pursued farther.

Solution 4:
Let $\varphi_1^2 = \varphi_2^2 = 0$. Equations (51) and (52) may now be solved simultaneously. However, the values that result for x_1 and for x_2 (and are shown in tabular form below) will not satisfy Equation (48), the functional constraint.

In tabular form, the results of the four solutions are shown below. For each solution, the starting values of the Λ's and the φ^2 values are shown on one line, and the results are shown beginning on the next line.

A graphical representation is shown in Figure 12.8. With no functional constraint, solutions are possible anywhere within the shaded region. Adding the functional constraint F (which is the equation of a circle), restricts the locus of possible solutions to that part of the circle that lies within the shaded area.

As pointed out earlier, the problem we have just solved falls into the category referred to as a general optimization problem, because it has a criterion function, at least one functional constraint, and at least one regional constraint. In addition, it falls into a category referred to as nonlinear programming because at least one of the constraints is not given by a linear equation. Obviously, such problems are considerably more difficult than linear programming problems, and various methods have been devised to make such problems more tractable. The interested reader may wish to consult Chapters 8 and 9 of Reference [1].

	λ	Λ_1	Λ_2	x_1	x_2	φ_1^2	φ_2^2	C
#1		0	0					
				1.5547	1.8321	0.4796	-0.5984	
				0.4453	0.1679	2.2920	2.1752	0.697
#2		0				0		
	-0.0812		-0.7426	1.9915	1.1298			3.6862
	-0.5077		-0.4359	1.1298	1.9915			4.1171
#3			0			0		
				x's are all complex.				
#4						0	0	
				0.6264	3.5920			Equation (48)
				2.8736	0.7830			not satisfied.

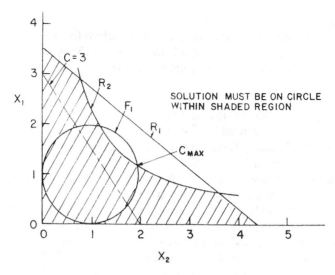

Figure 12.8 Optimization problem with nonlinear constraints.

12.7 SEARCHING

In this section, we will look at the general problem of finding the maximum or minimum value of a function of one or more variables by use of procedures known collectively as search methods. For ease of visualization, the discussion will be restricted to two variables, so that the variables can be displayed in two dimensions, and the magnitude of the function can be shown by contour lines and points at which extrema are reached, as shown in Figure 12.9.

In 1977, one of the authors (RHE) wrote a program for a TI-59 programmable calculator designed to obtain the roots of a polynomial function f(x), whether the roots were real, complex, or imaginary, up to the thirtieth order. The method used was a search process, beginning at the origin, making a first step along the imaginary axis, and then computing directional derivatives that would steer the search toward an extremum (a zero). Provisions were built in to reduce the step size as a zero was approached, and to declare that a root had been found when the magnitude of f(x) was less than an arbitrary value predetermined by the program user. If the root was real, the original function was reduced by the corresponding linear factor, and the process began again. If the root was imaginary or complex, a quadratic factor was formed and the original function reduced by that factor. Obviously, given the machine on which the algorithm ran, the solution time was long, especially if the original polynomial were of high order, but it had good accuracy if the original function was what the mathematician refers to as a well-defined function. In one test of a thirtieth-order polynomial for which the roots were known to twelve places after

OPTIMUM DESIGN

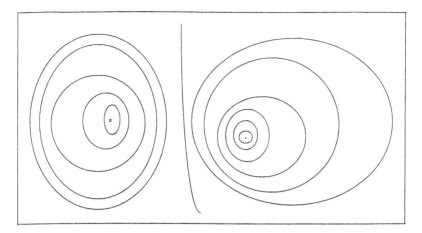

Figure 12.9 Graphical representation of a function having two different maxima.

the decimal for both the real and the imaginary parts, the largest difference in the roots found by the program from the actual roots was three in the last place, and for most there were no differences at all.

The search process used, despite the possible high order of the polynomial, was in many respects much easier than search processes used in optimization problems. For one thing, all of the extrema had the same value, zero, whereas in optimization problems there may be more than one extremum, and they frequently have different values. In Figure 12.9, the open contour is assumed to have the minimum value of the function, and the increment from one contour to another is the same everywhere. Hence the right-hand "peak" has a larger maximum value than the left.

A second difference between the high-order polynomial solution and the world of optimization problems is that there were no constraints, such as regional constraints. In Figure 12.10, a "fence" is shown enclosing part of the right-hand peak of Figure 12.9. If one is searching for the largest value of the function, a search beginning at point C will end at D, whereas one beginning at A will end in the vicinity of B, yielding a larger value of the function.

A third difference is that in the high-order polynomial problem, the function could be expressed in a fairly simple form. In many optimization problems, the functional form may involve e (the base of natural logarithms) raised to various powers of one of the variables, trigonometric or hyperbolic functions, integrals, or various other functions. It is even possible that the form of the function to be optimized may not be readily reducible to a mathematical expression at all.

In the brief descriptions of various methods that follow, we will assume that the extremum sought is a maximum. If a minimum is sought, the process is

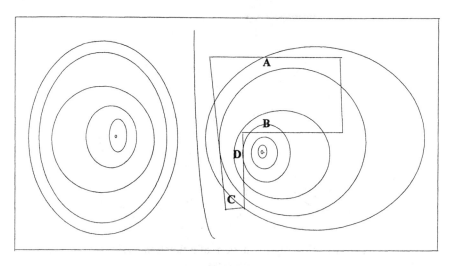

Figure 12.10 Function having two maxima, with the addition of regional constraints (that is, a "fence").

simply reversed, words like "greater" being replaced by "smaller," and "ascent" by "descent." It will also be assumed that the starting point for the search process is selected using as much *a priori* knowledge as the designer can bring to bear on the optimization problem. That is, the designer will make an initial "guess" as to where the solution lies, but does not do so blindly. In the case of inductor design, for example, one would not select a #40 AWG conductor as an initial value for wire size if the current to be carried is 10 amperes because the resulting power dissipation would melt the conductor.

A number of different methods may be used for the search process, and most of these have variations. One method, discussed in Reference 1, is known as a multivariate grid search, and is shown (again for two variables only) in Figure 12.11. In this method, a grid of predetermined size is placed over the selected region. A particular node of the grid is selected as a starting point, such as the one marked "A." The function is evaluated at those points of the grid that surround the selected node, and the point at which the greatest value is found becomes a new starting point. In this figure, the greatest value will be found at the node marked "3." For the new point, at least four of the values of the function will already be known and the remaining new ones must be calculated. The process is repeated until the central node yields the greatest value of f(x). This may be sufficiently close to the actual maximum that no further work is needed. However, if it is desired to improve the accuracy with which the location of the maximum is to be found, the grid size can be reduced, and the process repeated beginning with the node located in the first phase of the search process. In Figure 12.11, the search began at point A and it would lead rather directly to the actual maximum, or at least to a point very close to it.

OPTIMUM DESIGN

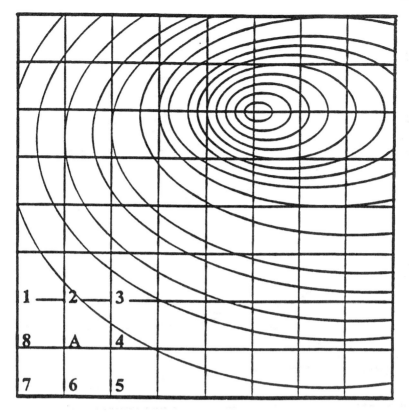

Figure 12.11 Two-dimensional grid search.

In the search shown in the figure, the solution would be reached rather rapidly. However, this process quickly becomes inefficient as the number of variables is increased. At each iteration of the calculation, we need $3^n - 1$ values of the function, where n is the dimensionality of the function. For the two-dimensional function illustrated, there are only eight values needed. If we are dealing with a three-dimensional function (n = 3), we need 26 values, and for a four-dimensional function we need 80. Although the procedure is tractable for two dimensions and possibly even for three, it is readily seen that it is highly inefficient for higher order functions.

Another method that has been used is referred to as a pattern search, originally proposed by Hooke and Jeeves [7]. This is known as a direct search method, and is illustrated in Figure 12.12. One begins at a preselected point, and makes a number of moves (called exploratory moves) using a step of small size in directions parallel to the two axes. From the results of these exploratory moves, a "pattern" is established that indicates in which direction a larger step (referred to as a "pattern" step) can be taken with a high probability of success in

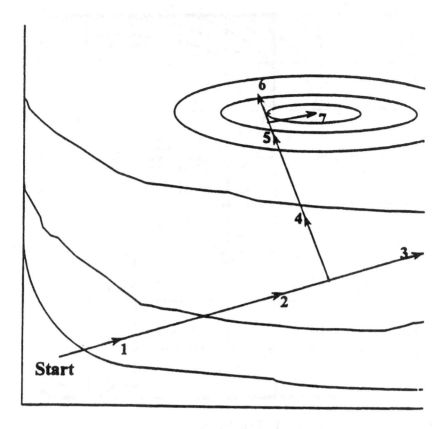

Figure 12.12 A two-dimensional pattern search.

reaching a larger value of the function. If this larger step does in fact result in a larger value of the function, the resulting point in the plane is used as a new starting point. The search pattern shown in Figure 12.12 illustrates a series of such steps. Step 3 is presumed to lead to a smaller value of the function (we do not know how it behaves off the graph as displayed). Since we do know that the direction of this step is correct, it is obvious that the size of the pattern step is too large. The pattern step is therefore reduced in size until f(x) shows an increase, and the process is repeated until the extremum is found. Note that step 6 also failed to yield an increase in the value of the function, and it was necessary to retreat.

The last method to be described is known variously as the method of steepest ascent or the gradient method. The basic concept is quite old, having been enunciated by Cauchy in 1847. In Figure 12.13, which exhibits the same function as that in Figure 12.12, we follow the direction of maximum slope until the extremum is reached. Because we take steps of finite size, the method as

OPTIMUM DESIGN

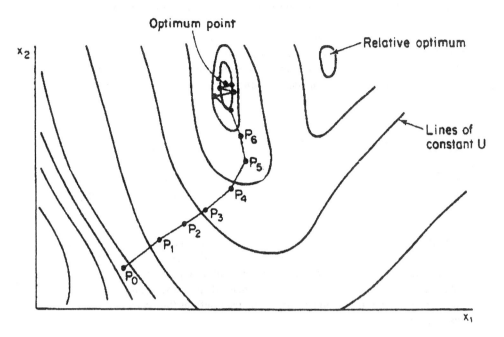

Figure 12.13 Optimum point found using the gradient method.

applied in practice does not result in a smooth curve from the starting point to the optimum, but produces a series of straight line steps. At its beginning, each step points in the direction of steepest ascent, but as one proceeds along the finite length of each step the direction along which one is proceeding will in general diverge from the actual direction of steepest ascent. The path followed, therefore, is only an approximation of the curve that should theoretically be followed, and it will certainly be different from that followed as the result of other search methods. It was a variation of this method that was used in the case of the root location polynomial problem described above. Detailed discussion of these methods will be found in References 1 and 5, as well as in 8.

Which is the best method to use? There is no definitive answer to this question. It may be necessary to try more than one method, because a method that works well on one class of problems will work poorly on another. Step size may have to be adjusted. Too large a step may result in overshooting an extremum, while too small a step size will result in long run times to reach a solution.

Suppose the criterion function has two (or more) extrema? Fortunately, in most applications in engineering, the function is unimodal. If it is suspected that it is not, it may be prudent to run the search three or four times, with searches after the first beginning from starting points well removed from previous starting points.

12.8 OPTIMIZATION USING COMPREHENSIVE CAE PROGRAMS

In Section 11.6, we looked briefly at one comprehensive CAE system, the I-DEAS™ system marketed by Structural Research Dynamics Corporation (SDRC™), and listed a number of programs available in that system. One that was omitted from that list was their I-DEAS Optimization software. This program allows the designer to use analysis results to directly drive design improvements, using finite-element analysis to simulate structural performance. "Performance" as used here may be any one of a number of measures related to the part in question, such as mass, stress, displacement under load, natural frequency of vibration, and so forth. The user defines the necessary information, as follows:

1. Design parameters: These are the physical features to be modified, such as hole dimensions, thickness of a part, and so forth. Limits on allowable modifications can be specified.
2. Design limits: These are the physical quantities that measure performance, such as displacement under load or natural frequency.
3. Design goals: The goals define the objective of the optimization. For example, the objective may be to minimize the mass, while still staying inside the design limits set in (2), or it may be to limit stress or displacement under load.

Once these entities have been specified, the program will modify the design parameters in order to reach the extremum of the specified design goal while keeping inside the design limits.

For example, refer to the final design of the fixed portion of the door hinge in Figure 11.10 and the stress and deflection patterns of Figure 11.12. Suppose that the stress concentrations that appear in Figure 11.12 are beyond the allowable design limits. Will the stress be reduced sufficiently to meet the design limits if the inside corner of the front flat surface (that closest to and facing the viewer) in Figure 11.10 is filleted?

Suppose that the stresses shown in Figure 11.12 are well within design limits, and the deflection under load is also within limits. The thickness of the stock from which the part is formed can be reduced, but by how much before one

OPTIMUM DESIGN

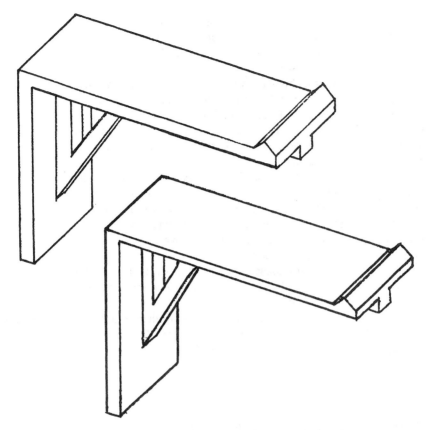

Figure 12.14 Shelf brackets. Upper: Original design.
Lower: Optimized design. (Courtesy of SDRC.)

encounters either a deflection or a stress limit? The I-DEAS Optimization program gives definitive answers to questions such as these.

Figure 12.14 shows a bracket, intended to be used to support shelves carrying heavy loads. The upper part of the figure shows the bracket as originally designed. The lower part of the figure shows the design as optimized by the I-DEAS Optimization software so as to minimize the mass while still keeping stresses under load below a specified upper limit. Note that there has been a considerable reduction of the thickness of the horizontal part of the bracket, and that the shape of the triangular web between the horizontal and vertical parts has been modified. Both changes have reduced the mass of the bracket.

This discussion is intended only to illustrate a few of the capabilities of the I-DEAS Optimization program and to answer only a few of the questions that can be asked about the program. The program has many other capabilities of value to the designer. For example, design limits can be applied to the material strength, specified as a maximum Von Mises stress, and multiple material strengths can be used in the same analysis for different loading conditions or for different regions of the structure. Some loading conditions, such as high cycle fatigue loads, will probably have a lower allowable stress than other loads, such as a static load. Different allowable displacements may be specified at different locations for a variety of loading conditions. Dynamic vibration performance can be modified by changing natural frequency values, with the objective of keeping natural frequencies well away from excitation frequencies, such as those dependent on engine speed.

Another capability is that of sensitivity analysis, that is, determination of the degree to which a change in a design parameter influences a change in the structural performance. Parameters having a high sensitivity are the ones the designer is most interested in because these generally lead to the most cost-effective design changes. The parameters for which sensitivity analysis may be done include thickness of shell elements, spring constants, beam parameters (including cross section dimensions and moments of inertia), and the material properties of isotropic, orthotropic, and anisotropic materials, such as Young's modulus and Poisson's ratio.

As noted at the end of Section 11.6, further information is available directly from SDRC at "http://www.sdrc.com/pub/catalog/ideas" or at the address given in Reference 11 of Chapter 11.

12.9 SUMMARY

There are three major points with respect to optimization that the designer should remember. The first of these is that an essential part of the activity of a skillful designer is the selection of a set of specifications, such as dimensions and material characteristics, that result in the best product possible for the particular design configuration chosen. That is, the designer is searching for the optimum design. We have looked at a number of examples in this and previous chapters, such as the spring of Chapter 10 and, in this chapter, the generator slot dimensions, the inductance coil optimum shape, the open-top container, and the cylindrical tank. These have all been design problems. During each new design or even redesign project, the designer should be alert to the opportunity to optimize the design.

Secondly, there is usually more than one way to find the optimum. The dimensions of the cylindrical tank were found in two different ways in this chapter, and the optimum ratio of dimensions of the inductor coil could also have been found in more than one way.

Finally, keep in mind that the available optimization procedures have certainly not been exhausted by the coverage in this chapter. Our intent has been to present the philosophy of optimization, with enough examples to demonstrate

OPTIMUM DESIGN

the utility of the concept. References 1 and 5-8 (or any of a number of other books on the subject) may be consulted if you are confronted by problems that cannot be handled by the methods discussed in this chapter.

REFERENCES

1. Cooper, Leon, and David Steinberg. *Introduction to Methods of Optimization,* W. B. Saunders, Philadelphia, 1970.
2. Engelmann, Richard, and William Middendorf. *Handbook of Electric Motors,* Marcel Dekker, New York, 1995.
3. Pender, H., and K. McIlwain. *Electrical Engineer's Handbook,* Third Edition, John Wiley & Sons, New York, 1941, pp. 4-14.
4. Brooks, H. B. Design of standards of inductance and the proposed use of model reactors in the design of air-core and iron-core reactors, *National Bureau of Standards Research Journal,* 1931; August:289-328.
5. Siddall, James N. *Analytical Decision-Making in Engineering Design,* Prentice-Hall, Englewood Cliffs NJ, 1972.
6. Pike, Ralph W. *Optimization for Engineering Systems.* Van Nostrand Reinhold, New York, 1986.
7. Hooke, R., and T. A. Jeeves. Direct search solution of numerical and statistical problems, *Journal of the Association of Computing Machinery,* 1961;8.
8. Siddall, J. N. *Optimal Engineering Design,* Marcel Dekker, New York, 1982.

REVIEW AND DISCUSSION

1. What is the important prerequisite to obtaining an optimum design?
2. How does one develop a criterion function that involves a number of desirable characteristics, at least some of which are measured in different units?
3. Does piecemeal optimization of parts result in optimum design of the product?
4. Can you give an example that confirms your answer to question 3? In searching for examples, broaden your base by considering products you have had considerable experience with, products used in the home, or products described in advertising.
5. Name and describe the three types of functions that comprise the general optimization problem, including limitations on numbers of variables, numbers of functions, and any other limitations of significance.
6. Does the use of Lagrangian multipliers have an advantage as compared to variable elimination using the function constraints? What is the advantage?

7. What is a regional constraint? Give an engineering example. Do regional constraints (possibly under another name) exist in other disciplines, such as economics? Can you give a simple example?
8. Where is the solution to a linear-programming problem almost always found?
9. What are slack variables, and where are they used?

PRACTICE PROJECTS

1. Many power-handling devices have one set of losses that are independent of the power output and another set of losses that depend on the square of the output. Designate the output by the symbol P, the first set of losses by the symbol K_1, and the second set of losses by K_2P^2. The efficiency of any device is given by the equation $\eta = P/(P + K_1 + K_2P^2)$. That is, the efficiency is the output divided by the output plus losses, which is the input. As output is varied, the efficiency varies. Under what conditions will maximum efficiency be obtained? Examples of devices for which there are losses that are fixed plus other losses that vary as the square of the output are transformers, electric motors, and internal combustion engines (at least to a first approximation).

2. In Figure 12.15, assume that R_G and X_G are fixed, but that R_L and X_L are adjustable. R_L may only be positive, but X_L may be either positive (if chosen as an inductance) or negative (if a capacitance is used). The power to R_L is given by the equation

$$P = \frac{E^2 R_L}{(R_L + R_G)^2 + (X_L + X_G)^2}$$

Determine the value of R_L and X_L for maximum power in R_L.

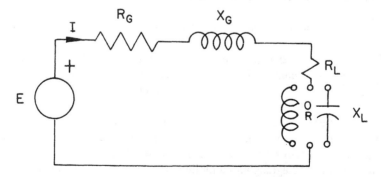

Figure 12.15 Circuit to be investigated for maximum power transfer.

OPTIMUM DESIGN

3. Repeat problem 2 if R_L and X_L are to be fixed, and R_G and X_G are adjusted so as to maximize the power dissipated in R_L.
4. Find the conditions required to maximize the power in R_L in the circuit of Figure 12.15 if only R_L and X_L are adjustable, but there is the added constraint that $R_L = X_L/2$, and X_L is always positive.
5. Four square segments can be cut from a rectangular panel to form an open box as shown in Figure 12.16. What should the relationship be among a, b, and c to result in minimum material (including the off-fall) for a given volume of box?

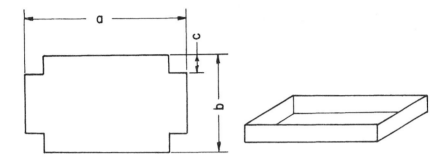

Figure 12.16 Optimum design of an open box.

6. Consider the lidless box problem used as an example in Section 12.3.1, and begin with a square base, that is, a = b. What is the optimum value of h from the criterion function written as a function of h? Is this solution easier than the one in Section 12.3.1, in which a rectangular base was assumed?
7. A typical shipping carton is made with top and bottom formed by flaps extending from each of the four sides as shown in Figure 12.17. The flaps are half as long as the shorter side, so that two opposite flaps completely close the top or bottom of the carton and form double thicknesses of the material. After being folded, the flaps are held by glue or tape. Suppose the

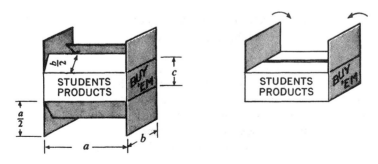

Figure 12.17 Optimum shipping carton.

product you make is quite small so that it can accommodate to any shape of carton without wasting space, each carton being capable of holding a few hundred. What should the relationship be among a, b, and c to result in minimum material for a given volume? Solve using Lagrangian multipliers.

8. Many design decisions are driven by a compromise between functions that vary as shown in Figure 12.18. For example, if we use x to designate reliability, the cost of repair service for a television set varies as k_1/x, whereas the cost of improved reliability to avoid customer service problems varies as $k_2 x^2$. Reliability is defined here as the probability of 5 years of use without a major repair. Suppose that $k_1 = 7.0$ and $k_2 = 20$. Find the optimum reliability that results in minimum total cost (that is, original cost of the set plus repair charges) to the customer.

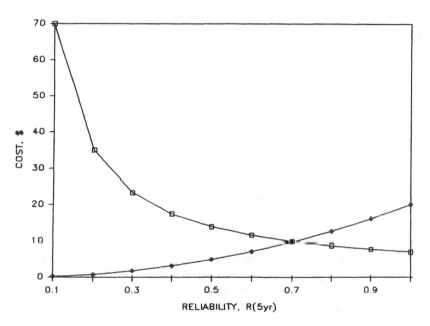

Figure 12.18 Average cost of repair (□) and incremental cost of reliability (◊) versus reliability.

9. In production of a product, set-up costs are minimized if the production runs produce as many units as possible. On the other hand, storage costs and the foregone earnings on funds not used for production are minimized by making many runs, each with a small number of units. Assume that the number of units in stock varies with time as shown in Figure 12.19. Determine the optimum lot size in terms of the following symbols:

OPTIMUM DESIGN

C_s = cost of setting up for a production run in dollars
C_i = cost of storing one widget per unit time
D = demand rate, the number of widgets sold per unit time
R = total number of widgets sold during the total time considered, so that R = tD
t = total time considered
N = lot size, the number of widgets produced per run
N_o = optimum lot size
t_s = time between runs
t_{so} = optimum time between runs
K = total cost during time t
K_s = total cost of set-ups during time t
K_i = total inventory (storage) cost during time t
n_s = number of set-ups during time t
n_i = average number of widgets in inventory during time t

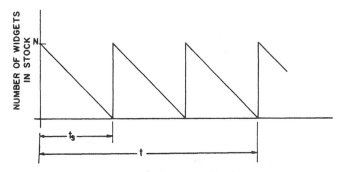

Figure 12.19 Optimum production runs.

10. The load impedance of the network shown in Figure 12.20 on the next page is connected to the source by an ideal transformer. The only effect of an ideal transformer is to multiply the load impedance (R_L and X_L) by the turns ratio (designated by the symbol a) squared, so that the network model may be replaced for analysis purposes by the equivalent network in the right half of the figure. What value of a^2 should be chosen to give the maximum power transfer to the load? $R \neq R_L$ and $X \neq X_L$. Problem 2 shows the form for the power equation.

11. A 7.5-hp, 900-rpm (nominal), three-phase wound rotor induction motor has a per phase equivalent circuit as shown in Figure 12.21. The mechanical output power is equal to three times (because of the three phases) the electrical power dissipated in the resistor labeled $R = R_2(1-s)/s$ in the figure. The symbol s stands for the slip, and $s = (900 - N)/900$, where N is the actual motor speed in rpm. (The motor will run at almost 900 rpm under no-load conditions, and will slow as load is applied.) Determine the slip and the actual speed at which maximum power output occurs. Again, the power

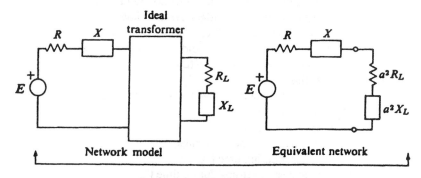

Figure 12.20 Maximizing power transfer using a transformer.

equation of Problem 2 applies. The parameters of the equivalent circuit have the following values:

$R_1 = 0.223\Omega$ $R_2 = 0.400\Omega$
$X_1 = 0.338\Omega$ $X_2 = 0.338\Omega$
$R_{H+E} = 200\Omega$ $X_m = 13.0\Omega$

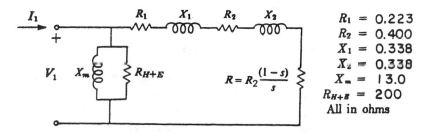

Figure 12.21 Induction motor equivalent circuit.

12. The shaft torque produced by an induction motor is proportional to the electrical power dissipated in the sum of R_2 and R in Figure 12.21, which is equal to R_2/s. Determine the slip at which maximum torque occurs in the motor of Problem 11. What is the actual speed?
13. A printed circuitboard assembly department processes two types of boards, which we shall call X and Y. Each board requires three operations: Component insertion, soldering, and inspection. The hours required for each board in each operation and the maximum work hours available per day are:

… OPTIMUM DESIGN 451

	X	Y	Person-hr/day
Insertion	8	4	80
Soldering	3	4	60
Inspection	1	3	24

   a) Determine the number of X boards and the number of Y boards that should be produced to maximize output.

   b) Assume that each X board contributes $120 profit and each Y board $60 profit to the company. Determine the number of each board that should be made to maximize profit.

14. For quite a few years, one gasoline marketer sold various grades of gasoline using only two basic components and blending the two at the pump. Call the basic components "regular" and "octane." There were five blends with the following proportions of the components:

Grade #	Regular (%)	Octane (%)	Price of grade (Cents/liter)
1	100	0	30
2	90	10	33
3	80	20	36
4	70	30	38
5	60	40	40

Regular costs the dealer 25 cents/liter; octane costs 35 cents/liter. The selling price is shown in the table. On a certain weekend the dealer has 64,000 liters of regular and 16,000 liters of octane. He wishes to maximize his profit. Set up the linear-programming problem to determine which of the five grades he should "push."

15. Find the maximum value of C under the following conditions:

$$C = x_1 + 1.5x_2$$
$$R_1 = x_1 + 0.8x_2 \le 3.5$$
$$R_2 = x_1 x_2 \le 2.25$$

16. Find the minimum value of C under the following conditions:

$$C = x_1 + 0.5x_2 + 1.5x_3$$
$$R = \frac{1}{x_1} + \frac{1}{x_2} + \frac{1}{x_3} \le 1$$

13

RELIABILITY

> *But the Deacon swore (as Deacons do,*
> *With an "I dew vum," or an "I tell* yeou"*)*
> *He would build one shay to beat the taown*
> *'N' the keounty 'n' all the kentry raoun';*
> *It should be so built that it* could n' *break daown.*
>
> Oliver Wendell Holmes, *The Deacon's Masterpiece*

Reliability can be defined in an engineering sense as the probability that a component, device, or system will perform without failure under a given set of conditions for a given period. If a system is in continuous or nearly continuous operation, the period is usually given in terms of time. For example, the motor driving the blower of a residential furnace/air conditioner may be set to run continuously, regardless of any signals from the thermostat to turn on the burner to supply heat or to turn on the air conditioner to supply cooling. In this case, the period will obviously be given in terms of time. On the other hand, liquid level controls used in industrial operations frequently incorporate reed switches that are used to open or close valves. In this situation, the period can be stated in terms of number of operations of the switches.

An alternate definition of reliability is the frequency with which failures occur as a function of time. The frequency at which failures occur is also referred to as the mean failure rate, and may be measured in terms of failures per operating hour. Intuition tells us that the two definitions are related, and indeed they can be related mathematically, as will be seen below.

Regardless of how reliability is defined, it is certain that a reliability of 100% (if the probability definition is used) or of zero failures per operating hour (if the frequency of failure definition is used) can never be achieved. Everything eventually wears out, fails because a component incorporated into a product was substandard, or as a result of some random combination of adverse effects.

RELIABILITY

The subject of reliability is of great importance because the designer must accept the primary responsibility for the reliability of a product. Reliability must be designed into the product; it cannot be added later. The discussion of reliability presented in this chapter is in sufficient depth that the engineer will be able to estimate or perform tests to determine the reliability of simple products or systems. It also includes guidelines by which the reliability of a product can be improved while in the design phase. For those readers interested in pursuing the subject in greater depth, References 1, 2, and 3 will be of value.

13.1 THE NATURE OF FAILURES

In practice, the units of a population of products are usually placed in service over an extended period of time. If the length of time is known for each unit from placement in service until occurrence of failure, one can track failures of the entire population as a function of time from the date of placement into service. This is the equivalent of placing all units in service simultaneously. If data are collected on the number of failures during a given time interval and plotted against time, a histogram such as that shown in Figure 13.1 will result. In this figure the total population was 60 units, and none survived more

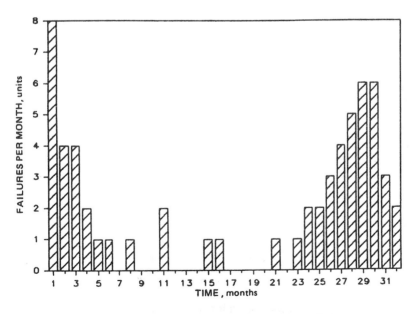

Figure 13.1 Typical failure history of a product.

than 32 months. The specific numbers are irrelevant; the general shape of the histogram is important. We see that early in the failure history there were a fairly large number of units that failed, but that the number of failures per month decreased rather rapidly, reaching a rather small value (zero to at most two per month) after 4 months. The failure rate then stayed fairly low until the twenty-fourth month, at which time 33 units still survived. At this point, however, the failures began to increase in number so that when the thirty-first month was reached almost all of the remaining 33 units had failed.

There are three distinct regions that can be noted in this histogram. In the first of these, the failures are referred to as "early failures" or "break-in failures." The failures in the middle region are referred to as "chance failures," and those in the third region are referred to as "wearout failures," although some chance failures are still occurring in that region. Each of these regions will be discussed in more detail in succeeding sections. For the moment, however, consider a normalization process that can be applied to the data displayed in the histogram.

The rate at which failures occur is directly proportional to the number of failures in a given time period and inversely proportional to the length of the time period. The number of failures is also directly proportional to the number of units in service at the time the failures occur. In order to have a measure of

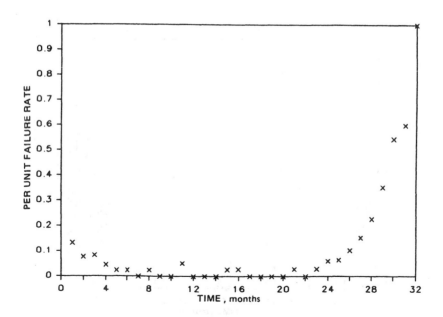

Figure 13.2 Per-unit failure rate versus time.

RELIABILITY

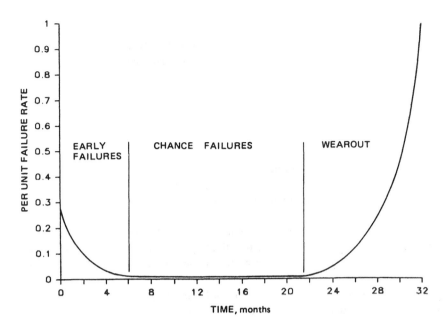

Figure 13.3 The idealized failure rate curve.

failure rate that is dependent only on the size of the present population, the number of failures in the given time period is divided by the number of surviving units. That is, we may define the failure rate by the equation

$$\lambda = \frac{1}{N_s} \cdot \frac{\Delta N_f}{\Delta t} \tag{1}$$

where λ = per-unit failure rate
  N_s = number of surviving units
  ΔN_f = the number of units that failed during time Δt

Application of this equation to the data of Figure 13.1 results in the data points of Figure 13.2, which may be smoothed to yield the idealized curve shown in Figure 13.3. As mentioned above, each part of the failure rate curve is discussed below, but it must be pointed out that the "chance failure" portion of the curve is the most important because it represents the useful life period of the product. To make discussion of the various parts of the curve more meaningful, the necessary mathematical relationships will first be developed.

13.2 BASIC RELATIONSHIPS

Most of the basic concepts of reliability theory follow from a rather simple mathematical investigation that begins with the definition of reliability as a probability. That is,

$$R(t) = \lim_{N_p \to \infty} \frac{N_s(t)}{N_p} \qquad (2)$$

where
$R(t)$ = the reliability for any time t
$N_s(t)$ = the number of survivors at time t
N_p = the total number of units in the population

Although it is not possible to have an infinitely large population of units, probability concepts are used as if this limiting condition were met. However, because the limiting condition is not met, reliability numbers applied to actual products are, at best, approximations.

A certain number of units $N_f(t)$ will fail between t = 0 and any time t thereafter. Since units must either survive or fail,

$$N_p = N_s(t) + N_f(t) \qquad (3)$$

Substituting Equation (3) back into Equation (2),

$$R(t) = 1 - \frac{N_f(t)}{N_p} \qquad (4)$$

which on differentiation yields

$$\frac{dR(t)}{dt} = \frac{-1}{N_p} \frac{dN_f(t)}{dt} \qquad (5)$$

If Equation (1) is placed in differential form (rather than the incremental form above),

$$\lambda = \frac{1}{N_s(t)} \frac{dN_f(t)}{dt} \qquad (1a)$$

On substitution of Equations (5) and (2) into (1a), we obtain

RELIABILITY

$$\lambda = \frac{-1}{R(t)} \frac{dR(t)}{dt} \tag{6}$$

an expression that involves only the reliability, the per-unit failure rate, and time. Rearranging,

$$\lambda dt = -\frac{dR(t)}{R(t)}$$

On integrating and noting that ln R(0) = 0 because R(0) = 1 (the probability of surviving for an infinitesimal period of time being unity),

$$-\int \lambda dt = \ln R(t)$$

or

$$R(t) = e^{-\int \lambda dt} \tag{7}$$

In some cases, it may be desirable to think in terms of unreliability, U(t). Since reliability and unreliability are mutually exclusive concepts,

$$R(t) + U(t) = 1 \tag{8}$$

In the following, we will use the equations that have been developed in discussing the early failure and chance failure portions of the failure curve. The mathematical approach to the wearout portion of the curve will be developed later.

13.3 CHANCE FAILURES

This portion of the failure curve is discussed first because it corresponds to the useful life period of the product. The early failures of the break-in period have all occurred and the wearout stage has not been reached. The per-unit failure rate is low and is constant, a characteristic that has been verified by experimental data for large populations of products. If the per-unit failure rate is constant, the failures are said to have a Poisson distribution.

What causes failure during this period? The reasons are obscure, but are probably related to randomly occurring stresses, the fact that the material properties of the components (such as resistance to heat, ability to withstand a certain voltage gradient, or other similar properties) have a probability distribution of their own, and environmental conditions that may cause failure also exhibit a probability distribution. When adverse combinations of these factors occur simultaneously, failure occurs. Because of the obscure nature of the causes for failure, such failures are difficult to prevent.

With the per-unit failure rate, λ, constant, Equation (7) becomes

$$R(t) = e^{-\lambda t} \qquad (9)$$

That is, for chance failures the reliability decreases exponentially, as shown in Figure 13.4. In this figure, the first part of the curve, for which $R(t) \cong 1$, is also shown with the time scale expanded by a factor of 100.

In order to be able to plot a reliability curve or to evaluate Equation (9), the value of λ must be known. In practice, data are usually obtained for the reciprocal of λ. The reciprocal is called the mean time between failures (MTBF), and has the units of time. That is,

$$\text{MTBF} = m = 1/\lambda \qquad (10)$$

It is necessary only to determine either λ or m for a given population of a product in order to be able to predict the reliability at any time during the chance failure period. An experiment to determine reliability usually involves taking a number of units that have survived the early failure stage, operating them in the

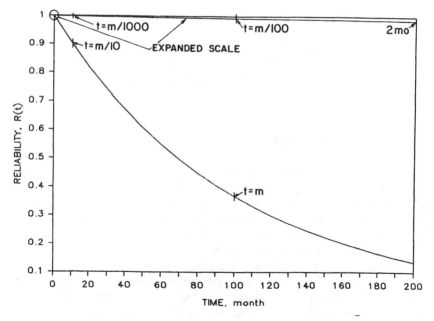

Figure 13.4 Reliability versus time for chance failure with $\lambda = 0.01$.

RELIABILITY

intended manner for some chosen time, and replacing units as soon as they fail with other units (which had also survived the early failure period) held in reserve for that purpose. The reason that replacement units can be tested along with those that have been on test for some time is that failure during this stage is due only to chance. Care must be taken to use a sufficiently large sample. If the test is prolonged into the wearout period, that fact must be recognized and data from the wearout phase discarded in making the calculation for m. If these conditions are met,

$$m = \frac{Nt}{N_f} \qquad (11)$$

where N = the number of units under test (a constant)
        N_f = number of units that failed during the test
        t = the duration of the test

The test may also be run without replacing the failed units. (The hypothetical product for which data are shown in Figures 13.1 and 13.2 was tested using this protocol.) This results in a decreasing number of devices on test, and the MTBF is given by

$$m = \frac{\sum N_i \Delta t_i}{N_f} \qquad (12)$$

where N_i = the number of units operating during time Δt_i
        Δt_i = the incremental times during which N_i units were operating

If one applies Equation (12) to the data in Figure 13.1 for months 5 through 23, the value of λ is found to be 0.0129/month, for which m = 77.5 months. For this device, it is clear that chance failure will not be the dominant factor in the life of the product, but that wearout will. Of the 42 units that survived the early failure region, 33 are still in operation when the wearout portion of the curve is reached.

Sample size becomes important, especially for components having low per-unit failure rates. If a test program is set up which runs continuously, it can run for only 8,760 hours in a year. If per-unit failure rates are to be quoted in terms of failures per million hours, one either runs the test on a single unit for 114 years (which makes no sense for any one of several reasons, including the fact that the unit may fail very early), or one runs the test on a sufficiently large number of units that early and late failures tend to offset one another, and the million hours can be reached cumulatively in a reasonable time. Testing of 114 units for one year on a continuous basis will yield the million hours, and larger samples will clearly result in improved precision or allow for shorter test runs.

The tests used must be designed to subject the devices or systems to all failure modes that are likely to occur. Simultaneous application of different stresses is more likely to cause failure than stresses applied singly and should be avoided unless such a combination is consistent with normal in-use conditions. For example, electrical insulation is used to support electrical conductors, and is therefore subject to known voltage gradients, mechanical stress, and elevated temperature.

It is also important to insure that stress is not applied with an intensity that is certain to cause complete or even partial failure. (These are not tests to destruction.) The objective of the tests is to determine what small fraction of the product population is likely to fail because of overstressing that occurs by chance in normal operation, not the limit of endurance. The failure modes that occur during the test should be the same as those expected in service before wearout is reached. If accelerated life tests are used, care must be taken that the conditions imposed during those tests are not in conflict with the conditions to be expected in normal service.

As noted above, wearout usually becomes important long before the MTBF point (m) is reached. Hence values of time (t) that are fractional parts of m give realistic reliability values. Values that are likely to be of interest are:

$$t = m/10^3 \quad R = 0.9990$$
$$t = m/10^2 \quad R = 0.9900$$
$$t = m/10 \quad R = 0.9048$$

If a reliability of 99.90% is desired, say for a military mission, then the MTBF must be 1000 times the expected mission time. Even if a reliability as low as 90% is satisfactory for consumer products, the mean time between failures must be ten times the expected life. To put this in perspective, suppose that a shaver is expected to have a useful life of 5 years; for 90% reliability the MTBF must be 50 years. Note that "to have a useful life of 5 years" is equivalent to operation for only 304 hours if the shaver is used for as much as 10 minutes per day. Hence the MTBF in real time is 3,040 hours, and a realistic program for testing a representative group of sample shavers is quite feasible.

How does one improve the reliability of a given design? The first thing to find out is the principal cause of failure. Examination of failed units will usually enable the designer to identify the point at which failure most often occurs. In the shaver, for example, this is most likely to be the brushes or the commutator of the motor. Having located the principal weakness, steps are taken to increase the capacity of the relevant parts. For example, substitution of a different grade

RELIABILITY

Table 13.1 Per-Unit Failure Rates (λ) of Selected Components in Units of Failures/Million Hours

Mechanical Components	λ	Electrical Components	λ
Accelerometer	43	Ammeter/voltmeter	26
Accumulator	30	Battery	
Actuator	52	Lead acid	0.5
Bearing		Mercury	0.7
Ball	13	Brush (electric motor)	5
Roller	200	Circuit board	0.3
Sleeve	23	Circuit breaker	1.2
Bellows	5	Connector	
Brake	13	Power	0.2
Clutch	2	Printed circuit board	0.1
Compressor	65	Generator	
Counter	5	AC	2
Differential	15	DC	40
Fan	6	Heater	4
Gyroscope	31	Lamp	
Heat exchanger	4	Incandescent	10
Gear	0.2	Neon	0.5
Pump	12	Lamp socket	0.1
Shock absorber	3	Motor	
Spring	5	Fractional horsepower	8
Tachometer	37	Integral horsepower	4
Thermostat	17	Relay	9
Valve	14	Solder joint	0.001
		Solenoid	1
		Switch	6

of brush may improve the MTBF. Derating of components prone to failure will also improve reliability.

Per-unit failure rates vary widely from component to component, and from manufacturer to manufacturer. For specific values for components being considered in a particular design, the manufacturer should be contacted. However, some typical values are shown in Table 13.1.

13.4 EARLY FAILURES

Early failures (the break-in stage) are usually caused by substandard parts, poor manufacturing techniques, or inadequate quality control. Another mechanism, one that results from lack of attention to all the facets of the design process, may also lead to early failure, as discussed below. In any case, the units that fail early on were not obviously inferior or they would have been rejected at the factory. In use, a variety of conditions can occur other than those used to test the product as part of the manufacturing process. Some of the operating conditions may result in electrical, mechanical, or thermal stresses that can easily be withstood by parts of normal strength, whereas parts which are marginal will fail.

The failure mechanism is not essentially different from that of the chance failure (normal life) failure mode. It is a matter of chance when a combination of conditions that will cause failure will occur. The major difference is that, since great overstressing is not required, the adverse conditions will occur more frequently. In the usual situation, only a small number of units will embody the deviations from normal that make them subject to failure. After these have been removed, the healthy majority of the population remains.

Since failures in the break-in stage are due to chance, a per-unit failure rate λ_d for these defective units may be identified. This will be several times larger than the per-unit failure rate λ of the useful life portion of the failure curve. That is, $MTBF_d$ will be so small that none of the defective units can be anticipated to reach the wearout stage. However, chance failures at rate λ are occurring simultaneously with early failures. The failure rate of the entire population during the break-in period can then be expressed as

$$\lambda_b(t) = \frac{N_g(t)\lambda + N_d(t)\lambda_d}{N_p} \tag{13}$$

where $\lambda_b(t)$ = the per unit failure rate during break-in at any time t
$N_g(t)$ = the number of good units in the population at time t
$N_d(t)$ = the number of defective units in the population at time t
λ = the per unit failure rate during the useful life stage
λ_d = the per unit failure rate of defective units, and
N_p = the total population

This elegant equation is, as it turns out, of only limited value. Neither λ_d nor N_d is known *a priori,* although their values may be determined after the fact if the devices or systems are life tested.

What is wanted is some way of getting rid of the problem of early failures. The simplest way of doing this is a method generally adopted for most consumer products: Provide the user with a warranty that the manufacturer will bear the cost of repair (or possibly replacement) if the product fails within a specified period of time. Many small consumer items carry a one-year warranty. The warranty on new automobiles, which was once as short as 90 days, is now routinely one year for a complete warranty, and limited warranties for periods of as long as seven years are not uncommon. The warranty approach is a tempting

RELIABILITY

one, but the designer must realize that there are repair expenses connected with this approach, and there will be attrition of customer good will if the number of early failures is high. It is better to identify the causes of as many of the early failures as possible and modify the design to eliminate those causes.

A second way of getting rid of the problem of early failures is simply to force those failures to occur before shipment by use of a procedure known as burn-in, break-in, or debugging. This approach is used for critical applications, such as the control for an industrial process which, if shut down, causes substantial financial loss. It is also used for applications in which there might be considerable hazard to life as the result of a failure, to say nothing of financial consequences. Many military and space applications fall into this category.

The burn-in procedure subjects the entire population to operation under rated conditions (and possibly somewhat more stringent conditions) for a predetermined period of time, a time that is presumably sufficiently long that all of the defective units will have been culled from the population. As an example, the amplifiers used by one well-known manufacturer of jet engines for afterburner control have been routinely tested at an ambient temperature of –20 C for 24 hours, followed by operation at 120 C for 50 hours. The purpose of the burn-in procedure is to lop off the left tail of a "strength" probability density function as indicated in Figure 13.5, so that no "stress" that can occur, assuming the stress probability density function of that figure, can possibly exceed the smallest value of "strength" of the remaining units. Further discussion of this concept appears in Section 13.6.

Regardless of the approach taken to deal with the problem of early failures, the designer should have some way of determining the duration of the early failure period. It can be shown [3, p. 66] that the mean burn-in time is given by

$$\text{BIT} = \frac{1}{\lambda_d}\left(1 + \frac{1}{2} + \frac{1}{3} + \ldots + \frac{1}{N_d}\right) \qquad (14)$$

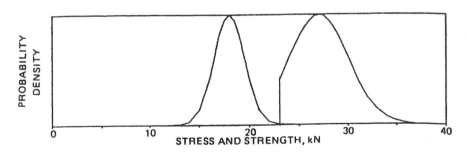

Figure 13.5 Ideal strength and stress density functions after burn-in.

This equation suffers from the same deficiencies as those pointed out in the discussion following Equation (13), that neither N_d nor λ_d is known before a test is run. Fortunately, it is somewhat insensitive to the number of defective units. If there are 5 substandard units, $BIT = 2.3/\lambda_d$; with 10, $BIT = 2.9/\lambda_d$; and with 100 substandard units $BIT = 5.2/\lambda_d$. That is, one can compensate for having only a small sample (and presumably a small number of substandard units) by increasing the burn-in time. Keep in mind also that, since the number of defective units decreases exponentially, only 63.2% of the defective units will fail during the mean burn-in time. That is, the equation provides at most a useful guide.

Now λ will be known to a considerable degree of accuracy after life tests are run. If one decides that any unit having a failure rate five to ten times the average for the entire population is to be eliminated by burn-in, then use of $\lambda_d = 5\lambda$ in Equation (14) will give a rough estimate of mean burn-in time. The value of BIT calculated must be increased by at least 300% to give reasonable confidence that the early failures will all be culled.

13.5 WEAROUT

The last stage of in-use life depicted in Figure 13.3 is the stage in which wearout becomes the determining factor leading to failure. The mechanisms causing failure can be described as general deterioration of parts due to wear, fatigue due to cyclical torques or bending moments, cyclical thermal cycles, or any other normal activity of the product or system that results in gradual weakening of or wear on various components. If life tests are run on a sufficiently large sample of units and if the early and chance failures are disregarded, it will be found that the remaining failures cluster around some mean lifetime M, and that the distribution of failure times is such that they approximate a normal or Gaussian distribution. That is, the lifetime to wearout f(t) may be given by

$$f(t) = \frac{1}{\sigma\sqrt{2\pi}} e^{-(t-M)^2/2\sigma^2} \tag{15}$$

where

M = mean wearout life
σ = the standard deviation
t = the age of the unit

Now f(t) is theoretically defined for $-\infty < t - M < \infty$, but if σ is small (as it will be in practical cases), the lower limit of t may be taken to be zero without appreciable error. Since reliability is the probability that a unit will perform without failure over a given period of time and because (ignoring early and

RELIABILITY

chance failures) the probability of performing without failure for small values of t is unity, we may write R(t) as

$$R(t) = 1 - \int_0^t f(t)dt \qquad (16)$$

Since $\int_0^\infty f(t)dt = 1$ provided only that σ is small, Equation (16) may be rewritten as

$$R(t) = \int_t^\infty f(t)dt = \frac{1}{\sigma\sqrt{2\pi}} \int_t^\infty e^{-(t-M)^2/2\sigma^2} dt \qquad (17)$$

Figure 13.6 shows the reliability for wearout failure versus time for Figure 13.1. Note that wearout reliability depends on the product's total in-use life. Wearout failures are related to the past history of the unit; early and chance failures have no relationship to history.

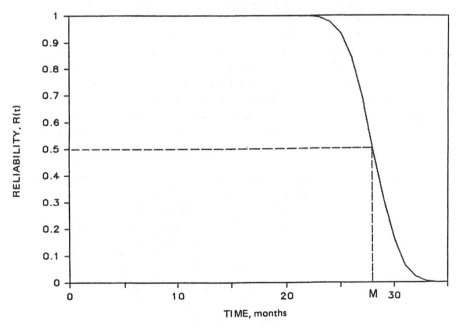

Figure 13.6 Reliability versus time for wearout failures.

One of the main concerns of a designer is to extend the mean wearout life, M, and to reduce the standard deviation, σ. That is, the designer may be interested in emulating the achievement of the deacon in Holmes's "The Deacon's Masterpiece," which was to build a chaise (shay) so "that it *could n'* break daown." If you have read the Holmes piece, you will recall that the shay lasted one hundred years, at which time a slight shock caused the entire shay to disintegrate into a pile of sawdust. Had the deacon actually built several units to the same standards as the one said in the poem to have been built, those units would have had a mean wearout life of $M = 100$ years and $\sigma = 0$. They would also have had the desirable characteristic that no part would have worn out significantly earlier than any other part. Practically, of course, the designer must also be interested in balancing first cost against the benefits of reduced in-use repair. Moreover, one must consider what the useful life of a product is likely to be, not in terms of wearout life, but in terms of years to obsolescence. Many products are replaced today, not because they are worn out, but because a new product is available that is so substantially better than the old one that replacement can be justified on economic grounds, or simply because the owner would like something new.

Nevertheless, there are large numbers of devices and systems that will be expected to operate well beyond their mean wearout life. For example, the drive systems for rolling mills represent enormous capital expenditures, and are replaced as infrequently as possible. To extend the life of equipment which must continue to be useful and productive beyond normal wearout, routine maintenance procedures can be used to identify parts that are beginning to deteriorate, scheduled periodic lubrication will extend the life of bearings, and the detailed history of the system will identify components that have a tendency to fail at a certain rate and which might be replaced on a routine basis. The designer may find it desirable to propose that the installation be equipped with vibration measuring devices, either for the purpose of triggering alarms or for the purpose of establishing a history of normal vibration patterns for comparison against those existing at any later time, significant deviations being an indication of incipient problems. Care must be taken that replacement components are of the proper quality and that they are correctly installed; failure to observe these precautions may lead to an early failure following routine maintenance. Discussion of some of these topics will be found in Chapter 14 of Reference 4, in which a long list of other references on maintenance will be found.

Although the three stages of in-use life have been discussed separately, they are obviously not really separable. It was pointed out earlier that chance failures of the type that occur during the normal life of a unit may occur during the early failure portion of the curve, and some will also occur during the wearout phase. On the other hand, a substandard unit may survive both the early and the chance failure regions, eventually failing by wearout. A reliability curve for all three regions appears in Figure 13.7 with the label "a." If an effective

RELIABILITY

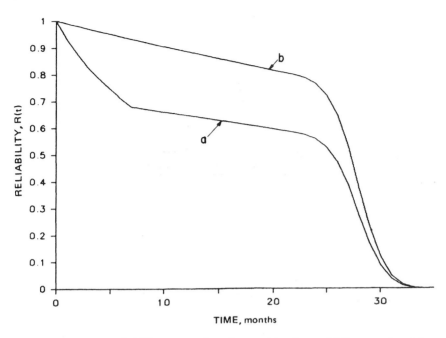

Figure 13.7 Reliability versus time for combinations of failure modes, (a) including early failure and (b) with early failures eliminated.

burn-in procedure has been used, the initial sharp reduction in reliability of curve "a" disappears, as shown by curve "b," and there is a significant improvement in reliability during the normal life of the product, although the wearout stage is unaffected.

13.6 STRESS/STRENGTH INTERFERENCE

Device or system design is usually done by using single-valued real numbers in the equations comprising the mathematical model. Yet we know that no two supposedly identical components are actually identical. There will always be some variations in dimensions and deviations from the nominal values of bulk properties for any component from the values used in the design. Moreover, we know that the stress (whether mechanical, electrical, or thermal) will also exhibit variations around the nominal value used in the design. Both the variations in bulk properties and the variations in stress are frequently Gaussian (or normal) in nature, as shown in Figure 13.8(a), in which the stress distribution is to the left and the strength distribution to the right, both distributions being normalized. If one concentrates attention on the mean values only, it appears that the mean strength exceeds the mean stress by a comfortable margin. In what follows, we

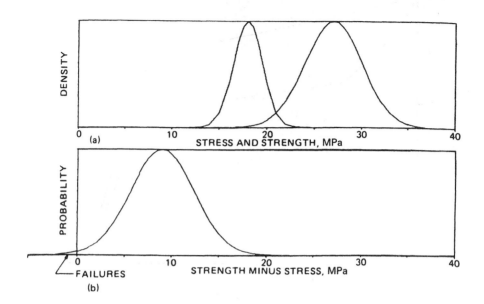

Figure 13.8 Effect of normal distributions of parameters on reliability.

shall see that the margin is not as comfortable as one might wish. Suppose that we know that the part represented by these curves is to be subjected to a mean stress μ_{ss} of 18,000 MPa, with a standard deviation of 1,500 MPa. The mean strength μ_s is computed to be 27,000 MPa, with a standard deviation of 3,000 MPa. Will the parts designed be reliable?

It is clear from the fact of overlap of the stress and strength curves of Figure 13.8(a) that some parts will lack sufficient strength to be able to withstand stresses that happen to be at the upper end of the stress distribution. One question of interest to the designer is "What proportion of the parts made will fall into this category?" If one subtracts the stress distribution from the strength distribution, the resulting curve will have a mean value of 9,000 Mpa, and the left tail will extend beyond the origin. The standard deviation can be computed using

$$\sigma = (\sigma_{ss}^2 + \sigma_s^2)^{1/2} = (1500^2 + 3000^2)^{1/2} = 3354 \text{ Mpa}$$

where the subscript ss was used to designate stress and s to designate strength, as above. Now we can compute the probability of nonfailure (survival) using

RELIABILITY

$$z = \frac{\mu_s - \mu_{ss}}{\sigma} = \frac{27{,}000 - 18{,}000}{3354} = 2.68 \qquad (18)$$

From Table 8.3, we see that the probability of nonfailure lies between 99.38% and 99.87%; more detailed tables show that the probability of nonfailure for this value of z is 99.63%. The probability of failure is therefore 0.37%, and the designer should expect that four units (approximately) out of each 1,000 made (or 3,700 units of each million) will fail. The burn-in procedure discussed earlier is intended to eliminate these units by lopping off the left side of the strength curve so that the strength distribution of Figure 13.5 is achieved.

O'Connor [5] defines z as a safety margin and shows that intrinsic reliability—which is defined as the probability of failure being negligible—requires a value of z of 4 or 5. For z = 4, the probability of failure is 30 units per million. The designer will have to decide whether this meets his or her definition of "negligible."

Lewis [6] shows that Equation (18) can be expressed in terms that may be more convenient to use. Defining the factor of safety as

$$FS = \frac{\mu_s}{\mu_{ss}}$$

and the coefficients of dispersion about the mean as

$$\rho_s = \frac{\sigma_s}{\mu_s}$$

and

$$\rho_{ss} = \frac{\sigma_{ss}}{\mu_{ss}}$$

it can be shown that

$$z = \frac{FS - 1}{\left(\rho_s^2 FS^2 + \rho_{ss}^2\right)^{1/2}} \qquad (19)$$

Using the data from the example above,

$$\rho_s = \frac{3{,}000}{27{,}000} = 0.1111$$

$$\rho_{ss} = \frac{1{,}500}{18{,}000} = 0.0833$$

$$FS = \frac{27{,}000}{18{,}000} = 1.50$$

Then

$$z = \frac{1.50 - 1}{\left(0.1111^2 \times 1.50^2 + 0.0833^2\right)^{1/2}} = 2.68$$

as before. That is, one can anticipate the failure rates computed earlier.

Equation (19) has the advantage that it can be rearranged so as to be able to solve for the value of FS once a value of z has been chosen. On rearranging, the quadratic form

$$\left(z^2 \rho_s^2 - 1\right) FS^2 + 2FS + z^2 \rho_{ss}^2 - 1 = 0 \tag{20}$$

is obtained. Using the values of the coefficients of dispersion already obtained and letting z = 4, we find that the factor of safety, FS, must be 1.91. If we can achieve this value, we can have confidence that there will be only 30 early failures per million units produced.

13.7 CALCULATION OF PRODUCT RELIABILITY

The best the designer can do if asked for the mean time between failures, m, the mean wearout life, M, and its standard deviation, σ, is to provide estimates. The mean time between failures for a complete device or system can be computed from data for the component parts, such as the typical values given in Table 13.1, or data from test or from the manufacturer of the component. While making the calculations, the designer must decide for each part whether or not its failure will cause failure of the device. For example, a leak in the heater core of an automobile does not render the automobile inoperable, whereas failure of any one of the tires on an automobile will cause "failure" of the automobile. Failure of one of the rear tires on a truck with double sets of rear tires will not cause "failure" of the truck if the truck is empty or lightly loaded, although a fully loaded truck would probably fail if one of the rear tires was punctured.

In addition to making decisions as to whether failure of a component will cause failure of the product or not, the designer must also take into consideration the configuration of the entire system, as will be seen in the following sections.

13.7.1 Series Reliability

Consider first the situation in which all components of the product are independent and in which the failure of any one element will cause failure of the entire system. For example, a mechanical drive system for an industrial mixer will consist of an electrical supply, circuit breakers, contacts on relays, an electric motor, a drive train that may involve gearing or a belt drive with variable pitch pulleys, and the final drive shaft and mixing head. Failure of any one of these elements will cause the mixer to be inoperative. In this example, the elements of the system are obviously arranged in a series. It is not necessary, however, that the elements be arranged in a series in order for the series concept to apply. The four tires on an automobile are in series for reliability calculations because failure of any one of them makes the car inoperative.

The total reliability is the joint probability of survival of all elements:

$$R(t) = R_1(t)R_2(t)\cdots R_n(t) = \prod_{i=1}^{n} R_i(t)$$
$$= e^{-(\lambda_1+\lambda_2+\cdots+\lambda_n)t} \quad (21)$$

That is, when the failure of any element will result in failure of the product, the per-unit failure rates of all elements are added to obtain the failure rate of the product. Designating the product failure rate for series reliability as λ_s,

$$\lambda_s = \lambda_1 + \lambda_2 + \cdots + \lambda_n = \sum_{i=1}^{n} \lambda_i \quad (22)$$

The series model for reliability is identical, except for notation, to the model used to determine the probability of occurrence with AND gates in Chapter 2. It also applies to a single component which may have more than one failure mode. A gear in a gear train may be secured to its shaft by a set screw, not by a key. There are two easily identifiable modes of failure: The gear may slip on the shaft because the set screw loosens or because it was not tightened originally, and one or more teeth may break off the gear. Either event constitutes a failure, and the per-unit failure rate will be the sum of the failure rates for each of the modes.

For the purpose of estimating reliability, the assumption is often made that all components of a product are in series; this yields a conservative assumption and is certainly the simplest mathematical form. As noted above, the per-unit failure rates as obtained from test, from technical periodicals, or from the manufacturer are added to obtain an overall failure rate and, from that number, the reliability. Admittedly, this is not accurate, but it is an often used measure.

13.7.2 Parallel or Redundant Systems

An empty truck with a double set of truck tires illustrates another possible relationship among elements. This is called parallel reliability, and is defined as the situation in which multiple units must fail before the device fails. A machine tool may have two control stations, each with a START pushbutton. Failure of the contacts on either pushbutton will still allow the machine to be started from the other station, but failure of the contacts on both pushbuttons renders the machine inoperative—there has been a failure of the machine.

In dealing with this problem, it is convenient to use the idea of unreliability, which is the probability of failure. The total probability of failure for n elements in parallel is given by

$$U(t) = U_1(t) U_2(t) \cdots U_n(t) = \prod_{i=1}^{n} U_i(t)$$

$$= (1 - e^{-\lambda_1 t})(1 - e^{-\lambda_2 t}) \cdots (1 - e^{-\lambda_n t}) \tag{23}$$

From Equation (8),

$$R(t) = 1 - U(t)$$

$$= 1 - \prod_{i=1}^{n} U_i t \tag{24}$$

For the special case of all elements being (supposedly) identical—as the truck tires or the START pushbuttons on the machine tool might be thought to be—

$$U(t) = (1 - e^{-\lambda t})^n$$

and

$$R(t) = 1 - (1 - e^{-\lambda t})^n$$

For $\lambda t \ll 1$, expansion of the exponential in series form and clearing of constants enables one to calculate the mean time between failures, m_p, for n identical elements in parallel as

$$m_p = \frac{1}{\lambda} + \frac{1}{2\lambda} + \frac{1}{3\lambda} + \cdots + \frac{1}{n\lambda} \tag{25}$$

RELIABILITY

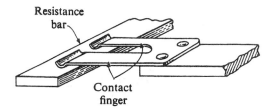

Figure 13.9 Redundant electrical contacts for a potentiometer.

The equations used to calculate reliability for parallel or redundant systems are identical, except for notation, to those used to determine the probability of occurrence with OR gates. Equation (25) may be used to calculate the MTBF for the truck tires. Assuming that the mean time between failures for a truck tire is 1 year, then the MTBF for the dual set is 1.5 years.

Redundant systems are essentially systems in parallel. These systems may or may not be "identical," as the truck tires were assumed to be, but they are operative whenever the product is in use. A simple example of redundancy is given by the sliding contacts on a potentiometer as shown in Figure 13.9. The two fingers shown are not identical so that they will have substantially different natural frequencies of vibration, thus increasing the probability that at least one finger is in contact with the resistance bar even if the potentiometer is subjected to vibration. The fingers not being identical, it is expected that the failure rates of the fingers will not be the same, and the reliability would have to be calculated using Equations (23) and (24), rather than Equation (25). On the other hand, suppose that redundancy is provided in an electric circuit by connecting both poles of a double-pole relay in parallel in the supply line; the contacts and the fingers supporting them are presumed to be identical, and the MTBF would be calculated from Equation (25).

13.7.3 Standby Systems

Another technique for increasing reliability is to provide standby systems. In this approach, the standby units are not used and hence suffer no wear until the primary unit fails. The standby technique is clearly different from the parallel or redundant technique. For this approach to be successful, there must be some means of sensing that a failure has occurred and a control system that brings the secondary unit into operation. Such systems may have a tertiary system that is activated if the secondary unit fails. Figure 13.10 shows in block diagram form a three-layer system for opening an escape hatch of a space capsule. The units shown are completely different in character; the primary unit is electrical, the

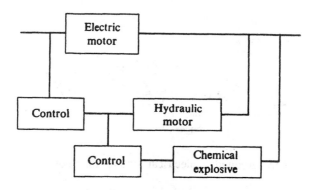

Figure 13.10 Standby systems for opening an escape hatch.

first standby is hydraulic, and the second standby is chemical (explosive bolts). The use of standby systems can result in very high reliability if the standby and control units receive careful periodic maintenance to assure that chance failure has not occurred while in the standby mode. Hospitals, for example, are required to have standby electrical supply systems for operating rooms and other critical areas, and these must be tested on a weekly basis. The lifetime of the complete system is the sum of all of the lifetimes of the primary, secondary, and (if present) tertiary components, since each operates in sequence until the final unit has failed. Assuming a system in which all units are kept in good repair and for which the failure rate of the units are identical, the reliability is given by [3, pages 112-114]

$$R_{stby}(t) = R(t) \sum_{i=0}^{n-1} \frac{(\lambda t)^i}{i!} \tag{26}$$

where $R(t) = e^{-\lambda t}$

and for which we keep in mind that $0! = 1$.

For the system of Figure 13.10 having $\lambda = 0.001$ failures/mission and with $X = 200$ missions, λt is replaced by $\lambda X = 0.001 \times 200 = 0.2$, so that

$$R(200) = e^{-0.2} \left(\frac{0.2^0}{0!} + \frac{0.2^1}{1!} + \frac{0.2^2}{2!} \right) = 0.9989$$

From this result, it may be shown that the mean number of missions between failures is 174,000, which should surely make the astronauts comfortable.

RELIABILITY

13.8 OTHER STATISTICAL DISTRIBUTIONS

In the treatment so far of reliability, it has been assumed that early failures, chance failures, and wearout can always be accurately modeled by the exponential, Poisson, and Gaussian distributions. This is not true; test data may show significant variations from the shapes dictated by these distributions. Furthermore, as is obvious from the section on stress/strength interference, the tails of the distributions, which are easily ignored, may have considerable importance in the determination of the probability of failure. As a consequence, it is often necessary to choose a distribution empirically by fitting a curve to the data. Goodness-of-fit tests such as the chi-square test may be used to decide whether a particular distribution fits a given set of data to the engineer's satisfaction.

The distribution usually used for more accurate reliability analysis is the Weibull distribution. Whereas the distributions considered so far have fixed shapes, the Weibull distribution can be made to model a wide range of life distribution characteristics as shown in Figure 13.11. In addition, it appears to be particularly useful for mechanical failures and in the studies of brittle materials, such as ceramics [6, Section 3.4]. The two-parameter Weibull distribution assumes a failure rate

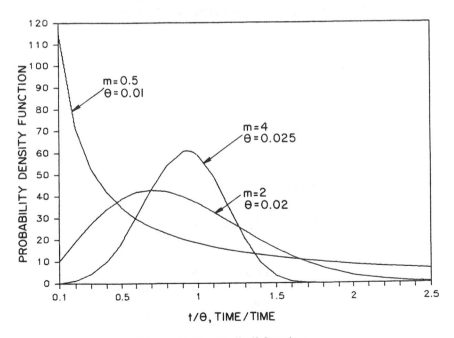

Figure 13.11 Weibull functions.

$$\lambda(t) = \frac{m}{\theta}\left(\frac{t}{\theta}\right)^{m-1} \tag{27}$$

m and θ being the two parameters.

The probability density function for unreliability is

$$\frac{dU(t)}{dt} = \frac{m}{\theta}\left(\frac{t}{\theta}\right)^{m-1} e^{-(t/\theta)^m} \tag{28}$$

giving the cumulative density function

$$U(t) = 1 - e^{-(t/\theta)^m} \tag{29}$$

Then, from Equation (8),

$$R(t) = e^{-(t/\theta)^m} \tag{30}$$

The values of m and θ may be estimated by plotting failure times on the special graph paper shown in Figure 13.12, for which the scales may be obtained by the following derivation. (See also Reference 7.) Using Equations (8) and (29),

$$R(t) = 1 - U(t) = e^{-(t/\theta)^m} \tag{31}$$

Taking the reciprocal, we have

$$\frac{1}{1 - U(t)} = e^{(t/\theta)} \tag{32}$$

and taking the natural logarithm twice results in

$$\ln\left[\ln\frac{1}{1-U(t)}\right] = m \cdot \ln t - m \cdot \ln \theta \tag{33}$$

which is the equation of a straight line on the axes of Figure 13.12.

The ordinate value is the cumulative failure fraction (or percentage, if the scale is given in percentage units) and is calculated as

$$U(t) = \frac{i}{N+1}$$

RELIABILITY

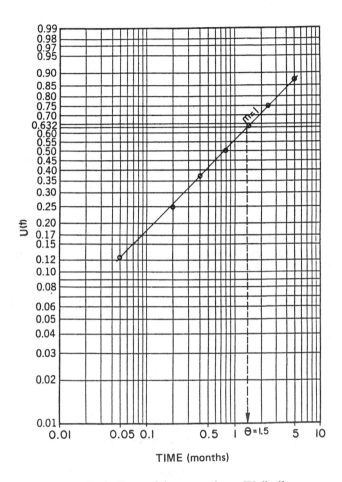

Figure 13.12 Data of the example on Weibull paper.

for the ith failure of the N units tested. The failure times are plotted on the abscissa using a log scale.

The two Weibull parameters, m and θ, may be estimated from the straight line as shown in Figure 13.12. Alternately, a scale showing m may be added to the figure, as shown in Figure 10.25 of Reference 7, and the plot of data points may be moved parallel to itself so that m is read directly. As an example, in Figure 13.12, seven data points are shown at (0.05,0.125), (0.2, 0.25), (0.4, 0.375),

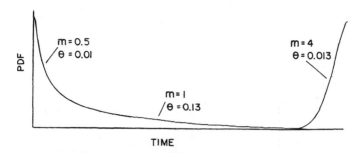

Figure 13.13 Reliability bathtub curve made from three Weibull distributions.

(0.8, 0.500), (1.6, 0.625), (2.5, 0.75), and (5.0, 0.875). The time is assumed to be in months. If the scales have been plotted using the same linear scale on each axis, the slope of the line through the data points is m, and the value on the abscissa for U(t) = 0.632 is θ, the left side of Equation (33) being zero at that point. For Figure 13.12 the slope, m, is 1 and θ is 1.5 months.

An important advantage of the Weibull distribution is that it is extremely flexible in nature. The disadvantage consequent on that flexibility is that the cumulative density function values cannot be given in tabular form as can be done with the exponential and Gaussian distributions. It becomes necessary to integrate the failure density function for each section of the reliability curve (early, chance, and wearout failures) with the relevant values of m and θ. In addition, because the chance and wearout sections of the curve are delayed in time, the probability density function becomes for those sections

$$\frac{dU(t)}{dt} = \frac{m}{\theta}\left(\frac{t-t_o}{\theta}\right)^{m-1} e^{-((t-t_o)/\theta)^m} \qquad (34)$$

where t_o is the beginning of the chance or wearout section of the curve. Figure 13.13 shows a typical reliability bathtub curve formed from three Weibull distributions, two of which were delayed.

The integration of the curve of Figure 13.13 is obviously not done using standard tables of integrals. Computer programs are available to carry out these calculations.

13.9 DESIGN CONSIDERATIONS

In the design of any product, one of the goals of the designer should be to attain the longest mean wearout life and mean time between failures that are compatible with the other design considerations. You will find that some substantial improvements can be made by following what appear to be rather

RELIABILITY

simple suggestions. You will also find that reliability improvement does not always result in increased cost, although the two are closely related when extremely high reliabilities are required, such as for space shuttles or satellite launch vehicles.

The simplest suggestion to improve reliability is to simplify the design as much as possible. That is, try to use the minimum number of parts and try to make those parts with the minimum number of fabrication operations. Keep in mind that every soldered connection, every welded or riveted joint, every extra resistor, and every extra access plate provides one more possible point of failure. Even considering a single part, if it can be made in one die with one stroke of a punch press you will find it to be more reliable than if it must be made by a series of operations using several different dies. Any attempt to simplify the design usually has the added advantage that cost is reduced. This is not a suggestion that performance be sacrificed to simplification; the suggestion is simply that there be a close look at every part of a system or device while questioning whether there is another and simpler way to do the same thing. In Section 6.2, there was discussion of great inventions of the past. In most of the cases mentioned, the great invention was simply a simplification of what had already been designed. See, for example, the work of Black on feedback amplifiers, and the simple way in which the need to have different fluid flows in heat pumps, depending on the direction of flow, was solved.

The use of standard, proven, components rather than the latest advance in the state of the art usually results in better reliability. The reason is that the manufacturer of the latest device is probably going through a learning period. There are exceptions, of course, especially when a new component that is inherently more reliable than the old is developed. Classic cases illustrating major improvements in reliability occurred in the electrical field, with the replacement of vacuum tubes by transistors, and later by the replacement of transistors and the discrete components associated with them in amplifiers, flip-flops, gates, and so forth by integrated circuits.

Another way of improving reliability is by choosing the most reliable components available. A particular function may be provided in several different ways, and not all of them have the same or even similar reliability. For example, an electric motor driving a ball screw to provide linear motion may be more or less reliable than a hydraulic cylinder that will provide the same travel. In many devices, the position of a certain element must be measured, and this can be done using either a potentiometer or a linear variable differential transformer (LVDT). The LVDT is far more reliable than the potentiometer, and simultaneously provides a much higher precision.

Parts should be designed and components chosen that have as small a variation within the population as is possible. As shown earlier in this chapter, a large deviation from the mean values of strength makes failure of at least a certain number of units virtually inevitable.

Still another way of improving reliability of a product is to derate components that are the major cause of unreliability. Figure 13.14 shows a typical set of derating curves. The reason for the effectiveness of derating is that components are rated to withstand certain combined values of electrical, mechanical, and thermal stress. If a component is used with one or more of these stresses significantly reduced, the chance of random overstressing is very unlikely. For example, to gain the advantage of derating, an electric motor is sometimes specified with normal voltage and having a horsepower rating greater than the horsepower required by the load. Although the electrical stress is unchanged, both mechanical and thermal stresses are reduced.

Reliability may be further improved by reducing the mechanical, electrical, and thermal stresses on components. A relay that is used in an environment in which there is a considerable amount of vibration can be shock-mounted. A bearing in an area in which a considerable amount of dirt is present should be a sealed unit. A wire passing through a hole in sheet metal should be protected by a grommet. Equipment in which a sizable amount of heat is dissipated should have adequate ventilation, possibly provided by a fan. Obviously it is not possible to list every possible problem area. The designer

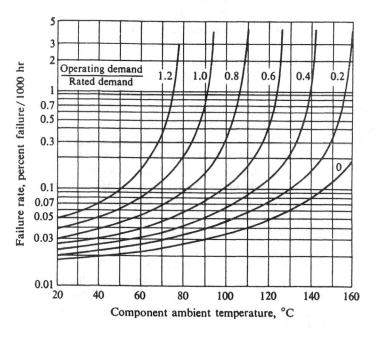

Figure 13.14 Typical derating curves.

RELIABILITY

should make a careful study of the physical model looking for potential problems. Practice the art of failure mode analysis, as discussed in Chapter 2.

Finally, reliability can be improved by proper testing. The designer should take the lead in suggesting to manufacturing personnel suitable performance or functional tests intended to discover those units likely to fail prematurely.

13.10 RELIABILITY GROWTH MODELING

Reliability is improved as the designer accumulates data on the various failure modes that are operative in a given product and makes the necessary changes in materials, manufacturing processes, or in the design to decrease the frequency of such failures [8]. The key to the improvement process is an analysis of the failures that occur during testing so that weaknesses can be identified and eliminated. In order to help precipitate failure, the equipment should be stressed to the limits of the specified operating environment, and may even be moderately overstressed. The process of growth in reliability is very similar to that of improving productivity, described in Section 8.3.4, which dealt with the learning curve. In fact, the learning curve is the model used for the analysis of reliability growth, with cumulative failure rate and cumulative operating time being the variables of interest.

The model is most accurate during the initial production runs. It is in this period that the product is likely to be subjected to untried or unanticipated environmental and stress conditions. It is also in this period that modifications in design and tooling are most acceptable. During later production, when dedicated tools or custom-formulated materials have been incorporated into the process, changes are less acceptable to manufacturing and purchasing, and the allowable remedies will probably result in only minor improvements in reliability. Moreover, unless the manufacturer maintains an in-house reliability improvement program, the impetus for improvements is likely to arise from customer complaints or products returned under warranty, and thus be related to the number of units produced rather than to time. For this reason, the cumulative failure rate and cumulative number produced after market introduction should be considered as appropriate variables during the later production phase of a product.

Using the learning curve model, the cumulative failure rate λ_c is

$$\lambda_c = K t_c^{-\alpha} = F_c / t_c \qquad (35)$$

where

F_c = cumulative failures $\qquad t_c$ = cumulative testing time
K = a constant $\qquad \alpha$ = growth rate

The unit failure rate at any time is

$$\lambda = \frac{dF_c}{dt_c} \tag{36}$$

From Equation (35),

$$F_c = Kt_c^{(1-\alpha)} \tag{37}$$

and hence

$$\lambda = (1-\alpha)Kt_c^{-\alpha} = (1-\alpha)\lambda_c \tag{38}$$

If the cumulative failure rate is plotted versus cumulative testing time on log-log paper, a straight line results, as shown in Figure 13.15. Hence

$$\alpha = \frac{\log \lambda_{co} - \log \lambda_c}{\log T_c - \log T_{co}}$$

from which

$$\left(\frac{T_c}{T_{co}}\right)^{\alpha} = \frac{\lambda_{co}}{\lambda_c}$$

or

$$\lambda_c = \lambda_{co}\left(\frac{T_{co}}{T_c}\right)^{\alpha} \tag{39}$$

or

$$T_c = T_{co}\left(\frac{\lambda_{co}}{\lambda_c}\right)^{1/\alpha} \tag{40}$$

The value of applying learning curve theory to reliability is that it gives a method of initiating and monitoring a reliability improvement program. Before beginning such a program, however, there should be a careful evaluation of the tests to be used, failure must be defined (Does one count a minor malfunction in the same way as complete shutdown?), and the failure rate to be attained should be agreed on.

The value used for the exponent α depends on the aggressiveness with which the improvement program is pursued. If elimination of all identified failure modes using accelerated (that is, overstress) tests under all foreseeable environmental conditions is vigorously pursued, the literature indicates that an α of 0.4 to 0.6 is appropriate. For programs in which reliability is given high priority but the tests used more nearly reflect expected performance under

RELIABILITY

normal environmental conditions, α of 0.3 to 0.4 should be used. If tests are to be performed without pushing the environmental limits, 0.2 is appropriate. If reports of failure from consumers are to be the major source of information, $\alpha = 0.1$ or less should be used.

An important unknown that must be determined is the failure rate at the beginning of the reliability improvement program. This rate can be based on past experience with similar products, on calculation of the system reliability, which in turn is based on component reliability as in Section 13.7, or on initial test data.

For example, suppose a product has been tested for 500 hours under the most aggressive testing category ($\alpha = 0.6$) and is found to have a cumulative failure rate of 0.01 units per hour. The goal is to reduce the unit failure rate to 0.0001 per hour by continuing to use these test methods. How long will the test program continue? Using the second form of Equation (38),

$$\lambda_c = \frac{\lambda}{(1-\alpha)} = \frac{0.0001}{(1-0.6)} = 0.00025$$

which is the desired cumulative failure rate. Then, from Equation (40),

$$T_c = 500\left(\frac{0.01}{0.00025}\right)^{1/0.6} = 233{,}921 \text{ hours}$$

This is an obviously impractical length of time, and management asks what value of unit failure rate could be attained if testing is discontinued at 8,000 hours, which is continuous testing for almost a year. Using Equation (39),

$$\lambda_c = 0.01\left(\frac{500}{8{,}000}\right)^{0.6} = 0.00189$$

and from Equation (38),

$$\lambda = (1-\alpha)\lambda_c = 0.00076$$

That is, the failure rate can be improved by a factor of 13 over the original value by extending the time for the testing program by a factor of 16. Another way of looking at these results is that the failure rate can be reduced to 7.6 times the original goal, and this can be done in about 1/30 the testing time that the original goal would have required. Regardless of the point of view taken, it should be evident that the major part of the reliability improvement will occur in the first part of an extended testing program, and that significant improvement becomes more and more difficult as time goes on.

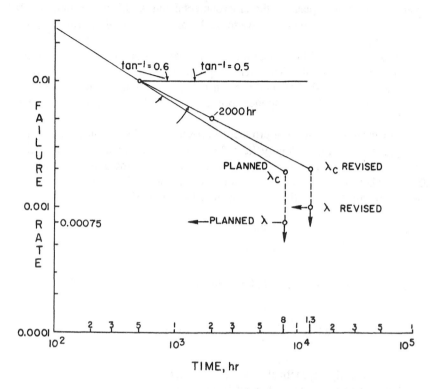

Figure 13.15 Graphical prediction of reliability growth.

Graphical solutions usually have sufficient accuracy for most purposes when using learning curve theory, especially in view of the imprecise data with which the analyst works. Figure 13.15 shows the solution of a more complex problem that may arise. Suppose that in the example above, testing has been continued, and at the 2,000 hour point if becomes obvious that the expected reduction of failure rate has not been achieved. In spite of the aggressiveness of the program, the experimental data shows that $\alpha = 0.5$ is the appropriate value to use rather than the $\alpha = 0.6$ that had been postulated after the initial 500-hour testing period. Management now wants to know how much longer the testing program will have to run if the target failure rate is raised to 0.001.

Using $\alpha = 0.5$ for the continuation of the test program, the target value for λ_c from Equation (38) is 0.002. Extending the line drawn through the data points for 500 and 2,000 hours, 0.002 is reached at 13,000 hours. A substantial increase in testing time is required.

RELIABILITY

13.11 HUMAN FACTORS

The most imprecise area for discussion under the topic of reliability is that of human error. Such errors often result in tragic events, or at least events that gain wide attention. The events at Three Mile Island and at Chernobyl were clearly the result of human error, and illustrate the fact that human error can defeat well-designed safeguards. It is estimated that 70% of all aircraft accidents are caused by human error; some of these errors are pilot errors, but others are errors on the part of maintenance personnel. The authors are reminded of other, more localized events, such as the case of the electrician who attempted to measure the current in a busbar in a switchboard with a clamp-on ammeter having uninsulated jaws, resulting in an arcover from one phase to another with molten copper being sprayed onto the electrician and his co-workers. There seems to be no end to the cases of individuals being shocked, sometimes fatally, by the use of singly polarized three-wire extension cords with the ground prong removed, and one frequently reads about the homeowner who, in the process of installing a TV antenna, was shocked or killed because the antenna and mast were allowed to contact a power line.

Warning signs seem to have little effect. A person may read the sign the first time it is encountered, but not the second and succeeding times, and even its presence is often forgotten. In suits that follow accidents such as those above, the victim is often unaware that the TV antenna and mast came with specific instructions and warnings about the danger from power lines, nor that aluminum ladders have similar warning labels. Virtually no one reads the tags on extension cords that specify limitations on the allowable size of the load.

Testing of people to determine their reliability using procedures analogous to the tests on hardware has been done, with results that give reliabilities ranging from lows of 0.01 to as high as 0.99. Mood, alertness, boredom, and fatigue are among the factors that determine reactions to situations that seem to be routine until an accident occurs. The degree of danger that would result from a mistake is frequently given little consideration, especially if the situation is such that the hazard is present much of the time. It is for this reason that we find electricians working on equipment that is energized, rather than having it disconnected and locked out.

There is no magic formula that can be used to solve the problem of human error. The designer must study the product with the intent of finding all of the ways in which misuse by a human being could initiate events leading to a disastrous result (an accident). Once these have been identified, the next step of course is to try to guard against their occurrence, or at least to mitigate the damage that might result.

In spite of the gloomy outlook of the previous paragraphs, progress is being made in the human reliability area. The reader may wish to refer to Section 6.3 of Reference 2 or to Reference 9. Some insight into the evaluation of a product as part of a device-plus-person system appears in Chapter 15.

13.12 SUMMARY

In this chapter, the following important points should be noted:

1. In-use life may be characterized by three stages, that of early failures, that of chance failures, and wearout.
2. The mean time between failures m, the mean wearout life M, and the standard deviation σ characterize the chance failure and wearout stages of product life.
3. Tests have been described to determine m, M, and σ experimentally.
4. The failure rates of some components are available in the technical literature, and the failure rates of others should be available from the manufacturers.
5. Methods available to improve reliability include redundant and standby systems. For most products, however, backup systems are not feasible. For these products, improvement of reliability during design, as discussed in Section 13.9, should claim the designer's attention.

What has been presented here constitutes a minimum of what every designer should know about reliability. For many products the designer can achieve an acceptably high level of reliability by using the contents of this chapter. If a specified value of reliability must be achieved, especially if it is a very high value, the designer should work with an engineer who specializes in reliability.

REFERENCES

1. Doty, Leonard A. *Reliability for the Technologies,* Industrial Press, New York, 1985.
2. Modarres, M. *What Every Engineer Should Know About Reliability and Risk Analysis,* Marcel Dekker, New York, 1993.
3. Bazovsky, Igor. *Reliability: Theory and Practice,* Prentice-Hall, Englewood Cliffs NJ, 1961.
4. Engelmann, Richard H., and William H. Middendorf. *Handbook of Electric Motors,* Marcel Dekker, New York, 1995.
5. O'Connor, P. D. T. *Practical Reliability Engineering,* 2nd Edition, John Wiley & Sons, New York, 1985.
6. Lewis, E. E. *Introduction to Reliability Engineering,* 2nd Edition, John Wiley & Sons, New York, 1996, pp. 182-188.
7. Siddall, James N. *Analytical Decision-Making in Engineering Design,* Prentice-Hall, Englewood Cliffs NJ, 1972, pp. 312-318.
8. Fuqua, N. B. *Reliability Engineering for Electronic Design,* Marcel Dekker, New York, 1987.

RELIABILITY

9. Kirwan, Barry. *A Guide to Practical Human Reliability Assessment,* Taylor & Francis, Bristol PA, 1994.

REVIEW AND DISCUSSION

1. What is the definition of reliability? Of unreliability?
2. What is the definition of per unit failure rate, and what is the defining equation?
3. How can tests be run to determine chance failure rate?
4. Draw the failure rate curve and the reliability curve for chance failures on the same set of axes.
5. What is the basic equation for $R(t)$?
6. What simplifying assumption is made for chance failure in computing reliability?
7. For a typical consumer product, which is greater, m or M? By how many orders of magnitude?
8. What is meant by burn-in time? How is it computed? How long is the actual burn-in period in terms of the calculated burn-in time? Why?
9. What well-known curve is used as a mathematical model for wearout?
10. Sketch a curve of failure rate due to wearout and a curve of the resulting reliability on the same set of axes.
11. What is meant by series reliability? Do the subsystems actually have to be connected in series for series reliability concepts to be applicable? How is overall reliability computed for series reliability?
12. What variable is convenient to use when computing parallel reliability?
13. What is a redundant system? What is a standby system? How do they differ?
14. Name three ways to improve reliability in the design process. Do any or all of these add to the product cost?
15. The curve of Figure 13.3 is sometimes referred to as a mortality curve. Can you relate this curve to the mortality of the human race? Which part of the curve are you on?
16. The owner's manuals of automobiles having a timing belt in the engine all recommend replacement at 60,000 miles. Timing belt failure in some engines can cause several thousand dollars of damage to the engine, and in any case failure will make the automobile inoperative until a new belt is installed. What do you believe to be the shape of the curve of timing belt failures versus mileage? Do the manufacturers anticipate any early failures? What is the order of magnitude of the mean number of miles to wearout?

PRACTICE PROJECTS

1. As you proceed through this chapter, design tests that will give data on the mean time between failures and the mean wearout time for the product you have chosen to develop.
2. One hundred valves were tested for 100 hours under normal operating conditions. One unit failed at each of the following times: 1, 6, 15, 20, 50, 70, 85, 85, 90, and 99 hours. What is the mean time between failures? What is the 100-hour reliability of these valves? Obtain values assuming (a) that failed units were immediately replaced so that the number of operating valves was constant, and (b) that failed units were not replaced.
3. A system is made of 100 *new* similar components. Ten are substandard, and have a mean time between failures of 10 hours. The standard units have a mean time between failures of 1,000 hours. All units have a mean wearout life of 100 hours. (a) In what time will 63% of the substandard units fail? (b) How many units (to the nearest integer) will fail in the first 100 hours of operation due to all modes of failure?
4. Classify the following failures as early, chance, or wearout: (a) An open circuit occurs in a solenoid coil in your 6-month-old washer. (b) Your bat breaks as you hit a ball. (c) You tear your trousers on a nail. (d) The heater core in your 8-year old minivan springs a leak. (e) An open circuit occurs in the starter motor of your 6-month-old car. [In (a) and (e), are your answers different? Would they be different if the age of each unit was 1 year rather than 6 months?]
5. In the following, assume that 100 new components of the same kind are used in each test. (a) The units have been debugged (they have gone through a burn-in process). They are operated for t = m hours, m being the MTBF. How many would you expect to survive chance failure? (b) The units have been debugged as in (a), but they are now operated for t = M hours, M being the mean wearout life. How many would you expect to still be operable neglecting chance failure? Taking chance failure into account? (c) The units have not gone through the burn-in process. They are operated for BIT (burn-in time) hours. What percent of the early failures would you expect to occur?
6. A simple resistor-capacitor differentiating network consisting of one resistor and one capacitor is used in a servo system. If the failure rate of the resistor is 0.21×10^{-6} units per hour and for the capacitor the failure rate is 0.23×10^{-6} units per hour, what is the reliability of the network for 5 years of operation at 2,000 hours per year?
7. A certain system has series-parallel relationships among its parts as shown in Figure 13.16. Reliability values for the subsystems for a given time t are: $R_1 = 0.8$, $R_2 = 0.7$, $R_3 = 0.6$, $R_4 = 0.5$, and $R_5 = 0.4$. What is the system reliability? Consider an electronic device composed of two subcircuits

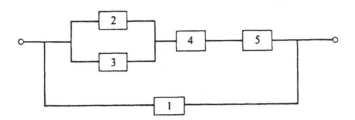

Figure 13.16 Subsystems in series and parallel relationships.

having 2,000-hour reliabilities of 0.90 and 0.80. Failure or nonfailure of one subcircuit has no effect on the life of the other, but failure of either part results in failure of the device. What is the reliability of the device for 500 hours?

8. Ten ball-point pens are tested to wearout failure—that is, they run dry. The failure times in minutes were measured as 1830, 2100, 1512, 1975, 1364, 1834, 1960, 1796, 2067, and 1839 minutes, respectively. Compute the mean wearout life M and the probability of wearout in 2000 minutes or less.

9. Assume that the reliability of a truck tire is 0.90 for 20,000 miles. The truck has double axles at the rear, each axle carrying two wheels on a side. Only two tires are needed on each side to carry the load. What is the reliability of the system of rear tires for 20,000 miles?

10. Reliability due to chance failure is given by Equation (9), λ being the per-unit failure rate. Unfortunately, this rate is generally not known with precision. Derive an equation for the tolerance of the reliability (T_R) in terms of the tolerance of the failure rate (T_λ). Note that the time (t) is specified in any reliability calculation, and therefore has zero tolerance.

11. Reliability associated with chance failure of a given product for a mission time of 1 year is 0.95. What is the reliability for 2 years?

12. Suppose that power lawn mowers have a mean wearout life (M) of 6 years and a standard deviation of 2 years. What is the probability that a new mower will wear out in 3 years?

13. A device to be operated at 80 C is composed of five components. To improve reliability, all components are derated by 20%. Assuming that the curves of Figure 13.14 apply to each component, what is the improvement in reliability associated with chance failure for a mission time of 5,000 hours?

14. To improve reliability, a certain military device has a primary and two standby devices used to initiate a certain critical operation, an operation that occurs once per mission. If the failure rate for the primary device is 1 per 100 missions, and 2 and 3 per 100 missions for the two standbys, what is the reliability of ten missions (a) assuming maintenance to return all devices to

prime operating condition after each mission, and (b) assuming no maintenance?

15. Strings of lights used for decoration typically contain a large number of small bulbs, all connected in series. If the per-unit failure rate of the bulbs is 1 per 1,700 hours and a typical string is turned on for 50 hours each year, what is the reliability of a string of lights having 36 bulbs for 1 year?

16. An electronic component is tested for 3 months and found to have a reliability of 0.97. It is expected that the component will have a constant failure rate. (a) What is the failure rate? (b) What is the MTBF? (c) For a life of 4 years, what is the reliability?

17. A manufacturer of home computers determines that its standard model has a failure rate of 0.035 per year in normal use. If the manufacturer is willing to repair 5% of the computers at no cost to the purchaser, what warranty period should be used?

18. The voltage to be applied to a stand-off insulator, such as that on an electrical distribution circuit, is 33,000 volts with a standard deviation of 1,500 volts. Insulators of this type have been found to have a standard deviation in their breakdown strength of 1/20 of the mean value. If the number of expected failures is to be limited to one per 1,000 insulators, what mean value of breakdown strength is needed?

19. Using the data of problem (18) and your value for mean breakdown strength, what factor of safety (FS) would be required to limit failures to one per 1,000 units?

14

ACCELERATED LIFE TESTING

Tempus edax rerum. [Time, the devourer of all things.]
Ovid, *Amores,* xv, 234

Throughout the design process, the engineer must test parts, subassemblies, or complete product models whenever it appears necessary to verify the functional integrity of a design. In addition, the product should be tested against comparable products from the leading competitors. The purpose of this latter test is to determine whether the proposed product is equal to, better than, or inferior to the product whose customers you want to attract.

The first problem faced by the designer at this stage is to define what "equal to, better than, or inferior to" means, and how this comparison should be accomplished. The test could conceivably be entirely quantitative in nature, but as the discussion in the introduction to Chapter 1 pointed out, although two products (for example, two refrigerators) will satisfy the same basic requirements, users or prospective users who compare the products will still judge one to be better or more suitable than the other. Such judgments, while subjective in nature, are frequently the basis for the decision to purchase one product and reject the other.

Even before a comparison test is done, tests should be performed to verify the functional integrity of the product. If functional integrity cannot be demonstrated, then the comparison test should not be done at all. For a product to be judged to have functional integrity, it must be shown that the product will perform acceptably to the end of its expected life. This means that the percentage of units that fail during the warranty period (presumably all early failures) will be less than a given acceptance value. It also means that the chance failure rate is sufficiently low and the mean wearout life is sufficiently long that customer satisfaction is assured. The tests needed are those described in the preceding chapter, but as indicated there it may be necessary to compress several years of usage into a few months. This is generally not an easy task. Furthermore, many products fail in complex ways, as will be discussed in an example later in the chapter.

Compression of years of use into a few months requires that tests be accelerated. There are four factors that govern the permissible degree of acceleration: The sample size, the environment, the testing time, and the anticipated failure mode. For example, as pointed out in the preceding chapter, in testing for chance failures the number of units on test can be increased with a corresponding decrease in testing time. The intensity of a selected aspect of the environment, such as temperature, voltage gradient, or humidity, can be increased and the testing time decreased. There are, however, limits to the increases in stress that can be used. If stress is increased too far, failures will begin to occur for reasons that are different from the reasons failures occur in normal operation. The test then yields results that are invalid as far as normal use of the product is concerned.

It is also important that the test not be accelerated to the point at which normal operation of the product no longer occurs. For example, if one is testing relays, such as that shown in Figure 8.4, after armature opening or closure the main spring will vibrate at some natural frequency, the amplitude of the vibrations gradually dying out. If the frequency with which the relay is cycled during testing is too high, the main spring's vibrations will not have died out before the next operation occurs, and the amplitude of the resulting vibrations may be larger or smaller than normal, resulting in either reduced life because of accelerated fatigue, or longer life because of fatigue reduction.

14.1 THEORY OF ACCELERATED TESTING

The simplest accelerated test is one that involves a repetitive action which can be done at a faster than normal rate, such as the relay testing of the preceding paragraph. Test of a shaver, referred to in the preceding chapter, is an example. The circuit breaker tester described in Section 10.4 is an example of a device used to cycle circuit breakers through a large number of repetitions in a short period of time. Many recognized standards specify the number of operations of a device, their frequency, and the duty cycle. For example, UL 943 Ground Fault Circuit Interrupters (GFCI's) specifies that the GFCI must pass a load test comprising 3,000 cycles, at the rate of 6 cycles per minute, each cycle consisting of 1 second ON time and 11 seconds OFF time, the load current being equal to the ampere rating of the device and the power factor being in the range of 0.75 to 0.80. Following this test, an additional 3,000 cycles are to be performed using the supervisory circuit (the TEST button, in the case of residential units) for automatic tripping. Such rigorous repetitive tests are commonly found in mandatory standards.

Most accelerated tests involve an increase in intensity of a particular environmental stress, such as an increase in temperature or humidity. Samples of

ACCELERATED LIFE TESTING

the test material (or component or device) are subjected to overstressing for relatively long periods of time. Specimens are withdrawn periodically from this test population and a selected characteristic is measured. When this characteristic has changed by 50% from its original value, the material is said to have reached the end of life. After data are obtained at several (usually four) high-stress levels, the end of life under normal stress can be estimated by extrapolation. This is an especially useful technique when the deterioration mechanism is such that the experimental end of life times fall along a straight line.

If time to end of life is now plotted versus stress level, a curve such as one of those in Figure 14.1 will frequently be the result. These curves can often be expressed analytically by an inverse power law equation

$$y = \frac{B}{x^m} \qquad (1)$$

Note that stress equal to unity corresponds to design stress, and the "Time to end of life" scale is therefore a normalized (percent) scale. If we now take the logarithm of Equation (1), we obtain

$$\ln y = -m \ln x + \ln B \qquad (2)$$

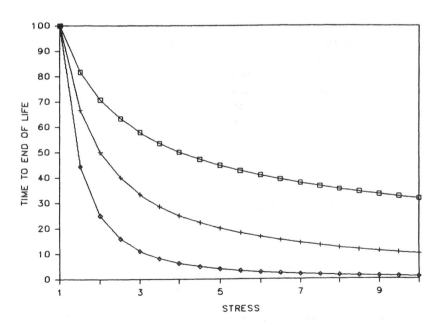

Figure 14.1 Deterioration expressed by the inverse power equation, where the symbol □ is for m = 0.5; + for m = 1.0; and ◊ for m = 2.0.

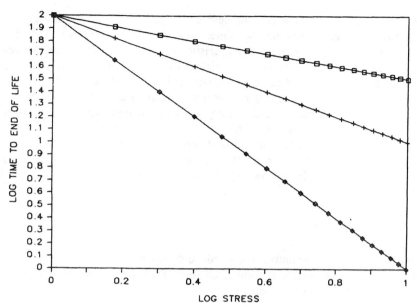

Figure 14.2 Logarithm of inverse power equation, where the symbol □ is for m = 0.5; + is for m = 1.0; and ◊ is for m = 2.0.

which is the equation of a straight line when plotted on log-log paper or on rectangular coordinate paper for which the axes are scaled in units of logarithms of the variables, as in Figure 14.2. A plot of this type will make it evident whether the given set of data fits the inverse power model, and, if so, what the values are of m and B. The reader may wish to consult Section 8.8 of Reference 1 for a more theoretical (and compact) discussion.

Another model that fits deterioration especially well when a chemical reaction is involved is the exponential equation

$$y = Ae^{-rx} \qquad (3)$$

This is known as the Arrhenius or single reaction rate equation. It is based on the following assumptions:

1. The failure rate is constant with time, and
2. The failure mechanisms are identical at each stress level used with those at the normal stress level.

Taking the natural logarithm of Equation (3) yields

$$\ln y = -rx + \ln A \qquad (4)$$

ACCELERATED LIFE TESTING

Whether data will fit this model or not can be determined by plotting the data on semilog paper. In the ideal case, data points will fall on a straight line.

It is obvious that with higher stress on a product failure will occur in less time. That is, every accelerated life curve of time plotted against stress will have a negative slope. However, accuracy of the forecast of expected life at normal stress levels requires that a theoretical basis be established that indicates that the functional relationship chosen was the correct one. In other words, it is not sufficient to note that the data generally fall along the inverse power law or along a negative exponential and then choose one or the other as a best fit. Extrapolation from closely spaced data points in an attempt to predict life under normal stress can result in significant error unless the model chosen truly fits the physical phenomena. Even if the model chosen is the correct one, minor variations in the distribution of the data points around a straight line can lead to significant error because of the uncertainty of the slope of the line. Some hints of the correctness of a given equation for a particular phenomenon can often be obtained by considering the behavior of the product at extremes, such as zero applied voltage, time approaching infinity, zero velocity, extremely high force, and so forth.

14.1.1 Accelerated Aging of Plastics

The use of plastic parts in all kinds of devices and products has been increasing rapidly in recent years. One has only to look at power hand tools, for example, to realize that the metal housings of previous manufacture have in many cases been replaced by plastic housings, and drive shafts that were previously made of steel must now be made of plastic for double-insulated tools. Plastic gears and gear trains have become the norm in light duty applications. FAX machines, printers, enclosures for monitors, and the cases of PC's are largely plastic. Because plastic has become such an important part of many products, the designer who intends to incorporate plastic parts into a product needs to know something about the way in which plastic ages. The following discussion therefore provides an introduction to the aging of plastics and also serves as an example of the way in which accelerated life tests may be conducted.

We tend to look at plastic as a simple product, but it is subject to aging in normal use to a much greater degree than other materials, such as metal. It degrades by a complex combination of effects due to scission of molecular chains in some cases, oxidation, changes in crystallinity, formation of dense cross-linked skin, and so forth. In 1948, T. W. Dakin [2] realized that there was a connection between aging phenomena and the Arrhenius law of chemical reaction rates, leading to formulation of Equation (3) as

$$k = Ae^{-E/RT} \qquad (5)$$

where
k = specific reaction rate
A = frequency factor of molecular encounter
E = activation energy (which is constant for a particular reaction)
R = universal gas constant
T = absolute temperature

Equation (4) then becomes

$$\ln k = -\left(\frac{E}{R}\right)\frac{1}{T} + \ln A \qquad (6)$$

Following Dakin's reasoning, if the physical property selected as an indicator of age is a function of the specific reaction rate, k, the aging test can be accelerated by subjecting specimens to various values of elevated temperature. Heat aging affects the properties of a plastic resin in various ways. The most adverse effect is on the tensile impact strength; the effect on tensile strength is smaller, as is the effect on electrical breakdown strength. When end-of-life of a material is being determined, the property to be measured is the one that corresponds to the failure mode to be expected in real-time. Tensile strength is therefore appropriate for evaluation in mechanical design applications; electrical breakdown strength is appropriate for the indication of aging for electrical properties.

The aging of plastics is carried out by placing a sufficient number of specimens (say 40 to 50) in each of a number of ovens operating continuously

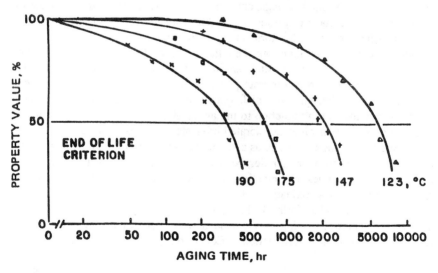

Figure 14.3 Aging according to the single chemical reaction rate theory.

ACCELERATED LIFE TESTING

at different fixed temperatures. Typically, four ovens are used, at temperatures of, for example, 190, 175, 147, and 123 C. The material at the highest temperature will change more quickly than will the others, and specimens will be withdrawn for testing from the 190 C oven first. Experience with the particular material to be tested or a similar material will serve as a guide as to how soon to withdraw the first samples, and how frequently to withdraw specimens, both from the 190 C oven and from the others. The specimens withdrawn from the ovens are tested to destruction for the property, such as tensile strength, that is relevant to the application intended for the material. The tensile strength as a percentage of the tensile strength of unaged specimens is then plotted as in Figure 14.3, and is used to determine the time necessary to reach the 50% level, which is defined as the end-of-life.

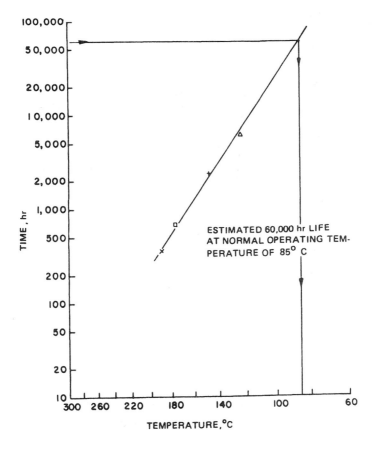

Figure 14.4 Prediction of life under normal operating conditions.

When data has been obtained for all of the temperatures, the data are plotted as shown in Figure 14.4, where the scale of the horizontal axis is a suitable multiplier on 1/T, thus conforming to Equation (6). The resulting curve can be used in a number of ways. One way is to extrapolate to normal operating temperature so that the life can be predicted. Another way, shown in the figure, is to determine from the required life (60,000 hours in the figure) a suitable operating temperature. The end-of-life point at the lowest temperature often requires that the test be run for over one year, but that year may represent 30 years of normal usage.

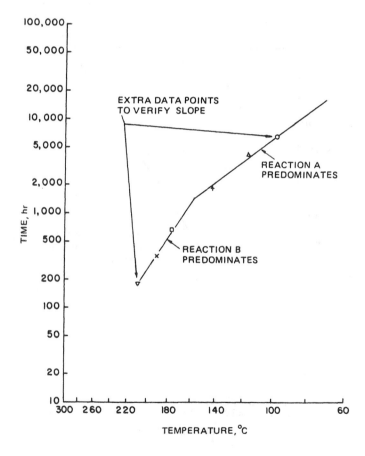

Figure 14.5 Aging when different reactions occur over the test range.

ACCELERATED LIFE TESTING

It is usually advisable to test a control material with the one for which a life rating is sought. The control material should have a life characteristic which is already known by previous tests, or possibly from field experience. If the control material performs as expected, one may be confident that the test was run correctly.

Not all plastic degradation involves a single reaction over the temperature range used during the test. The first clue that such a situation exists is likely to be a scattering of points around a straight line on the plot. The tendency to draw the best straight line should be resisted, even though the scattering may be the result of experimental error. This is especially true if the data for the control material fits a straight line. If the original four points of Figure 14.5 are examined on the assumption that more than a single reaction is involved, there is also the temptation to draw a broken curve, such as that shown there. However, neither segment of the curve is defined well with only two points, and additional test temperatures should be used as shown. The test at the higher temperature will consume a relatively small amount of additional test time, but the test at the lower temperature will require a very long period. If it is suspected that more than one reaction will be involved, one should allow for this possibility when selecting temperatures initially, even though more ovens will be required.

This example can be used as a model for the design of accelerated tests of any product. The steps are:

1. Identify the deterioration mechanism, if at all possible.
2. Determine the performance characteristic most critically affected by the deterioration.
3. Subject an appropriate number of specimens to several levels of increased stress of the variable that causes deterioration or subject them to numbers of operations beyond that to be expected in normal use.
4. After close examination, classify failures as early, chance, and wearout. Determine (1) suitable burn-in time at elevated stress (if applicable), (2) mean time between chance failures, and (3) an estimate of mean wearout time at normal stress using extrapolation, as in Figure 14.4.

One more step should be taken with most products. The failure should be classified according to the potential for causing property damage or injury, as discussed in Section 2.3.2. Product safety must be given careful consideration.

14.2 TESTING A FEW SAMPLES

Suppose that there are relatively few samples available for test and the aim of the test is to determine to some confidence level that the desired mean load to failure is exceeded by the actual mean load to failure of the population, a level that we shall designate by W_D. If a number of samples passed a test at a stress level

W_D you would feel that the design is close to that which you want. What if they failed significantly below W_D? Certainly you would be convinced that the device or system needs to be improved. On the other hand, if the test was passed at a stress significantly greater than W_D, you would probably conclude that there may be a degree of overdesign.

That is, although all of these samples may in fact be from the same population, consistent failure below W_D or consistent successful operation above W_D gives a strong indication that the samples are from populations whose means are either below W_D or above W_D. In fact, if a single sample is tested at a stress level of $W_O > W_D$ as shown in Figure 14.6 and passes, the shaded area P represents the probability that it is from a population whose mean is greater than W_D, but not necessarily equal to W_O. The unshaded area $(1 - P)$ represents the probability that the sample is from a population whose mean is less than W_D. The area P increases as higher stresses are used for the test and successfully passed, that is, as W_O in Figure 14.6 is moved to the right. Also, the unshaded area, the probability that the sample is from a population whose mean is less than W_D, decreases as the stress level is increased and the test is passed. The shaded area P (expressed in percent) is also known as the confidence level or degree of confidence.

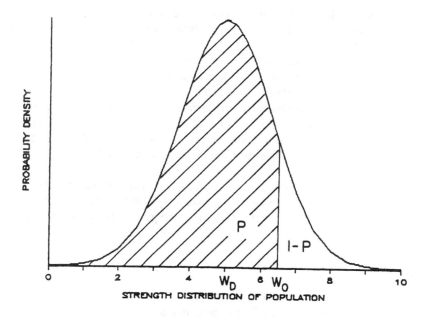

Figure 14.6 Normal distribution of test results.

ACCELERATED LIFE TESTING

An even more satisfactory way of increasing confidence that the product has a mean characteristic equal to or greater than W_D is to repeat the test on a number of samples at a stress level equal to W_O. If all survive, the joint probability that the population mean is less than W_D is

$$p(1,2,3,\ldots,n) = (1-P_1)(1-P_2)\ldots(1-P_n) \tag{7}$$

or, if each comes from the same distribution,

$$p(1,2,3,\ldots,n) = (1-P)^n \tag{8}$$

so that

$$p_n(W > W_D) = 1 - (1-P)^n \tag{9}$$

Testing a number of samples permits the use of lower stress levels, avoids the possibility that the stress introduces a different failure mode than occurs in normal use, and realistically introduces the variability of manufactured products into the test.

As an example of working with a limited number of samples, suppose a component must be designed to sustain an average load of 2,000 newtons. At what load should five units be tested to give a 95% confidence that the design goal has been reached? Assume that the standard deviation of the population is one-tenth of the average load W_D, or 200 newtons.

Table 14.1, constructed for k = 0 by use of Equation (9), is used to obtain the solution. In this table, N is the sample size and k is the number of test units that fail. $(W_O - W_D)/\sigma$ is the difference between the unknown load (the one we are to solve for) and the desired load (mean of the standard distribution) expressed in terms of standard deviations. That is, it is analogous to the z of Equation (18) of Chapter 13 or the z of Table 8.3. The degree of confidence is given in decimal form by the three-place values in the columns below k = 0, 1, or 2, respectively.

To solve for the load to be used in the situation cited above, we enter the table for N = 5 and k = 0, and interpolate between 0.935 and 0.954 to find that

$$\frac{W_O - W_D}{s} = -0.12$$

Hence

$$W_O - W_D = -0.12 \times 200 = -24$$

and

$$W_O = 2000 - 24 = 1976 \text{ newtons.}$$

Table 14.1 Confidence that the Mean Value of a Property Exceeds the Desired Value of a Normal Distribution

$\dfrac{W_O - W_D}{\sigma}$ (N=1)	k = 0	$\dfrac{W_O - W_D}{\sigma}$ (N=2)	k = 0	$\dfrac{W_O - W_D}{\sigma}$ (N=3)	k = 0	k = 1
0.3	0.618	0.0	0.750	−0.3	0.764	
0.4	0.655	0.1	0.788	−0.2	0.806	
0.5	0.692	0.2	0.823	−0.1	0.843	
0.6	0.726	0.3	0.854	0.0	0.875	
0.7	0.758	0.4	0.881	0.1	0.903	
0.8	0.788	0.5	0.905	0.2	0.926	
0.9	0.816	0.6	0.925	0.3	0.944	
1.0	0.841	0.7	0.941	0.4	0.959	
1.2	0.885	0.8	0.955	0.5	0.971	0.773
1.4	0.919	0.9	0.966	0.6	0.979	0.816
1.6	0.945	1.0	0.975	0.7	0.986	0.853
1.8	0.964	1.2	0.987	0.8	0.990	0.884
2.0	0.977	1.4	0.993	0.9	0.994	0.911
2.4	0.992	1.6	0.997	1.0	0.996	0.932
2.8	0.997	1.8	0.999	1.2	0.998	0.963
3.2	0.999	2.0	0.999	1.4	0.999	0.981

$\dfrac{W_O - W_D}{\sigma}$ (N=4)	k = 0	k = 1	$\dfrac{W_O - W_D}{\sigma}$ (N=5)	k = 0	k = 1	k = 2
−0.5	0.771		−0.8	0.696		
−0.4	0.815		−0.7	0.750		
−0.3	0.854		−0.6	0.799		
−0.2	0.887		−0.5	0.842		
−0.1	0.915		−0.4	0.879		
0.0	0.936		−0.3	0.910		
0.1	0.955	0.745	−0.2	0.935		
0.2	0.969	0.796	−0.1	0.954	0.759	
0.3	0.979	0.841	0.0	0.969	0.813	
0.4	0.986	0.879	0.1	0.979	0.858	
0.5	0.991	0.910	0.2	0.987	0.896	
0.6	0.994	0.934	0.3	0.992	0.926	
0.7	0.997	0.954	0.4	0.995	0.949	0.773
0.8	0.998	0.968	0.5	0.997	0.966	0.825
0.9	0.999	0.978	0.6	0.998	0.978	0.869
1.0	0.999	0.986	0.7	0.999	0.986	0.905

ACCELERATED LIFE TESTING

That is, if all five samples pass a loading test at 1976 newtons, there is a 95% confidence that the population from which the five samples came will, on the average, sustain a load of 2000 newtons.

Now, suppose there is only one sample to test. What load should be applied to state with 95% confidence that the average load sustained by the population will be 2000 newtons? Entering the table for N = 1 (and k = 0), we interpolate between 0.945 and 0.964 and find that

$$\frac{W_O - W_D}{\sigma} = 1.63$$

from which

$$W_O = 2000 + 1.63 \times 200 = 2325 \text{ newtons}$$

If several units are tested and one or more of the units fail during test, the confidence with which statements can be made as to the relationship between W_O and W_D will be sharply diminished. The method for computing the confidence level is the same as in the examples above, except that one enters the appropriate part of Table 14.1 with k = 1 or 2. The derivation of the equation used to calculate the confidence level with $k \neq 0$ may be found in Section 5.2 of Reference 3, which also contains in Appendix A tables for a number of samples up to 10 and with $-4.0 \leq \frac{W_O - W_D}{\sigma} \leq 4.0$.

14.3 GOODNESS OF FIT; CENSORED DATA

Because many natural phenomena display normal (Gaussian) distributions of one quality or another, we have a tendency to assume that data obtained from tests should somehow or other fit such a distribution. In the following, we shall see one way of testing that hypothesis, and of simultaneously gaining other useful information.

To illustrate the process, consider a test in which various values of voltage are applied to insulation and hours to failure are recorded, as shown in Table 14.2. The failure phenomenon is expected to behave in accordance with Equation (3), $y = Ae^{-rx}$, with x replaced by the reciprocal of the voltage and y replaced by the rate of deterioration. The design specification is that the product must have a 50% probability of surviving for 50 hours with a voltage of at least 250 volts.

Now it sometimes happens in testing a product with an accelerated life test that a few units are stubborn, and refuse to fail within a reasonable time. The test is finally terminated when the judgment is made that "They will never break down!" Even if prolonged testing would finally result in the failure of these stubborn units, there would always be doubt whether the resulting data should be included in the analysis because the stubborn samples do not appear to be truly representative of the entire population. They are referred to as "outliers." In such a situation, one *censors* the data by including the data for these units but designating it so that it is evident that the data does not represent time to failure. A simple way to do this, of course, is by the use of the symbol > to indicate that the unit was still operable after a certain number of hours, months, or number of operations. This device has been used in Table 14.2.

The analysis of the data requires that the raw data be plotted on log probability paper. We begin by placing the data for each voltage in ascending order. The raw data for 250 volts, when arranged in this way, are:

14, 22, 42, 44, >75

The censored datum (>75) thus appears at the end of the list. If there were repeated values, those values would go in their proper places, but would not be differentiated among themselves.

The next step is to count all values (five in this case) and, using Table 14.3, assign a "plotting position" for each value according to its rank. This table gives plotting positions for ranks 1 to 10 for sets of observations from 2 to 10.

The plotting positions are calculated from the equation

$$P = 100(R - 0.5) / N$$

in which P is the plotting position on the log probability paper and R is the rank of the observation, ranging from 1 (the lowest position) to N, the total number of observations, including any censored data. It will be noted that the plotting positions are centered on 50 (percent) and are distributed uniformly over a scale of 100 (percent) with left and right ends of the scale closed against each other (as, for example, by forming the scale into a ring). The plotting position is numerically equal to the cumulative percentage of the area under the normal distribution curve.

Table 14.2 Test Results Including Censored Data

Voltage (V)	Time to failure in hours for sample number				
	1	2	3	4	5
300	6	30	10		
250	>75	22	14	42	44
240	54	76	38		
225	>75	>75	>75		

ACCELERATED LIFE TESTING

Table 14.3 Plotting Positions for Test Data

N*	1	2	3	4	Rank 5	6	7	8	9	10
10	5.0	15.0	25.0	35.0	45.0	55.0	65.0	75.0	85.0	95.0
9	5.6	16.7	27.8	38.0	50.0	61,1	72.2	83.3	94.4	
8	6.2	18.7	31.2	43.7	56.2	68.7	81.2	93.7		
7	7.1	21.4	35.7	50.0	64.3	78.6	92.9			
6	8.3	25.0	41.7	58.3	75.0	91.7				
5	10.0	30.0	50.0	70.0	90.0					
4	12.5	37.5	62.5	87.5						
3	16.7	50.0	83.3							
2	25.0	75.0								

* N is the number of observations.

For the ranked data for 250 volts, the plotting positions become:

Value	Plotting Position
14	10.0
22	30.0
42	50.0
44	70.0

Note that the censored datum point enters into the determination of plotting positions, but the value itself is not used. Each usable value is now plotted on the log-probability paper of Figure 14.7, the plotting position defining the abscissa and the value being plotted on the ordinate. (This paper is available in many technical supply stores.) The data for the other voltages in the test are processed in the same way. Since none of the samples in the 225-volt test failed during the test, all of this set of data is censored. Obviously, there must be at least two uncensored data points for any kind of plot to be drawn.

The next step is to draw the best fitting straight line through the points representing time in hours and plotting positions for each of the voltages of the test. There are mathematical techniques for determining the best fitting line, but Shapiro in Section 6.2 of Reference 1 states that "It is helpful to use a clear ruler in evaluating whether the plot is linear. . . More formal techniques for finding the best line are usually not worth the effort." We note in Figure 14.7 that the

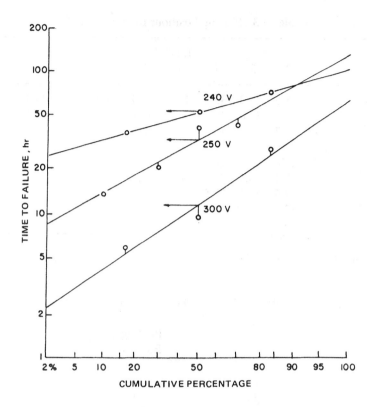

Figure 14.7 Determination of most likely failure times.

data for 240 volts lie almost exactly on a straight line, and that for 250 volts and 300 volts the data points are all very close to the straight lines on the plot.

Because of the straight-line plots, we can now conclude that the insulation failure times at each test voltage are probably represented very closely by a normal distribution curve. This conclusion may be drawn because the log-probability paper used has a horizontal scale that is dependent on the shape of the normal distribution curve. If we cannot draw a straight line through the points on the plot, we conclude that the distribution of failure times was not a normal distribution. We could then try other log-probability papers, such as one for an exponential distribution, until we found one for which a straight line could be drawn. At that point we would be able to identify the distribution to which the data conforms.

ACCELERATED LIFE TESTING

Having the straight lines of Figure 14.7, we can now read off the value of hours at the 50% cumulative percentage point for each of the test voltages, resulting in the following tabulation:

Volts	Hours to failure
240	54
250	34
300	12

Plotting these results, as on the log reciprocal paper of Figure 14.8, results in data points which lie on a straight line, and from which we can confirm that the

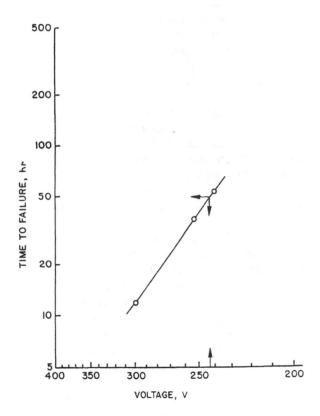

Figure 14.8 Insulation failure time versus applied voltage.

insulation will not meet the specification of a 50% survival probability at 250 volts.

As a bonus, we note that the slope and the standard deviation are related. However, rather than attempting the difficult task of determining the standard deviation from the slope, it is far easier to make use of the fact that σ is 40% of the difference between the values of the function at the 90% and 10% points on the probability scale.

The reader interested in various statistical models and means for testing them may wish to refer to Chapter 6 of Reference 1.

14.4 EVALUATING PROPOSED IMPROVEMENTS

In engineering practice, experiments on products are often performed in order to evaluate proposed improvements that may be made by minor changes in the materials, processes, or components used to produce the device or system. Such experiments are known as factorial experiments because they determine the effect of various factors (independent variables) on some characteristic of the product that is of special interest. For example, it may be desirable to determine the effect of temperature on the output of a chemical process, to compare the life of tires supplied by different manufacturers, or to determine the effect on surface wear of a sliding block if the viscosity of the lubricant is changed. Unfortunately, the effect of the change, if any, may appear to be lost when immersed in the data that reflect normal manufacturing variation or experimental error.

A very powerful method of analyzing data for such a situation is known as the analysis of variance. This method allows us to separate the changes in the measured variable due to the various changes introduced by the experimenter (these are frequently called "treatments" in the literature) from the changes due to normal variation in specimens, experimental error, and so forth. In this section, we will examine the analysis of variance very briefly and look at an example of what is called a single-factor experiment, that is, an experiment in which only one independent variable is changed. Those who have interest in the subject or a need to explore the area in greater depth may wish to refer to Chapter 6 of Reference 3 or to Reference 4.

The model that leads to a procedure for handling the data begins with the realization that every measurement X_{ij} may be treated as if it were the sum of three components. One component is the population mean of the variable, μ; a second component is the effect T_j of the treatment (or change of independent variable); and the third consists of uncontrolled effects, which may be symbolized by e_{ij}. In what follows, the subscript j is used to identify the treatment (the various values of the independent variable) and runs from 1 to k. The subscript i is used to identify the observations for a particular value of the independent variable. That is, X_{ij} is the ith observation of the quantity being measured while treatment j is in place. The subscript i runs from 1 to n_j.

ACCELERATED LIFE TESTING

If every measurement could be made with complete accuracy and precision and if every sample were identical with every other sample, the effects of the controlled change by the experimenter, that is, the treatment, would be obvious. However, as experiments are repeated, many changes occur. The specimens are not all identical, ambient temperature or humidity (or both) will change, supply voltages and air pressures fluctuate, and so forth. The analysis of variance gives the experimenter a method by which hypotheses regarding the data can be tested and conclusions expressed in terms of a degree of confidence.

As the first step after the data have been collected, the data are arranged as shown above the dashed line in Table 14.4. Each column contains the data observed during the first test, that is, when treatment (1) is in effect. The number of entries will vary from column to column.

The quantities below the dashed line are then calculated. A very effective shorthand notation is used in this area; the summation over a running subscript is shown by replacing the subscript with a period or dot. For example, $T_{.1}$ means

Table 14.4 Data Layout for Analysis of Variance

Treatment	1	2	...	j	...	k	
Original data	X_{11} X_{21}	X_{12} X_{22}	... ...	X_{1j} X_{2j}	... ...	X_{1k} X_{2k}	
	X_{i1}	X_{i2}	...	X_{ij}	...	X_{ik}	
	$X_{n_1 1}$			$X_{n_j j}$			
		$X_{n_2 2}$				$X_{n_k k}$	
Totals	$T_{.1}$	$T_{.2}$	...	$T_{.j}$	...	$T_{.k}$	$T_{..} = \sum_{j=1}^{k} T_{.j}$
Number	n_1	n_2	...	n_j	...	n_k	$N = \sum_{j=1}^{k} n_j$
Means	$\overline{X}_{.1}$	$\overline{X}_{.2}$	...	$\overline{X}_{.j}$	...	$\overline{X}_{.k}$	$\overline{X}_{..} = \sum_{j=1}^{k} T_{.j}$

the sum of all the data in the first column. $\overline{X}_{.1}$ means the average of the data in the first column. Moreover, $T_{..}$ is used to represent the sum of all the data, that is, $T_{..} = T_{.1} + T_{.2} + \cdots + T_{.k}$. Finally, $\overline{X}_{..}$ means the average of all the data.

Now consider the sources of variations. One obvious source is that the experimenter is changing the treatment (independent variable); this is being done deliberately and the eventual outcomes are what the experimenter wishes to examine. There are k different treatments used during the course of the experiment, but there are only k − 1 changes from one treatment to another. The factor (k − 1) is referred to as the number of degrees of freedom for this source of variations. Now it can be shown [4] that the sum of the squares due to the variations resulting from the treatments may be written as

$$SS_{treatment} = \sum_{j=1}^{k} \frac{T_{.j}^2}{n_j} - \frac{T_{..}^2}{N} \tag{13}$$

and the mean square for the variations due to the treatments is

$$MS_{treatment} = \frac{SS_{treatment}}{(k-1)} \tag{14}$$

This is the variance due to the treatments.

The remaining variations occur within treatments for the reasons cited earlier, such as variations from one specimen to another, random variations in environmental properties such as humidity, and power supply voltage or air pressure variations. These are the uncontrolled effects or errors, previously designated by the symbol e_{ij}. There are N data points altogether, but there are k treatments used in the experiment. The factor N − k is referred to as the number of degrees of freedom for these variations. Again it may be shown that the sum of the squares of the errors may be written as

$$SS_{error} = \sum_{j=1}^{k} \sum_{i=1}^{n_j} X_{ij}^2 - \sum_{j=1}^{k} \frac{T_{.j}^2}{n_j} \tag{15}$$

and the mean square for the variations due to errors is

$$MS_{error} = \frac{SS_{error}}{(N-k)} \tag{16}$$

This is the variance due to the errors.

ACCELERATED LIFE TESTING

Finally, we calculate a factor F given by

$$F = \frac{SS_{treatment}/(k-1)}{SS_{error}/(N-k)} \tag{17}$$

The factor F follows a statistical variation known as an F distribution. Table 14.5 gives values of this distribution for 75, 90, and 95% confidence levels. If F is greater than the number in the table corresponding to the same numerator and denominator degrees of freedom, the conclusion may be drawn that the treatments used significantly affect the variable measured during the test. Table 14.5 is an abbreviated version of more extensive tables which may be found in References 1, 3, and 4. In some of these tables $k-1$ and $N-k$ may be found with values up to ∞, with confidence levels to 99.9%.

This general discussion will be clearer if we use as an example a problem that arose some years ago in the electrical insulation industry. One of the tests

Table 14.5 F Distribution

| $N-k$ | $1-\alpha^*$ | \multicolumn{5}{c}{$k-1$} |||||
		1	2	3	4	5
1	0.75	5.83	7.50	8.20	8.58	8.82
	0.90	39.9	49.5	53.6	55.8	57.2
	0.95	161.4	199.5	215.7	224.6	230.2
2	0.75	2.57	3.00	3.15	3.23	3.28
	0.90	8.53	9.00	9.16	9.24	9.29
	0.95	18.51	19.00	19.16	19.25	19.30
3	0.75	2.02	2.28	2.36	2.39	2.41
	0.90	5.54	5.46	5.39	5.34	5.31
	0.95	10.13	9.55	9.28	9.12	9.01
4	0.75	1.81	2.00	2.05	2.06	2.07
	0.90	4.54	4.32	4.19	4.11	4.05
	0.95	7.71	6.94	6.59	6.39	6.26
5	0.75	1.69	1.85	1.88	1.89	1.89
	0.90	4.06	3.78	3.62	3.52	3.45
	0.95	6.61	5.79	5.41	5.19	5.05
6	0.75	1.62	1.76	1.78	1.79	1.79
	0.90	3.78	3.46	3.29	3.18	3.11
	0.95	5.99	5.14	4.76	4.53	4.39

Table 14.5 *(Continued)*

		\multicolumn{5}{c}{k − 1}
N − k	1 − α*	1	2	3	4	5
7	0.75	1.57	1.70	1.72	1.72	1.71
	0.90	3.59	3.26	3.07	2.96	2.88
	0.95	5.59	4.74	4.35	4.12	3.97
8	0.75	1.54	1.66	1.67	1.66	1.66
	0.90	3.46	3.11	2.92	2.81	2.73
	0.95	5.32	4.46	4.07	3.84	3.69
9	0.75	1.51	1.62	1.63	1.63	1.62
	0.90	3.36	3.01	2.81	2.69	2.61
	0.95	5.12	4.26	3.86	3.63	3.48
10	0.75	1.49	1.60	1.60	1.59	1.59
	0.90	3.28	2.92	2.73	2.61	2.52
	0.95	4.96	4.10	3.71	3.48	3.33
11	0.75	1.47	1.58	1.57	1.57	1.56
	0.90	3.23	2.86	2.66	2.54	2.45
	0.95	4.84	3.98	3.59	3.36	3.20
12	0.75	1.46	1.56	1.56	1.55	1.54
	0.90	3.18	2.81	2.61	2.48	2.39
	0.95	4.75	3.89	3.49	3.26	3.11
13	0.75	1.45	1.55	1.55	1.53	1.52
	0.90	3.14	2.76	2.56	2.43	2.35
	0.95	4.67	3.81	3.41	3.18	3.03

* α is the same as P in Figure 14.6; that is, it is the area under a normal distribution curve from $-\infty$ to a particular point on the population axis. Then $1 - \alpha$ is the confidence level in decimal notation.

performed on electrical insulation is an accelerated test designed to determine the tendency of the insulation to develop carbonized conductive paths (this is known as "tracking") under conditions of voltage stress and moisture. The test requires an electrolyte of approximately one-tenth of one percent ammonium chloride in distilled water. This is applied in measured drops at the rate of one drop every 30 seconds between two electrodes spaced 4 mm apart with specified potential difference. The electrolyte is specified as having a resistivity of 385 ± 5 Ω-cm at 25 C. The number of drops is counted until failure (tracking) occurs. Inconsistency in the results among several laboratories indicated that a study of the sensitivity of the test to variations in electrolyte resistivity should be made.

ACCELERATED LIFE TESTING

Because the test tends to give a wide variation in the data even under seemingly identical conditions, this was not an easy study to perform.

Three batches of electrolyte were prepared having resistivities of 340, 382, and 415 Ω-cm, and the test was run using each of the three batches of electrolyte on five specimens each, yielding the original data in Table 14.6. In order to reduce the size of the numbers and make the calculations easier, the data may be coded (that is, reduced by a fixed number), as shown in the center section of Table 14.6, without altering the results below. We note that there were three treatments; thus $k = 3$ and $k - 1 = 2$. Since $N = 15$, $N - k = 12$.

We now apply Equations (13) through (17) to these data with the objective of obtaining a value of F with which to enter Table 14.5.

Equation (13) yields

$$SS_{treatment} = 1470 \tag{18}$$

Then Equation (14) yields

$$MS_{treatment} = SS_{treatment}/(k-1) = 735 \tag{19}$$

Table 14.6 Number of Electrolyte Drops to Failure for Various Values of Electrolyte Resistivity.

Treatment (resistivity)	340	382	415	
Original data	33	37	58	
	38	71	100	
	15	33	54	
	40	39	40	
	56	33	47	
Coded data (Original data less 45)	−12	−8	13	
	−7	26	55	
	−30	−12	9	
	−5	−6	−5	
	11	−12	2	
Totals, $T_{.j}$	−43	−12	74	$T_{..} = 19$
Number, n_j	5	5	5	$N = 15$
Means, $\overline{X}_{.j}$	−8.6	−2.4	14.8	$\overline{X}_{..} = 1.27$

which is the variance due to treatments.
Equation (15) yields

$$SS_{error} = 4113 \tag{20}$$

and Equation (16) yields

$$MS_{error} = SS_{error}/(N-k) = 343, \tag{21}$$

the variance due to errors.
Finally,

$$F = MS_{treatment}/MS_{error} = 735/343 = 2.14 \tag{22}$$

Entering Table 14.5 with $k - 1 = 2$ and $N - k = 12$, we find that $F = 1.56$ for a confidence level of 0.75 and $F = 2.81$ for a confidence level of 0.90. With $F = 2.14$ for the experimental data, we conclude that the hypothesis that there is no difference among treatments is rejected with a confidence level greater than 0.75, and probably (assuming that linear interpolation is reasonably accurate) with a confidence level greater than 0.80. The practical effect of this test was that it became obvious to the laboratories that the resistivity of the electrolyte must be kept under very close control so that the results of tracking tests would be consistent from one laboratory to another.

More information can be gained from the data at hand by a study of contrasts. We will follow what is referred to in the literature as the method of orthogonal contrasts. In this method, a contrast, C_m, given by the equation

$$C_m = \sum_{j=1}^{k} c_{jm} T_{.j} \tag{23}$$

is a contrast if

$$\sum_{j=1}^{k} c_{jm} = 0 \tag{24}$$

for equal n_j's in the columns.

Two contrasts, C_m and C_q, are said to be orthogonal contrasts if

$$\sum_{j=1}^{k} c_{jm} c_{jq} = 0 \tag{25}$$

again for equal n_j's in the columns. The number of contrasts may not exceed the degree of freedom of the treatments, $k - 1$, which is 2 in this case.

The sum of the squares for a contrast is given by

ACCELERATED LIFE TESTING

$$SS_{C_m} = \frac{C_m^2}{n\sum_{j=1}^{k} c_{jm}^2} \qquad (26)$$

Since all contrasts have a degree of freedom of 1, the ratio of the contrast mean square is the same as the sum of the squares, and we may calculate

$$F_{1,(N-k)} = \frac{SS_{C_m}}{(SS_{error})/(N-k)} \qquad (27)$$

Careful choice of the contrasts allows us to learn more about the results of the experiments. For example, since we wish to concentrate our attention on the effect of the resistivity of the electrolyte, we can contrast the results at the two extremes of the resistivity range by making $c_{11} = -c_{31} = 1$ so that

$$C_1 = T_{.1} - T_{.3} = 0$$

If we wish to examine the linearity of the results due to the concentration of the ammonium chloride in the electrolyte, we may choose the second contrast so that we are comparing the average of the test results at the extremes of the resistivity range to the center result. (The average of the resistivities at the extremes is very close to the center value of resistivity used during the experiment.) We can do this by selecting $c_{12} = c_{32} = 0.5$ and $c_{22} = -1$. Then

$$C_2 = 0.5T_{.1} - T_{.2} + 0.5T_{.3} = 0$$

We note that these selections of c's satisfy both Equations (24) and (25).

Now, picking up the necessary values from Table 14.6, we obtain

$$C_1 = -43 - 74 = -117$$
$$C_2 = -21.5 + 12 + 37 = 27.5$$
$$SS_{C_1} = \frac{(-117)^2}{5\left[1^2 + (-1)^2\right]} = 1369$$
$$SS_{C_2} = \frac{(27.5)^2}{5\left[2\cdot(0.5)^2 + (-1^2)\right]} = 100.8$$

The first contrast is between test results at the extremes of the resistivity range, and for this contrast we use Equation (27) to calculate

$$F_{1,12} = \frac{1369}{343} = 4.0$$

where 343 is the mean square error shown in Table 14.6. Entering Table 14.5 with this value of F, we find that it lies between 3.18 for a confidence level of 0.90 and 4.75 for a confidence level of 0.95. We may thus say with a confidence level of about 0.92 (92%) that there is a strong correlation between test results and the ammonium chloride concentration in the electrolyte.

The second contrast was chosen so as to give some measure of the linearity of the test results with concentration. For this contrast,

$$F_{1,12} = \frac{100.8}{343} = 0.29$$

This is such a small value as compared to those in Table 14.5 that we can conclude that there is a linear relationship between the electrolyte resistivity in the neighborhood of 382 Ω-cm and the number of drops of the electrolyte to cause failure by tracking.

Section 3.3 of Reference 4 contains a discussion both on tests on means set prior to experimentation (the orthogonal contrasts) and tests on means when the decision to conduct the investigation is made after experimentation.

14.5 SUMMARY

This chapter is obviously not an exhaustive treatment of accelerated life tests nor of the statistical methods available to interpret the data of such tests. The main purpose is to emphasize the importance of statistical techniques and accelerated testing in product development.

Application of energy in any form is likely to cause deterioration and lead to eventual failure. Some of these applications are obvious, such as the impact force in a drop test. Others are more subtle in nature, and may well be overlooked. The gradual deterioration of many materials due to ultraviolet radiation is one such. Adverse reactions to acids are to be expected, but the adverse effect of moisture on certain products is often disregarded. Unfortunately, there is no single list of conditions that can cause deterioration, nor is there a set of guidelines for the development of experiments to measure deterioration. The best approach to development of relevant tests for evaluation is to make a careful study of how the product will be used, the conditions under which it will very likely be used, and what the user expects, both short term and long term.

The few statistical techniques given here were selected as being particularly relevant to accelerated life tests. References 1, 3, and 4 cover a much wider range. One of the important points to be made about statistical methods of data interpretation is that the method chosen and the tests run must be applicable to the situation at hand. Lipson and Sheth [3] pointed out in 1973 that an examination of test data drawn from several manufacturer's files indicated that fully 75% of the data were statistically not significant, and that the conclusions drawn from those data based on statistical analyses might or might not have been valid. It is better to plan your experiments so that the data are statistically significant rather than try to find statistical significance after the fact.

REFERENCES

1. Wadsworth, Harrison M., Jr., Editor. *Handbook of Statistical Methods for Engineers and Scientists*, McGraw-Hill, New York, 1990.
2. Dakin, T. W. Electrical insulation deterioration treated as a chemical rate phenomenon, *AIEE Transactions,* 1948, 67(1):113-22.
3. Lipson, Charles, and Narendra J. Sheth. *Statistical Design and Analysis of Experiments,* McGraw-Hill, New York, 1973.
4. Hicks, Charles R. *Fundamental Concepts in the Design of Experiments,* Holt, Rinehart and Winston, New York, 1964.

REVIEW AND DISCUSSION

1. What are the four factors that determine the extent to which a test can be accelerated?
2. What limits the stress level that can be applied to a product during an accelerated test?
3. What is the advantage of increasing the stress level to the maximum allowable level?
4. What level of performance is typically defined as the end of life for a product?
5. Sketch a typical curve of life versus stress of a product.
6. What mathematical manipulation is used to increase the accuracy of the extrapolation to normal stress conditions of the life versus stress curve?
7. How does one distinguish between test data that is best represented by $y = B/x^m$ and those which are best represented by $y = Ae^{-rx}$?
8. In some cases, the life data involving chemical reactions fall along two intersecting straight lines when plotted on semilog paper. What is the significance of this phenomenon?
9. What steps should be followed in developing an accelerated life test?

10. If only a few specimens are tested for some performance characteristic, is it possible to state the expected value of the population mean to some degree of confidence?
11. What are censored data? What causes them to occur? How does probability paper allow us to use censored data?
12. What is meant by "treatment" in analysis of variance?
13. What are the three parameters derived from the test data for a given treatment?
14. How are the values of F, $k-1$, and $N-k$, as calculated and as found in F-tables, used? What does one learn from comparison of the two values?
15. If test data are to be examined using contrasts, what are the two conditions that must be met for orthogonal contrasts?

PRACTICE PROJECTS

1. The following data were obtained on a certain electrical insulation by using a test procedure described in the American Society for Testing Materials standard ASTM D3638-77. What analytical expression best fits the data? What voltage calculated by the equation would cause failure to occur at 50 drops? Can orthogonal contrasts be used on these data?

Volts	Number of drops to failure
275	3, 4, 3, 7, 3
250	11, 12, 21, 3, 15
225	13, 24, 3
200	42, 9, 16, 83, 27
175	19, 51, 72

2. Test equipment is available that will apply 86,000 V (and no more) to electrical insulation to determine its dielectric strength. Material is to be tested to determine with 85% confidence whether or not its mean breakdown strength is 90,000 V. It is known that the standard deviation of the insulation breakdown strength is 10,000 V. Specify how the test is to be run.

3. The engineer running the tests of Problem 1 decides that additional data are needed for lower test voltages. She obtains the following additional values.

Volts	Number of drops to failure
150	>100, 60, 24, 52
125	>100, >100, 32, 71, 43

    Using censored data techniques, repeat Problem 1.

ACCELERATED LIFE TESTING

4. Underwriters Laboratories Inc. requires that 15- and 20-A circuit breakers operate satisfactorily after 10,000 on-off operations performed at the rate of six per minute. (See Figures 10.9 and 10.10.) There are 21 devices to be put on test, seven having no lubricant, seven having a petroleum-base lubricant, and seven having a silicone-base lubricant. On test, the circuit breakers yield the following data on number of operations to failure:

No lubricant	Petroleum base	Silicone base
10,721	12,363	13,357
12,915	14,218	17,023
10,876	11,420	16,008
11,573	12,123	15,666
12,519	14,830	13,537
11,203	11,283	10,732
13,840	13,480	12,120

Use analysis of variance to determine whether the apparent improvement in performance as a result of lubrication is statistically significant. Use contrasts to test the hypothesis that petroleum-base and silicon-base lubricants are not significantly different.

15

HUMAN FACTORS ENGINEERING

Some products are designed to be useful with little or no human intervention. Examples of such products are such widely diverse devices as space satellites and residential furnaces. The device at each end of this spectrum must have been designed by engineers, it must have been fabricated and assembled, in the case of a furnace it must have been installed, whereas the satellite was mounted on a launch vehicle and placed into orbit, but once these events have occurred, there is only infrequent interaction—sometimes never—between the device and a human being.

By contrast, most of the products we engineers design are intended to be used by people, many times on an almost continual basis. The lathe operator, the punch press operator, the over-the-road truck driver, and the computer operator are all intimately related to the product being used. Good product design demands that the user and the product be compatible with each other, and since we are not in the business of redesigning the human body, this means that the product must be adapted as necessary. It further means that we must take into account the limitations of the human body, such as dimensions, weight, strength, and comfortable seating positions. In addition, the capabilities of the body, such as its ability to rotate some joints only in a hinge mode while others have a ball and socket mode, must be factored into the design process. All of this should be done as early as possible so that no design features are adopted simply because a few decisions have already been made, and the inertia of the process simply carries them along.

If the discussion is limited to the factors just enumerated, the subject of this chapter can be summed up in the word "ergonomics," which is composed of two Greek words meaning "work" and "laws." Ergonomics considers the ability of the human to do physical work, including the effects of fatigue due to

HUMAN FACTORS ENGINEERING 521

repetitive motions. Professionals who are engaged in such activities are known as ergonomists, and perform a valuable function in the design field. However, ergonomics in the narrow sense fails to consider the effects of such other factors as reaction time, noise, illumination, and perception limitations. For this reason we prefer the phrase "human factors" or "human factors engineering." Whether one uses the term "ergonomics" or "human factors engineering," the field requires a great deal of knowledge, and we can do no more than give a broad-brush treatment. The references at the end of this chapter cover a spectrum ranging from very practical approaches [1,2] to human reliability assessment [3] and the simulation of human beings in the workplace [4].

15.1 HUMAN-MACHINE INTERACTIONS

When considering human-machine interactions, a number of considerations must be taken into account. These include:

1. The "fit" between human and machine. The product must be convenient to use and the relationship between human and machine must be comfortable. To enhance this "fit," the ergonomist makes extensive use of published data on body dimensions, as discussed below.
2. When using some products, as on production lines, parts must be manipulated or moved, force must be applied, or work done. This requires that speed of movement, accuracy of movement, force and power available from the body, and the effects of repetitive actions be understood. Understanding of skeletal and muscular capabilities and limitations is therefore important. The limitations of the sensory organs must also be included in the knowledge pool of those engaged in design.
3. Many devices are controlled in a very direct manner by the user. The lathe operator and the over-the-road truck driver were cited above. The controls used in operating the lathe and the truck must be unambiguously identified with the function to be performed and the identity must be maintained under conditions of operator stress and fatigue. In these situations the user must perform actions based on information received from the product, all of which is normally nonverbal in nature.

No matter how well we engineer a product and regardless of the attention we pay to the human-machine interaction, human errors will occur. The consequences of these errors depends on the circumstances attendant on the situation. The incorrect key stroke on a word processor is easily correctable, either by the person using the processor, or by the spell-checker if one is used. (But even the spell-checker will not identify the word "from" when it should be "form.") But there are three other classes of errors which may not be so innocuous, especially in large and complex systems, and which may lead to disastrous results such as the Challenger disaster or the chemical plant catastrophe at Bhopal, India. These are errors of omission, that is, failing to carry out an act that is required; errors of commission, carrying out an act that should not have been performed; and extraneous acts, which are acts which are not related to the required act. Considerable attention has been paid in recent years to "a means of properly assessing the risks attributable to human error and for ways of reducing system vulnerability to human error impact" [3, page 1].

15.2 ANTHROPOMETRY

Ergonomics began to emerge as a discipline about the time of World War II, but anthropometry—the science of body measurement—is far older. That study began as early as the time of the ancient Greeks, who were interested in body proportions so that sculptures would represent life-like images, even when smaller or larger than real life. Subsequently, the officers of many European armies became interested because of the desire to have cadres of tall soldiers. Regardless of the source of the interest, however, it must be pointed out that the data do not remain constant from one generation to the next, and over the course of a few centuries there have been major changes in human body dimensions, although not in proportions. Anyone who has looked at the armor worn by full-grown Englishmen during Elizabethan times may conclude that many twelve-year-old American males would fit into that armor rather well today. Even casual observation is convincing that there are large populations of individuals from one ethnic or racial background whose anthropometric data are at variance with those from another population.

There are a number of sources for anthropometric data. One that has been cited many times is a 1964 study of U.S. Army males by the U.S. Army Human Engineering Laboratories [5], but there have been others, some studies having been done with very specific tasks and occupations in mind. One source of information is SAE J833, *Human Physical Dimensions* [6], which defines the sizes of people from all over the world. Another source is ISO Std. 3411, *Human Physical Dimensions of Operators* [7]. Other sources are military standards. Even NASA has entered into anthropometric data-generation with NASA STD 3000, published as *Human-System Integration Standards* [8] and with NASA Reference Publication 1024, *Anthropometric Source Book,* published in three volumes [9].

It must be noted that the data in the tables will differ from one study to another, some of the differences being due to such minor decisions as to whether to measure height with or without shoes, more substantive differences being due to the longitudinal effect of dietary and nutrition changes, but the major differences being due to selection of different groups of individuals on which to

HUMAN FACTORS ENGINEERING

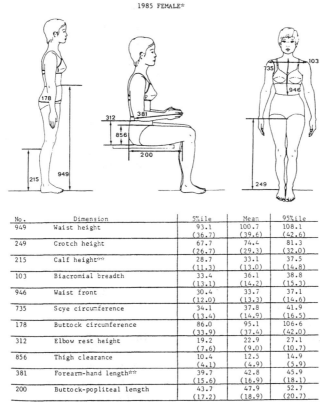

Figure 15.1 Figures used for females (1985), with some dimensions, from NASA Ref. Pub. 1024; *Anthropometric Source Book* [9].

make measurements. The 1964 data were entirely on U.S. Army males, other military studies were on ground troops and on aviators as two distinct classes, and still other studies include American males (as being representative of large individuals) and Japanese females (as being representative of some of the smaller individuals in the developed world). The point is that there is a large amount of data available, but one must be certain to use a source that is appropriate for the design at hand or that is required by the intended purchaser of the device or equipment, such as one of the branches of the military. Figure 15.1 shows the human figure in some of the positions used when the NASA measurements were made. Table 15.1 (from the 1964 study) is representative of the kind of measurement that is valuable to the designer.

Table 15.1 Nude Body Dimensions: U.S. Army Males (1964)

Description	Design values (percentiles) 5th [cm (in.)]	95th [cm (in.)]
A. Standing		
1. Stature	165.6 (65.2)	185.7 (73.1)
2. Eye height	154.4 (60.8)	174.2 (68.6)
3. Chest depth	20.3 (8.0)	26.4 (10.4)
4. Waist breadth	23.8 (9.4)	31.2 (12.3)
5. Elbow height (arms hanging)	103.1 (40.6)	117.9 (46.2)
B. Seated		
1. Sitting height	85.9 (33.8)	96.5 (38.0)
2. Shoulder height	54.1 (21.3)	63.8 (25.1)
3. Knee height	51.1 (20.1)	59.2 (23.3)
4. Elbow rest height	18.8 (7.4)	27.4 (10.8)
5. Eye height	74.7 (29.4)	85.1 (33.5)
C. Reach		
1. Functional reach	75.4 (29.7)	88.9 (35.0)
2. Arm reach from wall	81.0 (31.9)	94.7 (37.3)
3. Maximum reach from wall	89.9 (35.4)	105.9 (41.7)
4. Span	167.4 (65.9)	192.0 (75.6)
D. Hand		
1. Grip diameter (outside)	9.4 (3.7)	11.2 (4.4)
2. Grip diameter (inside)	4.1 (1.6)	5.3 (2.1)
3. Hand length	17.5 (6.9)	20.3 (8.0)
4. Thumb thickness	1.8 (0.7)	2.0 (0.8)
5. Thumb length	5.1 (2.0)	6.6 (2.6)

The partial listing of static dimensions in Table 15.1 presents only part of the information necessary to the designer. There are dynamic measurements to be considered, such as the rotations pointed out earlier as being available in many of the joints of the human body and of the torso as a whole. Some of these rotations may be readily allowable under normal working conditions, but others may involve awkward rotations that, on a repetitive basis, can lead to damage to wrists, for example. Functional reach as listed in Table 15.1 (75.4 to 88.9 cm) represents the maximum reach if the arm is extended horizontally, but the geometry of the situation will cause the functional reach to be reduced to about 40.0 cm if one is seated at a desk and working on the surface of the desk or reaching toward a control knob mounted above head height. There will be still more decrease because of the necessity to consider whether the fingertips are simply to touch the work, or whether the work is grasped.

Differences in body dimensions in the expected population are accommodated in many products by making parts adjustable. Automobiles are routinely equipped with seats that can be moved backward or forward, some seats can be raised or lowered, and many can be tilted. Chairs used at desks in offices are typically adjustable as to height of the seat, and arm height may also be adjustable. If there is a computer terminal display, provision may be made for adjusting its height, or at the very least for fitting it with a tiltable platform so as to provide a comfortable viewing angle or to reduce glare.

In spite of the fact that the range of sizes of the human body has been taken into account in many ways and provisions have been made to accommodate to that range, it must also be pointed out that most of this effort has been expended in making improvements in situations that are frequently encountered. Events that occur only rarely have drawn less attention and effort. A current example is the controversy over airbags in automobiles. Although there is sound evidence that these devices have saved many lives (although not nearly as many as predicted early on), there is also evidence that these same devices have caused the deaths of children (who should probably have been in rear seats) and of small adults. One automobile manufacturer recognized when airbags were first being introduced on a large scale that the small adult represented a different problem than larger adults, but could not convince the industry and governmental authorities that a new (although smaller) problem was being created.

Although the main thrust of product design should be on designing to ensure safety, productivity, and avoidance of errors, comfort is also important. Lack of a comfortable ambiance will lead to frustration and fatigue, which lead on to human errors and a reduction in safety. It may also lead on to trauma disorders, as will be seen later.

15.3 BONES AND MUSCLES; FORCE AND WORK

The human skeleton is comprised of 200 to 210 bones (depending to some extent on age). Some of these bones, such as the skull, are protective, but the bones of interest to us are those that move relative to each other by means of a variety of joints, such as in the fingers, wrists, elbows, and shoulders, or by a slight tilting action, as in the vertebrae. Regardless of the joint, all movements are controlled by systems of muscles which are attached to the bones by tendons. Muscle action, whether contraction or expansion, requires the expenditure of energy, which in turn produces waste products. Continued action by a muscle requires that the energy supply be replenished and the waste material be removed. If muscular activity is too intense, the rates at which the replenishment and removal processes proceed will be too low for a high level of muscular activity to be sustained, and muscle control will be degraded and may even be temporarily lost.

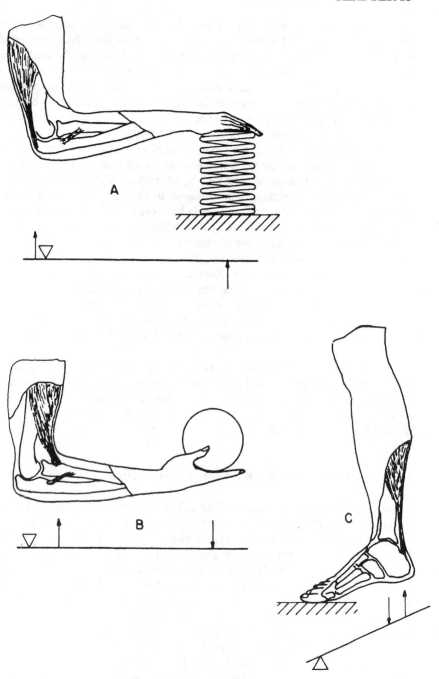

Figure 15.2 Mechanical structure of the human arm and leg.

HUMAN FACTORS ENGINEERING

Level of muscular activity is thus of importance, as is the repetitive nature of some tasks.

In an engineering sense, bones and muscles may be viewed as mechanical elements, as shown in Figure 15.2, and conclusions may be drawn as to possible performance of the body by drawing (or sketching) diagrams of forces such as those used in the study of statics. For example, examination of Figures 15.2A and 15.2B leads us to conclude that the placement of bones and tendons in Figure 15.2B creates a much higher mechanical advantage than that of Figure 15.2A, and we can therefore conclude that it should be much easier to hold a weight in one's hand than it is to push down on a spring, the load forces being assumed equal in the two cases. Practical observations confirm this conclusion. As a matter of fact, the "push-down" situation seems to be largely neglected in the literature, but the support situation of Figure 15.2B has been the subject of much investigation, and the National Institute for Occupational Safety and Health (NIOSH) has developed a method to assess such lifting conditions [1, pages 29-31].

The size of the muscle is also a major factor. Shoulder muscles and thigh muscles, when pushing, provide the largest forces. The phrase "put your back into it" is a common one. If one wants to slide a heavy load standing fairly close to a wall and if the body can be placed between the load and the wall, it is often possible to start the load into motion while bracing the back against the wall and pushing with the hands. Data on forces available in both the push and pull directions are available. A reasonable limit seems to be about 200 N, although both men and women are capable of considerably more force production on a one-time basis, especially when standing and using two hands to push or pull.

It is not only the amount of force that can be exerted that is of interest, but the rate of energy consumption (power) of which the human body is capable on a sustained basis also enters into the design of some products. It is estimated that a healthy young male can, for a few seconds, produce as much as 1500 W of power, or about 2 horsepower. For purposes of longer term work activity, the physical work capacity of European males has been expressed [10] by a variety of equations, among them

$$Hp = 0.39 - 0.104 \log t \quad \text{at age 20,}$$
$$Hp = 0.35 - 0.092 \log t \quad \text{at age 35, and}$$
$$Hp = 0.29 - 0.077 \log t \quad \text{at age 60,}$$

where $4 < t < 480$, and is in minutes. For short bursts (6 to 120 s),

$$Hp = 2.5 t^{-0.33}$$

t being in seconds. If one considers the 35-year-old and assumes a 4-hour work period (240 minutes), the equation gives a power (converted into watts) of almost 100. Since some of the developed power is used for life support functions (possibly 75 W), we see that the sustained effort capability of human beings is quite small. For heavier work, effort on the part of males may be at the level of 150 to 250 W (exclusive of life support functions), but only for relatively short periods. Consider, however, the pattern of work for the usual task. It rarely, if ever, consists of such tasks as rowing a racing shell, a task that requires sustained effort for many minutes. It is more likely to be a task that requires considerable effort for a few seconds to a minute, followed by a period of activity during which effort is expended at a very low rate, thus averaging out to values such as those suggested by the equations above.

In ergonomic studies, there is rarely a way to measure power directly in watts or any other units except in such set tasks as pedaling or cranking. It is common practice to make indirect measurements, however, one such being the rate of oxygen consumption in liters per minute while doing a particular task. Since five nutritional calories are consumed per liter of oxygen, an indirect measure of energy rate is possible. Keep in mind, however, that the results must include the "housekeeping" functions of the human body as well as the useful work produced.

In addition to the factors of force, power, and energy, one must also take into account the effect of repetitive actions. Certain kinds of tasks tend to produce trauma disorders if repeated too frequently. For example, typing and cashiering may produce the disorders known as tension neck and the carpal tunnel syndrome. The assembly of small parts may also cause tension neck as well as tendinitis of the wrist and epicondylitis (tennis elbow), and manual material handling may cause shoulder tendinitis.

Various steps may be taken with the objective of avoiding the trauma that may otherwise result. These include scheduling activities so that an employee does not perform the same motions for more than a limited time, say an hour. If possible, rotate employees who perform very repetitious tasks. Most important of all, examine the task design so that it is less repetitious, and so that the motions required at skeletal joints are not such as to exacerbate the possible trauma.

It may also be possible to make use of auxiliary devices (such as the popular wrist pads for keyboards) or to substitute one form of device for another. One form of keyboard, referred to frequently as a "natural" keyboard, has the keys arranged along sweeping curves that are claimed to be easier to use than those on standard keyboards.

Finally, one needs to pay attention to the weight of tools that may be used in various tasks, especially on assembly lines. Even a small tool, lifted repeatedly, becomes a major burden. Many small power tools can be supported so that they remain suspended at the height at which they are released by the worker, making it easy to grasp a tool and move it to the next task with less effort than if it had to be picked up every time.

HUMAN FACTORS ENGINEERING

15.4 SPEED AND ACCURACY

Some products require an insignificant force or consumption of energy, but test human limitations in regard to speed of response and accuracy of motion. An automobile is one such product, especially when equipped with power steering. Not much power is required from the driver for the steering function, but in a collision avoidance situation on a highway the speed of response and the accuracy of the resulting action may very well make the difference between a "near-miss" and a fatal collision.

Reaction time to an expected stimulus applied at random varies from about 100 ms to 500 ms. This wide range occurs because of the variability among human beings, older individuals tending toward or even beyond the upper value. The process comprises receptor recognition of the stimulus, which is usually visual or auditory in nature, neural transmission to the brain, information processing within the brain, neural transmission to the actuator (muscle), and muscle activation. For simple tasks, the information processing requires 70 to 300 ms of the total of 100 to 500 ms. If, on the other hand, there are several choices that may be made as the result of the original stimulus, the reaction time can easily be several hundred percent greater. During this lengthened response time the brain must evaluate the choices and make a decision as to which to choose. Even in such a situation, however, the human brain is a remarkably fast information processor.

When comparing reaction times due to auditory or visual stimuli, it is generally agreed that an auditory stimulus will elicit a quicker reaction than a visual one, taking about 200 ms as compared to 300 ms for the visual stimulus. A suspicious or loud sound tends to cause a quick flow of adrenaline, leading to more rapid action than will occur with a bright light, even if it is flashing. A study of detectable flicker rates from rapidly flashing lights tends to show that visual stimuli longer than about 50 ms begin to merge with one another, and 50 ms is already half of the lower limit mentioned above for reaction time. The kind of response also has an effect on reaction time. A manual response (defined as a hand or a foot action) can be accomplished in about three-fourths the time for a verbal response. Thus a manual response to an auditory stimulus has a double advantage over a verbal response to a visual stimulus, reaction time being about 250 ms versus 400 ms for a simple task. Unfortunately, we cannot always select the type of stimulus to be used. The physical contact of one vehicle with another on the highway may be the stimulus to trigger the quicker reaction, but the visual stimulus of seeing another vehicle about to strike the right-front fender is much to be preferred.

Accuracy of response to a stimulus is also of importance. Here again there is a response time, although the response time should now be thought of as the time necessary for the hand (or the foot) to move from one position to another, the decision to make the movement having already been made as a result of the original stimulus. The times are on the order of 100 ms, but much of this time is taken up in acceleration from the original position and the deceleration as the fingers or hand approach the final position. It has been reported that, while it might take 100 ms to move the hand to a target 15 cm away, it requires only 25 ms more if the target is 45 cm distant.

Once the hand has reached the target, the task to be performed has an influence on the total response time. Actually, the total response time is adjusted before the hand ever reaches the target because the brain knows what action is to be taken, and has begun positioning the hand properly during the traverse from initial to final position. For example, if a four-position rotary switch (the positions being at 0°, 40°, 80°, and 120°) is to be moved after the hand reaches the switch, it takes longer to reach the switch if it is to be rotated to 120° than if it is to be rotated to only 40° or 80°. The pre-positioning of the hand and the wrist for the final switch rotation is occurring during the travel to the location of the switch.

The capabilities for control of the skeletal joints also enter into the accuracy with which hands or feet can reach desired targets. Of the joints of the arm, the least controllable with precision is the shoulder, followed by the elbow. The wrist joint, however, is eminently controllable, having several degrees of freedom, and the finger joints are also highly controllable. In general, the joints with least flexibility are the most difficult to control for accuracy, and this is true whether one considers the arm or the leg. The greater the flexibility, the more accurate the positioning can be. For example, the movement of an object from a position directly in front of the body to a point 45° away to the right requires the use of the elbow as the principal element controlling the direction of movement; if, on the other hand, it were desired to move the same object to a point farther away but directly in line with its original position, the principal element controlling the direction of movement would be the shoulder, a less controllable joint.

15.5 DESIGNING TO AVOID HUMAN ERRORS

Errors, other than casual lapses, were categorized in Section 15.1 as errors of omission, errors of commission, or extraneous acts. Human errors are not intrinsically different from other human actions; they are simply actions that should or should not have been taken at a particular time. Many times they are actions that have worked in the past or have worked in the past in a somewhat different situation. In any case, it would be highly desirable to be able to develop a model of human action that would be able to identify any errors that were likely to occur. As may be surmised, this has proved to be a very elusive target.

One of the best ways to avoid human errors is by use of the instruction manuals described in Chapter 2. The operator should learn from the operating manual how the product is to be used. Errors should be avoided if the instructions are written correctly and followed faithfully, but such an outcome cannot be relied on. Instead, the user frequently requires more instruction than will be found in the instruction manual, and sometimes this instruction can be better given by use of labels on the device itself. For example, the change from the foot-operated dimming switch of an automobile to a switch operated by movement of the turn-signal stalk on the steering column (which itself quickly acquired additional functions) was announced only by the change of wording in the owner's manual. It was true that most drivers became accustomed very quickly to the new situation, but the automobile manufacturers were not of great assistance in another aspect of the change. Some headlight switching was now done by a two-position switch, requiring the turn-signal stalk to be in one position for down beams and another for high beams, while other manufacturers used a relay operated by a momentary impulse from the stalk switch to switch between high and low beams.

Under some circumstances, a seemingly innocuous design can result in a hazardous situation. An industrial mixer was described in Section 2.2. The control station consisted of pushbuttons intended to start and stop the mixer, to open and close gates for introduction and removal of product, and other controls, such as a knob to control mixer speed. The layout of the control station was logical, and there seemed to be no reason for use of different sizes or shapes of the pushbuttons, or for barriers between one portion of the control station and another. However, as described in that section, an electrician was checking the gate operation with the access door open, his head inside the drum, and his fingers on the pushbuttons. An interlock failure resulted in starting of the mixer drive when he inadvertently pushed the START button for the mixer drive, resulting in decapitation. This death would probably have been avoided, in spite of the failure of the interlock, if barriers had been used between the drive section of the control station and the section that controlled the product gates, or if pushbuttons differentiable by touch had been used.

Design changes from one machine tool to a later model can also present hazardous situations. If a certain positioning of the pushbuttons is used on one model and the next model has a different positioning, any user having one of each model and having operators expected to use either machine can certainly expect errors of some sort. It would be far better to retain only one design in such a case, or to rebuild one control station so that their functions are identical.

When controlling devices or systems, errors are best avoided by allowing the operator to rely upon some expected way of operating the product or by making it possible to develop an easily remembered mnemonic device. Operations to be done in sequence are best performed if the initiating pushbuttons are aligned next to each other, in sequence, and probably left to right. Random placement of controls should be avoided. Moreover, there are certain control movement stereotypes that are ingrained into American culture. Among these are that a switch movement up or clockwise turns the switch ON, and that a knob movement in a clockwise direction tends to move a device, such as the slide on a lathe, to the right or to increase the speed of a motor. The clockwise movement of a valve handle, however, tends to CLOSE the valve, or turn it OFF. Even though this is contrary to the switch convention, American culture has no problem with it. If one is building equipment for use abroad, however, it is mandatory that the conventions used in the culture at the destinations be understood and observed. British practice with regard to switches, for example, is generally the opposite of the American practice, down being the ON position.

15.6 ILLUMINATION

One consideration in designing a product or a place where a product is to be used is the level of illumination needed, desirable, or mandatory. One of the problems that arises almost immediately is that of defining such seemingly obscure terms as lux, candles, foot-candles (fc), candelas (cd), lamberts, and foot-lamberts, most of which we shall ignore. Some of the difficulty arises from the fact that the terminology has changed, but old terminology persists in the literature. Difficulty also arises from the fact that the detector of illumination is the human eye, the eye's response to different wavelengths of light cannot be neatly characterized by an equation, and there are other psychological factors such as desirable and undesirable contrasts between adjacent surfaces that are important, but difficult to quantify. Before discussing levels of illumination, we need to have some understanding as to what the relevant terms mean and how we are to interpret them.

Let us review the fundamental process that results in our "seeing something." Illumination begins with the production of visual energy by a source, such as an incandescent or fluorescent lamp. This energy is referred to as flux; the units are the "international candle" (now outdated) and the candela. Because the flux from a source spreads out, the flux density becomes an important concept. This is expressed in lux or in foot-candles (theoretically outdated also). The flux reaches a surface, that which is to be illuminated, and some of the flux is reflected in various ways. That which reaches the human eye enables us to see various objects.

Although the mathematics would probably be extremely complex because of the fact that we are dealing with ill-defined fields and multiple reflections of various kinds, the process just described can be entirely analytical from the production of visual flux up to the last step, that at which the human eye enters. Physiological responses of the eye and of the brain are probably the most important of the entire process as far as human factors engineering is concerned. For our purposes, however, we need to back up two steps. We need to know something about the flux density on the surface of interest (lux or foot-candles) produced by the light sources being used. We also need to know the "reflectance" of the surface, that is, the ratio of reflected flux to the flux incident on the surface. The reason that we can use an abbreviated chain of investigation

HUMAN FACTORS ENGINEERING

is that the relevant data are in terms of lux or foot-candles needed for a particular task. We leave to others the task of determining which light fixtures to use, and how to arrange them. Our starting point is that we need a certain level of illumination on the task at hand.

There are some general rules that we can follow. If visual aspect is not critical, such as in a hallway in a building or in the usual stockroom, a light intensity of 10 to 200 lux, or in terms of foot-candles, about 1 to 20, is sufficient. (The conversion factor is 10.76 to 1.) Reasons for choosing the upper level are that there may be some instructions on labels or notices posted on bulletin boards that are not to be missed, or individuals are expected to be in the area for only a short time, and will then be returning to a higher level illumination area. Use of the higher illumination level reduces brightness adaptation when going from one area to another.

Most normal activities require a light intensity of 200 to 800 lux. These are activities such as operating machines, doing assembly work, or reading files or drafts of proposals. Which end of the light intensity range to choose depends on a closer examination of the tasks involved. If print is small or the parts to be assembled are medium to small in size, the higher level will be needed. If there is an adjacent space that has a high level of illumination, it may be necessary to raise the level of illumination in the space under consideration simply to reduce brightness differences where the spaces merge. For some special applications,

Table 15.2 Illumination Levels for Various Tasks

Work area or task	Illumination levels, lux (fc) Recommended	Minimum
Assembly		
Coarse	540 (50)	325 (30)
Medium	810 (75)	540 (50)
Fine	1075 (100)	810 (75)
Precise	3230 (300)	2155 (200)
Electric equipment testing	540 (50)	325 (30)
Office work, general	755 (70)	540 (50)
Reading		
Large print	325 (30)	110 (10)
Small type	755 (70)	540 (50)
Prolonged	755 (70)	540 (50)
Transcribing and tabulation	1075 (100)	540 (50)

still higher light intensities are need, running as high as 3,000 lux or more. Table 15.2 gives a few representative levels of illumination for various tasks.

There are a number of considerations other than absolute light levels. For example, it is necessary to avoid substantial brightness differences in the visual field. A desk with a very dark surface is not the place to read a paper with black printing on a white background; the luminance ratio (ratio of light reflected from the paper as compared to that reflected from the desk) can easily create an unpleasant condition.

Improper use of contrast can also make the visual task very difficult. For example, the use of a pale yellow heading over a black tabulation on white paper is to be avoided; the heading is virtually unreadable. The effect becomes worse if the paper is calendered (that is, having a high gloss, which leads to specular reflection). Glare is frequently a contributor to difficulty in the visual field. If one is reading a calendered sheet of paper, lying flat on a table, with a single desk lamp directly in line with the paper, there will be much picking up of the paper and rotating it so as to direct the specular reflection from the paper elsewhere. The presence of reflections on the screen of a computer terminal can contribute to errors, which may be especially serious if the terminal is being used to control a machine tool. A copying machine, built by a prominent manufacturer and currently in use, has displays under a reflective cover that are small in size, but are probably readable if there are no reflections on the cover. But when the machine is placed in a typical office environment with fluorescent lighting in the ceiling directly over the machine or with natural illumination from a window, the displays become virtually unreadable.

15.7 INFORMATION FEEDBACK FROM THE PRODUCT

The user's interaction with a product demands that information be transmitted to the human senses in an unambiguous way. Of the five senses, only hearing, sight, and touch can be used. The only use of smell as a feedback mechanism of which the authors are aware is the introduction of a compound into natural gas that is easily smelled if a leak occurs, and taste seems to be an unheard-of possibility.

Sound has the advantage of being pervasive. The user does not need to be oriented toward the source in order to receive a signal. As noted earlier, sound signals also have the advantage of initiating the shortest reaction times. They are easily made annoying, demanding an early response. Wailing sirens, horns, and buzzers are universally identified as warning devices. Unusual sounds, such as the noise generated by disk brakes when the pads become too thin, will attract the driver's attention and send him or her to the repair shop. For many years, turn signals in automobiles gave a clicking sound that reminded the driver to reset the signal if a turn was made that did not automatically clear the setting. For a time, the automobile industry seemed to make an effort to reduce the level of this sound or to eliminate it entirely, probably with the misguided objective of trying to make the interior of the car as quiet as possible. More recently, the sound has been restored as automotive engineers realized that the sound sent a signal that was desirable because it signaled the need to reset the signal under certain circumstances.

Single sounds of short duration may be used to signal events of importance to the user. The "end of cycle" tone from an electronic timer or from a microwave oven alerts the user that a point has been reached at which some further action may be taken. The pop-up toaster emits a distinctive sound when the toast is done. The distinctive beep used by television transmitters to alert the viewer (who may be engaged in some other activity, such as doing counted cross-stitch work) that supposedly important announcements are being run across the bottom of the screen is common. The bell announcing the arrival of an elevator car is effective in alerting the user to the fact that the car has arrived, even if attention has been distracted by casual conversation with a colleague. Most of these sounds are now designed into the product, but the designer should not ignore the fact that "natural" signals may also be of value. In general, instruction as to the significance of the signal is not necessary. One or two experiences with a signal are sufficient to "give the instruction."

The development of voice synthesizers gives sound additional possibilities for feedback to the user. Synthesizers are now very small units, and appear in many consumer products. In one version of a home security system, the user presses buttons in an appropriate sequence to arm the system, and the system itself then announces the level at which it has been set and by a later beep that the system is actually armed. If a door, for example, is opened while the system is armed, the system announces "INTRUSION! INTRUSION!" loudly, clearly, and persistently until the system is disarmed, as one of the authors has discovered to his chagrin. To make use of a voice synthesizer, it will undoubtedly be necessary to incorporate additional sensors into a system, but the advantage of audio feedback (as compared to a visual display, for example) of not requiring the user to direct his or her attention in a specific direction enhances the communication channel. Some automobiles have warning chimes that sound if the headlamps are left on or the key is left in the ignition, but it is then left to the driver to discover which of several possible errors has been committed. However, other automobiles have their own voice synthesizers that announce that a particular door is open (or is ajar), that headlamps have been left on, or that the fuel level is low. The same techniques can be used with machine tools. Speech messages can be given directing the user to take action to correct an error or informing the user of a certain condition that will require attention. One must, of course, be able to foresee the congeries of possible errors, and make provision for sensing each of these so that the appropriate message can be initiated.

Although sound signals are no doubt the better choice (as compared to visual signals) for feedback to the user, one must keep in mind that the ambient noise level may mask the signal. The environment in which the system is to be used must be understood, and possibly modified. In addition, there is usually a tacit assumption that the user has normal hearing.

Light signals require that the user's attention be directed generally toward the display. The location of the display should be near the central part of the user's zone of vision (without interfering with the visual area needed to perform the task at hand), although it may be placed in the user's peripheral zone provided that the intensity of the display when a message is placed on it is sufficiently high that the user's attention is drawn to it or the user is alerted in some other manner. The remark about normal hearing above may be recast to the assumption that the user has normal vision. If the user suffers from tunnel vision, placement of a display in what is conventionally understood to be a peripheral zone may result in the display being unnoticed by a sight-impaired person. In order to accommodate such possibilities, the idea of a sound signal produced simultaneously with the visual signal must not be neglected. Even if a person has normal vision, the attention-getting function of the sound signal makes it a worthwhile backup, alerting the user to the fact that a message is on the display.

Light has the advantage over sound of being more versatile. Shape, size, color, and position may all convey different information. We are accustomed to this with signs, for example, but advantage may be taken of the same flexibility if a screen on a computer terminal is the display mechanism. A single 72-point word, especially if in a flashing mode, will quickly draw the attention that a 10-point nonflashing word never will. It is not, of course, mandatory that a computer terminal be used. Passive displays, using all kinds of display devices, such as LED's or liquid crystals, will probably be less expensive, but the potential versatility of the computer screen is lost.

Regardless of the device on which the visual display appears, there are a number of additional considerations to keep in mind. Among these are decisions as to whether to use all capitals or not, which type faces are least likely to generate confusion between two characters, contrast, the use of different colors (if the display has that capability), whether or not to right-justify text, placement of various messages on the display, preferable types of chart displays (line, bar, pie?), and so forth. We conventionally read left to right and top to bottom, giving us a rule-of thumb as to what we consider to be a logical placement if there is more than one message to be displayed. Before making the assumption that this is a logical placement everywhere in the world, one must look into the conventions used wherever the device may reasonably expected to be sold and adapt the placement as necessary. Reference 1 has a fuller discussion of some of these points.

One of the principal methods of feedback to the user is by means of the readings on various instruments, such as those that indicate voltage, speed, feed rates, temperatures, and so on. The choices here are between analog and digital readouts. Analog readout devices are generally less expensive, but digital readouts are becoming so inexpensive that the choice between the two is seldom one of cost. On what basis should a choice be made? This depends on the use that is to be made of the information, and the accuracy with which the information must be known. In general, the dial (analog) indication is better held in the user's memory than the digital readout. Because humans have the ability to detect slight angular differences, if one is controlling a function the needle on the dial can be instantly read as being below, above, or "right on" the desired value if that value has been marked on the dial. The user's adjustment of the input will, after some learning runs, be proportional to the angular differences exhibited on the dial. The precision of the control will, however, be less than the precision possible with a digital readout. The user will probably find it more difficult to make fine adjustments of the input if a digital display is used because the sense of how far the actual value is off from the desired value will depend on a mental subtraction process, which may lead to errors.

When devising displays and selecting sensors, the entire range of possible sensors should be considered, even though many will be rejected as unnecessary or unsuitable. Among sensors sometimes overlooked is the video camera, especially when used with fiber optics. Visual surveillance of components or ongoing processes inside a system can now be accomplished, giving the user more information than could previously be obtained. One must, of course, be wary of providing so much information to the user that information overload occurs.

The third sense used for feedback is the sense of touch, and this always involves the complete hand or the fingers individually. Considering the hand as a unit, touch enables the user to obtain feedback principally by the size of an object grasped, by its shape, or by its placement. Feedback may also be obtained from the ease or difficulty with which a handle may be moved. If the product is an aircraft, the shape of the throttle, afterburner, flap, gun control, and other knobs will all be dictated by the U.S. Air Force in the case of aircraft procured for Air Force use. The intent is to avoid unintentional operation of the wrong control. The information is so ingrained into the pilot by many hours of practice and use that the pilot need not look at the control during high-stress maneuvers.

Some tactile sensing is so common that it is unnoticed by the user. That is, the user does not realize that his or her action has been affected by a form of feedback from the device being used. Consider the key action of typewriters or keyboards, such as those used with computers. The keys must always return to their "home" position, and this requires some spring return action, thus determining the initial force by the finger that is to depress the key. As the key is depressed, the spring will compress, requiring increasing force until contact closure (or type bar movement in the case of a manual typewriter) occurs. The problem is, how much force and how much movement is required to achieve closure? With too little force, closure will not occur. One of the authors has a touch-tone telephone with this kind of force-motion relationship for the keys. The problem with it is that, when "dialing," there is only one way to make certain that closure has occurred, and that is to hold the receiver to the ear as the buttons are pushed. If one does not do this, numbers may be missed, and redialing must be done.

If you depress the keys of a computer keyboard very carefully, trying to sense the relationship between force and movement, you will find that the force increases in the very early part of the travel, and then suddenly the key "falls away" in the sense that the force required becomes smaller as travel increases. The user is not even aware of this phenomenon, but the practical result is that once the fall-away region has been reached the key travel will be completed to the point of closure of the contacts. The user "knows" that closure has occurred, and proceeds quickly to the next keystroke.

15.8 FATIGUE, BOREDOM, AND VIGILANCE

Fatigue, boredom, and vigilance are interdependent topics. They are also alike in that the degree of each cannot be quantified with any precision. The first two are negative characteristics, whereas the third is positive. In the following, we attempt to differentiate between the first two and suggest some steps that can be taken to alleviate them.

Fatigue is usually associated with prolonged periods of heavy work, which drain the muscles of their energy reserves. However, fatigue can also result from various forms of physical and mental stress. It can also result from monotony on the job. Stressful situations include attempting to perform work beyond the skill of the worker, attempting to absorb information requiring background knowledge the worker does not possess, or just plain dissatisfaction with one's occupation. A low level of physical activity coupled with monotony can also lead to fatigue, as a full day of interstate driving in certain parts of the country will clearly show. Whatever the cause, fatigue is counterproductive and increases the probability of operator errors. Distaste for the task at hand and a slowing of reactions will result in a general decline in body and mental performance.

Fatigue is quantifiable, although not with high precision. One way of quantifying fatigue is for the person to record his or her own impression of the degree of fatigue; this is a highly subjective measure. Electroencephalography has been used in an attempt to reach a more objective measure, but the difficulty is that no one is quite certain how to interpret the electroencephalograph for this purpose. Perhaps the most quantitative measure is the flicker-fusion frequency test. In this test the subject views a two-part illuminated area, one part being illuminated with light of constant brightness. This is the control or reference. The other part is illuminated by a light whose intensity is varied at slowly increasing frequencies until the subject can no longer detect the flicker. The frequency at which flicker detection inability occurs decreases after a period of arduous mental tasks, that is, when the subject is fatigued. Regardless of the method used, none is an on-line, real time method.

HUMAN FACTORS ENGINEERING

The negative effects of fatigue require that remedies be sought. One remedy is to provide a work environment that is challenging but not physically or mentally exhausting or that exhibits monotony. Tasks beyond the capability of the user can be automated, the tasks done on the job can be varied, and suitable rest periods provided. The difficulty with doing this well is that none of the aspects mentioned can be quantified; they are all subjective in nature. Nevertheless, the directions to take are relatively clear, even though the size of the steps to be taken is nebulous.

If fatigue is physical or mental exhaustion, then boredom may be characterized as a state of weariness and dissatisfaction resulting from inactivity or lack of interest. A boring task is one in which the person is not challenged, either mentally or physically, by the task at hand. Prolonged periods of repetitive work, especially if no thought has to be given to what is being done, are a certain prescription for boredom, but may not be quite as bad as repetitive work whose nature does not allow the worker to think of other things. Attitude and motivation are keys to whether the person will be bored or not. Counted cross-stitch work done on the basis of a contract to produce a certain number of pieces per month may be the most boring job in the world, but the same work done because the person loves to do it and simply enjoys seeing the finished work develop will not be boring at all. If a variety of subtasks can be introduced into the main task, the risk of boredom is far lower. However, these subtasks cannot be simply "make work" in nature, but must have value that is perceived by the worker.

The flicker-fusion frequency test has also been found to be a reliable indicator of boredom. A more direct approach is to include some activities in a given task that require some direct responses by the operator. The operator's condition may be assumed to be reflected by the accuracy or inaccuracy of the responses, thus allowing for an indirect measurement of boredom.

Vigilance is the state of being alert or watchful. Someone who is fatigued or bored cannot simultaneously be vigilant. Vigilance seems to be more amenable to quantification, and some findings may be summarized as follows:

1. Alertness begins to decrease after the first 30 minutes.
2. Alertness to signals improves with the frequency of the signals up to the point at which the operator can comfortably respond.
3. Stronger signals improve alertness.
4. Informing the operator about observed performance improves alertness.

Taken together or individually, fatigue, boredom, and vigilance are qualities that we understand, but that we cannot measure with any precision. They are largely subjective in nature.

15.9 INCLUDING HUMAN FACTORS IN DESIGN

Although we have addressed human factors very late in this work, they must be woven into the design of a product very early in the process. This is not the inclination of most engineers, who tend to proceed to hardware as fast as possible. That is a tendency to be resisted, because early decisions may result in hardware that is difficult to manufacture, to install, or to use. The advice to delay physical design while setting up specifications is rarely well received, but it is sound advice.

In Chapter 3 need analysis and specifications were addressed. Some of the sample specifications, such as "convenient to install," are motivated by sensitivity to human factors in design. Other specifications relating to human factors could be "easy to assemble," or "easily read display panels." The photocopy machine having unreadable displays under fluorescent lighting referred to in Section 15.6 is a case in point as to failure to take human factors into account. Many VCR's are built so that, with the VCR at any height other than eye level when standing, the programmer cannot read the top of the display without bending over. If it is placed on the bottom shelf of a TV cart, one must almost get down on the floor to do the programming. A popular form of small entertainment center is one that contains an AM-FM radio, CD player, and one or two tape decks. These units are obviously designed so that they may be placed on shelves. One such unit has two tape decks on top of the main unit, with the usual set of pushbuttons to control the various functions. To verify what each pushbutton controls, it is necessary to peer over the top of the unit. The edge of the pushbutton facing toward the front of the unit has sufficient space for a brief legend describing its function, but that face is left blank.

The determination of needs should be preceded by a four-step procedure. First, the operational events should be defined. Without any knowledge of the eventual form of the product, it is possible to write a scenario describing the probable interaction between the user and the proposed product.

The second step is for the events most difficult for the operator to monitor or control to be identified. Operator limitations must be taken into account when making this assessment so that there are no unreasonable expectations as to speed of response, memory capabilities, or operator strength.

Third, evaluate the expected environmental conditions, especially if the product being designed is expected to be used in an industrial environment. What temperatures may be expected? What is the probable noise level? What illumination level is likely?

Finally, the designer should decide which operations can be automated or must be automated, which require special operator training, and which can be made easier to do by appropriate design of the equipment.

The results of these four steps should be worked into the need analysis and specification list proposed in Chapter 3. Only after this is done can the designer have confidence that the product being designed will be convenient to use and most likely to be error-free in use.

15.10 SUMMARY

As noted at the beginning of this chapter, we have given a broad-brush treatment of the entire area of human factors engineering. We have given general directions and observations, trying to point in the best direction for improvements. Much of what has been said is qualitative in nature simply because of the fact that quantification of the various factors is virtually impossible, humans being very adaptable creatures.

REFERENCES

1. Dul, J., and B. A. Weerdmeester. *Ergonomics for Beginners,* Ninth Edition, Taylor & Francis, Bristol PA, 1991.
2. Ostrom, Lee T. *Creating the Ergonomically Sound Workplace,* Jossey-Bass, San Francisco, 1993.
3. Kirwan, Barry. *A Guide to Practical Human Reliability Assessment,* Taylor & Francis, Bristol PA, 1994.
4. Badler, Norman I., Cary B. Phillips, and Bonnie Lynn Webber. *Simulating Humans,* Oxford University Press, New York, 1993.
5. Hedgcock, R. E., and R. F. Chaillet. *Human Factors Engineering Design Standard for Vehicle Fighting Compartments,* HEL STD.S-2-64, U.S. Army Human Engineering Laboratories, Aberdeen Proving Ground MD, 1964.
6. SAE J833, *Human Physical Dimensions,* Society of Automotive Engineers, 400 Commonwealth Avenue, Warrendale PA 15096.
7. ISO 3411, *Human Physical Dimensions of Operators,* American National Standards Institute (ANSI), 11 West 42nd Street, New York NY 10036.
8. NASA STD 3000, *Human-System Integration Standards,* MSIS Custodian, NASA-Johnson Space Center, Houston TX 77058.
9. NASA Reference Publication 1024, *Anthropometric Source Book* (3 vols.), NASA Scientific & Technical Information Office, Yellow Springs OH 45387.
10. Baumeister, Theodore, Eugene A. Avallone, and Theodore Baumeister III, Editors. *Marks' Standard Handbook for Mechanical Engineers,* 8th Edition, McGraw-Hill Book Company, New York, 1978.

REVIEW AND DISCUSSION

1. People and products interact in various ways. Three ways are listed early in the chapter. What are they? Are there other interaction mechanisms?
2. What is the difference between "ergonomics" and "human factors engineering"? Is this more than a semantic difference? Is a difference in outlook implied?
3. You are designing a machine to be used in Japan. Suggest some possible sources of ergonomic data. Can you suggest some ways of learning about conventions (such as switch conventions) that are in effect in Japan?
4. Why are the data in Table 15.1 given for the 5th and 95th percentiles?
5. You are designing a lathe. At what height should hand controls be located, assuming that the data of Table 15.1 are relevant? At what height should a display screen be placed? What information from the table and from the text leads you to these decisions?

16

THE ART OF DESIGN

The main purpose of this book has been to develop various techniques that are useful in design. These techniques may be characterized as engineering techniques or scientific approaches. Design, however, involves a considerable amount of art, that is, a body of knowledge that is acquired principally by experience or by observation and for which no engineering rationale can be advanced. In certain senses it is intuitive in nature.

Placing this discussion in the final chapter should not lead you to infer that the art of design is to be considered only toward the end of a product development project. Like reliability and human factors engineering, it should be considered as the development of the product proceeds, entering into the thinking about the design at the beginning. If a final design is so unacceptable that a major redesign is necessary, the changes needed are almost invariably expensive. Moreover, if product evaluation and test results have been used in the original work, all those results are placed in doubt.

16.1 DESIGN FOR PRODUCTION

We cannot emphasize too strongly that design is an iterative process. Figure 1.2, reproduced as Figure 16.1, illustrates numerous feedback paths that may be used during the design process. It is important to understand that the feedback paths shown are not intended only for the designer. As the design moves forward, designers must avail themselves of the expertise of others within the organization. These include engineers and technicians, especially those with special training or experience in areas different from that of the designer, and appropriate manufacturing, quality control, purchasing, and marketing personnel. As noted earlier, the term that is now frequently used for the interactive relationship just described is concurrent engineering.

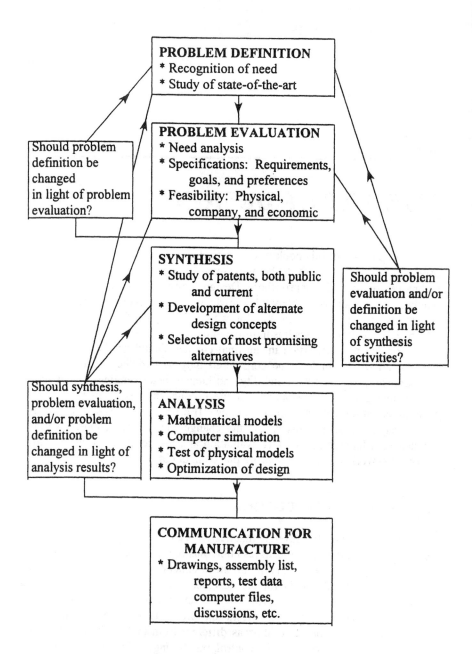

Figure 16.1 A design procedure.

THE ART OF DESIGN

New materials and new manufacturing processes are two of the driving forces in design. This reinforces the comments of the previous paragraph as to the relationship among the designer, manufacturing engineers, sales force, purchasing agents, and others. Productivity begins with the design function and ends with the safe delivery of a reliable product to the consumer. There are many seemingly insignificant design decisions, and most are probably insignificant, especially when considered individually. Some, however, can make a difference between a product that is inexpensive to manufacture and one that is not. As each part is designed, ease of manufacture and ease of assembly should be prime considerations, and ease of installation must not be neglected.

Production may be divided into four activities:

1. Purchasing materials and components.
2. Fabrication of parts.
3. Assembly of parts and components.
4. Packaging the finished product.

16.1.1 Materials and Components

The specifications will presumably determine the values of the physical properties needed for the material of each part or component. These values are obtained using the techniques for determining the design parameters as discussed in Chapter 10. When these parameters are known, the designer can search for materials or components whose physical characteristics meet or exceed the requirements. One function of the purchasing agent is to assist the designer in this search. Usually there will be many materials and components available having the required properties. You must then make selections based on a combination of considerations. Quality is of importance, but cost may not be neglected, including the cost of assembly of purchased components or fabrication of materials.

When making these selections, there is a tendency to use what was used in previous applications that are similar in nature. Although this is a good starting point, it must not limit the materials and components that are reviewed. Plastics are available that are as strong as steel, others are electrical conductors, and metal oxides can serve as insulators. The constant development of new materials is one reason for continual examination of designs, even in what are normally thought of as mature industries.

As a case in point, consider the induction motor. United States Patent Number 382,279 was issued to Nikola Tesla on May 1, 1888, for this invention. Its operation depended on the availability of alternating current supplies having more than one phase, a technology that was readily available. Because of its simplicity, production costs of this motor were much lower than that of the direct current motor. Moreover, the low maintenance of the induction motor as compared to the direct current motor made it popular in industrial applications. Between the time of its invention and World War II, gradual design evolution brought the motor to a form that is quite similar to that of today's machine. Yet manufacturers mounted major redesign efforts just after World War II, again in the 1950's, again in the 1960's, and yet again in the 1970's. The first three of these redesign efforts came about because of improved characteristics of insulation. The last of them came about because there was an energy crisis; the redesign effort was directed toward increasing the efficiency of the motor. During the entire evolution from 1888 until now, there has been no change in the way in which the motor operates, but so great have been the advances in insulation and in the magnetic properties of the steels used that the size and weight have been reduced while the reliability and range of operating characteristics have been constantly improved.

One characteristic of a good designer is that he or she is continually collecting data on new materials and components, and thus has the data readily available to aid in making decisions. Suppliers will always furnish information on their products on request, and many will supply the information on a routine basis once they know that there is an interest. Periodicals that are sent without subscription cost to qualified engineers and designers are another source of information on materials and components.

The cost of materials is very important information. Keep in mind that material is usually sold on a weight basis but is used on a spatial basis. For example, a chair has a certain size, as does a link connecting two elements of a mechanism, and weight is a consequence of the dimensions and the material. As a result, a material that is more expensive than another on a weight basis may turn out to be less expensive on a spatial basis. For example, a pound of aluminum costs about 7-1/2 times as much as a pound of steel. However, a cubic inch of aluminum costs only about three times as much as a cubic inch of steel. Moreover, the link mentioned above, if made of steel, may have to be machined or forged from the raw stock, whereas it may be possible to extrude the aluminum so that it is usable with little machining, thereby offsetting the increased material cost. As a result of this kind of analysis, aluminum (and other materials) has replaced steel in many applications in which steel was traditionally used. The use of aluminum conductors, in spite of the lower conductivity of aluminum as compared to copper, rather than copper is another illustration of the fact that the end cost may be reduced by substituting nontraditional materials.

Plastics are another group of materials which will need careful investigation, and the situation is even more complex than when considering the various metals. The basic electrical properties of the various types of plastics are surprisingly similar, but the mechanical properties, such as density, show great differences. If one made three designs of a simple insulator to support an electrical conductor, each using a different plastic, the designs would probably

THE ART OF DESIGN

result in three different cross sections, three different weights, and three different costs.

Finally, consider the purchase of components. There are certain primary design parameters that must be satisfied, but the designer must consider the reliability of components from various suppliers and their quality control. Components that are functionally equivalent from different manufacturers may have different configurations, making substitution impossible. What is the manufacturer's reputation for prompt and reliable delivery? The formal decision procedures of Chapter 8 can help to evaluate suppliers, but intangible factors, such as the degree of rapport between your company's purchasing agent and the supplier's managers, may not be neglected.

16.1.2 Fabrication

The fabrication of parts is the second manufacturing activity to be considered in the design process. This is undoubtedly the largest body of art the design engineer needs to draw on, and he or she should not hesitate to consult with the company's manufacturing engineers and technicians. Consider metal parts as an example. In many instances, functionally equivalent parts can be made by metal removal, cold forming, extruding, casting, or sintering. Each of these general techniques is further divided by specific methods of production. Cold forming, for example, can be done by successive punch press operations or by a four-slide machine, or in some cases by impact techniques. Sometimes a slight change in the configuration of a part can greatly reduce the required manufacturing operations. An example is the production of an arc of metal from a strip. If the arc to be formed is greater than 180°, two strokes of a punch press are required. If the design can be modified so that the arc is less than or equal to 180°, the arc can be formed with one stroke, thus speeding production. The designer needs to balance the increase in speed of production (and hence lower cost) against the fact that the design must be modified to accommodate the change. Such detailed analyses, annoying as they may be, are important when many parts and operations on parts are involved. Design for ease of fabrication is one of the most important skills of the designer because it always results in reduced costs.

We cannot cover the details of all manufacturing techniques, and the designer does not need global comprehension either. For each type of product, there are some techniques that must be understood quite well, others that may be understood in general, and many that may be neglected. If you design earth-moving equipment, you must understand castings and casting processes, but you have absolutely no need to understand how to diffuse impurities into a semiconductor. However, you need to keep informed of new manufacturing techniques as they are developed to the extent that you can judge applicability or nonapplicability to the products your company manufactures. A good rapport with your manufacturing personnel is of great help in deciding which of the fabricating techniques requires more in-depth knowledge, and where to look for the needed information.

One of the first steps a new designer should take is to make a technical tour of the company's manufacturing facilities so that he or she has current information on the fabrication techniques in house. A second step is to examine the products of competitors in order to ascertain, if possible, the fabrication techniques they are using. These techniques, both in-house and those of competitors, should receive in-depth study in order to have a firm understanding of their capabilities and limitations. Keep an open mind about techniques described in advertisements or articles in trade journals, as well as those proposed by visiting sales engineers.

Information relative to a manufacturing technique can be subdivided into four categories:

1. A general description of the technique.
2. Limitations on the shape or size of a part imposed by the technology.
3. The number of parts in a run necessary to achieve economy.
4. The tolerances required on part dimensions.

For example, a study [1] in 1963 of a particular type of casting called an *investment casting* revealed the following information under each of these points.

1. The technique consists of the following basic steps:
    a. Forming expendable patterns out of wax or other suitable material.
    b. Gating and assembling the patterns.
    c. Placing the assembled patterns in a mold (usually of silica and plaster of Paris).
    d. Melting out the wax.
    e. Pouring the metal to be cast.
    f. Removing and cleaning the castings.

    The technique has a major advantage over other casting techniques in that there is virtually no restriction on the complexity of the part. Any piece whose pattern can be made in wax can be cast. Parts are removed from the mold after casting by breaking the silica away from the part. Small parts are usually cast in groups attached to a gating core, whereas large parts are cast singly.
2. Parts may range in size from "very small," where each pattern has many parts on a common core, to parts having a major dimension of 2-1/2 feet or so and weights to 200 lb. Generous radii should be allowed at all edges.
3. The cost of the part is rather insensitive to the number of parts cast. This is because the cost of the equipment is low compared with that of other casting

THE ART OF DESIGN

techniques, and labor is about the same cost per piece regardless of the number made. Hence this technique is used in applications ranging from parts for prototypes to high-production parts.

4. The commercial tolerance on linear dimensions is 0.010 inch per inch with a 0.010 inch minimum tolerance. The general tolerance on radii is 1/64 inch. Long, thin parts are made to a straightness tolerance of 0.005 inch per inch and surfaces to a flatness tolerance of 0.005 inch per inch. All tolerances can be reduced to half the values given if special care is used. (Note: Castings are routinely made today with much smaller tolerances.)

This is the kind of information that can be found in the literature. If you make parts using the investment casting technique, you will learn much more about the process than is contained in the brief statements above, but there is a considerable amount of information available in less than a page. If you have similar information on other casting techniques, first comparisons are rather easily made.

16.1.3 Assembly

This is the third of the manufacturing activities that must be considered by the designer. One of the best ways to approach the challenges of assembly while the product is still in the design phase is for the designer to place himself or herself mentally in the position of the person who will actually carry out the assembly. Can the parts be held properly while being fastened together? Will the assembly require a special jig or a special tool? Have I provided for the necessary alignment of parts? What is the possibility of misassembly, especially if the assembly worker becomes fatigued? Are there changes possible that will reduce the number of steps in the assembly, thus reducing assembly costs?

Separate parts must always be fastened together in some way. Traditionally this has been by machine screws, sheet metal screws, rivets, screws designed to hold well in plastic, "Christmas trees" used in the automotive field to fasten interior trim in place, welding, and so forth. Mechanical fastening technology has been improved at a rapid rate, so much so that adhesives are now frequently used for fastening parts together that were formerly fastened by one of the traditional methods above. Another method that is extremely popular is the spring tab on plastic moldings which, when inserted into a slot, pops out behind the mating piece after the tab has been inserted to its full depth.

Electrical connections must also be considered. Mating plugs are very common, especially with computers, but you will find them in numbers in automobiles. Printed circuit boards with edge-mounted connectors are ubiquitous. Soldering is used for components on printed circuit boards. Pressure connectors are replacing soldered joints otherwise, and have shown a high degree of reliability.

Although assembly and its costs are important, one should also consider the possibility of disassembly. If a product may need repair or maintenance, can it be disassembled relatively easily, and without special tools? Some products are not intended to be repaired but to be replaced because the repair costs could very easily far exceed the cost of a replacement. The modern residential smoke detector is one example, and no one would even consider repairing a defective ground fault circuit interrupter. If a product should not be disassembled for repair, has the assembly method assured that a disassembly attempt is reasonably certain to fail?

16.1.4 Packaging

The most important purpose for packaging is to protect the product. Products often receive rough treatment between the time they leave the manufacturer and the time they reach the consumer. A product which in and of itself is properly made for an intended application may be damaged when it is inadvertently dropped 5 feet off a loading dock. But whether this unusual event occurs during shipment or at the hands of a wholesaler or retailer is of interest only if one is going to claim damages, and this is frequently an unprofitable venture, the compensation received often being eaten up by the cost of recovery of damages. The key question to be asked is "What is the normal amount of abuse that the product will be subjected to between the time it leaves the manufacturer and the time it reaches the consumer?" If a satisfactory answer can be obtained to this question (and an answer can frequently be inferred from an examination of returns), then the designer can make recommendations as to the next step. One alternative is to strengthen whatever part of the product has been shown to have failed, and retain the present packaging. A second alternative is to improve the packaging so that the product is isolated from the shocks causing the damage. Either approach represents an expenditure, and the less expensive solution is obviously the one of choice.

Packaging serves other purposes as well. Attractive designs are common on the packages of many consumer products; they are part of the advertising. (If costs are accounted for by department, then part of the cost of packaging should probably be charged to the sales department.) The packages in which many toys come have openings covered by a clear plastic so that the toy may be seen but not touched or removed. Other packaging is designed to be reusable. For example, jet engines are shipped in this way, and shipping pallets (which are really part of a package) are also intended to be reused.

There is a general rule concerning packaging that should be remembered: Make it as inexpensive as is consistent with the needs for protection because it adds nothing to the intrinsic value of the contents.

THE ART OF DESIGN

16.2 INDUSTRIAL DESIGN

Industrial design was first mentioned in 1913, but the first person to style himself as an industrial designer was Norman Bel Geddes, in 1927. The concept of industrial design became generally accepted during the Great Depression of the 1930's through the work of Raymond Loewy, who had been a freelance illustrator for fashion magazines, and a little later of Henry Dreyfuss, who had designed theater sets [2]. These men built their clientele by product redesigns that were largely aesthetic in nature, enhancing the appearance of a product without necessarily making an improvement in its function. One of Loewy's first commissions was the redesign of trash cans in Pennsylvania Station in New York. So successful was he in this commission that the next task he received from the Pennsylvania Railroad was the redesign of a locomotive. This resulted in the first of the streamlined locomotives which soon became the norm for passenger trains.

The objective of industrial design is to improve a product's appearance so as to make it more attractive to a discriminating public. The aesthetic value of a product depends upon its ability to evoke a positive response from the person looking at it. In the process of meeting its functional purpose within economic restraints, many products have, under the guidance of industrial designers, evolved a characteristic form that has become meaningful to the public. People have learned that the form represents the ability to do the job for which the product is intended even if they do not understand the physics that impose the form requirements. The shape of a racing car denotes speed to the average person, even though relatively few may be able to understand aerodynamic drag, know nothing about laminar flow, and do not understand the relationship between engine torque and acceleration. That inability does not diminish people's ability to judge in a heuristic sense that the sleek, low-slung racer is built just right.

The job of the industrial designer is to give the product the outward form that expresses unmistakably, and without the need for a salesman to speak, that the product has the qualities the potential customer wants. For that reason, a bank safe must project security and impenetrability by its size, structure, and workmanship; a set of dishes must project cleanliness and bounty; and a micrometer must symbolize precision, reproducibility of measurements, and sturdiness.

There are four guidelines [3] that can help the engineer select the appropriate form for a product:

1. The law of organic form. This is based on the observation that a natural form (that is, one that satisfies a desired function) has a tendency to respond to environmental changes in the most efficient way without changing its basic form. For example, the kitchen gas stove of the 1930's had four burners for heating skillets and pots, an oven was located at about the same

level to one side, and there was a considerable amount of waste space underneath. The amount of wall space required could easily run to 60". As kitchens became smaller, the stove evolved into a unit having an oven below the burners, and requiring only 30" of wall space. The functions are unchanged, but there is more efficient use of space.

2. The law of conformation. This law states that there is a characteristic form that is common to all products within the class. We expect, for example, that all hand-held telephones in a residential setting will be basically the same in form, having both microphone and receiver in one unit. This is true for the most basic residential unit, as well as for phones with built-in answering machines, cordless phones, and cellular phones. (The variations come in the base units.)

3. The law of adaptation. A class of products will preserve itself through evolution. The telephone evolved from the receiver-on-the-hook and the microphone (mouthpiece) attached to a large base on the wall (vintage 1910) to the receiver-on-the-hook and the microphone at the top of a small stand on a desk or table (vintage 1930) into the units we have today.

4. The law of essentials. Do not incorporate unnecessary embellishments. The difficulty is to determine what is necessary and what is not. There is a constant urge to add gadgets or additional functions to every product. In some cases these additions turn out to be very welcome, such as the timers and the power level controls on microwave ovens and the speed controls on automobiles. In other cases, there is so much complication built in by these additions that it becomes necessary to resort to a very careful reading of an instruction manual if one inadvertently pushes the wrong button on a dishwasher, for example. How does one get the machine out of an unwanted cycle, such as a 6-hour delay in initiating the wash cycle when one wants to start the dishwasher now?

At some point in the design procedure, a review of the appearance of the proposed product must be made. An effective practice to follow is to sketch the outward appearance (possibly in consultation with an artist or industrial designer) before the design is so sufficiently fixed as to cause problems if it must be changed. As the design progresses, appearance similar to the sketches is one of the goals to be achieved by the designer. Sometimes, of course, trade-offs must be made to satisfy critical specifications. Then, after a working model is available, a more serious effort is made to improve appearance. The model and sketches help to point out difficulties that may arise if changes are made for the sake of appearance. The operation of the model also gives a far more accurate idea of control and operational problems related to choice of form than premodel thinking can possibly yield.

In carrying out the industrial design activity, the designer must be concerned with the form, materials, and color and texture of the materials that will be seen by the user.

16.2.1 Form

The search for the proper form is probably the most difficult. The best guidance that can be offered is that aesthetically acceptable products tend to follow the form of similar products of their class. Those few typewriters still being produced have a near sculptured look, but the basic form is still the same as its very early predecessors. The reason, of course, is that a product is designed to perform one or more functions, and these do not change with time. However, major changes in outward appearance may follow significant changes in life style, as was true in the development of the washing machine from a device that looked like a barrel on legs with a wringer, always relegated to the basement, to the automatic washer in a square cabinet, which may now be found in a kitchen or even in an upstairs closet in a modern home.

In addition to being pleasing to the eye, a product should look as if it is capable of doing its job. The racing car has already been mentioned as having this characteristic. Another example is the off-the-road vehicle, intended for use in the back country or in underdeveloped areas. These vehicles are almost purely utilitarian in design, and make it easy to visualize one crossing extremely rough country because it looks rugged and dependable, just the qualities that are of interest when considering purchase of such a vehicle.

16.2.2 Materials, Texture, and Color

The basis for selecting materials to improve product appearance is usually related to the shape, color, finish, or texture of the part under consideration. Aluminum may be chosen over steel for the front of an enclosure because it can be formed with rounded edges, or because it can be anodized to give a durable colored finish. Plastics may be chosen because they can be molded in a wide variety of colors and shapes, and cutouts are easily incorporated to hold other parts.

Consumers associate (sometimes incorrectly) various attributes with certain materials, and these associations should be considered in the material selection process. For example, plastics have long been thought to be inexpensive and breakable, aluminum or stainless steel to be long-lasting, steel to be strong but tending to rust, fine-grained wood to be expensive but a mark of quality, and leather to be elegant. In fact, the association of wood and leather with quality is so strong that the plastics industry has spent considerable effort in developing imitations that are quite good. As a result, there are now available on the market from six or seven manufacturers flooring materials made of a high-pressure melamine laminate, having the appearance of wood, but having far greater resistance to wear and scuffing than the wood for which they are a substitute.

The choice of texture is usually made on the basis of implied qualities, although sometimes it is made for contrast or for ease of cleaning. If a fine finish is needed for parts, the customer's choice, if confronted by two competitive models of precision grinders, will certainly be influenced by the finish on each. There is an implication that, if the finish of the machine is excellent, it will be capable of producing finer finished work.

Smooth surfaces are more easily cleaned than textured surfaces, but they have the disadvantage that they will show fingerprints and scuffs more easily. Hence many kitchen appliances, such as refrigerators, now have textured exterior surfaces so that fingerprints are not noticeable. At the same time, the interior surfaces of refrigerators are quite smooth. The designer has in mind the user's subconscious assumption that a smooth surface, because it is more easily cleaned, is inherently more sanitary (which it no doubt is).

The selection of color or combinations of colors for a product is usually made on the basis of the association that a color has with certain attributes or on the basis of color preference of the expected customers. Industrial equipment is generally painted gray. This is a neutral color, it blends well with others, and it implies strength by association with the colors of aluminum and steel. Green suggests coolness or restfulness by association with vegetation; red denotes danger or excitement; blue is associated with calmness or serenity. There is a difference in preference for color between different groups of people. Studies have shown that American men prefer blue, red, violet, green, orange, and yellow in that order; women have the same order of preference except that blue and red are interchanged. Color preferences in other parts of the world are different, and some colors in other cultures will produce a negative response. The implications for export trade are obvious.

Finally, it should be noted that color preferences or the acceptance of various colors changes with time. Kitchen appliances and laundry equipment have been available at one time or another in the past thirty years or so in a variety of colors. Those that come to mind are a kind of bronze or brown color, yellow, green, and even black, in addition to white, off-white, and almond. Only the last three are at all common today, and appliances having any of the other colors are perceived as being "dated." The assumption cannot be made, of course, that we will not see the outmoded colors become popular again.

16.2.3 Industrial/Human Factors Engineering

The demands of industrial engineering and human factors engineering must often be considered together. Besides appearance, color selection must enhance the human-machine interaction discussed in Chapter 15. For example, control stations on industrial equipment are usually painted gray, but they will almost always have a red button (probably oversize as compared to others) or a red handle. The use of red in this situation has nothing to do with appearance; the

THE ART OF DESIGN

color is chosen so that the appropriate control can be quickly identified in case of an emergency. Hence red will not be used on other controls on the station. Other colors are used, however. Green is rather common, and has the connotation that pressing a green pushbutton is the control station equivalent of taking one's foot off the brake and placing it on the accelerator—we expect some desired function to begin to be performed, and this is frequently accompanied by the appearance of a green indicator light on the station.

Industrial design helps a product to be immediately attractive to the potential buyer. This is an important attribute of a product, but it is not a substitute for a good design, which must be a good one to begin with. One of the major manufacturers of portable power tools, intended for use by the homeowner, has recently introduced a completely new line of battery-powered tools. One of the authors has two of these tools. Both are attractively styled, the colors chosen are easy on the eye, and both perform the functions for which they are intended. On the store shelf they are immediately attractive. One of them is extremely easy to use. It fits the hand well, has a good balance, and the batteries take a charge that keep the tool in a good performance range for rather extended tasks. But the other is awkward to use because the hand must be placed in a position that is not at all natural in order to hold the tool and to operate the on-off switch.

The point is that care must be taken not to sacrifice ease of use to appearance.

16.3 SUMMARY

This chapter is intended to strike a balance between design engineering as an art and as a science. Design is sometimes practiced as an art in areas in which it could be practiced as a science. The use of computers, the development of synthesis procedures, optimization, and the myriad other topics included in this book can put design on a more scientific basis. However, design will always be partly an art, and this aspect is no less important than the science of design. A designer must be able to contribute skillfully in both aspects.

REFERENCES

1. Bolz, R. W. *Production Processes—The Producibility Handbook,* Penton, Cleveland OH, 1963.
2. Petroski, Henry. *The Evolution of Useful Things,* Alfred R. Knopf, New York, 1992.
3. Pulos, A. J. The meaning of product aesthetics, *Machine Design,* 1967, June 22:162-167.

INDEX

Accelerated testing, 42
   censored data, 503
   evaluation of improve-
      ments, 508
   of a few samples, 499
   plastics, 495
   theory, 492
Accuracy, human movement, 529
Activity, CPM or PERT, 278
Activity, dummy, 280
Adaptation, 198
   law of, 552
Aging, accelerated tests, 42
Aircraft, human-powered
   flight, 101
Alternatives
   dependent, 124
   independent, 124
   mutually exclusive, 124
American National Standards
   Institute (ANSI), 85
American Society for Testing and
   Materials (ASTM), 86
Analogs, 200
Analysis, 2, 12
   dimensional, 231
   economic, 112
   failure, 43, 45
   fault tree (FTA), 49
   finite element, 383
   needs, 77, 90
   repeated, 340, 341, 342
   spreadsheet, 379
   of variance, 508
   worst case, 43
Annual worth method
   (AW), 124, 128
Anthropometry, 522
Apollo 13, 62
Application, divisional, 166

Area thinking, 202
Arrhenius equation, 494
Assembly, 549
Authority, groups with, 80

Benefit/cost method
   (B/C), 124. 128
Bisociation, 184
Block diagrams, 216
   functional, 217
Bones and muscles, 525
Boredom, 538
Brainstorming, 203
Brooks coil, 426
Burn-in time (BIT), 463

CAD, 392
CAE, 399
Cash flow diagrams, 117
Censored data, 503
Certification marks, 172
Challenger disaster, 21
CIM. 389
Chance failures, 457
Circuit, equivalent, 218
Claims, patent, 155
Classification of patents, 159
Collective marks, 172
Color of product, 553
Communication for
   manufacture, 13
Community-level project, 333
Company restrictions, 89
Compatibility, company, 109
Competition, product, 136
Computers
   expert systems, 396
   general purpose programs, 383

557

hardware, 374
languages, 376
software, 376
Computer-aided design (CAD), 214, 390, 392
Computer-aided engineering (CAE), 389, 399
Computer-aided manufacturing (CAM), 390, 392
Computer integrated manufacturing (CIM), 389
Computer program, procedure for, 378
Computer software, patents, 169
Conception, invention, date of, 166
Concurrent engineering, 9
Confidence level, 500
Conformation, law of, 552
Constants having dimensions, 237
Constraint
    functional, 415, 420
    regional, 415, 426
Consumer Product Safety Commission (CPSC), 87
Consumers' Research, 88
Consumers Union (*Consumer Reports*), 81, 88
Consumption, 16, 78
Continuation-in-part, patent, 167
Copyright, 171
Cost/benefit method (See benefit/cost method.)
Cost estimating, 132
Creative destruction of capitalism, 136
Creativity
    blocks to, 187, 191
    characteristics conducive to, 189
Criterion function, 294, 414
Critical path, 280
    method (CPM), 276
Cross-examination, 68
Customer needs and wants, 79

Curvilinear square plot, 385
Customer desires, 79

Danger, 45
Data, censored, 503
Decision analysis, economic, 110
Decision, 4, 264
    combination of models, 312
    characteristics, 265
    criterion, choice of, 311
    elements of, 265
    make/buy, 135
    matrix, 265, 305
    models, 264
    tree, 301
    types of, 266
    under certainty, 267
    under risk, 301
    under uncertainty, 308
Decision tree, 301
Defendant, 31
Dependency among product parts, 269
Dependent matrix rows, 238
Depletion allowance, 122
Deposition, 32, 67
Depreciation, 117
    double declining balance, 110
    straight line, 120
Design
    alternative, 177
    avoiding human error, 530
    breakthrough, 412
    checking, 62
    competitor's, 80
    definition, 4
    by evolution, 340
    human factors in, 540
    industrial, 551
    levels of, 6
    optimum, 411
    procedure, 8
    by repeated analysis, 340, 342

INDEX

by spreadsheet, 379
by synthesis, 340, 348
unacceptable, 15
Design patent, 171
Design procedure, 5, 8
Design for production, 543
  assembly, 549
  fabrication, 547
  materials and components, 545
  packaging, 550
Design for reliability, 478
Design review, 62
Design of systems, 327, 332
Designer, duties of, 14
Diagrams, block, 216
Diligence, invention pursuit, 167
Dimension normalization, 294, 296
Dimensional analysis, 231
  choice of variables, 236, 237
  to develop equations, 237
  need for experimentation, 239
  π terms, 232
  use in design, 242
Disclaimer, 32
Disclosure, patent (See patent specification.)
Discounting, 113
Distribution, 17, 78
Dominance in decision matrix, 308
Door closer, 226
Drawings,
  engineering, 213, 390
  patent, 155
Duals, 200
Duty to warn, 58

Early failure, 454, 461
Economic decision analysis, 110
Education, limited, causing mental blocks, 188
Elegant solutions, 183
Engineering certainty in legal sense, 68

Engineering method, 5
Environmental concerns,
  benefit/cost analysis, 128
Environmental Protection
  Agency (EPA), 88
Equations, development of, 239
Ergonomics, 520
Essentials, law of, 552
Estimates of cost, 132
Ethics, 17
  exercises in, 20
  ideal behavior, 17
  unethical actions, 18
Event, CPM or PERT, 278
Event, decision tree, 301
Evidence, Federal Rules of, 64
Evolution, product, 340
Examination
  cross-, 68
  direct, 67
Expected time to complete
  β-distribution, 285
Experience, in design decisions, 12
Expert systems, 396
Expert witness, 64
  Federal rules of evidence, 64
  guidelines for, 69
Experts, wrong opinion, 109
Express warranty, 55
Extremum, 416

Fabrication, 547
Factor of safety, 469
Factory automation, 391
Failure
  break-in, 454
  chance, 457
  early, 461
  nature of, 453
  to warn, 58, 62
  wearout, 464
Failure modes and effects
  analysis (FMEA), 46

Failure rate, substandard units, 461
Fatigue, 538
Fault tree analysis (FTA), 46, 49
F-distributions, 511
Feasibility, 97
    economic, 110
    example of, 101
    lessons from, 107
    physical, 98
    requirements for, 97
Federal Trade Commission (FTC), 56, 87
File wrapper, patent, 168
Files, design, 329
Finite-element analysis, 383
Fit, goodness of, 503
Float (CPM or PERT)
    free, 283
    independent, 283
Flywheel design, 343
Food and Drug Administration (FDA), 87
Form of a product, 553
Force, human capability, 527
Foreseeable use and misuse, 36
Functional constraint, 415
Functional requirements, 12
Functional synthesis, 204
Functions, criterion, 294
Fund, sinking, 117

Gate
    AND, 49
    OR, 49
Gazette, U.S. Patent Office, 165
Goal, 12, 77, 94
Gossamer Condor, 107
Government agencies, 86

Hazard
    analysis of, 54
    definition, 45

Human errors, designing to avoid, 530
Human factors, 520
    fatigue, boredom, vigilance, 538
    inclusion in design, 540
    reliability, 485
Human-machine interaction, 521
Human power capability, 104, 527
Human speed and accuracy, 529
Hurwicz criterion, 309

Illumination, 532
Improvement, evaluating, 508
Index to the U.S. Patent Classification System, 163
Inductor design, 363
Industrial design, 551
    form, 553
    materials, texture, color, 553
Industrial/human factors engineering, 554
Information feedback, 534
    audible, 534
    instrument readings, 536
    tactile (touch), 537
    visible, 536
    voice synthesizers, 535
Infringement, patent, 149
Inspection, 44
Installation instructions, 44
Instruction manual, 61
Interest, 113
Interference
    patent, 167
    stress/strength, 467
Intuitive approach, 8
Invent, improving the ability to, 194
Invention
    adaptations, 198
    general method, 198
    involvement, 204

INDEX

Patent Office criteria, 146
stimulation, 197
strategy, 191
theory of, 184
Inventions of great inventors, 179
Inventors
nature of, 189
visualization skills, 190
Investment tax credits, 122

Key, desirable action of, 537
Keyboards, 528

Labels, warning, 44, 58
Lagrange, J., 411
Lagrange multipliers, 422
extension of, 430
Landmark case, 31
Laplace criterion, 309
Latest event time, 281
Learning curve, 290
Levels
of design, 6, 327
of mental activity, 188
Liability, product, 31
strict, 32
Life cycle, product, 17
community level products, 330
Life testing, 491
Linear dependence, 238
Linear programming, 427
Literature, technical, 54

Magnuson-Moss Warranty Act, 55
Maintenance, 44
Manual of Classification, 159
Manuals, instruction, 61
Manufacturing resource planning, (MRP), 391
Marks, service, certification, and collective, 172

Materials, used in product, 553
Mathematical models, 221
Matrix
coefficient, 233
of a community-level project, 331
decision, 265, 305
payoff, 307
precedence, 268, 337, 362
rank, 234
regret, 310
Maximax criterion, 309
Maximin criterion, 309
Mean time between failures (MTBF), 458
Mental block, 187
Method
engineering, 5
scientific, 5
Minimax criterion, 310
Minimum attractive rate of return (MARR), 123
Misuse, 32, 36
Models, 212
analytical, 341
block diagrams, 216
combinations of, 225
decision problems, 264
distorted, 245
mathematical, 221
network, 218
off scale, 232, 244, 245
physical, 224, 341
scaled, 242
sketches and drawings, 213
small signal, 220
system, 332
theory, 242
using existing products, 250
Money, time value of, 113

National Electrical Code (NEC), 81, 86, 89, 90, 138

National Electrical Manufacturers
  Association
  (NEMA), 57, 81, 83
National Fire Protection
  Association (NFPA), 86
National Safety Council, 86
Nature, states of, 306
Need(s) analysis, 12, 77, 90
  technology, 111
Negligence, 33
Network models, 218

Oath, patent application, 151, 159
Objective function, 414
Occupational Safety and Health
  Act (OSHA), 81, 87
Office action, patent, 168
Office automation, 391
Optimist criterion, 309
Optimization
  Lagrange method, 422
    extended, 430
  problem, general, 414
  searching, 436
  using CAE programs, 442
  with a criterion function, 416
  with functional constraints, 420
  with functional and regional
    constraints, 430
  with regional constraints, 426
Organic form, law of, 551
Organizations, standards writing,
  85
Outcomes, of decisions, 265
  quantifying, 296
Overhead, 133

Packaging, 550
Patent Gazette, 165
Patents, 146
  basics of patent law, 165
  Cassis CD-ROM, 163

commercial databases, 163
constitutional basis, 146
continuation-in-part, 167
classification system, 159
computer software, 169
design, 171
diligence in pursuit of, 167
drawings, 155
infringement search, 149
interference, 167
maskwork for
  semiconductors, 173
novelty search, 149
oath (declaration), 151
office action, 168
pending, 146, 168
plant, 171
search methods, 159
specification, 151, 154
state-of-the-art search, 148
structure, 151
types of searches, 148
validity search, 150
Path, critical, 280
Payoff, decision matrix, 265
Payoff (payback) time, 114, 131
People, creative, 189
PERT, 276, 285
Pessimist criterion, 309
Physical models, 224
π terms, 232
  model theory, 242
  number of, 233
Pilot run, 44
Plaintiff, 31
Plastics aging, 495
Power, human capability, 104, 527
Precedence
  table, 269
  matrix, 268, 271
Preferences, 12
  color, 554
Present worth method
  (P/W), 116, 124, 125

INDEX

Press, printing, 186
Price, selling, 136
Privity, 31
Probability
    element of decision
        matrix, 302
    of failure, 53
    of job completion, 286
Problem definition, 10
Problem evaluation, 10
Product
    competitive, 80
    definition of, 7
Product/component/system, 7
Product history, 43
Product evolution, 341
Product liability, 31
    risk, 40, 41
Product reliability,
    calculation of, 470
    parallel/redundant, 472
    series, 471
    standby systems, 473
Production, 15, 77, 545
Program evaluation and review
    technique (PERT), 276
Programming, linear, 427
Public domain, 148
Purchasing, 77

Rank, matrix, 238
Rate of return, minimum acceptable
    (MARR), 123
Reaction rate, Arrhenius, 495
Realizability,
    human-powered flight, 101
    physical, 98
Reduction to practice,
    invention, 166
Regional constraint, 415
Relay
    precedence matrix for, 276
    precedence table for, 275

Reliability
    affected by human error, 485
    basic relationships, 456
    definition, 452
    design considerations, 478
    growth, 481
    human factors, 485
    product, 470
    stress/strength
        interference, 467
    Weibull functions, 485
Reliability calculations (see
    Product reliability)
Repair service, 79
Requirements, 77, 94
Res ipsa loquitur, 36
Resistor design, 350
Responsibility, designer's, 14
Restrictions, company, 89
Retirement, 17, 78, 332
Return-on-investment method
    (ROI), 124, 130
Review, design, 13, 62
Risk, 45
    decisions under, 301
Risk/utility considerations, 44

Safety, 41, 44
Sales department, 79
Salvage value, 121
Savage criterion, 310
Scaling, 242
Scheduling, 285
Scientific method, 5
Searches, patent, 159
Searching, 436
    gradient, 440
    multivariate grid, 438
    pattern, 43
Secret, trade, 173
Selling price, 136
Service marks, 172
Sinking fund, 117

Sketches and drawings, 213
Slack, CPM or PERT, 285
Small signal model, 220
Smell, use in feedback, 534
Solutions, graphical, 484
Sound, as signal, 534
Specification, patent, 151
Specifications
    absolute, 346
    other, 346
    performance, 82
    prescriptive, 82
Speed and accuracy, 529
Spreadsheet analysis, 379
Spring design, 355
Standard deviation, 286
Standards, 80
    locating, 83
    mandatory, 81, 82
    product, 81
    voluntary, 81
    work place, 81
Standards-writing organizations, 85
States of nature, 306
Strategies, of decision matrix, 306
Stress/strength interference, 467
Strict liability, 32
Surveys of prospective
    customers, 79
Synthesis, 2, 12
    design by, 348
    functional, 204
    necessary condition for, 342
    system, 332
System
    characteristics of, 328
    design of, 327, 332
    designer, 329
    expert, 396
    illustration of, 328
    need analysis, 332
    reduction to components, 359
    redundant, 472
    standby, 473
    synthesis, 332
    task matrix, 331

Table, precedence, 269
Tactile sensing, 537
Task matrix, 331
Technical literature, 54
Testing, accelerated, 491
    censored data, 503
    design of, 499
    few samples, 499
    by independent laboratory, 43
    theory of, 492
Time
    burn-in, 463
    earliest event, 280
    expected, 285
        deviation, 286
    mean, between failures, 458
    payback, 131
Tolerance, 287
Tort, 32
Touch, 531 (see Tactile sensing)
Trademarks, 172
Trade secret, 173
Tree, decision, 301
Trier of fact, 32

Ultimate issue, 32
Underwriters Laboratories (UL), 85
Uniform Commercial Code, 56
Units, SI system, 257
Unreliability, 457
Use, foreseeable, 36
Utility curves, 297

Variance, analysis of, 508
Vigilance, 538
Visualization, need for, 189

INDEX

Warning, 61
Warning labels, 44, 58
Warranty, 32, 54
    express, 32, 55
    fitness, 56
    implied, 32, 55, 56
    limited, 55
    of merchantability, 56
Wearout failure, 454, 464
Weibull functions, 475
Weighting factors, 295
Whistle blowing, 19
Witness, expert, 64
Work hardening, 37
Worst case analysis, 43

Years-to-payback, 124, 131